Photochemistry and Photophysics

Volume IV

Editor

Jan F. Rabek, D.Sc., Ph.D.
Professor of Polymer Chemistry
The Royal Institute of Technology
Stockholm, Sweden
and
Polymer Research Group
Department of Dental Materials and Technology
Karolinska Institutet
Stockholm, Sweden

CRC Press
Boca Raton   Ann Arbor   Boston

Library of Congress Cataloging-in-Publication Data
(Revised for volume 4)

Photochemistry and photophysics

    Includes bibliographical references and indexes.
    1. Photochemistry. 2. Optics. I. Rabek, J. F.
QD714.P46 1990   541.3'5   89-17254
ISBN 0-8493-4041-1 (v. 1)
ISBN 0-8493-4042-X (v. 2)
ISBN 0-8493-4043-8 (v. 3)
ISBN 0-8493-4044-6 (v. 4)

Direct all inquiries to CRC Press, Inc., 2000 Corporate Blvd., N.W., Boca Raton, Florida 33431.

International Standard Book Number 0-8493-4041-1 (Vol. 1)
International Standard Book Number 0-8493-4042-X (Vol. 2)
International Standard Book Number 0-8493-4043-8 (Vol. 3)
International Standard Book Number 0-8493-4044-6 (Vol. 4)

Library of Congress Card Number 89-17254
Printed in the United States

PREFACE

The aim of these volumes is to present a highly coherent critical review of photochemistry and photophysics including inorganic, organic, atmospheric, environmental, material, bio, and polymer fields to a general collection of papers discussing new developments in these areas. Photochemistry and photophysics have increasing scientific and technological importance, e.g., solar energy utilization, photosynthesis, photochemistry of the atmosphere, photopolymerization and photocuring, photoimaging, or highly photoresponsive systems (photosensors). A mechanistic approach to photochemistry gave practical application of a number of photochemical processes in reprography, microelectronics, and holography. The application of lasers on an industrial scale solved problems of such complicated and expensive technologies as isotope separations. Also, photobiology and molecular photochemistry gave new information on processes which occurred in organisms under light irradiation or during which light is produced (emission of light by living matter). These and other interesting problems will be reviewed in these volumes.

Jan F. Rabek
November 1990

THE EDITOR

Dr. Jan F. Rabek is Professor of Polymer Chemistry, working in the field of polymer photochemistry and photophysics since 1960. His research interests lie in the photodegradation, photooxidation, and photostabilization of polymers, photocuring, singlet oxygen photooxidation, spectroscopy of molecular complexes in polymers, and photoconducting polymers.

Dr. Rabek obtained his D.Sc. in Polymer Technology at the Department of Polymer Technology, Technical University, Wroclaw, Poland (1965) and his Ph.D. in Polymer Photochemistry at the Department of Chemistry, Silesian Technical University, Gliwice, Poland (1968). He is employed at the Royal Institute of Technology, Stockholm and is head of the Polymer Research Group in the Department of Dental Materials and Technology, Karolinska Institute, Stockholm. He has published more than 130 research papers, review papers, and is author, co-author or editor of ten books on the photochemistry of polymers.

CONTRIBUTORS

M. V. Encinas
Departamento de Química
Facultad de Ciencia
Universidad de Santiago de Chile
Santiago, Chile

Horst Kisch
Institut für Anorganische Chemie II
Universität Erlangen-Nürnberg
Erlangen, Germany

Ronald Künneth
Institut für Anorganische Chemie II
Universität Erlangen-Nürnberg
Erlangen, Germany

E. A. Lissi
Departamento de Química
Facultad de Ciencia
Universidad de Santiago de Chile
Santiago, Chile

Ulrich E. Steiner
Fakultät für Chemie
Universität Konstanz
Konstanz, Germany

Chen-Ho Tung
Institute of Photographic Chemistry
Academia Sinica
Beijing, China

Hans-Joachim Wolff
Fakultät für Chemie
Universität Konstanz
Konstanz, Germany

Cheng-Ba Xu
Institute of Photographic Chemistry
Academia Sinica
Beijing, China

TABLE OF CONTENTS

Chapter 1

MAGNETIC FIELD EFFECTS IN PHOTOCHEMISTRY

Ulrich E. Steiner and Hans-Joachim Wolff

TABLE OF CONTENTS

I. INTRODUCTION

Magnetic field effects (MFEs) in chemical kinetics require interaction of a magnetic field with reaction intermediates having unpaired electron spins.[1] Since photoreactions start from electronically excited states that correspond, as a rule, to configurations with unpaired spins, it is not surprising that most of the reactions known today to exhibit magnetic field-dependent rates or product yields are photochemical reactions. Although the interest in magnetic field-dependent reactions has quite a long tradition (see, for example, Reference 2), there are few reports of reproducible effects with a clear molecular mechanism before the mid-1960s. At that time a number of magnetic field-dependent phenomena in the photoluminescence and photoconductivity of molecular crystals (cf. References 3 to 7) and unusual line intensities in magnetic resonance spectra recorded during chemical reactions with radical pair intermediates were discovered (cf. References 8 to 10). The molecular mechanisms, particularly of the latter effects (CIDNP, CIDEP), provided the conceptual basis for a systematic and successful search for suitable conditions to observe magnetic field-dependent reaction yields and kinetics.

In 1972 Molin and co-workers[11] found the first example of magnetic field-dependent reaction yields when studying thermal reactions between lithium alkyls and benzylchloride in homogeneous solution. In 1974 Brocklehurst et al.[12,13] published stimulating work on the magnetic field dependence of radioluminescence originating from the recombination of spin-correlated radical ion pairs, which was followed, in 1976, by the first reports of corresponding effects with photochemically generated radical ion pairs by the groups of Weller[14] and Michel-Beyerle.[15] At the same time the first reports on magnetic field-dependent reaction yields observed under conditions of continuous photolysis[16,17] and the first successful experiments of photochemically induced magnetic isotope enrichment also appeared.[18,19]

From the historical point of view it may be of interest to note that actually the first example of a photochemical MFE, the magnetic fluorescence quenching of iodine vapor[20] that is due to a magnetic field-induced predissociation,[21,22] dates back to 1913. The mechanism of this MFE seems to be restricted to small molecules in the gas phase and has not stimulated the development of photochemical magnetokinetics in a way comparable to the impact that the above-mentioned magnetic field-dependent spectroscopic phenomena in condensed phases have had.

Since the experimental and theoretical basis of chemical MFEs has been largely developed during the 1970s the field has continuously expanded during the last decade and a variety of reactions and systems have been studied so far, mainly in liquid solutions, but also in the gas phase, solid state, and on interfaces (for a previous comprehensive survey cf. Reference 1).

When assessing the potential influence of a magnetic field on the course and kinetics of a chemical reaction one must consider the order of magnitude of the interaction energy involved. The energy for a spin-flip of 1 mol of electrons corresponds to

$$\Delta E_{spin} = N_A g \beta B \tag{1}$$

where N_A is Avogadro's number, g the electronic g-factor, β Bohr's magneton, and B the magnetic induction.* In a strong magnetic field of 1 T* $\Delta E_{spin} = 5.6\ J*mol^{-1}$, corresponding to $2.2*10^{-3}$ RT at room temperature. On the other hand, creating (or compensating) 1 mol of unpaired spins that have a thermal equilibrium distribution of orientations in a magnetic field of strength H is associated with a Gibbs energy change of

$$\Delta G_{spin} = (\mp) \frac{1}{2} \frac{(N_A \beta)^2}{RT} H^2 \tag{2}$$

In a field of 10^4 Oe (B $= 10^4$ G) this corresponds to $\Delta G_{spin} = 6.4*10^{-3}\ J*mol^{-1}$ or $2.6*10^{-6}$ RT at room temperature. From the orders of magnitude of both ΔE_{spin} and ΔG_{spin} it is clear that at normal temperatures significant kinetic MFEs cannot be expected for electronically allowed adiabatic reactions, since possible changes of activation barriers by a magnetic field are too small in comparison to RT.

The situation is quite different in the case of nonadiabatic reactions. If the magnetic field can induce (or prevent) transitions between close lying states that differ significantly in their electronic coupling to product states, sizable kinetic effects can ensue. (Note that for B $= 1$ G the angular frequency corresponding to Δ_{Espin} is $17.6*10^6\ rad*s^{-1}$). There are two types of mechanisms where this principle applies. The first type involves reactive intermediates consisting of pairs of weakly coupled paramagnetic particles as two doublets (DD or radical pairs; this is the most important case in connection with chemical reactions), two triplets (TT pairs), or a triplet and a doublet (TD pairs). The second type is the so-called triplet mechanism. In the case of the pair mechanisms the various spin substates of the pair represent a near-degenerate set of states that may be divided into subsets according to their quantum number of total spin. Usually the reactivity strongly differs among the pair states of different multiplicity so that the whole set of spin states does not decay uniformly. In general, transitions between the slowly and the fast reacting spin substates will be involved in the overall reaction. It is by interference with these transitions that the external magnetic field can exhibit its kinetic effects.

In the case of the triplet mechanism a differentiation of reactivity exists within a set of spin substates belonging to the same multiplicity (triplet). This differentiation results from spin selection rules associated with spin-orbit coupling (SOC). In an external magnetic field a coupling among the triplet substates is introduced that changes the overall kinetics of their decay.

---

*   In dealing with magnetic quantities we will use the Gaussian system of equations together with the Gaussian CGS system of units,[23] where the magnetic induction B is measured in *gauss* (1 G $=$ 1 $dyn^{1/2}$ cm) and the magnetic field strength H in *oersted* (1 Oe $=$ 1 $dyn^{1/2}$ cm). For convenience we will also use the conversion 10^4 G $=$ 1 T. One should note, however, that the unit *tesla* belongs to the MKSA system of units and that when applying the conversion 1 T $=$ 1 $Vs*m^{-2}$ one must use modified versions of the magnetic equations in compliance with the system using four basic quantities.

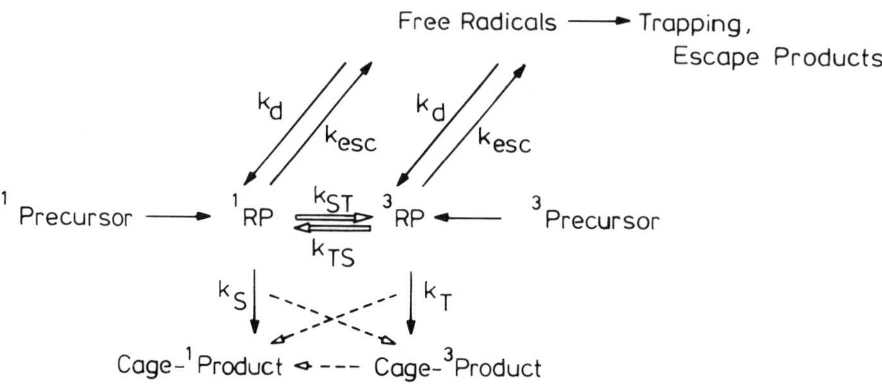

FIGURE 1.   General reaction scheme for the radical pair mechanism (RPM).

In this review, which is complementary to the one by H. Hayashi in Volume I of this series, we will first deal in some detail with the theoretical concepts of the radical pair mechanism (RPM) and the triplet mechanism (TM) and will then discuss in the light of these concepts a number of representative experimental examples of MFEs on photochemical reactions. Thereby we will confine ourselves to reactions in homogeneous and micellar solution, since most of the work dealing with MFEs in the gas phase and in the solid state is concerned with photophysical processes not actually leading to chemical change. MFEs according to the RPM have found important applications in the study of primary processes in photosynthetic reaction centers. This topic has been reviewed recently by Boxer et al.[24] and will not be taken up here.

II. THEORETICAL PRINCIPLES

A. THE RADICAL PAIR MECHANISM

The basic kinetic scheme underlying MFEs according to the RPM is presented in Figure 1. Radical pairs (RPs) that may be generated in thermal or in photochemical reactions originate with specific spin correlation (^{1}RP = S-RP, ^{3}RP = T-RP) according to the multiplicity of their direct precursors. This rule of spin conservation in an elementary chemical reaction is, in general, also obeyed in subsequent processes of the RPs so that S-RPs will be only allowed to react to singlet recombination products, whereas T-RPs may only recombine to (excited) triplets, provided they have sufficient energy. These rules may be relaxed in some cases due to the effect of SOC, but for an explanation of the principles we may neglect such effects in the first place.

The RPs may diffuse apart to yield free uncorrelated radicals that may be trapped or may recombine in a statistical fashion to form so-called escape products. Of course, if the radical lifetime is long enough, the original kinds of RPs may be formed again from the free radicals, too, however with a statistical probability of 1:3 of forming S-RPs or T-RPs. (Such pairs with random spin are usually termed F-pairs.) One must keep in mind, however, that the decay of the initially generated, so-called geminate RPs by cage recombination and escape takes place on a much shorter time scale (usually in the range of 1 to 100 ns) than formation of F-pairs from the homogeneous recombination (usually >1 μs).

It is important to realize that the notion of a RP is not specific about the separation of the two radicals, which could actually vary between contact distance (*circa* 5 Å) and several tens of Å. The variation in time of RP separation caused by diffusion is considered as an internal coordinate of motion of the RP. During its "diffusional lifetime" a RP may undergo several separations and re-encounters. As has been shown by Noyes[25] for the case of Brownian diffusion the probability f(t) per unit time for a first re-encounter of the RP occurring after a contact at t = 0 is approximately given by

$$f(t) = mt^{-3/2} \exp(-\pi m^2/p^2 t) \tag{3}$$

with

$$m = [27/(8\pi)]^{1/2}(1 - p)^2(a/\sigma)^2\tau_h^{1/2} \tag{4}$$

and[26]

$$p = \int_0^\infty f(t)dt \approx 1 - \left(\frac{1}{2} + \frac{3}{2}\frac{a}{\sigma}\right)^{-1} \tag{5}$$

Here a, σ, and τ_h are the encounter distance, the hopping length, and the hopping time, respectively. The latter two quantities are related to the sum of the diffusion coefficients of the radicals by

$$\sigma^2 = 6D\tau_h \tag{6}$$

Since at long times the time dependence of f(t) is determined by the factor $t^{-3/2}$ one cannot uniquely assign an average "cage time" to the diffusive RP decay. The effective cage time or its inverse, the rate constant k_{esc}, will depend on the rate constants of competing processes. The essential magnetokinetic aspect of the RP reaction scheme is the change of RP multiplicity. Here, again, for reasons of simplifying the arguments, rate constants have been assigned to processes that are beyond ordinary reaction kinetics and require the terminology of quantum mechanics for an exact description. In dealing with the time dependence of the RP spin state it has become customary to speak of "spin-motion" or "spin-evolution".[27] These processes may be influenced by external magnetic fields whereby, depending on the magnetic properties of the RP, the effective rate constants k_{TS} and k_{ST} may decrease (this is rather the normal case) or increase in a magnetic field.

Magnetic field effects on the overall kinetics and on product yields follow from the magnetic field dependence of k_{TS} and k_{ST}. For example, in many cases RPs are produced from a triplet precursor and k_{TS} and k_{ST} are decreased in a magnetic field. In these cases the yield of escape products will increase, unless the RP has enough energy to produce an electronically excited triplet state. In other typical cases investigated, S-RPs are generated of high energy so that recombination with formation of excited triplets is possible. In these cases the triplet yield decreases and that of singlet cage products increases in a magnetic field if k_{TS} and k_{ST} are slowed down. The yield of escape products may increase or decrease depending if k_T or k_S is larger.

To describe spin-correlated states of RPs that may be defined by their total spin and their spin projection onto one selected direction, in a pictorial way, use of the vector model as shown in Figure 2[8,28] has become customary. This should, however, be interpreted not too "literally" because of a shortcoming that arises in the vectorial representation of the spin states T_0 and S. The algebraic representation of T_0, for example, is

$$T_0 = \frac{1}{\sqrt{2}}(\alpha_1\beta_2 + \beta_1\alpha_2) \tag{7}$$

It corresponds to a coherent superposition of two situations, $\alpha_1\beta_2$ and $\beta_1\alpha_2$, having α spin on radical 1 and β spin on radical 2, or the converse, respectively. Physically, each of these

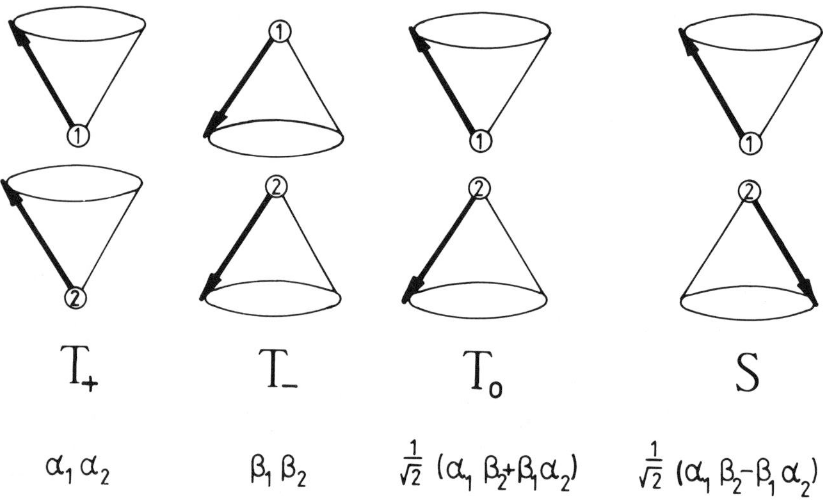

$$\alpha_1\alpha_2 \qquad \beta_1\beta_2 \qquad \tfrac{1}{\sqrt{2}}\,(\alpha_1\beta_2+\beta_1\alpha_2) \qquad \tfrac{1}{\sqrt{2}}\,(\alpha_1\beta_2-\beta_1\alpha_2)$$

FIGURE 2.   Vector representation of radical pair spin states T_+, T_-, T_0, and S. (Adapted from Turro and Kräutler.[28])

products of single spin functions corresponds to a so-called spin-polarized situation, where in an ESR experiment one of the radicals would appear in emission and the other in absorption. In the T_0 and S RP spin states, however, α and β spins appear with equal probability on both radicals. The difference between T_0 and S lies in the phase factor by which the *products* $\alpha_1\beta_2$ and $\beta_1\alpha_2$ are added. In the vector model, however, it is attempted to represent phase factors by the relative azimuth angle of transversal orientation of the *individual* spins. Translated into the algebraic representation, a rotation of a spin function by an angle φ around the z-axis of quantization is represented by a phase factor $e^{i\varphi}$. In this sense the graphical representations of T_0 and S as given in Figure 2 would be expressed as $\alpha_1\beta_2$ and $e^{i\varphi}\alpha_1\beta_2$, which is actually the same physical state.

On the other hand, the states T_+ $(\alpha_1\alpha_2)$ and T_- $(\beta_1\beta_2)$ do not cause problems to the imagination since they are both spin-polarized (definite spin projection on each radical) *and* spin-correlated (one can tell the spin orientation on radical 2 if one knows it for radical 1). The imaginational problem arising with T_0 and S results from the fact that they are spin-correlated but not spin-polarized. Thus it is undetermined if in a measurement on radical 1, say, one finds α or β spin. What is difficult to perceive, particularly if the radicals are so far apart that they do not interact any longer, is how radical 2 "knows" about the decision of radical 1 to show up either with α or β spin. This difficulty is related to the Einstein-Podgolsky-Rosen paradox.[29]

Although one prefers to think about pair spin states in the spin-polarized basis, algebraically they have no principal priority over spin-correlated states. In many cases it may be favorable to express spin-polarized pair states by the spin-correlated ones:

$$\alpha_1\beta_2 = \frac{1}{\sqrt{2}}\,(T_0 + S) \tag{8}$$

$$\beta_1\alpha_2 = \frac{1}{\sqrt{2}}\,(T_0 - S) \tag{9}$$

Another aspect important for understanding spin motion in a RP in terms of a semi-classical picture (cf. Reference 30) is the axis of spin quantization. Actually a spin $S = \tfrac{1}{2}$, for example, may be found in only two orientations with respect to the selected direction.

However, it may be quantized in this way with respect to any direction. In particular, the two spins on two radicals may be quantized along two different directions, determined by some local magnetic field differences in the two radicals. Using the α,β basis defined by quantization along the z-axis of a Cartesian coordinate system, a spin, quantized parallel to a direction given by the direction cosines u,v,w, would be represented by the spin function

$$\chi_{u,v,w} = (2 + 2w)^{-1/2}[(1 + w)\alpha + (u + iv)\beta] \tag{10}$$

or by the vector

$$\chi_{u,v,w} = (2 + 2w)^{-1/2}\begin{pmatrix} 1 + w \\ u + iv \end{pmatrix} \tag{11}$$

Conversely, from the coefficients a and b of a general spin function

$$\chi = a\alpha + b\beta \tag{12}$$

the direction (u,v,w) along which the spin is quantized can be uniquely determined.

Combining two radicals, one with spin quantized along (u,v,w) and the other along (u',v',w'), it is a straightforward matter to assess the RP spin function in terms of a coherent superposition of spin-correlated states:

$$
\begin{aligned}
\chi_{RP} &= \chi(1)\chi'(2) = aa'\alpha_1\alpha_2 + ab'\alpha_1\beta_2 + ba'\beta_1\alpha_2 + bb'\beta_1\beta_2 \\
&= aa'T_+ + bb'T_- + \frac{1}{\sqrt{2}}(ab' + ba')T_0 + \frac{1}{\sqrt{2}}(ab' - ba')S \\
&= AT_+ + BT_- + CT_0 + DS
\end{aligned}
\tag{13}
$$

This relation will be required when composing the spin motion of the pair from the spin motion of the individual radical spins. The reverse decomposition of a general pair spin state characterized by the set of coefficients (A,B,C,D) in terms of two sets (a,b) and (a',b') is, however, not generally possible, since the latter situation corresponds to a pair of two spin-polarized radicals. Such situations represent only a subset of all possible RP states.

In a macroscopic sample the RPs are statistically distributed over a manifold of pair spin states. Let us, for simplicity, consider the subset spanned by combinations of T_0 and S only. Then one might specify the state of the macroscopic sample by giving the probabilities p_i at which the RP spin state $C_iT_0 + D_iS$ occurs. Actually one is interested in calculating expectation values of observables O, obtained as

$$
\begin{aligned}
\langle O \rangle &= \sum_i p_i \langle C_iT_0 + D_iS|O|C_iT_0 + D_iS \rangle \\
&= \sum_i p_iC_i^*C_i\langle T_0|O|T_0 \rangle + \sum_i p_iC_i^*D_i\langle T_0|O|S \rangle \\
&\quad + \sum_i p_iD_i^*C_i\langle S|O|T_0 \rangle + \sum_i p_iD_i^*D_i\langle S|O|S \rangle
\end{aligned}
\tag{14}
$$

Thus it is shown that the complete physical information on the ensemble of RPs is contained in the set of statistical averages of the products $C_i^*C_i$, $C_i^*D_i$, $D_i^*C_i$, $D_i^*D_i$, which may be represented as a matrix

$$\rho = \begin{pmatrix} \overline{C^*C} & \overline{C^*D} \\ \overline{D^*C} & \overline{D^*D} \end{pmatrix} \equiv \begin{pmatrix} \rho_{T_0T_0} & \rho_{T_0S} \\ \rho_{ST_0} & \rho_{SS} \end{pmatrix} \tag{15}$$

This matrix is called the density matrix. It is the most suitable and consistent instrument to deal with statistical ensembles of quantum systems.[31] Its diagonal elements correspond to the ensemble averaged populations of the corresponding basic states, its off-diagonal elements to the corresponding average of phase relations between the basis states. (Note that non-vanishing off-diagonal elements indicate the presence of spin correlation in a spin-polarized basis and vice versa.)

1. Spin Motion

The time dependence of quantum mechanical systems with a Hamiltonian H is described by the Schrödinger equation

$$\dot{\chi} = -\frac{i}{\hbar} H\chi \tag{16}$$

For a statistical ensemble described by a density matrix the corresponding equation is the Liouville equation:

$$\dot{\rho} = -\frac{i}{\hbar} (H\rho - \rho H) \equiv -\frac{i}{\hbar} [H,\rho] \tag{17}$$

The spin Hamiltonian of a noninteracting RP is

$$H = H_1 + H_2 \tag{18}$$

with

$$H_k = g_k \beta B_o S_{z,k} + \sum_{i(k)} a_i I_i S_k + \text{anisotropic terms} \tag{19}$$

In its isotropic part (independent of radical orientation) it includes the electronic Zeeman interaction $g\beta B_o S_z$ (with the electronic g-factor, Bohr's magneton β, magnetic induction B_o, and the spin operator component S_z parallel to the magnetic field) and the hyperfine interaction terms $a_i I_i S_k$ (with a_i the isotropic hyperfine coupling [hfc] constant between spin k and nucleus i with spin operator I_i). Both Zeeman interaction and hyperfine interaction also comprise anisotropic contributions that depend on the orientation of the radicals and therefore are modulated by rotational diffusion. In liquid systems the correlation times of molecular tumbling are usually much shorter than the time scale of spin motion induced by H, so that to a first approximation they have no effect on spin motion.

In higher order, however, the fluctuating forces caused by the anisotropic interactions induce incoherent spin motion, leading to relaxation toward canonical equilibrium populations of the spin substates (cf. below).

a. Coherent Spin Motion

The time-independent isotropic part of H induces a coherent spin motion that may be derived in a straightforward manner from Equation 16 or 17. If the RP spins are noninteracting the solution to Equation 16 simply corresponds to the combination of the independent motions of the individual electron spins.

Let us, for the moment, adopt the semiclassical point of view where the isotropic hyperfine interaction is considered as equivalent to a static magnetic field and nuclear spin changes due to the hyperfine interaction are neglected.[30] Then we have for electron spin 1

$$H_1 = g_1 \beta B_{(1)} S_{(1)} \equiv \hbar \omega_{(1)} S_{(1)} \tag{20}$$

where the vectorial angular frequency $\boldsymbol{\omega}_{(1)}$ may be decomposed as:

$$\boldsymbol{\omega}_{(1)} = \omega_{(1)}\begin{pmatrix} u \\ v \\ w \end{pmatrix} = \begin{pmatrix} A_{(1)x} \\ A_{(1)y} \\ A_{(1)z} + \omega_0 \end{pmatrix} \tag{21}$$

Here $\mathbf{A}_{(1)}$ is the angular frequency vector contributed by the effective hyperfine field. To correspond most closely with the actual hfc pattern in a radical, $\mathbf{A}_{(1)}$ should be chosen from a probability distribution[30]

$$f(\mathbf{A}_{(1)}) = [\tau_{(1)}^2/(4\pi)]^{3/2} \exp\left(-\frac{1}{4} A_{(1)}^2 \tau_{(1)}^2\right) \tag{22}$$

characterized by a mean square of

$$\overline{A_{(1)}^2} = 6\tau_{(1)}^{-2} = \sum_{(1)} a_i^2 I_i(I_i + 1) \tag{23}$$

Solving Equation 16 with the Hamiltonian (Equation 20) one obtains for the time dependence of a spin that is α at $t = 0$:

$$\alpha_1(t) = [\cos\left(\frac{1}{2}\omega_{(1)}t\right) - iw\sin\left(\frac{1}{2}\omega_{(1)}t\right)]\alpha_1 - [i(u + iv)\sin\left(\frac{1}{2}\omega_{(1)}t\right)]\beta_1 \tag{24}$$

or for β at $t = 0$:

$$\beta_1(t) = -[i(u - iv)\sin\left(\frac{1}{2}\omega_{(1)}t\right)]\alpha_1 + [\cos\left(\frac{1}{2}\omega_{(1)}t\right) + iw\sin\left(\frac{1}{2}\omega_{(1)}t\right)]\beta_1 \tag{25}$$

These functions represent the classical spin precession with angular frequency $\omega_{(1)}$ around the effective magnetic field vector $\mathbf{B}_{(1)}$.

The result for a RP is obtained by substituting the time-dependent single spin functions (Equations 24 and 25) into the spin function for the initial state of the RP and finally determining the time-dependent coefficients of T_+, T_-, T_0, and S according to Equation 13. The somewhat lengthy results can be found in the book by Salikhov et al.[32]

Schulten and co-workers[30,33] have derived results for the evolution of the RP spin state averaged over an ensemble with statistically distributed hyperfine fields (Equations 22 and 23). The results are given for the triplet probability (^Sp_T) of an ensemble of RPs generated with singlet spin correlation. It may be useful to note a general relation, whereby various conditional probabilities $^xp_y(t)$ (probability at time t to find the RP in state y after being generated in state x at $t = 0$) can be interconverted:

$$^Sp_T = 1 - {}^Sp_S(t) = 3{}^Tp_S(t) = 3[1 - {}^Tp_T(t)] \tag{26}$$

As an example of RP spin motion we show in Figure 3 the spin evolution $[^Tp_S(t)]$ in RPs produced by electron transfer between aniline and triplet excited thionine (cf. Section IV.D).

The growing-in of singlet probability at short times corresponds to the quadratic time dependence[35]

$$^Tp_S = ct^2 \tag{27}$$

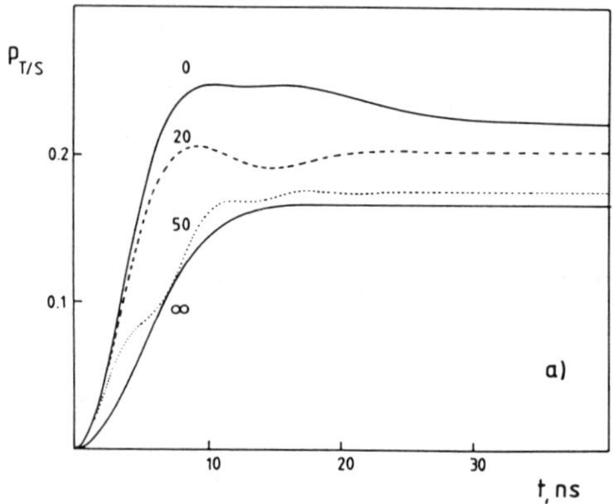
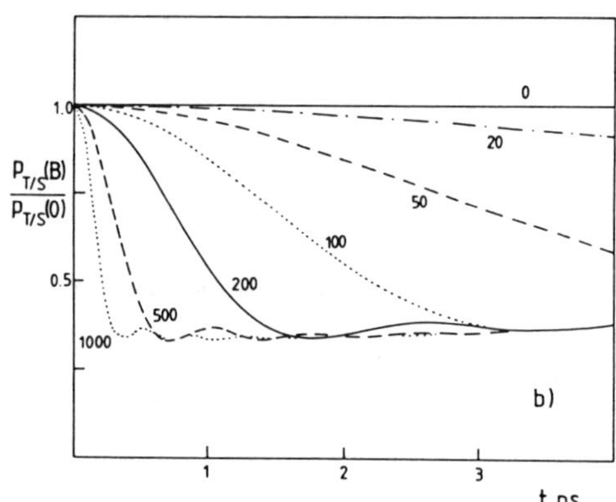

FIGURE 3.   (a) Evolution of singlet character $p_{T/S}$ (^Tp_S) in an original triplet RP. The curves were calculated according to the semiclassical method of Schulten et al.[30,33] using time constants of $\tau_1 = 13$ ns and $\tau_2 = 8$ ns, corresponding to a TH·/An$^+$· RP (cf. Section IV.D and Reference 34). (b) Details of the evolution of $p_{T/S}(B_o)$ at short times, represented as the ratio $p_{T/S}(B_o)/p_{T/S}(0)$. (Reprinted with permission from Schlenker, W., Ulrich, T., and Steiner, U. E., *Chem. Phys. Lett.*, 103, 118, 1983. ©1983, Elsevier Science Publishers B.V.)

with

$$c = \frac{\epsilon}{36} \left(\overline{A_{(1)}^2} + \overline{A_{(2)}^2} \right) \tag{28}$$

and $\epsilon = 3$ for $B_0 = 0$, but $\epsilon = 1$ for $B_0 \rightarrow \infty$. This ratio reflects the fact that in high fields, where $\mathbf{B}_{(1)}$ and $\mathbf{B}_{(2)}$ are essentially parallel, coherent transitions from $T_\pm$ to T_0,S are no longer possible because both electron spins precess around the same direction. The only difference that may be left is in the precession frequency $(A_{(1),z} - A_{(2),z})$ by which only T_0/S transitions can be achieved.

At long times ^Tp_S approaches the values of 2/9 at zero-field and 1/6 at high field. The long time limit at zero-field does not exactly correspond to the spin statistical average of 1/4. This is a consequence of the coherent nature of the spin evolution process, where not all of the initial state information is wiped out. Another feature specific to the coherent nature of the process is notable at intermediate times. Here the function ^Tp_S passes through two shallow maxima reflecting the characteristic times $\tau_{(1)}$ and $\tau_{(2)}$ of the two radicals involved. In the high field limit the coherence is only notable in the t^2 dependence at short times. The limiting value of 1/6 corresponds to the spin statistical ratio of $T_0/S = 1:1$.

The curves corresponding to spin evolution at intermediate fields are all enclosed between the zero-field and high field limits. At short times they start exactly as the zero-field case and then cross over to the high field curve at times varying inversely with B_0 (cf. Figure 3 b). This feature is important for understanding observed magnetokinetic $B^{1/2}$ values. From the characteristic field region where the field dependence of yield effects is most pronounced, one can draw conclusions about the relevant crossover times from low field to high field behavior. These reflect the time constants of processes that are themselves independent of spin evolution (e.g., escape or scavenging processes) but compete with chemical processes depending on it.

The exact solution of the spin motion problem requires quantum mechanical treatment of the nuclear spins, too. In a general analytical form the solution may be readily expressed in terms of the eigenvalues ϵ_k and eigenstates $|k>$ of H:[35]

$$^Tp_S = \frac{2}{3N} \sum_{k,l} |p_{kl}^S|^2 \sin^2\left(\frac{1}{2}\omega_{kl}t\right) \tag{29}$$

Here

$$N = \prod_{i(1)j(2)} (2I_i + 1)(2I_j + 1) \tag{30}$$

is the number of possible combinations of nuclear spin states,

$$\omega_{kl} = (\epsilon_k - \epsilon_l)/\hbar \tag{31}$$

and

$$p_{kl}^S = \langle k|Q_s|l \rangle = \langle k|\frac{1}{4} - \mathbf{S}_1\mathbf{S}_2|l \rangle \tag{32}$$

where Q_S is the operator projecting onto the electronic singlet manifold of electron-nuclear spin states.

Due to the large number 4N of electron nuclear spin states, closed-form expressions even for the most favorable cases of equivalent nuclear spins are rather clumsy. Nevertheless, with the use of computers fairly large nuclear spin systems can be treated. An example is the case of the pyrene$^-$·/dimethylaniline$^+$·RP.[36]

The situation with a general number of nuclear spins is much simpler to treat in the limit of high fields, where, unless resonant B_1 fields are present, only $T_0 \leftrightarrow S$ processes can occur by the coherent mechanism. In this case the nuclear spin states remain, in fact, unchanged and the problem may be decomposed into N independent 2 * 2 problems for the T_0/S process. The general result is

$$^{T_0}p_S = \frac{1}{N} \sum_n \sin^2\omega_n t \tag{33}$$

with

$$\omega_n = \frac{1}{2} [(g_1 - g_2)\beta B_o/\hbar + \sum_{(1)} a_i m_i - \sum_{(2)} a_j m_j] \tag{34}$$

where n denotes a specific combination $\{m_i\},\{m_j\}$, the m_i, m_j referring to the longitudinal spin projections of the individual nuclear spins of the two radicals, respectively. In view of this analytical simplification many theoretical studies of the RPM have been restricted to the case of T_0/S coupling. Although this situation prevails only at high fields the probability $^{T_0}p_S$ (or $^Sp_{T_0}$) may still be variable with the external field due to the term $(g_1 - g_2)\beta B_o/\hbar$. This mechanistic contribution to T_0/S transitions is called the Δg-mechanism or Zeeman mechanism.

b. Incoherent Spin Motion

As mentioned above the anisotropic terms in the spin Hamiltonian are averaged to zero over the typical time scale of coherent spin motion. However, their time-integrated effect on spin motion is not completely averaged out. Due to the uncorrelated fluctuations of the orientations of different radicals the spin-state distribution of an ensemble of RPs starting out from exactly the same initial spin state and having the same nuclear spin configuration will broaden about the state reached by the purely coherent motion. From this very reason the theoretical treatment of the combined effect of coherent and incoherent spin motion requires use of the density matrix. Relaxational terms may be added in a consistent way to the Liouville equation, now representing a stochastic Liouville equation (SLE) (cf. below).

As above for the coherent case we will consider the effects that independent processes in the separate radicals of a pair have on the overall spin evolution of the RP. For this purpose we can make use of the theoretical results obtained for single electron spin relaxation in connection with ESR spectroscopy. A $S = \frac{1}{2}$ system is characterized by two relaxation times, T_1 and T_2, corresponding to the longitudinal and transversal relaxation of magnetization. In terms of the spin density matrix, T_1 is the time constant at which the populations of α and β spins are equilibrized and T_2 the corresponding one for the disappearance of any phase relations between α and β spin states.

From the values of T_1 and T_2 of the radicals constituting the RP, rate coefficients for the time-dependent changes of the RP density matrix elements can be obtained.[37] Here we will give, as an example, the rate coefficients connecting the diagonal elements of the density matrix that may be directly interpreted as transition rate contants between the RP spin substates:

$$k_{T_\pm \to T_0} = k_{T_0 \to T_\pm} = k_{T_\pm \to S} = k_{S \to T_\pm} = \frac{1}{4T_1'} \tag{35}$$

$$k_{T_0 \to S} = k_{S \to T_0} = \frac{1}{2T_2'} - \frac{1}{4T_1'} \tag{36}$$

with the definitions

$$\frac{1}{T_1'} = \frac{1}{T_{1(1)}} + \frac{1}{T_{1(2)}} \quad \text{and} \quad \frac{1}{T_2'} = \frac{1}{T_{2(1)}} + \frac{1}{T_{2(2)}} \tag{37}$$

The relaxational part of the SLE corresponds to a coupled system of first-order rate equations with real coefficients. The separate solution of it would lead to a multiexponential decay of the density matrix toward the thermodynamic equilibrium value. In combination with the coherent part of the SLE the resultant solution corresponds to a damped oscillatory behavior.

In the following we will quote some convenient expressions for T_1. The effect of T_2 processes provides for a $T_0 \to S$ mechanism only. Usually it contributes much less than the coherent $T_0 \to S$ process due to the isotropic hfc or the Δg mechanism.

Expressions for T_1 are based on the general result derived by the first-order perturbation theory (cf. Reference 38).

$$\frac{1}{T_1} = \frac{2}{\hbar^2} \int_{-\infty}^{\infty} \overline{V_{\alpha\beta}^*(t + \tau)V_{\alpha\beta}(t)}\, e^{i\omega_0\tau} d\tau \tag{38}$$

Here $V_{\alpha\beta}(t)$ is the time-dependent perturbation matrix element. The relation expresses the relaxation rate constant $1/T_1$ as the probability of a resonant transition induced by the corresponding Fourier component in the power spectrum of the perturbation. Thus T_1 relaxation may be conceived as a kind of internal magnetic resonance transition by which energy is exchanged between the bath and the spin system (spin lattice relaxation). For an exponential correlation function

$$G(\tau) = \equiv \overline{V_{\alpha\beta}^*(t + \tau)\, V_{\alpha\beta}(t) V_{\alpha\beta}^*(t) V_{\alpha\beta}(t)}\, \exp(-|\tau|/\tau_c) \tag{39}$$

as is usually assumed, the general result of the integral in Equation 38 is

$$\frac{1}{T_1} = \frac{2}{\hbar^2} \overline{|V_{\alpha\beta}|}^2 \frac{2\tau_c}{1 + \omega_0^2\tau_c^2} \tag{40}$$

with τ_c the characteristic correlation time of the particular type of molecular motion controlling the time dependence of V(t).

The most important anisotropic interactions contributing to electron spin relaxation in radicals are anisotropic hyperfine interaction, Zeeman interaction, and spin-rotational interaction.

For *anisotropic hyperfine interaction* of one magnetic nucleus with spin I at a distance r from the radical electron, one has

$$W_{ahf} \equiv \frac{2\overline{|V_{\alpha\beta}\ (ahf)|^2}}{\hbar^2} = \frac{2}{3}\ I(I\ +\ 1)\gamma_I^2 g^2 \beta^2 \langle r^{-6}\rangle$$

$$= 1.22*\ 10^{17}\ \langle\frac{r^{-6}}{\mathring{A}^{-6}}\rangle\ s^{-2} \qquad \text{for } I = \frac{1}{2} \tag{41}$$

The corresponding expression for *anisotropic Zeeman interaction* (g-tensor anisotropy) is

$$W_{gta} \equiv \frac{2\overline{|V_{\alpha\beta}\ (gta)|^2}}{\hbar^2} = \frac{1}{10}\ \overline{\Delta g^2}\ \beta^2 \hbar^{-2} B_o^2 \tag{42}$$

For *spin-rotational interaction* $1/T_1$ is given by [39]

$$\left(\frac{1}{T_1}\right)_{sri} = \frac{\overline{\delta g^2}}{9\tau_c} \tag{43}$$

The latter two mechanisms are due to SOC effects that make the electronic g-tensor anisotropic:

$$\overline{\Delta g^2} = g_{xx}^2 + g_{yy}^2 + g_{zz}^2 - 3g^{-2} \tag{44}$$

or make it deviate from the spin value of a free electron:

$$\overline{\delta g^2} = (g_{xx} - 2.0023)^2 + (g_{yy} - 2.0023)^2 + (g_{zz} - 2.0023)^2 \tag{45}$$

The correlation times τ_c correspond to the mean molecular reorientation time of the radical.

According to Equation 40 the various mechanistic contributions to $1/T_1$ should exhibit characteristic field dependences (cf. Figure 4). For ahf it is constant at low fields and varies as B_0^{-2} when $\omega_0^2\tau_c^2 \gg 1$, i.e., when many Larmor precession cycles occur during the radical's orientational correlation time. For the gta mechanism $1/T_1$ increases as B_0^2 at low fields, due to the field dependence of the perturbation, and it becomes constant at high fields ($\omega_0^2\tau_c^2 \gg 1$) where the B_0^2-type increase of W_{gta} is compensated by the decrease of perturbational power density at ω_0.

The expression for the sri contribution (Equation 43) seems to be of a different type than Equation 40. Actually, in the corresponding expression for the sri mechanism[40] there appears the rotational correlation time τ_R, i.e., the average time at which angular momentum correlation is lost. This is usually by several orders of magnitude smaller than τ_c and the condition $\omega_0^2\tau_R^2 \ll 1$, rendering the power density of sri perturbation independent of the field, holds up to very high field values. The orientational correlation time τ_c and the factor $\overline{\delta g^2}$ appear in Equation 43 because of the relation

$$\tau_R = \frac{\Theta}{6kT\tau_c} \tag{46}$$

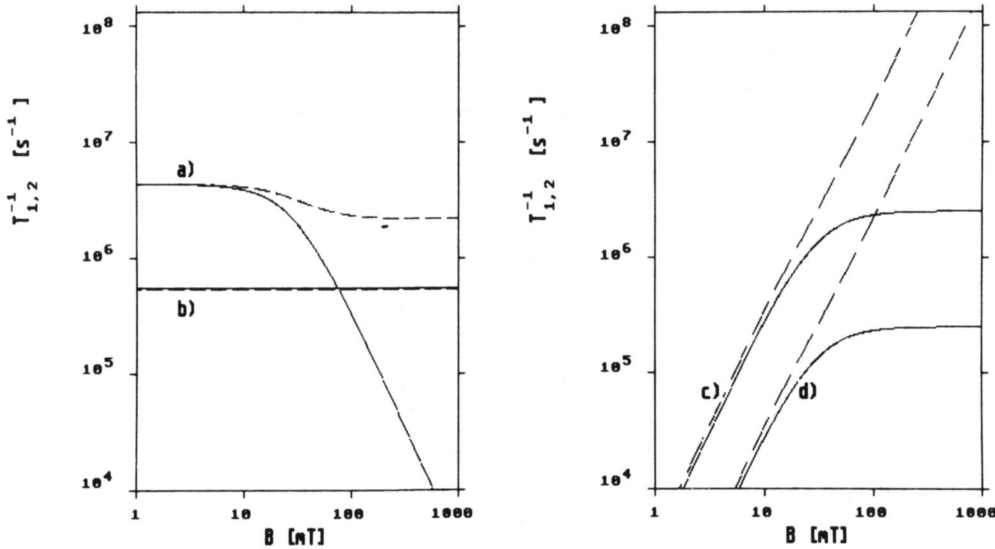

FIGURE 4.   Typical field dependences of electron spin relaxation rate constants $1/T_1$ (solid lines) and $1/T_2$ (dashed lines) according to various mechanisms ($\tau_c = 2*10^{-10}$ s). (a) Anisotropic hyperfine interaction (Equation 41, I = $^1/_2$, $<r^{-6}> = [1.5\ \text{Å}]^{-6}$); (b) Spin-rotational interaction (Equation 43, $\delta g^2 = 10^{-3}$); (c,d) g-tensor anisotropy (Equation 42, [c] $\Delta g^2 = 10^{-2}$, [d] $\Delta g^2 = 10^{-3}$). (Adapted from Lüders and Salikhov.[37])

and the possibility to express the spin-rotational coupling constant originally appearing in $V_{\alpha\beta}$ (sri) in terms of the moment of inertia Θ and δg, whereby the dependence on Θ is canceled.[39]

Apart from individual contributions of the separate radicals to the overall spin evolution in the RP, there are also contributions arising from direct interaction between the radicals in the pair. These are dipolar electron spin-spin interaction and exchange interaction.

Dipolar spin-spin interaction conserves total electron spin of a RP and thus induces $T_\pm \rightarrow T_0$ relaxation only. If the distance r_{12} between the radicals is assumed to be fixed the corresponding rate constants can be expressed as[38,41]

$$k_{T_\pm \rightarrow T_0} = k_{T_0 \rightarrow T_\pm} = \frac{3}{10}\frac{g^4\beta^4}{r_{12}^6\hbar^2}\frac{\tau_{12}}{1 + \omega_0^2\tau_{12}^2} \tag{47}$$

Here τ_{12} is the orientational correlation time of the vector $\mathbf{r}_{12}$.

Spin relaxation may contribute to RP spin evolution on a time scale $\gtrsim 0.2\ \mu$s, and therefore it is thought to play a key role in micellar RP kinetics[41-43] where reencounter periods of this order of magnitude are to be assumed. At fields where coherent $T_\pm \rightarrow S$ transitions caused by isotropic hfc are suppressed the situation in micellar supercages (cf. Section IV) can be adequately described by a classical kinetic scheme[41,44] (cf. Figure 5). Here k_{esc} is the rate constant of micellar exit, and k_S and k_T are the effective first-order rate constants for direct recombination of S- and T-RPs, respectively. The possibility of a direct SOC-assisted reaction of T-RPs to form singlet ground-state recombination products has been emphasized by Levin and Kuzmin[44] (cf. also Section IV). The rate constants $k_{r,1}$, $k'_{r,1}$, and $k_{r,0}$ refer to transitions between $T_\pm/T_0$, $T_\pm/S$, and T_0/S, respectively. Actually, the latter will be usually dominated by the coherent hfc process at any field. However, since $k_{r,0}$ ($\gtrsim 10^8$ s^{-1}) is usually much larger than k_S and k_T ($\lesssim 10^7$ s^{-1}) the actual value of $k_{r,0}$ is not important for the observed kinetics. In the limit $k_{r,0} \rightarrow \infty$ the general time dependence of the RP concentration corresponds to a biexponential decay. For an initial T-RP the solution is[44]

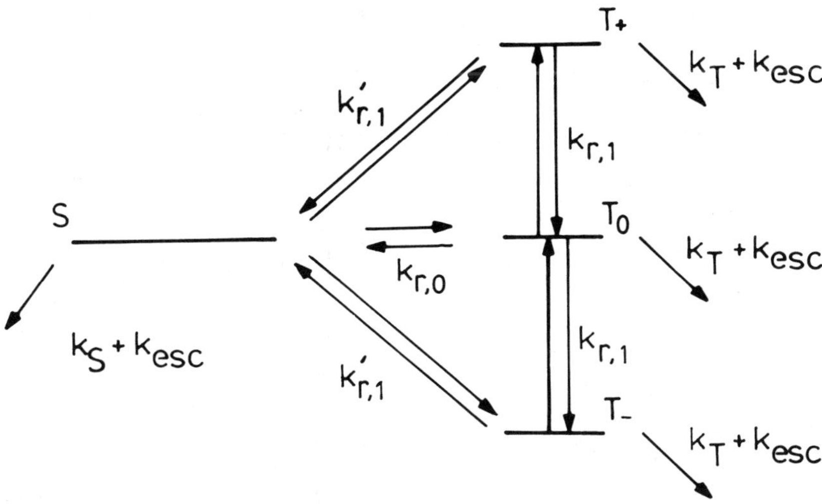

FIGURE 5. Kinetic scheme applicable for describing reactions and interconversions of radical pair spin substates under high-field conditions. For definition of rate constants see text.

$$c_{RP}(t) = c_{RP}(0)(A_+ e^{-k_+ t} + A_- e^{-k_- t}) \tag{48}$$

with

$$k_\pm = \frac{1}{4}(k_S + 3k_T) + k_{esc} + k_{r,1} + k'_{r,1} \pm D \tag{49}$$

$$A_\pm = \frac{1}{2} \mp \frac{1}{6D}\left[\frac{1}{4}(k_S - k_T) + 3(k_{r,1} + k'_{r,1})\right] \tag{50}$$

$$D = \left[(k_{r,1} + k'_{r,1})^2 + \frac{1}{16}(k_S - k_T)^2\right]^{1/2} \tag{51}$$

These relations allow to determine the magnetic field dependence of the sum of relaxation rate constants $k_{r,1} + k'_{r,1} = k_{T_\pm \to S} + k_{T_\pm \to T_0}$ from the observed radical decay kinetics.

C. Exchange Interaction

Exchange interaction is actually an orbital correlation effect and the relation with spin functions arises from the antisymmetry principle obeyed by many-electron wave functions. Wave functions for pure spin states can be represented as single products of a spin part and an orbital part. In the case of a two-electron function by which a RP may be represented one has, for example,

$$\Psi_S = \frac{1}{2}(\alpha_1\beta_2 - \beta_1\alpha_2)[\Phi_a(1)\Phi_b(2) + \Phi_b(1)\Phi_a(2)] \tag{52}$$

$$\Psi_{T_0} = \frac{1}{2}(\alpha_1\beta_2 + \beta_1\alpha_2)[\Phi_a(1)\Phi_b(2) - \Phi_b(1)\Phi_a(2)] \tag{53}$$

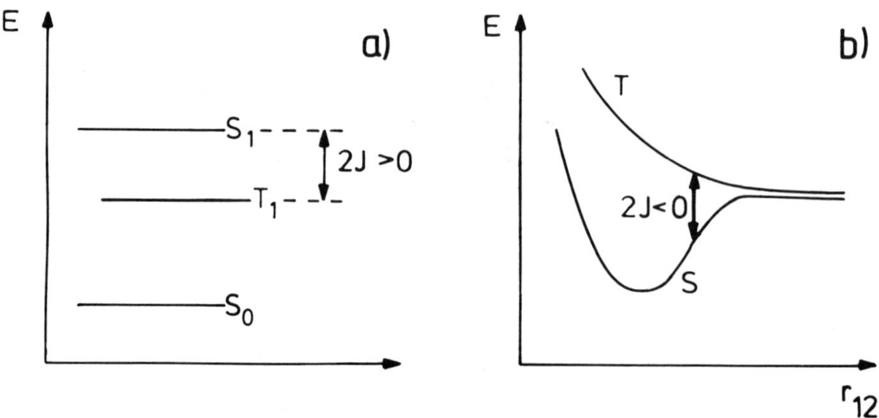

FIGURE 6.   Sign definition of exchange energy J. (a) Positive J is typical for situations where RP spin states correlate with S$_1$ and T$_1$ of the unified electronic system; (b) negative J is observed if RP spin states correlate with bonding and antibonding states of the unified electronic system.

Whereas, with respect to electron exchange, the spin part is antisymmetric for the singlet wave function, it is symmetric for the triplet wave function, the opposite being the case for the orbital parts. Calculating the Coulombic repulsion energy of the electrons places the triplet state at lower energy than the singlet state, provided there is nonzero differential overlap between Φ_a and Φ_b, i.e., if the electrons may "exchange". The situation $E_S > E_T$ is usually encountered if Φ_a and Φ_b are orbitals localized on a single molecule and is the typical situation for electronically excited states (S$_1$/T$_1$ energy gap in molecules; cf. Figure 6a).

If Φ_a and Φ_b are the lowest singly occupied orbitals in two separate radicals there will occur a bonding/antibonding interaction and the degenerate T,S states of the separate pair will correlate with antibonding and bonding states of the recombination product (cf. Figure 6b). In this case $E_T > E_S$ because the orbital wavefunctions of T and S differ not only by exchange symmetry.

Within the spin formalism of a RP one disposes of the explicit representation of the orbital wavefunctions by defining in a phenomenological way "exchange interaction" as a formal spin interaction:

$$H_{ex} = -\hbar J\left(\frac{1}{2} + 2S_1S_2\right) = -\hbar J(S^2 - 1) \tag{54}$$

with $J(r_{12})$ depending on the interradical separation (since no ambiguities can arise, the index 12 will be dropped in the following). With this form of the exchange energy operator S- and T-RP states are eigenstates with energies $E_T = J$ and $E_S = -J$, so that $\Delta E_{TS} = -2J$. The sign convention in Equation 54 is such that in the situation typical for RPs capable of forming σ-bonds, $J < 0$ ("antiferromagnetic interaction", cf. Figure 6b).

The dependence of J on r is usually expressed as (cf. References 45 to 47)

$$J(r) = J_o \exp(-r/r_J) \tag{55}$$

with $J_0 = 10^{17}$ to 10^{23} rad · s^{-1} and $r_J = 0.35$ to 0.5 Å. Due to the diffusional motion of the radicals whereby the interradical separation r is stochastically modulated, the effect of H_{ex} is a dynamic one. Before taking up this point, however, let us shortly consider the case of time-independent exchange interaction to see the principle of its action most clearly.

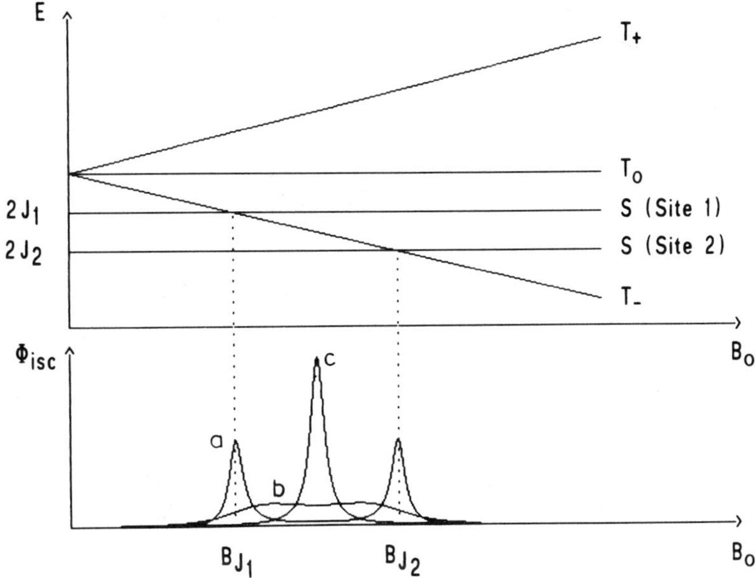

FIGURE 7.   Top: T_-/S level crossing depending on value of J at various RP separations (sites). Bottom: MARY spectra to be expected in the case of dynamic exchange between sites 1 and 2. Parameterization of model calculation: $J_2 = 2J_1$, hopping rate constants between the sites correspond to (a) $k = 0.1\ J_1$, (b) $0.6\ J_1$, (c) $5\ J_1$.

If, for example, one starts with a spin-polarized RP state $\alpha_1\beta_2$ the time-dependence obtained when solving Equation 16 with $H = H_{ex}$ is[363]

$$\chi_{\alpha\beta}(t) = (\cos \frac{J}{\hbar}\, t)\alpha_1\beta_2 - (i \sin \frac{J}{\hbar}\, t)\beta_1\alpha_2 \tag{56}$$

This corresponds to a periodic change between the oppositely polarized states $\alpha_1\beta_2$ and $\beta_1\alpha_2$, i.e., the spins are coherently exchanged between the two radicals, the time for a full round trip being $\pi\hbar/J$.

What is the interaction between H_{ex} and the local hfc contributions H_1 and H_2? If J is much larger than the individual hfc at radical 1 and 2, then the fast spin exchange has the effect that both spins are subject to the same average hyperfine field and spin correlation will be maintained even if local spin motions at both radicals may be quite different.

Simple analytical results are obtained for the effect of exchange interaction if only T_0/S transitions are considered (cf. Reference 48). For example, with ω_n given by Equation 34,

$$^{T_0}P_S = \frac{\omega_n^2}{\omega_n^2 + J^2} \sin^2[(\omega_n^2 + J^2)^{1/2}t] \tag{57}$$

As J increases the T_0/S oscillation frequency is increased but the singlet amplitude is concomitantly decreased. At $J^2 \gg \omega_n^2$ T_0/S transitions are suppressed. In general, RP spin states belonging to different multiplicities can be efficiently mixed by hfc only if their splitting does not exceed the order of magnitude of $\overline{A}_{eff} \approx (\overline{A_{(1)}^2} + \overline{{}_{(2)}^2})^{1/2}$.

The T_0/S energy gap 2 J does not depend on a magnetic field. On the other hand, with T_+ or T_- (depending whether J is positive or negative) an interesting level crossing situation occurs (cf. Figure 7). At the magnetic field $B_J = 2J\hbar/g\beta$ where the level crossing occurs $T_-/$

S (or T_+/S if $J > 0$) transitions induced by hfc will be particularly efficient. Magnetokinetic effects depending on ISC between such crossing levels will exhibit peaks positioned at $B_0 = B_J$, their width being determined by the hfc coupling strength and the chemical lifetime of the T_- and S level.

Dealing with an ensemble of RPs where the distances r, and hence $J_r(r)$, are fixed but distributed according to a probability density function $p(r)$, the observed magnetokinetic effect will be represented by an integral over the distribution:

$$\Phi_{isc}(B_o) = \int dr\, p(r)\Phi_{isc}(B_o,J_r) = \int dJ\, I_o(J)\Phi^o_{isc}(B_o,J) \tag{58}$$

with

$$I_o(J) = p[r(J)][dr(J)/dJ] \tag{59}$$

The function $\Phi_{isc}(B_0)$ may be regarded as a spectrum (so-called MARY spectrum, cf. Section V) representing a superposition of independent MARY lines $\Phi_{isc}(B_0,J_r)$ belonging to different "sites" r. It is a well-known phenomenon, for example in magnetic resonance, that such rigid-solution-type spectra undergo characteristic changes when the medium becomes less rigid and dynamic exchange between different sites takes place on the typical time scale of the experiment.

In a very instructive analysis of MARY spectra of RPs subject to dynamically varied exchange interaction, Schulten and Bittl[46,49] have shown the close analogy to the dynamic line shape problem as it is encountered for randomly modulated harmonic oscillators treated by Kubo.[50] The actual dynamic line shape was shown to be representable to a very good approximation by an expression of the form

$$\Phi_{isc}(B_o) = \int dJ_r\, I(J_r)\Phi^o_{isc}(B_o,J_r) \tag{60}$$

that is formally equivalent to a modified "quasi-statistic" distribution $I(J_r)$ of J_r sites, called the exchange energy spectrum. Thus the effect of dynamic exchange is visualized as an effective change of the static distribution, but without changing the homogeneous line shapes $\Phi^o_{isc}(B_0,J_r)$. If the homogeneous lines are very narrow the MARY spectrum $\Phi_{isc}(B_0)$ corresponds essentially to the exchange spectrum plotted against $B_0/2$ instead of J_r.

In the case of a dynamic equilibrium characterized by the discrete distribution $p_0(r_m)$, a stochastic operator $\hat{I}(r)$ for hopping between the sites, and the r-dependence $J_r(r)$, the general expression for $I(J)$ is[49,51]

$$I(J) = \frac{2}{\pi}\,\text{Re}\sum_{m'}\sum_{m}\langle m'|[2i(\hat{J}_r - J\hat{1}) - \hat{1}]^{-1}|m\rangle p_0(r_m) \tag{61}$$

Here $\hat{J}_r$, $\hat{1}$, and the unit operator $\hat{1}$ operate on vectors $[p(r_1),\ldots,p(r_N)]$ representing the distribution of the ensemble over the different sites $r_1,\ldots,r_N$. The symbol $|m\rangle$ represents the *m*th unit vector $[0,\ldots,p(r_m) = 1,0,\ldots]$. The operator $\hat{J}_r$ is diagonal representing the function $J_r(r)$:

$$\hat{J}_r = \text{diag}[J(r_1)\ldots.J(r_N)] \tag{62}$$

The application of Equations 60 and 61 shall be demonstrated with an example dealing with two sites only, that have equal equilibrium populations, viz.,

$$\hat{J}_r = \text{diag}(J_o, 2J_o) \tag{63}$$

$$\hat{1} = k\begin{pmatrix} -1 & 1 \\ 1 & -1 \end{pmatrix} \tag{64}$$

The solution to Equation 61 is

$$I(J) = \frac{k/\pi}{4(J_o - J)^2(2J_o - J)^2 + (3J_o - 2J)^2k^2} \tag{65}$$

which for a narrow homogeneous line shape of

$$\Phi_{isc}^o(B,J) = c_\Phi \delta(B_o - 2J) \tag{66}$$

would yield a MARY spectrum given by

$$\Phi_{isc}(B_o) = \frac{2}{\pi} c_\Phi \frac{k}{(2J_o - B_o)^2(4J_o - B_o)^2 + 4(3J_o - B_o)^2k^2} \tag{67}$$

In Figure 7 the resultant spectrum is plotted for various values of k, the rate constant of hopping between the two sites. For slow hopping rates the spectrum corresponds to the static distribution of J_r values. With k increasing the two lines broaden, show a coalescence, and finally motional narrowing is observed with one sharp line centering at the average J_r-value. Instructive experimental examples of dynamic exchange spectra have been reported for chemically linked radical ion pairs (cf. Section III.A).

2. Combining Spin Evolution, Diffusion, and Reaction

Two different approaches have been followed in quantitative theories of the magneto-kinetic behavior of RPs. In the first one, aiming for evaluation of time-integrated yields only, one considers the statistics of re-encounters, summing the averaged yield contributions of an infinite series of re-encounters. In this case the partial sums cannot be assigned to definite times and therefore this ensemble-averaging procedure may be termed "asynchronous".

In the second one the statistics of the complete ensemble of RPs is treated synchronously by solving a SLE for the distance- and time-dependent RP spin-density matrix.

The *re-encounter method* approach introduced into CIDNP and CIDEP theory of RPs by Adrian[52-54] and by Kaptein[26] has been further developed by Pedersen[55] and by Salikhov.[56]

The spin motion in the separate RP may be conveniently expressed in a form independent of the initial state of the ensemble by using a time-dependent super operator F^x (cf. References 32 and 56) so that

$$\rho(t) = F_t^x \rho(0) \tag{68}$$

For a RP separating from a nonreactive contact at t = 0, undergoing free spin evolution according to F_t^x, and having a first re-encounter at time t, the probability of recombination in that encounter is described by

$$p_t^{(1)} = \text{Tr}\{P^x F_t^x Q^x \rho(0)\} \tag{69}$$

where P^x is a suitable super operator describing spin-selective reaction and the effect of an unreactive encounter on the RP density matrix may be expressed by another super operator Q^x. Averaging over the distribution f(t) of first re-encounter times results in

$$p^{(1)} = \mathrm{Tr}\{P^x F^x Q^x \rho(0)\} \tag{70}$$

with

$$F = \int_o^\infty F_t^x(t) f(t) dt \tag{71}$$

The probability that the RP will react in the n*th* encounter is

$$p^{(n)} = \mathrm{Tr}\{P^x (F^x Q^x)^n \rho(0)\} \tag{72}$$

Summing an infinite series of reencounters finally yields the general result:

$$p = \mathrm{Tr}\{P^x (1 - F^x Q^x)^{-1} \rho(0)\} \tag{73}$$

Solutions of this equation for general fields have been worked out by Salikhov[56] utilizing the semiclassical representation of spin motion to express the super operator F^x. These calculations involved numerical averaging of analytical integrals over the distribution of all possible hyperfine fields.

For the high-field situation where only T_0/S mixing is to be considered, the closed form results for individual hyperfine states of RPs can be given. We will specify their solution[55] in some more detail since this allows us to comment on some problems arising with the form of the super operator Q^x and since in a recent extension of this case by Vollenweider and Fischer[57] it was shown how to incorporate the effect of exchange interaction into it.

The relevant part of the RP spin density matrix may be represented as a vector

$$\rho \equiv [2I_m(\rho_{ST_o}), \rho_{T_oT_o}, \rho_{SS}]^t \tag{74}$$

For the case of singlet selective reactivity P^x is represented as

$$P^x = \mathrm{diag}(0,0,\lambda_S) \tag{75}$$

For Q^x Pedersen[55] chose the form

$$Q^x_{(P)} = \mathrm{diag}(1,1,1 - \lambda_S) \tag{76}$$

It should be pointed out that this form of Q^x does not strictly conserve all the properties that the density matrix should have in order to represent real physical situations (cf. Reference 58). Assuming, for example, $\lambda_S = 1$, application of Q^x to ρ would leave us with a density matrix $[2I_m(\rho_{ST_o}) \neq 0, \rho_{T_oT_o}, 0]$ with zero singlet probability but still nonvanishing S/T_0 phase contributions. A physically sound form of Q^x would be

$$Q^x_{(H)} = \mathrm{diag}(\sqrt{1 - \lambda_S}, 1, 1 - \lambda_S) \tag{77}$$

corresponding to the reaction operator justified by Haberkorn[58] (cf. Equation 102) and implying the required minimum loss of S/T_0 phase correlation during a spin-selective reaction. Another possibility is

$$Q^x_{(S)} = \mathrm{diag}(0,1,1 - \lambda_S) \tag{78}$$

implying complete loss of S/T phase correlation in an encounter. The latter can be physically justified as an effect of strong exchange interaction notable during a RP contact. Equation 78 corresponds to the form of Q^x used by Salikhov.

The matrix of F^x is given by

$$F^x = p \begin{pmatrix} C & -S & S \\ {}^1\!/_2\, S & {}^1\!/_2\,(1 + C) & {}^1\!/_2\,(1 - C) \\ -{}^1\!/_2\, S & {}^1\!/_2\,(1 - C) & {}^1\!/_2\,(1 + C) \end{pmatrix} \tag{79}$$

with

$$C = \frac{1}{p} \int_o^\infty f(t)\cos(2\omega_n t)dt \tag{80}$$

$$S = \frac{1}{p} \int_o^\infty f(t)\sin(2\omega_n t)dt \tag{81}$$

and p the total probability of at least another contact after separation

$$p = \int_o^\infty f(t)dt \tag{82}$$

The frequency ω_n is obtained according to Equation 34. The result, for example, for a RP originating in T_0, may be cast in the form

$${}^{T_o}p_S = \frac{\Lambda\,{}^{T_o}p_S^*}{1 + (1 - \Lambda)\,{}^{T_o}p_S^*} \tag{83}$$

where Λ corresponds to the hypothetic recombination yield obtained if both T_0 and S spins would react with the same probability λ_S, and ${}^{T_o}p_S^*$ denotes the hypothetic recombination yield for S-RP recombination with $\lambda_S = 1$. Thus Λ and ${}^{T_o}p_S^*$ may be considered as measures of spin-free reactivity and effective spin conversion, respectively. The result is

$$\Lambda = \frac{\lambda_S}{1 - p(1 - \lambda_S)} \tag{84}$$

and

$${}^{T_o}p_S^* = \frac{p(1 - C) + p^2 X(C^2 + S^2 - C)}{2 - p(1 + C + 2CX) + p^2 X(C^2 + S^2 + C)} \tag{85}$$

Here the value X depends on the form chosen for Q^x. For $X = 1$ it corresponds to the original Pedersen result ($Q_{(P)}^x$). For $X = 0$ and $X = \sqrt{1 - \lambda_S}$ it corresponds to the versions $Q_{(S)}^x$ and $Q_{(H)}^x$, respectively. Actually, for $X = {}^{T_o}p_S^*$ the expression (Equation 85) for ${}^{T_o}p_S^*$ is no longer independent of λ_S. In that case complete separation of spin-motion-dependent terms and reactivity-dependent terms is not possible since, by the choice of $Q_{(H)}^x$, a reactivity-dependent effect on S/T_0 phase evolution has been introduced.

Employing the Noyes' re-encounter probability function (Equation 3) C and S are expressed as[57]

$$C = \exp(-Z)\cos(Z) \qquad (86)$$

$$S = \exp(-Z)\sin(Z) \qquad (87)$$

with

$$Z = \frac{3}{2}[(1 - p)^2/p](a/\sigma)(a^2/D)^{1/2}\omega_n^{1/2} \qquad (88)$$

In an extension of Pedersen's treatment it has been shown by Vollenweider and Fischer[57] that the effect of exchange interaction may be incorporated into the re-encounter formalism by replacing in Equations 86 and 87 C and S by C' and S' defined as

$$C' = (C + \varkappa)/(1 + \varkappa) \qquad (89)$$

$$S' = S/(1 + \varkappa) \qquad (90)$$

with

$$\varkappa = [5r_J/(3a)]\ln(1 + J_a ar_J/D)\ln(1 + \delta^{0.9}) \qquad (91)$$

where

$$\delta = (\omega_n a^2/D)^{1/2} \qquad (92)$$

and

$$J_a = J_o\exp(-a/r_J) \qquad (93)$$

The result of Equation 85 is greatly simplified when reducing it to the limit $\sigma \to 0$, corresponding to continuous diffusion.

Thus for $X = 0$ (case $Q_{(S)}^x$) one obtains

$$^{T_o}p_S^* = \frac{\delta}{\delta + 2} \qquad (94)$$

which together with Equation 83 corresponds exactly to the result obtained from a direct solution of the continuous diffusion equation with suitable boundary conditions by Salikhov et al.[59]

For $X = 1$ (case $Q_{(P)}^x$) the continuous diffusion limit is

$$^{T_o}p_S^* = \frac{\delta(1 + 2\delta)}{2 + 3\delta + 2\delta^2} = \frac{\delta}{\delta + 2}\frac{1 + 2\delta}{\dfrac{2 - \delta}{2 + \delta} + 2\delta} \qquad (95)$$

As can be seen from the rearranged second expression, a recombination requiring spin inversion is always more efficient when treated according to conditions (P) than according to (S). The difference disappears as $\delta \ll 2$.

If the Vollenweider and Fischer treatment of exchange interaction is included the result has the same form as Equation 95 but δ has to be replaced by

$$\delta' = \frac{\delta}{1 + \varkappa} = \left(\frac{a^2}{D} \frac{\omega_n}{(1 + \varkappa)^2}\right)^{1/2} \tag{96}$$

Thus the effect of exchange interaction expressed in $\varkappa$, is seen to correspond to a reduction of the effective S/T_0 transition frequency.

Finally, for $X = (1 - \lambda_S)^{1/2}$ (case $Q^x_{(H)}$) the continuous diffusion limit is

$$^{T_o}p^*_S = \frac{\delta[2 - \Lambda + 4(1 - \Lambda)\delta]}{2(2 - \Lambda) + (6 - 5\Lambda)\delta + 4(1 - \Lambda)\delta^2} \tag{97}$$

As can be easily verified, for $\Lambda \to 0$ this expression approaches the case (P) result and for $\Lambda \to 1$ the case (S) result.

For small $\delta < 1$ the expressions in Equations 95 to 97 yield results that coincide to first order in δ:

$$\lim_{\delta \to 0} {}^{T_o}p^*_S = \frac{\delta}{2} \tag{98}$$

In this limit the total recombination yield is

$$^{T_o}p_S = \frac{1}{2} \Lambda \delta \tag{99}$$

implying a proportionality to the square root of the S/T_0 transition frequency ω_n and the inverse of the diffusion constant D.

To illustrate the effect of the various forms of Q^x and of exchange interaction the relative field dependence of the free radical yield $(1-{}^{T_o}p_S)$ has been calculated for RPs where ISC is due to the Δg-mechanism only (cf. Figure 8). The parameters J_0 and r_J used for the exchange interaction in curve 4 correspond to those that have been used in a successful modeling of the magnetokinetic behavior of chemically linked RPs[45,46] (cf. Section III.A). As can be seen in Figure 8 effects of such an exchange interaction are also noteworthy in the behavior of freely diffusing RPs, where they cause a considerable decrease of the MFE that would be expected on the basis of reaction term $Q^x_{(P)}$ with no extra exchange (curve 1). For a $= 5$ Å the MFE with the Salikhov-type reaction term $Q^x_{(S)}$ (curve 3) is intermediate between parameter cases 1 and 4. It must be noted, however, that the exchange effect due to the $Q^x_{(S)}$ reaction term increases relative to that of case 4 ($Q^x_{(P)}$,J) if the reaction diameter a is increased. For a $= 7$ Å, for example, a value that may apply to fast electron transfer reactions, the exchange effect of $Q^x_{(S)}$ is even stronger than that of case 4 with the parameters of J as in curve 4.

As to be expected, for $\Lambda = 0.7$ the relative position of curve 2, corresponding to reaction term $Q^x_{(H)}$, is intermediate between cases $Q^x_{(P)}$ and $Q^x_{(S)}$. For $\Lambda \to 1$ it approaches the result for $Q^x_{(S)}$; for $\Lambda \to 0$ it corresponds to $Q^x_{(P)}$.

It should be noted, though, that so far there are no experimental examples of magnetokinetic effects with freely diffusing RPs, where all the magnetokinetically relevant parameters are known with such an accuracy as to allow for an unambiguous demonstration of the correct reaction/exchange interaction term.

a. Stochastic Liouville Equations

The maximum of information on an ensemble of RPs is expressed by the spin density matrix $\rho(\mathbf{r},\mathbf{\Omega}_1,\mathbf{\Omega}_2,t)$ characterizing the spin situation at any time t for RP subensembles, with specified separation vector $\mathbf{r}$ and radical orientations $\mathbf{\Omega}_1$ and $\mathbf{\Omega}_2$. The description of

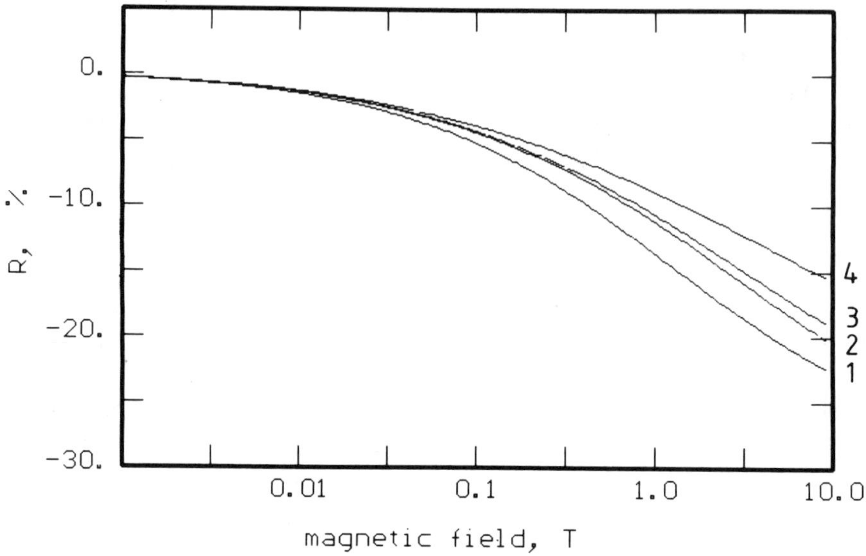

FIGURE 8.   Theoretical magnetic field dependence (according to Equations 83 and 85) of relative change R in free radical yield if the Δg-mechanism is operating in a T-RP that may undergo singlet recombination only. Parameters: $D = 10^{-6} \text{ cm}^2 \text{ s}^{-1}$, $a = 5$ Å, $\Delta g = 0.01$, $\Lambda = 0.7$, σ: continuous diffusion limit. The reaction terms used were (1) $Q^x_{(P)}$, (2) $Q^x_{(H)}$, (3) $Q^x_{(S)}$ with $J_0 = 0$, and (4) $Q^x_{(P)}$ with $J_0 = 1.67*10^{17} \text{ rad*s}^{-1}$, $r_J = 0.47$ Å (the exchange energy parameters correspond to those used in References 45 and 46).

simultaneous motion in configuration space $(\mathbf{r}, \mathbf{\Omega}_1, \mathbf{\Omega}_2)$ and Liouville space of the spins may be cast into the general form

$$\dot{\rho} = (H^x + R^x + K^x + L^x)\rho \qquad (100)$$

where the super operators describe the effects of coherent spin motion (H^x), spin relaxation (R^x), reaction (K^x), and stochastic motion in configuration space (L^x).

The Liouville operator H^x is represented by the spin Hamiltonian H as

$$H^x\rho = -\frac{i}{\hbar}[H,\rho] \qquad (101)$$

The typical spin Hamiltonian has been specified above. It comprises isotropic and anisotropic Zeeman and hyperfine interaction that are generally independent of $\mathbf{r}$. Distance-dependent terms are due to exchange interaction and electron spin-dipolar interaction. Consideration of the latter is, however, rather exceptional (cf. References 60 and 61).

The super operator of spin relaxation R^x may be included to dispose of an explicit treatment of the rotational diffusion that is usually much faster than the other processes of interest. Then, only the spin density matrix $\rho(\mathbf{r},t)$ averaged over $\mathbf{\Omega}_1$ and $\mathbf{\Omega}_2$ must be explicitly considered in Equation 100. The elements of R^x are conveniently expressed in terms of T_1 and T_2 of the individual radicals (cf. References 37 and 62). It should be noted, however, that inclusion of spin relaxation is usually considered as unimportant on the time scale of coherent spin motion and of the typical cage time of RPs in homogeneous solution. Therefore the $(\mathbf{\Omega}_1, \mathbf{\Omega}_2)$ dependence of ρ as well as its substitution by a relaxation super operator R^x is usually omitted when treating the SLE.

The reaction super operator K^x is usually expressed as

$$K^x = K_S^x + K_T^x = -\frac{1}{2} k_S(r)[Q_S,\rho]_+ - \frac{1}{2} k_T(r)[Q_T,\rho]_+ \tag{102}$$

with Q_S and Q_T the projection operators on S and T manifold of spin states.

Use of the anticommutator $[A,B]_+ \equiv (AB + BA)$ warrants conservation of all physically necessary properties of ρ.[58] This does not generally apply to the form

$$K_{S(T)}^x = -k_{S(T)}Q_{S(T)}\rho Q_{S(T)} \tag{103}$$

as employed, e.g., in References 55 and 63.

The distance dependence of the rate constants $k_S(r)$ and $k_T(r)$ is usually represented as a step function, nonzero only at $r = a$, when the RP is in "contact". (For more general treatments cf. References 51 and 64.)

From the solution $\rho(\mathbf{r},t)$ to Equation 100 the yields into the various recombination channels are obtained. For example,

$$p_S(t) = \int_o^t \int_{(r)} k_S(r)\mathrm{Tr}\{Q_S\rho\}dt \; d^3\mathbf{r} \tag{104}$$

The stochastic operator L^x describes the time evolution of the RP probability distribution over $(\mathbf{r},\mathbf{\Omega}_1,\mathbf{\Omega}_2)$, but is usually considered explicitly for translational diffusion $(\mathbf{r})$ only. L^x may be represented as a Smoluchowski operator, e.g.,

$$L^x\rho(\mathbf{r},t) = D[\nabla^2\rho + \frac{1}{kT} \nabla(\nabla U(\mathbf{r}))\rho] \tag{105}$$

where D is the sum of the diffusion constants of both radicals and $U(\mathbf{r})$ denotes an inter-RP potential, e.g., due to Coulomb interaction. Usually, as in Equation 105, L^x does not operate on the spin part of $\rho(\mathbf{r},t)$. If, however, the effect of exchange energy is included in $U(\mathbf{r})$, allowance for a different diffusional behavior of S- and T-RPs can be made (cf. Reference 65).

Mathematically the SLE corresponds to a set of N coupled linear partial differential equations, where N is the dimension of the electronic-nuclear spin space. As mentioned above, except for some solid-state situations[61,66,67] or several investigations concerned with orientation-dependent reactivity,[68-72] the Ω-dependence is usually neglected. Furthermore, the $\mathbf{r}$-dependence is assumed to be isotropic so that only one configurational coordinate (r) remains. The main obstacle against a closed-form analytical solution is that, in general, the discrete quantum problem cannot be separated from the continuous diffusional problem if H^x or K^x is considered as $\mathbf{r}$-dependent. In this general case, without any drastic simplifications, only numerical solutions employing discretization in the $\mathbf{r}$- and eventually also in the t-domain are possible. Examples may be found in the work of Pedersen and Freed[63,73-75] and of Schulten and co-workers.[36] This method has been particularly important in providing information on the role of exchange interaction,[76] noteworthy above all in the case of biradicals or chemically linked radical pairs (cf. References 64 and 77 to 81).

A limiting case of separability of spatial and quantum variables is given if the $\mathbf{r}$-dependence of K^x is limited to RP contact and exchange interaction is represented by a δ-function at some $r_J \geq a$. Then the $\mathbf{r}$-dependent coupling of quantum states may be accounted for in the form of boundary conditions and an analytical disentangling of the diffusional and the quantum problem is possible.[59,82-87] Still, however, for the case of low magnetic fields the solutions are only available as series expansions. Closed form expressions have been given for the case of S/T_0 coupling at high fields only.[59,82,84,85]

If the spatial discretization as applied in numerical methods is reduced to two positions (contact and irreversible separation) or three positions (contact, reversible separation, irreversible separation), the so-called exponential one- or two-site RP models are obtained. These seem particularly adequate for RPs generated in photosynthetic reaction centers.[60,88-93] They have also been applied to the situation of RPs in micelles.[37] Even with the simple one-site exponential model an analytical closed-form solution is available only for the high field case (with general nuclear spin system), and for the zero-field case (with a general number of nuclear spins but coupled to one electron spin only and all with the same coupling constant[94]).

B. OTHER PAIR MECHANISMS

The RPM is the most important case of the general paramagnetic pair mechanism comprising also the cases of a pair of triplets (TT pairs) and of a triplet and a doublet (TD pairs). TT and TD pairs are, for example, encountered in the spectroscopy of molecular crystals or in electrochemically generated luminescence (cf. Reference 1).

Thus, triplet-triplet annihilation (in crystals also called triplet exciton fusion) may lead to excited singlets,

$$T + T \rightleftarrows (TT) \rightarrow S^* + S_o \tag{106}$$

a mechanism producing delayed fluorescence (Equation 106). The reverse of this process does also occur (so-called singlet exciton fission, Equation 107):

$$S^* + S_o \rightleftarrows (TT) \rightarrow T + T \tag{107}$$

The interaction between a triplet and a doublet species may lead to triplet deactivation with dissipation of electronic energy or, eventually, electronic excitation of the doublet molecule (Equation 108):

$$T + D \rightleftarrows (TD) \rightarrow S_o + D \text{ (or } D^*) \tag{108}$$

We will not go as deeply into the theoretical details here as in the case of the RPM (for details cf. References 1 and 95), but will only introduce the basic model applied to these cases. Also, it will be interesting to compare the underlying concept with that of the usual description of the RPM.

The general form of the pair spin Hamiltonian is

$$H = H_1 + H_2 + H_{12} \tag{109}$$

As for the DD pairs (RPM) the MFEs are essentially due to the independent contributions of H_1 and H_2 to "spin motion". The influence of the interaction term H_{12} on spin motion has to be considered in quantitative treatments only. For a molecular triplet the spin Hamiltonian is of the form

$$H = H_{zfs} + H_Z \tag{110}$$

The zero-field splitting operator H_{zfs} describes the effects of electron spin-spin dipolar interaction but can also account for effects of SOC. The energetic splitting of the triplet substates due to H_{zfs} is typically by two orders of magnitude larger than the energy of hfc, the most important interaction in radicals. Therefore, in TT and TD pairs the influence of the hfc may be neglected in a first-order approximation. In each of the eigenstates (T_x, T_y, T_z)

of H_{zfs} the electronic spin is oriented perpendicular to one of the principal axes of the ZFS tensor. As the Zeeman interaction increases the direction of spin quantization is more and more dominated by the magnetic field direction, through the Zeeman Hamiltonian H_Z.

As different from the typical situations where the RPM applies, TT and TD pairs are usually not generated as pairs with a definite spin correlation (an exception is the so-called singlet fission process in molecular crystals, cf. Reference 95), but originate from random encounters of free T and D species. The second-order rate constant of homogeneous recombination may be written as:

$$k_{TT} = {}^{TT}p_S k_{diff} \tag{111}$$

$$k_{TD} = {}^{TD}p_D k_{diff} \tag{112}$$

where ${}^{TT}p_S$ denotes the probability that on an encounter a pair will react to a singlet product and ${}^{TD}p_D$ is the corresponding quantity for a TD pair reacting to a product with overall doublet multiplicity.

According to Merrifield[96] the reaction probabilities may be expressed as

$$^{TT}p_S = \frac{1}{9} \sum_i \frac{k_S P_i^S}{k_{esc} + k_S P_i^S} \tag{113}$$

and corespondingly

$$^{TD}p_D = \frac{1}{6} \sum_i \frac{k_D P_i^D}{k_{esc} + k_D P_i^D} \tag{114}$$

The concept underlying these expressions is that the eigenstates $|i\rangle$ of the spin Hamiltonian (not the eigenstates of the total electron spin of the pair) are equally populated in a re-encounter and that they are kinetically independent during the encounter, where they undergo recombination with a rate constant $k_S P_i^S$ or $k_D P_i^D$ in the TT pair and TD pair case, respectively. During an encounter spin-selective recombination competes with spin-independent dissociation (rate constant k_{esc}).

The reaction probabilities ${}^{TT}p_S$ and ${}^{TD}p_D$ are magnetic field-dependent, because the sums in Equations 113 and 114 depend on the way that the singlet probability (P_i^S) of a TT pair or the doublet probability (P_i^D) of a TD pair is distributed among the different energy eigenstates $|i\rangle$.

Let us, for example, consider the two extreme cases $P_i^S = 1/9$ and $P_i^S = \delta_{i,1}$. For these one obtains from Equation 113

$$^{TT}p_S(P_i^S = 1/9) = \frac{1}{9} \frac{k_S}{k_{esc} + 1/9 k_S} \tag{115}$$

$$^{TT}p_S(P_i^S = \delta_{i,1}) = \frac{1}{9} \frac{k_S}{k_{esc} + k_S} \tag{116}$$

The results indicate that the reaction probability ${}^{TT}p_S$ is the higher the more even the singlet character is distributed among the energy eigenstates. The general effect of a magnetic field is to concentrate the singlet character on fewer energy eigenstates so that ${}^{TT}p_S$ decreases in a magnetic field. Analogous rules hold true for TD pairs.

Why is the Merrifield model not applicable to the case of RPs? The essential point here is not the presence or absence of an initial spin correlation in the pair, since the Merrifield

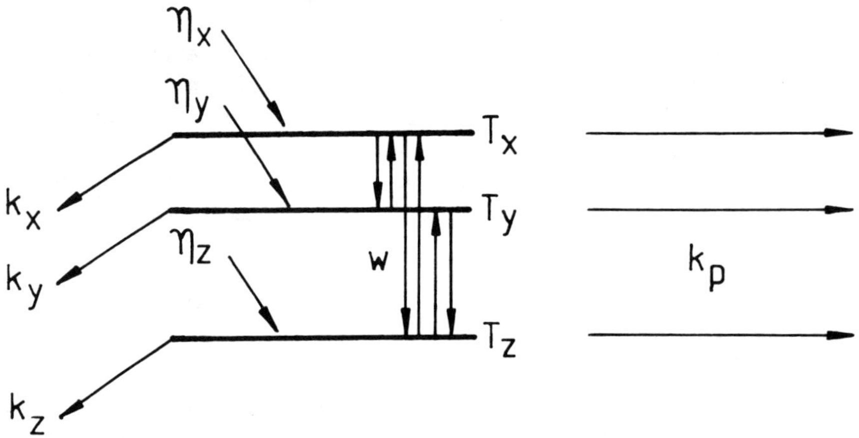

FIGURE 9.   General reaction scheme for the triplet mechanism (TM).

approach may be adjusted to the case of initially spin-correlated pairs by multiplying the terms in Equations 113 and 114 by corresponding population factors, for example, for a $^1(TT)$ pair:

$$^{1(TT)}P_S = \sum_i \frac{k_S(P_i^S)^2}{k_{esc} + k_S P_i^S} \tag{117}$$

However, in the RPM one does not start with incoherently populated eigenstates but with spin-correlated states that, in an energy eigenstate basis, would generate nonvanishing average phase relations between the basis states. These phase relations decay on a time scale determined by the average spacing of the energy eigenstates. For a RP these are on the order of the hfc constants; for TT and TD pairs they are on the order of the ZFS parameters and therefore the phase relations between the energy eigenstates are lost on largely different time scales. For RPs the time scale is in the order of $1/k_S$ or $1/k_{esc}$ and therefore spin motion must be treated explicitly. For TT and TD pairs the picture of quasi-stationary energy eigenstates is, however, quite appropriate. (An exception is the case of degenerate eigenstates of the spin Hamiltonian. Such situations may occur as level crossing phenomena giving rise to typical magnetokinetic resonances when varying the magnetic field or the orientation of single crystals.[95])

It shall be mentioned that in micellar supercages k_S and k_{esc} are greatly reduced in comparison to homogeneous solution. Here the Merrifield picture should be also appropriate for RPs as far as the role of the coherent spin processes is concerned. Corresponding treatments have been recently suggested.[97,98]

C. THE TRIPLET MECHANISM (TM)

Magnetokinetic effects according to the TM are based on the general reaction scheme shown in Figure 9. It represents population and decay of an excited molecular triplet state with a detailed consideration of the triplet substates T_x, T_y, T_z to be specified below. In contrast to the RPM (or other pair mechanisms) where the principle of spin conservation determines the selection rules for RP spin substate decay, in the TM the selectivity of various reaction channels open to the triplet substates[99-101] is due to symmetry selection rules governing SOC.[102]

In Figure 9 η_x, η_y, η_z denote the population efficiencies of the triplet sublevels. These will, in general, be different if the population is due to ISC from an electronically excited singlet state, but may be equal if it is due to energy transfer from or to chemical transformation of a relaxed precursor triplet. The decay of the triplet may occur by substate-selective processes, usually ISC to the singlet ground state, with rate constants k_x, k_y, k_z and by nonselective processes (k_p) that we will identify here with a chemical product formation channel. The triplet substates may be interconverted by transitions represented in Figure 9 as a stochastic process with rate constant w. Magnetic effects, e.g., on the product yield Φ_p, arise from a MFE on the effective value of w. All the other parameters are considered to be magnetic field-independent. As will be shown below w increases monotonically with the field.

Writing for short

$$x = k_x + k_p + 2w \qquad y = k_y + k_p + 2w \qquad z = k_z + k_p + 2w \tag{118}$$

the product yield can be expressed as

$$\Phi_p = \sum_{i=x,y,z} \eta_i{}^{T_i}\Phi_p = k_p \sum_{(ijk)} \eta_i \frac{jk + (j + k)w + w^2}{xyz - (x + y + z)w^2 - 2w^3} \tag{119}$$

where (ijk) means a cyclic permutation of (xyz). In the case that population is not sublevel-selective the result simplifies to

$$\Phi_p = \frac{k_p}{3} \frac{xy + yz + zx + 2(x + y + z)w + 3w^2}{xyz - (x + y + z)w^2 - 2w^3} \tag{120}$$

From the analysis of Equations 119 and 120 it becomes clear that magnetokinetic effects on the effective rate w of triplet substate interconversion can be borne out in any product yield only if the *decay* is sublevel selective. Selective *population* of the substates is neither necessary nor sufficient to produce magnetic effects on product yields. This is a difference to the role of the TM in producing electron spin polarization in radical products (CIDEP, cf. Reference 103), where selective population is a sufficient condition.

For $k_x = k_y = k_{x,y}$ Equation 120 is greatly simplified yielding explicitly[104]

$$\Phi_p = k_p \frac{\frac{1}{3} k_{x,y} + \frac{2}{3} k_z + k_p + 3w}{(k_{x,y} + k_p)(k_z + k_p) + (2k_{x,y} + k_z + 3k_p)w} \tag{121}$$

To see the principal effect due to a magnetic field let us specify Φ_p for $w \to 0$ and $w \to \infty$.

For $w \to 0$

$$\Phi_p = \frac{1}{3} \frac{k_p}{k_z + k_p} + \frac{2}{3} \frac{k_p}{k_{x,y} + k_p} \tag{122}$$

For $w \to \infty$

$$\Phi_p = \frac{k_p}{\frac{2}{3} k_{x,y} + \frac{1}{3} k_z + k_p} \tag{123}$$

For $k_{x,y} \neq k_z$ the product yield is always larger in Case 122, where the levels are completely

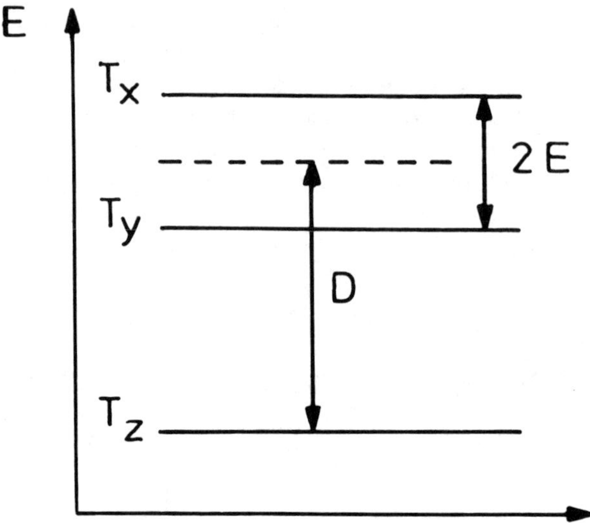

FIGURE 10.   Definition of zero-field parameters D and E of a triplet state.

decoupled, than in Case 123 where they decay as one level with averaged decay constants. From Equation 121 one can deduce that significant magnetic field effects will ensue only if

$$3w(B = 0) \lesssim \bar{k}_i \quad \text{and} \quad k_p < \bar{k}_i \tag{124}$$

implying that in liquid solutions the decay of the triplet must be rather fast (cf. Section III.A.2).

To describe the spin orientation in the triplet excited molecule from the kinetic point of view it is most appropriate to use as a basis the eigenstates of the zero-field spin Hamiltonian. These may be labeled according to the axes x, y, z of the ZFS tensor, fixed within the molecular frame.

$$\chi_{T_x} = \frac{1}{\sqrt{2}} (-\alpha_1\alpha_2 + \beta_1\beta_2) \tag{125}$$

$$\chi_{T_y} = \frac{1}{\sqrt{2}} (\alpha_1\alpha_2 + \beta_1\beta_2) \tag{126}$$

$$\chi_{T_z} = \frac{1}{\sqrt{2}} (\alpha_1\beta_2 + \beta_1\alpha_2) \tag{127}$$

In each of these states the direction of the spin ($S = 1$) is quantized perpendicular to the respective axis. The ZFS of the three H_{zfs} substates is usually described in terms of the quantities D and E (cf. Reference 102) defined in Figure 10. For aromatic π-systems, where the ZFS is determined by dipolar spin-spin interaction of the two unpaired electrons, the main splitting parameter D is in the order of 0.1 cm^{-1}. Both D and E decrease with the average separation of the electrons.

In an external magnetic field T_x, T_y, and T_z are no longer the energy eigenstates. We will, however, still keep to this basis because the substate selectivity of the triplet decay processes is fixed within the molecular frame. The effect of an external magnetic field will then have to be described as a time-dependence ("spin motion") in the basis (T_x, T_y, T_z) of states which have the right spin orientation in the molecular frame.[105] This approach is completely equivalent to that followed in the RPM, where spin selectivity of the reaction suggests to use S and T substates as a basis and describe the effect of the magnetic field-dependent spin Hamiltonian as inducing spin motion in this basis.

Since here we are mainly interested in photochemical reactions in liquid solution, the first problem when considering spin motion in the molecular frame is to account for the effect of rotational diffusion. Two extreme cases are conceivable. The first is that the spin orientation does not follow the molecular tumbling motion at all (rotational jumps as sudden perturbations), the second, that it follows the molecular motion as if rigidly coupled to the molecular frame (rotational jumps as adiabatic perturbations). In the first case where, during molecular rotation, the spin remains essentially stationary in a laboratory-fixed frame, it rotates with respect to the molecule-fixed frame of reference, whereby its motion can be described as a rotational diffusion with the same rotational diffusion constant as that of the molecule. During such a motion the spin energy will fluctuate between extremes separated by the ZFS parameter D. From time-dependent perturbation theory the following condition for the applicability of the quantum mechanical "sudden perturbation" approximation can be derived:

$$\frac{\hbar \dot{E}}{D^2} \gg 1 \tag{128}$$

where $\dot{E}$ is the rate of spin energy change during rotation. Let us, for example, consider the spin state T_y at $t = 0$ and a molecular rotation around the x-axis. Then, in the molecular frame the spin would oscillate between T_y and T_z and

$$\langle \hbar \dot{E} \rangle \approx D \frac{2 \dot{\varphi}}{\pi} \approx \frac{2}{\pi} D \left(\frac{3kT}{\Theta} \right)^{1/2} \tag{129}$$

where $\dot{\varphi}$ is the angular frequency that may be expressed by the thermal average of rotational energy. Substituting typical values $D = 0.1 \text{ cm}^{-1}$, a rotation constant $B = \hbar/(2c\Theta) \approx 0.2$ cm^{-1}, and $T = 300$ K, the quantity $\hbar\dot{E}/D^2$ is about 50. Thus the sudden approximation should provide a good basis to describe the effect of rotational motion on the orientation of a triplet spin in solution.

It can be shown[105] that by the influence of isotropic rotational diffusion any nonuniform substate population in the molecular frame will decay monoexponentially with a decay constant equal to $6D_r$, where D_r is the rotational diffusion coefficient. In terms of a rate constant of transitions between the different triplet substates this corresponds to (cf. Figure 9) the equation

$$w = 2D_r \tag{130}$$

In an external magnetic field the electronic spin precesses around the laboratory-fixed direction of the magnetic field. Combining this precession with the rotational-diffusion-type spin motion in the molecular frame corresponds to a damped oscillational behavior. Starting, for example, at $t = 0$ with a pure substate T_α ($\alpha = x,y,z$) in randomly oriented molecules, the average probability $p(T_\alpha,t|T_\alpha,0)$ to find any of these molecules still in the substate T_α at time t is given by (cf. Figure 11)

$$p(T_\alpha,t|T_\alpha,0) = \frac{1}{3} + \frac{2}{15} [1 + 2\cos(\omega_o t) + 2\cos(2\omega_o t)]\exp(-3wt) \tag{131}$$

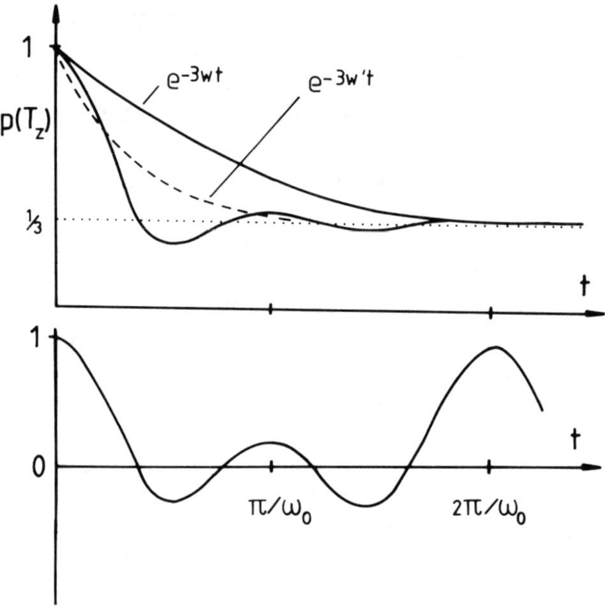

FIGURE 11.   Decay of nonuniform triplet substate population (case $p[T_z] = 1$ at $t = 0$) due to rotational diffusion ($w = 2D_r$) and Larmor precession (ω_0) of the spin. Bottom: Oscillating behavior of excess spin population of T_z due to Larmor precession in a randomly oriented ensemble of molecules. Top: Solid lines, decay of $p(T_z)$ in zero-field and in a magnetic field ($\omega_0 = 3\pi w$). Dashed line, effective monoexponential process adapted to $k_0 = 20w$ (cf. text).

where ω_0 is the Larmor frequency. In deriving this equation[105] the ZFS energy has been neglected. As will be seen, however, for low viscous solutions and $D < 0.1$ cm^{-1}, this will be a reasonable approximation when calculating magnetokinetic effects.

In order to model the effect that spin motion according to Equation 131 has on the product yield, one can take the approach to substitute the oscillating part in Equation 131 by an effective monoexponential process

$$p'(T_\alpha,t|T_\alpha,0) = \frac{1}{3} + \frac{2}{3}\exp(-3w't) \tag{132}$$

and substitute w in the yield expressions (Equations 120 and 121) by w′ which will be a function of the Larmor frequency ω_0. As has been corroborated by the more exact treatments described below the suitable condition to determine w′ is that the time average of p′(t) during the decay of the triplet with an effective rate constant

$$k_o \equiv \frac{1}{3}(k_x + k_y + k_z) + k_p \tag{133}$$

should be equal to the average of p(t). From this condition one obtains

$$w' = \frac{k_o(1 - k_o\tilde{p})}{3k_o\tilde{p} - 1} \tag{134}$$

where $\bar{p}$ is the Laplace transform $\bar{p}(s)$ taken at $s = k_o$:

$$\bar{p} = \int_o^\infty p(t)e^{-k_o t}\, dt = \frac{1}{15}\left(\frac{5}{k_o} + \frac{2}{k} + \frac{4k}{k^2 + \omega_o^2} + \frac{4k}{k^2 + 4\omega_o^2}\right) \qquad (135)$$

with

$$k = k_o + 3w \qquad (136)$$

Magnetokinetic effects on Φ_p calculated by using Equations 121, 134, and 135 are in very good agreement with exact treatments as long as

$$|D| \ll k_o + 3w + \omega_o \qquad (137)$$

In cases where $k_0 + 3\,w \gg |D|$ (cf. the experimental examples in Section III.A.2), this condition holds well even for low fields.

Our description of the TM presented so far corresponds to the "spin motion picture" of magnetokinetic effects. Application of the "Merrifield picture" of kinetically independent quasi-stationary spin energy eigenstates is not suitable for most cases of interest here, since the condition of its applicability, that phase randomization between the spin energy eigenstates be fast on the time scale of triplet decay, is not met. Of course, in a general quantum mechanical treatment the choice of the basis states is quite arbitrary and only a matter of convenience. Such general treatments have been based on the use of SLEs.

1. Treatment of the TM by SLEs

In the case of the TM a 3 * 3 spin density matrix is used to describe the ensemble of reacting triplets. The spin Hamiltonian is

$$H = g\beta \mathbf{B}_o \mathbf{S} + \mathbf{SDS} \qquad (138)$$

where **D** is the ZFS tensor.

Substate selective reactivity may be expressed using the projection operators (Q_x, Q_y, Q_z) onto the respective substates. A convenient form of the SLE is

$$\dot{\rho}(t) = -\frac{i}{\hbar}[H,\rho] - \frac{1}{2}[K_T,\rho]_+ - k_p\rho \qquad (139)$$

with

$$K_T = k_x Q_x + k_y Q_y + k_z Q_z \qquad (140)$$

describing selective depopulation. The selective population of the substates may be expressed by suitable initial conditions for $\rho(0)$.

Since $\mathbf{B}_0$ and **D** are at rest in a laboratory-fixed and in a molecule-fixed system, respectively, in any set of basis states the spin Hamiltonian H depends on molecular orientation Ω. The operator K_T depends on Ω if a laboratory-fixed basis is used.

An appropriate analytical solution to Equation 139, employing the time-dependent perturbation theory to account for the stochastic modulation of H and K_T, has been worked out by Atkins and Evans[103] in an application to CIDEP effects. Whereas they considered spin-selective population only, Serebrennikov and Minaev[106] applied this method to the case of spin-selective decay. For a uniform initial population of the triplet substates their result for the relative magnetic field effect on the product yield is

$$R(B) = \frac{\Phi_p(B_o) - \Phi_p(0)}{\Phi_p(0)} = -\frac{2}{3}\left(\frac{\omega_o}{k}\right)^2 \frac{D_k^2 + 3E_k^2}{D^2 + 3E^2}\frac{1}{1 + k_oT_1} \tag{141}$$

with

$$D_k = \frac{1}{2}(k_x + k_y) - k_z \tag{142}$$

$$E_k = -\frac{1}{2}(k_x - k_y) \tag{143}$$

and

$$T_1^{-1} = \frac{2}{15k}(D^2 + 3E^2)\left(\frac{k^2}{k^2 + \omega_o^2} + \frac{4k^2}{k^2 + 4\omega_o^2}\right) \tag{144}$$

It should be noted that, contrary to the case of spin polarization effects (CIDEP), magnetic field effects on the reaction yield do not depend on a nonvanishing ZFS. Letting $(D^2 + 3E^2) \rightarrow 0$ the result for $R(B)$ is

$$R(B) = -\frac{4}{45}\left(\frac{\omega_o}{k}\right)^2\frac{(D_k^2 + 3E_k^2)}{k_o}\left(\frac{k}{k^2 + \omega_o^2} + \frac{4k}{k^2 + 4\omega_o^2}\right) \tag{145}$$

These results for $R(B)$ are accurate to second order in the parameter ϵ, i.e., require the condition

$$1 \gg \epsilon^2 \approx [(|D| + |D_k|)/k]^2 \tag{146}$$

to be fulfilled for obtaining quantitatively reliable results. Unfortunately this condition does not hold well in cases where the magnetic field effects on Φ_p are large enough for precise experimental determination. Therefore they ought to be applied with care for quantitative interpretation of experimental results.[107]

A numerical solution to the SLE for the TM has been developed by Pedersen and Freed[108] for the CIDEP case and has been extended to the case of magnetic field-dependent reaction yields by Steiner and co-workers.[104] In these methods rotational diffusion is introduced into the SLE for $\rho(\Omega,t)$ the density matrix considered explicitly as a function of molecular orientation.

$$\dot{\rho}(t) = -\frac{i}{\hbar}[H(\Omega),\rho(\Omega,t)] - \frac{1}{2}[K_T(\Omega),\rho(\Omega,t)]_+ - k_p\rho(\Omega,t) + D_r\nabla_\Omega^2\rho(\Omega,t)] \tag{147}$$

The time-integrated solution to Equation 147 is obtained by Laplace transformation, letting $s \rightarrow 0$ and expanding the Ω-dependence of the Laplace transform $\tilde{\rho}(\Omega)$ in a series of Wigner rotation matrix elements, representing a complete orthonormal set in the space of Eulerian angles Ω. These Wigner functions are eigenfunctions of the Laplace operator ∇_Ω^2 so that the differential Equation 147 is transformed into an infinite set of coupled linear equations. This is truncated at some order, sufficient to guarantee convergence of the solution. The product yield is obtained from the zeroth order expansion coefficient, since the corresponding Wigner function is the only one that has a nonvanishing average when integrated over Ω. In Figure 12 results from such numerical calculations are shown, demonstrating some typical

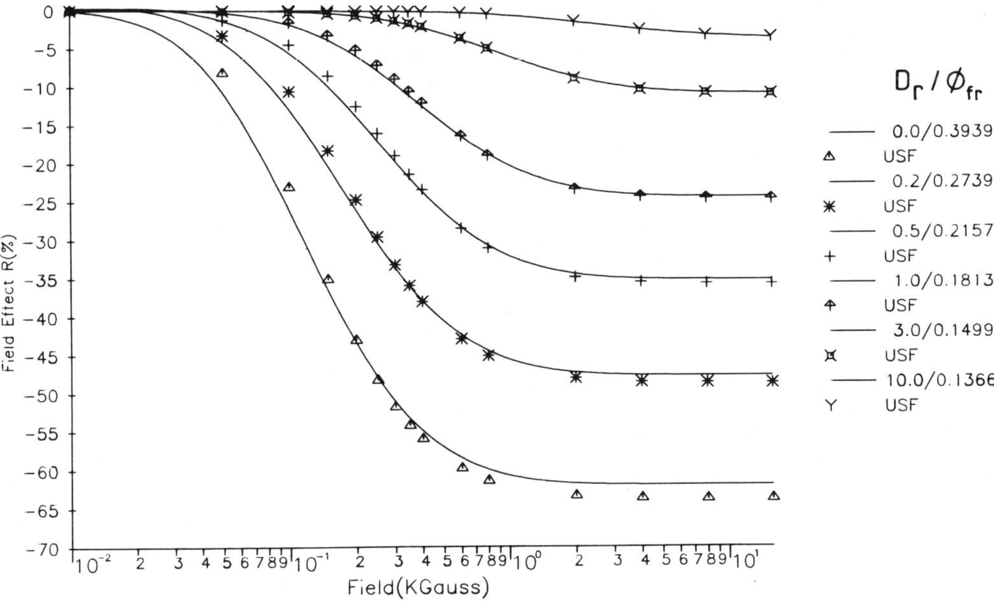

FIGURE 12. Theoretical field dependence of relative MFE R on product yield according to the TM. Solid lines: numerical solution of SLE (Equation 139); data points: from approximation USF (Equations 121 and 134). Parameters (in units of 10^9 s^{-1}) are $D,E = 0$, $k_S = 20/3$, $D'_k = 10$, $E'_k = 0$, and $k_p = 1$. D_r is varied as indicated in the figure, Φ_{fr} values given are absolute product yields in zero-field. (Reprinted with permission from Steiner, U. E. and Ulrich, T., *Chem. Rev.*, 89, 51, 1989. © 1989, American Chemical Society.)

features of MFEs according to the TM. For fixed rate constants of decay the magnitude of the effect increases with a decrease of D_r. For $6D_r \gg D_k$ the effect disappears. The effects saturate at fields about a factor of 10 larger than $B_{1/2}$ the field of half the saturation effect, for which one finds

$$\omega_o(B_{1/2}) \approx k + D \tag{148}$$

As is demonstrated in Figure 12 the results of the approximate solution (Equations 121, 134, and 135) denoted USF in Figure 12 are in good agreement with the exact numerical solution for the complete range of molecular mobility. The limited range of applicability of Equation 145 may be judged by comparing the approximate and exact high-field results obtained with the parameters used for the set of curves in Figure 12. Thus for $D = 0, 0.2, 0.5, 1.0, 3.0$, and 10.0, Equation 145 (in parentheses exact numerical solution of Equation 147) yields $R(B \to \infty) = -0.30\,(-0.62)$, $-0.26\,(-0.48)$, $-.22\,(-.35)$, $-0.17\,(-0.24)$, $-0.09\,(-.11)$, and $-0.034\,(-0.036)$, respectively.

III. REACTIONS IN HOMOGENEOUS SOLUTION

A. ELECTRON TRANSFER REACTIONS

Electron transfer reactions have figured prominently in the development of magnetokinetics. Actually, in the field of radioluminescence where oppositely charged radical ion pairs (RIPs) produced by high energy radiation recombine by electron transfer to yield electronically excited states, the idea of spin correlation and spin motion, characteristic of the RPM, has been suggested[109] independently and developed in parallel[12] to the ideas in CIDNP (for reviews of magnetokinetic effects in radioluminescence cf. References 110 and 111).

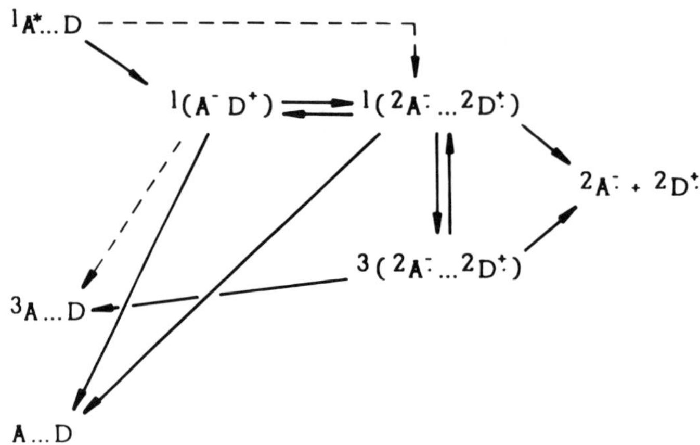

FIGURE 13.   General reaction scheme for singlet photoelectron transfer reactions.

1. Singlet Electron Transfer Reactions

A general reaction scheme entailing the important mechanistic features for magnetokinetic effects arising in photoelectron transfer reactions with excited singlet states is shown in Figure 13 (cf. Reference 112). The electronically excited state may accept (donate) an electron from (to) a suitable electron donor (acceptor). Such reactions occur at rates that are close to diffusion controlled if the free energy of RIP formation is negative.[113] In these cases electron transfer may occur between solvent-separated reactants so that a solvent-separated RIP is produced that can dissociate to uncorrelated radicals or eventually come into closer contact forming a contact RIP or fluorescing exciplex. The ratio of separation to contact formation strongly depends on the polarity of the solvent. Exciplexes may be also formed as the initial products of electron transfer during a close contact of the reactants. The extent to which this occurs has been a matter of controversy that cannot be regarded as generally settled. In any case, however, MFEs on the yield of exciplex fluorescence arise through exciplexes formed from solvent-shared RIPs only. Thus MFEs may yield important information on the mechanism of exciplex formation and on the exact pathway of electron transfer.

A singlet exciplex may decay to the singlet ground state either in a radiationless process or by fluorescence emission, the latter providing a sensitive probe of its detection. Otherwise it may undergo ISC to yield a locally excited triplet state. Such an ISC process is induced by SOC and not susceptible to MFEs (in this respect the situation is different for triplet exciplexes, cf. below).

The RIP formed either directly by the photoelectron transfer process or by dissociation of the singlet exciplex originates with singlet spin correlation. After separation to distances where exchange interaction can be neglected (>10 Å) hfc-induced spin motion will become efficient and eventually change the spin correlation to triplet so that in a subsequent re-encounter recombination to a locally excited triplet state may occur (if it is energetically below the RIP). This pathway of triplet formation is therefore magnetic field dependent. A spin conversion of the RIP will, of course, affect the singlet reaction channels, viz., regeneration of the singlet exciplex or direct recombination to the singlet ground state, in a sense opposite to the triplet yield. Effects on the yield of free radicals will only ensue if the recombination probabilities of S-RIPs and T-RIPs are different.

MFEs have been observed in numerous experimental studies on photoelectron transfer reactions, whereby the various transient species specified in Figure 13 have been probed by suitable techniques. A few selected references may serve to exemplify the variety of methods applied. Thus the kinetics of RIPs and the yields of free radicals have been probed by time-resolved laser flash spectroscopy.[15,114-116] Free radicals have also been detected under stationary illumination with modulated light utilizing photo conductivity measurements[117] or absorption spectroscopy,[118,119] whereby, in the latter case, enhanced sensitivity was achieved by magnetic modulation of the low stationary concentrations of the free radicals. The magnetic field dependence of free radical formation has also been detected by time-resolved photoconductivity measurements.[120] Generation of triplet recombination products has been monitored by transient absorption in laser flash experiments,[15,114,115,121] by delayed fluorescence due to triplet-triplet annihilation,[14,122] and also through photochemical quantum yields of olefinic *trans/cis* isomerization sensitized by the triplets generated by electron back-transfer.[123]

For the detection of exciplexes, fluorescence probing is, of course, the appropriate method. Corresponding observations have been made in plain stationary mode in the case of strongly fluorescent linked electron donor/acceptor systems,[124] or by the magnetic modulation technique[119,125] in the case of unlinked systems when exciplex emission in polar solvents is weak. Time-resolved MFEs on exciplex emission have been also reported.[126]

In what follows we will deal with magnetokinetic information that is available on several mechanistic aspects of reaction scheme Figure 13. Our examples will be selected from the field of photochemical reactions in solution. Many interesting results related to the problem of electron transfer have been also obtained from studies on photosynthetic reaction centers (cf. References 24, 127, and 128). These will, however, not be reviewed here.

a. Isotropic Hyperfine Coupling

If ISC occurs due to coherent hfc-induced spin motion in separate radicals, from what has been said in Section II.A, one can infer that in external magnetic fields several times higher than the effective hfc in the RP the S/T_+ transitions must be suppressed and any magnetic field-dependent effect must be saturated. To characterize the magnetic field dependence of such saturating effects one may define, as a so-called $B_{1/2}$ value, the field where the observed effect corresponds to the mean value of zero-field and a saturating field. If the lifetime of a RIP is long enough that the individual electron spin precession can evolve sufficiently, i.e.,

$$\tau_{RP} \gtrsim \tau_1, \tau_2 \tag{149}$$

where τ_1, τ_2 are the characteristic spin precession times of the two radical species according to Equation 23 it is to be expected that $B_{1/2}$ will be in some well-defined relation to the effective hyperfine fields of the radicals, defined as

$$B_{(1,2)}^{hfc} = (\overline{A_{(1,2)}^2})^{1/2} \tag{150}$$

A most representative model study to establish such a relation experimentally has been carried out by Weller and co-workers.[122] Here the magnetic field dependence of recombinant triplet formation from RIPs produced in electron transfer reactions between singlet excited pyrene (Py, (*1*) or Py-d$_{10}$ and several aromatic electron donors and acceptors (cf. Table 1) was investigated in methanol and acetonitrile.

TABLE 1
Hyperfine Coupling Effect on $B\,^{1/2}$-Value[a] of Magnetic Field-Dependent Triplet Formation from RIPs

Donor-acceptor system[b]		$B^{1/2}(exp)$		B_1^{hfc}	B_2^{hfc}	$B^{1/2c}$	$B^{1/2d}$	$B^{1/2e}$
(1)	(2)	In MeOH	In ACN					
Pyrene-d_{10}	DCNB-d_4	8	8	2.5	3.7	6.4	8.1	8.2
	DCNB	9	9	2.5	4.6	7.7	9.5	9.6
Pyrene	DCNB-d_4	18	16	10.1	3.7	16.8	19.5	19.8
	DCNB	16	17	10.1	4.6	16.8	20.1	20.5
	DMA-d_{11}	34	34	9.1	17.6	29.4	35.9	36.5
Pyrene-d_{10}	DMA-d_{11}	33	35	2.3	17.6	31.7	32.4	32.7
Pyrene	p-F-DMA-d_6	48	49	9.1	26.4	43.9	50.7	51.5
Pyrene-d_{10}	p-F-DMA-d_6	51	45	2.3	26.4	48.9	48.4	48.9
Pyrene	DMDMA	52	51	9.1	29.6	49.6	56.3	57.1
Pyrene-d_{10}	DMDMA	55	55	2.3	29.6	55.3	54.2	54.8
Pyrene	DMA	58	60	9.1	32.5	54.8	61.4	62.2
	DMT	59	59	9.1	34.5	58.4	64.9	65.8
Pyrene-d_{10}	DMA	59	63	2.3	32.5	61.0	59.5	60.1
Pyrene	p-F-DMA	67	66	9.1	37.4	63.7	70.1	71.0
Pyrene-d_{10}	DMT	60	62	2.3	34.5	65.0	63.1	63.8
	p-F-DMA	74	72	2.3	37.4	70.7	68.4	69.1

[a] Magnetic field values are in G.

[b] MeOH, methanol; ACN, acetonitrile; DCNB, p-dicyanobenzene; DMA, N,N-dimethylaniline; p-F-DMA, p-fluorodimethylaniline; DMDMA, 3,5-dimethoxy-N,N-dimethylaniline; DMT, N,N-dimethyl-p-toluidine

[c] Calculated according to Equation 152.

[d] Calculated according to Equations 153 and 154.

[e] Calculated according to Equation 155.

After Weller et al.[122]

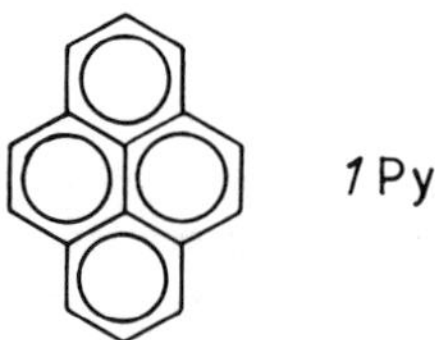

The MFE on the triplet yield was monitored by the intensity of delayed fluorescence due to triplet-triplet annihilation (I_{df}), using the relation

$$\frac{\Phi_T(B)}{\Phi_T(0)} = \left(\frac{I_{df}(B)}{I_{df}(0)}\right)^{1/2} \tag{151}$$

While there was practically no effect of the solvent, the $B_{1/2}$ values exhibited a characteristic dependence on the effective hyperfine fields B_1^{hfc} and B_2^{hfc} of the radicals involved. (cf. Figure 14 and Table 1).

Weller suggested the relation

$$B_{1/2} = \frac{B_1^{hfc}}{\overline{B}^{hfc}} B_1^{hfc} + \frac{B_2^{hfc}}{\overline{B}^{hfc}} B_2^{hfc} = 2\frac{(B_1^{hfc})^2 + (B_2^{hfc})^2}{B_1^{hfc} + B_2^{hfc}} \tag{152}$$

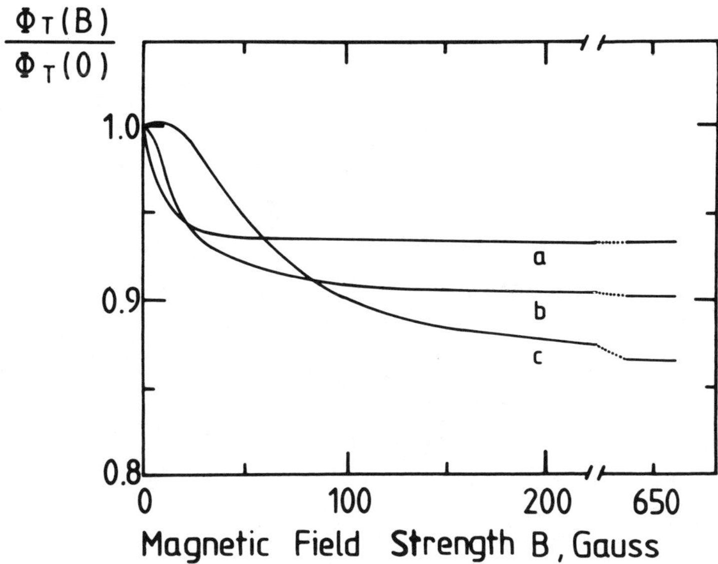

FIGURE 14.   Magnetic field dependence of relative yield of (*1*) triplets derived from delayed fluorescence measurements of pyrene in acetonitrile with addition of (a) *p*-dicyanobenzene-d₄, (b) *p*-dicyanobenzene-h₄, (c) dimethyl-*p*-toluidine. (Adapted from Weller et al.[122])

As is demonstrated in Table 1 this equation holds quite well. It is noteworthy that the functional dependence chosen in Equation 152 accounts for the fact that $B_{1/2}$ does not necessarily increase (cf. the cases Py/*p*-F-DMA and Py-d₁₀/*p*-F-DMA: although B_1^{hfc} is smaller in the latter case $B_{1/2}$ is found to be larger, in agreement with the predicton from Equation 152.

Schulten[129] has attempted to derive a suitable functional relation for $B_{1/2}(B_1^{hfc}, B_2^{hfc})$ from the semiclassical picture of spin motion. Since in this picture the transition from zero-field to high field corresponds to the transition from the case of uncorrelated axes of spin motion ($\langle\cos\Psi\rangle = 0$, if Ψ denotes the angle between the two local precession axes) to the case of parallel precession axes ($\langle\cos\Psi\rangle = 1$) one could attempt to assign the $B_{1/2}$ situation to some universal value of $\langle\cos\Psi\rangle$. From the semiclassical model one can derive for an arbitrary external magnetic field B[129]

$$\langle\cos\psi\rangle = \left[1 - \frac{1}{3}\left(\frac{B_1^{hfc}}{B}\right)^2\right]\left[1 - \frac{1}{3}\left(\frac{B_2^{hfc}}{B}\right)^2\right] \tag{153}$$

Averaging over the experimental $B_{1/2}$ values one obtains

$$\overline{\langle\cos\psi(B_{1/2})\rangle} \approx 0.9 \tag{154}$$

The correlation between $B_{1/2}$ and B_1^{hfc}, B_2^{hfc} based on Equation 153 and 154 is not quite as favorable as that based on Equation 152 (cf. Table 1). In particular it does not account for the increase of $B_{1/2}$ between Py/*p*-F-DMA and Py-d₁₀/*p*-F-DMA. If subtle features as the latter are neglected an even simpler relation than Equation 152 may be suggested,

$$B_{1/2} = \sqrt{3}\,[(B_1^{hfc})^2 + (B_2^{hfc})^2]^{1/2} \tag{155}$$

giving a comparably good correlation as Equation 152 (cf. Table 1).

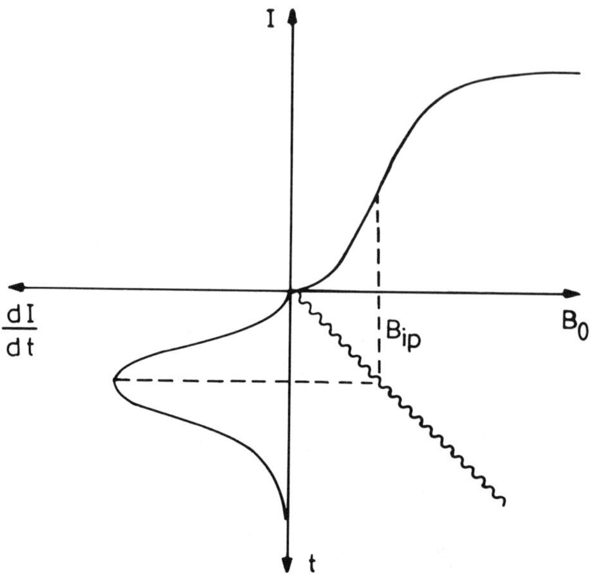

FIGURE 15.   Detection of first derivative MARY spectrum by low amplitude modulation of linear field sweep as applied in Reference 119.

MARY spectra, as plots of the magnetic field dependence of reaction yields are sometimes denoted (cf. Section V), are particularly easy to record if magnetic field-dependent stationary concentrations of transients can be detected under steady illumination conditions. An example is the observation of singlet exciplexes, whereby fluorescence emission represents a most sensitive probe of the magnetic field-dependent product yield. Since this fluorescence intensity may be modulated by a magnetic field, phase-sensitive detection can be applied for a direct reading of the MARY intensity as the amplitude of a locked AC signal. Such methods have been developed by Nath and Chowdhury[125] and by Hamilton et al.[119] Similar techniques had been previously applied to magnetic field-dependent chemiluminescence by Frankevich and co-workers.[130,131] In the method described by Nath and Chowdhury a sine wave-type alternating magnetic field of variable amplitude is applied and the AC signal is detected at twice the modulation frequency. In this way a somewhat distorted MARY spectrum is obtained. The technique followed by Hamilton et al. employs a sine wave modulation of low but constant amplitude superimposed on a linear scan of the magnetic field (cf. Figure 15). The AC signal amplitude detected at the modulation frequency corresponds to the first derivative of the MARY spectrum. This technique is completely analogous to that used in cw ESR spectroscopy. It is very sensitive for inflection points (B_{ip}) of the MARY spectrum.

Examples of first derivative MARY spectra recorded in this way with the exciplex fluorescence of Py/dicyanobenzene (DCB) donor/acceptor pairs are shown in Figure 16. The B_{ip} values obtained with 1,2-DCB (B^{hfc} = 6.3 G), 1.3-DCB (B^{hfc} = 10.4 G), and 1,4-DCB (B^{hfc} = 4.6 G) are 15, 19, and 15 G, respectively, and are close to the $B_{1/2}$ values according to Equation 152 (17, 20, 17 G). In other cases the method has allowed to assess another feature of the MARY spectrum: an initial antiphase of the MARY slope at very low fields corresponding to an enhancement of the ISC process in the RIP. These observations will be dealt with in more detail below (cf. Section III.A.1.d).

It shall be mentioned that magnetic isotope substitution in the solvent has been reported to affect the $B_{1/2}$ value, too. Thus Chowdhury and co-workers[132-134] found a decrease of 6 G

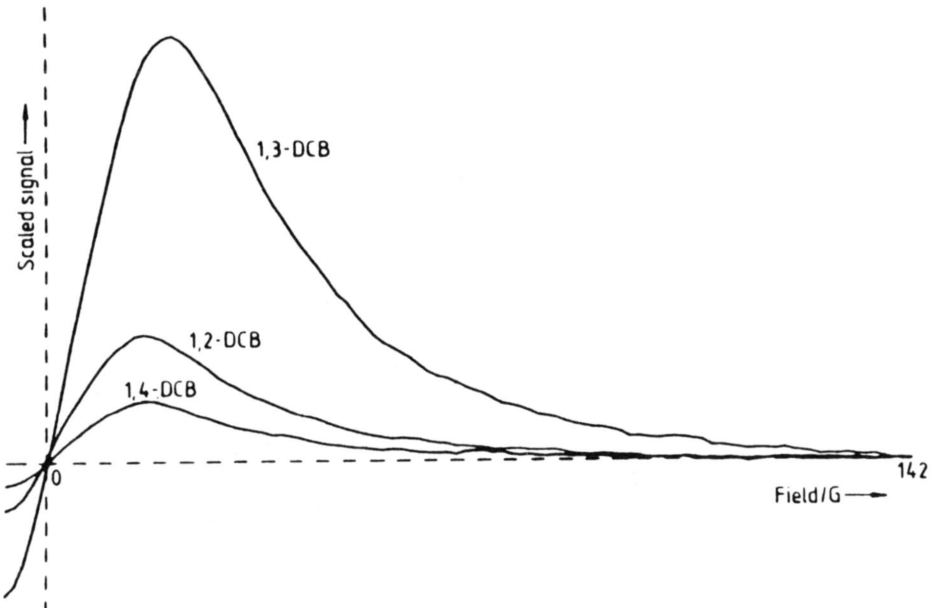

FIGURE 16. Field-modulated MARY observations of the fluorescence from cyclohexanol/acetronitrile (5:3 v/v) solutions of pyrene (*1*) with 1,2-dicyanobenzene, 1,3-dicyanobenzene and 1,4-dicyanobenzene, all sweeps starting from slightly negative values while the vertical scale is in the same arbitrary units for each isomer. (Reprinted from Hamilton, C. A., Hewitt, J. P., McLauchlan, K. A., and Steiner, U. E., *Mol. Phys.*, 65, 423, 1988. With kind permission of K. A. McLauchlan; copyright 1988, Taylor & Francis.)

in $B_{1/2}$ of the MARY spectrum for Py/*N,N*-dimethylaniline (DMA) exciplex fluorescence when replacing the solvent CH_3OH by CD_3OD. The suggestion that this effect be due to a delocalization of the radical spins to the solvent, so that they may sense some hfc from it, seems less acceptable than relating the effect to other changes in the mechanism as become apparent from an increase of the exciplex lifetime.

b. Time-Dependent Effects

In Section II (cf. Figure 3) it has been pointed out that the size of magnetokinetic yield effects depends on the time available for unperturbed spin motion. There have been some efforts to assess such spin evolution time effects theoretically and experimentally.

Using the semiclassical picture Schulten and co-workers[30,33,129] developed the theory of spin motion in RPs subject to stochastic hopping of one of the radical spins between different sites of the same chemical type and studied its effect on the magnetic field dependence of geminate recombination yields. Although the changes of the MARY spectra, resulting as the hopping rate is progressively increased, cannot be completely characterized by one single phenomenological parameter, as the $B_{1/2}$ value, this may serve as a rough characteristic here. It is found that as the hopping rate ($1/\tau_h$) increases $B_{1/2}$ increases first, too, but then decreases. At hopping rates $1/\tau_h \gg 1/\tau_1$, where τ_1 is the characteristic precession time for the radical undergoing paramagnetic/diamagnetic exchange, a final limit of $B_{1/2}$ is reached corresponding to spin motion on radical 2 only. Before approaching the fast-exchange limit the MARY curves, although exhibiting a relatively fast drop at low field, are characterized by a very slow saturation behavior.

The changes of $B_{1/2}$ as a function of hopping rate variation are related to the line broadening and motional narrowing effects known from other fields of spectroscopy. At fairly low hopping rates $B_{1/2}$ is expected to change as $1/\tau_h$ in conformity with the uncertainty relation.[135] In the limit $1/\tau_h \gg 1/\tau_1$ the stochastic changes of the effective hyperfine field at radical site 1 average to zero on the time scale of a typical precession period.

TABLE 2

Probe Time and Donor Concentration Effects on $B_{1/2}$ Values[a] for Anthracene (Ac) Triplet Yields from Ac⁻· . . . DMA⁺· RIP Recombination in Acetonitrile

Donor	Delay Time (ns)		
Concentration	7	11	20
$2.3 * 10^{-2}\ M$	61 ± 1	55 ± 1	54 ± 1
[b]	57	53	52
[c]	54	51	49
$8.0 * 10^{-2}\ M$	67 ± 1	58 ± 1	57 ± 1
$7.5 * 10^{-1}\ M$	≥ 110	≥ 84	≥ 78

[a] In G.

[b,c] Theoretical results (without hopping) from quantum mechanical (b) and semiclassical (c) model calculation.

Data from Reference 138.

Experimentally, controlled changes of the time interval of spin evolution in RPs have been made by applying variable delay times for probing the magnetic field dependence of recombination yields or by using high electron donor concentrations in order to speed up the diamagnetic/paramagnetic exchange process:[135-139]

$$D + (^2A^-\!\cdot\ldots^2D^+\!\cdot) \to (^2D^+\!\cdot\ldots^2A^-\!\cdot) + D \tag{156}$$

In Table 2 are shown some characteristic data obtained by Michel-Beyerle and co-workers[138] in the observation of magnetic field-dependent triplet recombination yields from S-RIPs of anthracene⁻·/DMA⁺·. The triplet yields have been obtained at definite delay times after generation of the RIPs by applying a two-laser pulse sampling technique with a second (weaker) laser pulse serving as an absorption probe pulse. The effect of delay time reduction is clearly borne out as an increase of $B_{1/2}$ and the results are in good agreement with theoretical expectation. Corresponding measurements have also been carried out with perdeuterated anthracene[138] exhibiting the expected reduction in $B_{1/2}$.

A significant increase of $B_{1/2}$ by increasing the donor concentration is only observed at very high donor concentrations. From model calculations a hopping rate $1/\tau_h \approx 3$ ns⁻¹ has been assigned to the MARY spectrum for [D] = 0.75 M, whereas $1/\tau_h < 0.1$ ns⁻¹ for [D] = $8*10^{-2}\ M$. This seems to indicate that no linear relation between $1/\tau_h$ and [D] exists. An inhomogeneous distribution of donor molecules has been suggested as an explanation.

Donor concentration-dependent effects in MARY spectra have been also observed with exciplex fluorescence. A low-field phase inversion feature of the MARY spectrum detectable with the Py/DMA system[119] is broadened and disappears as the concentration of donor is increased (cf. Figure 20, below).

A time-dependent study of MFEs on exciplex fluorescence with pyrene/diethylaniline has been reported by Lavrik and Nechaev.[140] In this investigation the fluorescence was integrated over time windows of selected width and delay. A maximum MFE was observed for a delay of about 100 ns. Investigated as a function of donor concentration the field effect at B = 300 G ($\gg B_{1/2}$) passed through a maximum at [D] $\approx 0.05\ M$. Such an effect is in line with the qualitative behavior of the average singlet characters attained after long times

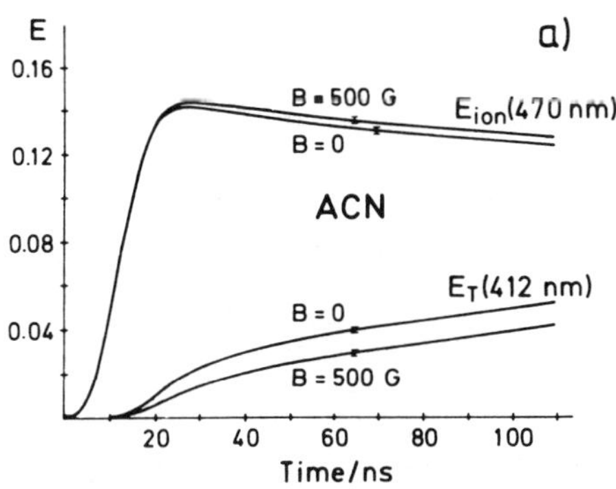

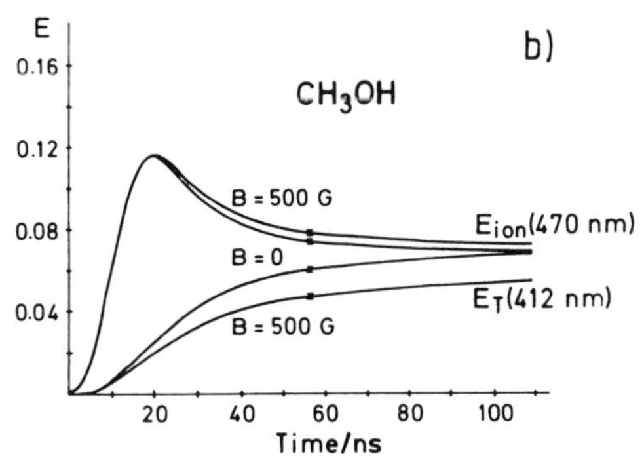

FIGURE 17. Time evolution of the radical ion (E_{ion}) and triplet (E_T) absorption of the system pyrene (*1*)/dimethylaniline. (a) In acetonitrile; (b) in methanol. (Reprinted from Werner, H.-J., Staerk, H., and Weller, A., *J. Chem. Phys.*, 68, 2419, 1978. With kind permission of A. Weller; copyright 1978, American Institute of Physics.)

by RPs generated with singlet spin. Model calculations based on the formulas of Schulten and Wolynes[30] for electron hopping under zero-field or high-field conditions have allowed to extract a hopping rate of about 0.5 ns^{-1} for [D] = 0.05 *M*.

The interpretation of $B_{1/2}$ variation with donor concentration as being due to electron hopping effects has been questioned by Nath et al.[134] on the basis of a study wherein the correlation between exciplex lifetime and $B_{1/2}$ was investigated. In these experiments the exciplex fluorescence lifetime was decreased at a fixed donor concentration by adding familiar fluorescence quenchers as KI or CsCl. Even though the exciplex lifetime was reduced from 9 to 6 ns no increase of $B_{1/2}$ was observed. (Rather it decreased by a few Gauss.) One should note, however, that this finding is at variance with the "hopping-interpretation" of [D] effects on $B_{1/2}$ only, if it can be taken for granted that the separate RIP, where spin evolution takes place, is similarly affected in its lifetime as the fluorescent exciplex. In our opinion this seems doubtful, however, since the effect of the applied quenchers on the exciplex results from its property as an electronically excited state, wherein deactivation pathways (ISC, IC) not available in the separate radical ions may be catalyzed by the quenchers. Lifetime shortening of the fluorescent exciplex only cannot be expected to affect $B_{1/2}$ since it does not limit the time intervals available for the spin evolution process.

In order to achieve high accuracy the time-resolved studies mentioned so far have been restricted to selectively sampled delay times. Reports on continuously time-resolved observations of magnetic field-dependent RIP recombination are few. The first investigations by Weller and co-workers[114] applied continuous transient-absorption probing after laser flash excitation (cf. Figure 17). In these experiments with the Py/DMA system the rise time of the radical ion signal amounted to about 20 ns and the MFE started to develop after about 10 ns and was complete after 30 to 60 ns, depending on the solvent (cf. below). The RIP decay times observed in MeOH and EtOH have been analyzed by Shushin,[141] who showed analytically that escape from the Coulomb cage should exhibit a time dependence that is exponential at short times and follows a t$^{-3/2}$ law at long times (cf. also the review by Mauzerall and Ballard[142]). The exponential part of the decay taken from the experimental results of Reference 114 corresponds to time constants of 37 ns in MeOH and 100 ns in EtOH. It is interesting to note that these values are about 20 times longer than expected from the Coulomb cage model. It has been suggested[141] that this disrepancy is due to deviations of the actual potential from the Coulomb potential at small distances. In fact, Werner et al.[114] maintained that in these alcoholic solvents the RIPs are formed by dissociation of long-lived fluorescing exciplexes that, in absorption, cannot be spectrally distinguished from separate radical ions.

Recently Basu et al.[126] have investigated the MFE on Py/DMA exciplex fluorescence in THF/DMF mixed solvent by the single-photon counting technique. According to their findings the MFE grows in during about 12 ns, a time that exceeds the theoretical value[36] by about 5 ns.

Finally it shall be noted that the growing-in, and even oscillations of MFEs due to hfc-dependent spin evolution have been time-resolved earlier in magnetic field-dependent scintillation processes.[143-145] It is particularly noteworthy that in selected systems quantum beats have been possible to follow over several S/T oscillation periods of RIPs.[146,147]

C. Solvent Effects

MFEs according to the RPM depend essentially on the diffusive motion of the RP which has to bring about efficient separation in order that magnetic field-sensitive spin evolution can ensue, but also reencounters in order that the result of spin evolution may become manifest as a product yield. From this point of view it is clear that solvent properties, like polarity and viscosity, should be important factors in controlling magnetokinetic effects and that vice versa, from such effects valuable pieces of information on the solvent dependence of reaction dynamics may be gained.

A detailed investigation and analysis of such solvent-dependent MFEs has been provided by Weller and co-workers.[112,114] In Figure 17 are shown examples of transient absorption from radical ions and pyrene triplet observed after laser flash-induced photoelectron transfer between DMA and singlet excited pyrene in MeOH and in ACN. The signals shown are for zero-field and a saturating field of 500 G. The geminate phase of RIP recombination can be assigned as the time interval during which the MFE develops. In MeOH it is much longer ($\lesssim$ 60 ns) than in ACN ($\lesssim$ 35 ns). Furthermore, in MeOH the radical signal shows a sharp drop during this period. An interpretation of this special kinetic feature in MeOH and other alcohols and lower polarity solvents is given by assigning it to the decay of a singlet exciplex through which the RIP is formed and which cannot be distinguished in absorption from the RIP. After the geminate recombination is complete the decrease of the radical signal and the increase of the triplet signal continues slowly due to homogeneous recombination processes. No further increase of the MFE is observed here.

In the quantitative analysis of the MFEs in Reference 114 it is assumed that in each solvent RIPs are formed through the singlet exciplex (route *a* in Figure 18), the dissociation lifetime of which being, however, strongly solvent-dependent. Fast triplet formation may occur either by SOC-induced ISC in the exciplex (route *b* in Figure 18) or by hfc-induced ISC in the RIP (route *b'*). Only the latter is magnetic field-dependent. On the basis of a comparison of the observed MFE on the yield of fast triplet formation and the theoretically predicted saturation value of R = -43% it was possible to assess the yields of triplets formed by SOC in the exciplex (Φ_T^{exc}) and the distribution $\Phi_{S\to T}^{gem}$, $\Phi_{S\to S}^{gem}$ over the spin-selective recombination channels of the RIP. The pertinent results are listed in Table 3.

In ACN and DMF the yield of triplets formed by SOC in the eciplex is zero (within the limits of theoretical predictability of R_∞). In MeOH and EtOH the values of Φ_T^{exc} increase in line with the exciplex lifetime. Actually these observations are consistent with a fairly solvent-independent rate constant k_{isc}^{exc} of about 10^7 s^{-1}.

In the solvents investigated in Reference 114 the exciplex lifetime is mainly determined by ionic dissociation. From a collection of data on more solvents Weller[112] established the following useful empirical relation

$$k_{isc}^{exc} = \frac{2.3 * 10^9 [cP*s^{-1}]}{\eta} \exp\left[- \frac{e_o^2}{\epsilon kT} \left(\frac{1}{d} - \frac{1}{a} \right) \right] \tag{157}$$

where η is the solvent viscosity, ϵ the dielectric constant, and d and a are the radical

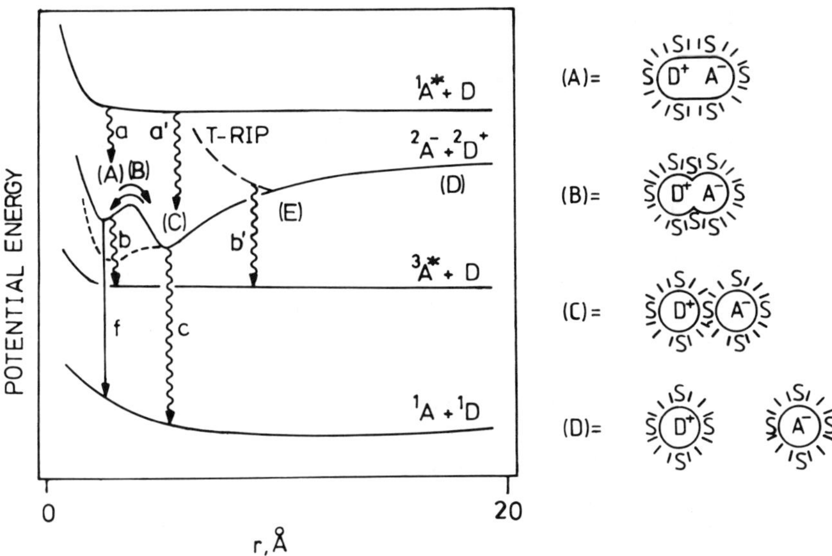

FIGURE 18.   General scheme of potential curves for spin-dependent electron transfer processes. Symbols denote (A) contact RIP or exciplex, (C) solvent-separated RIP, (B) transition state between (A) and (C), (E) point of near-degeneracy of T-RIP and S-RIP, (D) RIP outside the Debye radius. The processes a, a′, b, b′, c, and f are assigned in the text. (Adapted from Basu et al.[148])

TABLE 3
Solvent Dependence of Spin-Selective Reaction Parameters[a] in the Pyrene/DMA RIP and Exciplex

Solvent	ϵ[b]	η[c]	τ^{exc}[d]	Φ_T^{exc}	$\Phi_{S \to T}^{gem}$	$\Phi_{S \to S}^{gem}$
ACN	37.5	0.39	2.0	—	0.07	<0.1
DMF	37.0	0.80	—	—	0.06	0.4
MeOH	32.7	0.54	9.4	0.08	0.09	0.4
EtOH	24.5	1.20	19.8	0.14	0.07	0.7

a Φ_T^{exc} triplet yield from ISC in the exciplex, $\Phi_{S \to T}^{gem}$, $\Phi_{S \to S}^{gem}$ geminate RP recombination to triplet and singlet ground-state product.
b   Dielectric constant at 25°C (20°C for ACN).
c   Dynamic viscosity, cP, at 25°C (20°C for EtOH).
d   Exciplex lifetime in ns.

Data from References 112 and 114.

separations in the exciplex ($\approx$3 Å) and the solvent-shared RIP ($\approx$7 Å), respectively. As it stands, Equation 157 is valid for an exciplex dissociating into radical ions with $z_+ = z_- = 1$. Setting the exponential factor equal to 1 this relation is expected to hold also for systems that lack Coulombic attraction (cf. Section III.A.2).

The different solvent dependence of spin-selective RIP recombination probabilities (cf. Table 3) seems surprising at first sight. Whereas the yield $\Phi_{S \to T}^{gem}$ is practically solvent independent, the singlet recombination yield $\Phi_{S \to S}^{gem}$ increases with a drop of solvent polarity. These findings are, however, in good agreement with theoretical predictions. They can be rationalized as follows. Singlet recombination does not require spin evolution and thus $\Phi_{S \to S}^{gem}$ simply reflects the ease of a RIP to dissociate which is favored by a high ϵ and a low

η value. Triplet recombination, on the other hand, does require spin evolution, implying that the radicals have to separate first but then have to reencounter again. Both demands depend on solvent polarity in an opposite way so that a change of these solvent parameters causes two counteracting effects that seem to balance each other rather perfectly.

Although spin evolution is not required for singlet recombination, the S $\rightarrow$ T process in the RIP allowing for triplet recombination must have a reducing effect on $\Phi^{gem}_{S \rightarrow S}$ that should be susceptible to a MFE. Transient absorption measurements allow determination of $\Phi^{gem}_{S \rightarrow S}$ only in a rather indirect and approximative way that is not accurate enough for a sensible assessment of MFEs on this quantity. Since, however, the singlet recombination process may — at least in part — regenerate the fluorescing exciplex, fluorescence detection (route f in Figure 18) provides a very sensitive probe for MFEs on the singlet recombination channel $\Phi^{gem}_{S \rightarrow S}$ (corresponding to the contribution of routes f, b, c). The first reports of such observations have come from Frankevich's group[149,150] who studied the pyrene/diethylaniline exciplex fluorescence in various solvents. While in their investigation changes of stationary fluorescence intensity were directly monitored as responses to the switching-on of a magnetic field, Nath and Chowdhury[133] applied a sine wave modulation of the magnetic field for more sensitive lock-in detection. The strongest MFEs on exciplex fluorescence have been reported for polymethylene-linked RIPs by Staerk et al.[124] and by Tanimoto et al.[151,152] Thus in Reference 124 the MARY spectrum could be simply measured by recording the fluorescence intensity while scanning the magnetic field strength.

All the MFEs with exciplex fluorescence provide unequivocal evidence of the formation of fluorescing exciplexes from separate geminate RIPs. Nevertheless, the question of to what extent the primary event in excited singlet quenching leads directly to the exciplex or the solvent-shared RIP (route a vs. route a' in Figure 18) cannot be regarded as definitely decided. Doubtlessly the corresponding ratio will depend on the solvent.

The MFE on exciplex fluorescence can provide important information on the role of solvent polarity on the behavior of RIPs. In the first study of solvent polarity dependence of $\Delta\Phi_f/\Phi_f$, the relative MFE in the saturating field limit, by Petrov et al.[149,150] a maximum effect for a solvent with a dielectric constant of $\epsilon \approx 26$ was found. Recently the effects have been studied in further detail by Chowdhury and co-workers,[148,153,154] who, apart from the familiar Py/DMA system, investigated also exciplexes formed between 9-cyanophenanthrene *(2)* and *trans*-anethole *(3)*.

CN

H_3CO — CH$_3$

2   *3*

Their results obtained with the latter exciplex system in various solvent mixtures are depicted in Figure 19. As has been known for a long time (cf. Reference 155), the quantum yield of exciplex fluorescence decreases strongly with increasing solvent polarity. It must be noted, however, that different correlation lines between Φ_f and ϵ are obtained for mixtures of aprotic solvents (cf. curve 1' in Figure 19) and alcohols (curve 2'). For alcoholic solvents the drop in Φ_f occurs at significantly higher values of ϵ than for the aprotic solvent mixtures. Such a peculiarity of alcoholic solvents is also borne out by the relative MFE $\Delta\Phi_f/\Phi_f$. Whereas aprotic solvents show a maximum value of 3 to 4% at ϵ values between 15 and 20 with alcoholic solvents, such a maximum is not obtained below $\epsilon = 35$ with a highest value of only about 1%. For the Py/DMA system the results are similar.[153] However, in this case a maximum for the alcoholic solvents, though again shifted to higher ϵ than for the aprotic solvents, is clearly borne out at $\epsilon \approx 29$.

The solvent polarity dependence of $\Delta\Phi_f/\Phi_f$ has been modeled theoretically on the basis

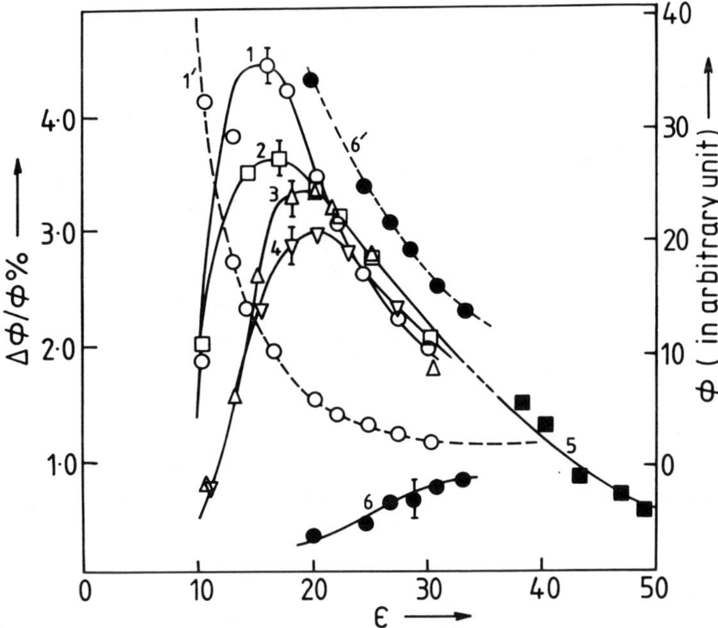

FIGURE 19. Quantum yield Φ (dashed lines) and relative MFE in saturating magnetic field $\Delta\Phi/\Phi$ (solid lines) on the quantum yield of exciplex fluorescence in the 9-cyanophenanthrene/*trans*-anethole *(2/3)* system as a function of solvent dielectric constant. Solvent mixtures employed are: $\circ$, THF-DMF; $\square$, THF-ACN; $\triangle$, benzene-ACN; $\triangledown$, ethyl acetate-ACN; $\blacksquare$, ACN-DMSO; $\bullet$, propanol-ethanol-methanol. (Reprinted from Nath, D. N., Basu, S., and Chowdhury, M., *J. Chem. Phys.*, 91, 5857, 1989. With kind permission of M. Chowdhury; copyright 1989, American Institute of Physics.)

of a RIP diffusing in a Coulomb potential and undergoing irreversible decay to the singlet channel described by some effective velocity of boundary crossing at reaction radius R into a sink representing those events where exciplex formation is irreversible. The treatment by Petrov et al.[150] was based on an analytical approximation with a one-proton model of spin motion as developed by Shushin. Application to the experimental results yielded activation energies of exciplex formation [process (C) $\rightarrow$ (A) in Figure 18] that were consistently about (7 to 8) *kT for all alcohols investigated. The calculation of Nath et al.[153] employing the semiclassical approximation of spin evolution nicely accounted for the occurrence of the ϵ-maxima. The fitting parameters of the model yielded reaction radii of about 6 Å for the nonalcoholic solvents but much smaller ones ($\sim$2.5 Å) for the alcoholic solvents. The interpretation of this result was that the activation barrier at (B) (cf. Figure 18) is present only in aprotic solvents, the reaction radius corresponding to that of the solvent-shared RIP (C). On the other hand, the activation barrier (B) was suggested to be absent in alcoholic solvents (cf. dashed potential curve in Figure 18). In the latter case the reaction radius corresponds to that of the contact RIP, or exciplex (A) and "reaction" means direct decay of this species.

The ϵ-dependence of $\Delta\Phi_f/\Phi_f$ has also been investigated with polymethylene-linked exciplex systems Py(n)DMA *(4)*[156] and Ph(n)DMA *(5)*[151] (cf. below). In these cases the ϵ-dependence was found to correspond to a monotonic increase of the relative MFE with increasing ϵ. Qualitatively, this behavior is to be expected because the initial separation of the RIP is supported by increasing the polarity, but the complete separation as occurring with unlinked RIPs is prevented in case of the chemically linked RIPs. Since the possibility of complete separation that controls the decrease of the MFE at high ϵ in the case of unlinked RIPs is absent in linked RIPs the ϵ-dependence must be monotonic in the latter case.

For unlinked RIPs it is possible to determine the relative order of triplet and singlet recombination rate constants from the sign of the MFE on the yield of free ions. Thus, for example, from Figure 17 it is apparent that the triplet recombination must be faster than singlet recombination because the free ion yield is increased when the S $\rightarrow$ T process is impeded by a magnetic field.

Observations of MFEs on free ion yields from photoelectron transfer reactions in the Py/*N*,*N*-diethylaniline (DEA) system have been made by Frankevich and co-workers[117,157,158] applying photoconductivity measurements under stationary illumination with modulated light. They obtained the interesting result that the MFE on photoconductivity was negative in isoamyl (-0.8%) and isopropyl alcohol (-2.7%), whereas it was positive ($+4.5\%$) in ACN. In mixtures of isopropanol and ACN a MFE sign inversion occurred when only about 1% of ACN was added to the alcohol. Obviously, in the alcoholic solvents the S-RIPs recombine faster than T-RIPs, whereas the opposite holds true with ACN present. An interpretation of these findings was given in terms of the energy difference between exciplex (contact RIP) and solvent-shared RIP, exciplex formation being energetically less favorable in ACN, or with even small amounts of ACN present in alcohols allowing for selective solvation after formation of the radical ions.

d. Exchange Interaction

Exchange interaction decreases rapidly with the separation of two radicals (cf. Equation 55). It is generally assumed that at distances of more than 10 Å it may be neglected. For freely diffusing RPs it is easy to move outside the region of strong exchange between successive reencounters, and with respect to spin-dependent recombination efficiency the lower probability of reencounter after a large separation is compensated by the fast hfc-induced ISC occurring outside the region of strong exchange. Therefore, as distinct from RPs with a fixed distance, the situation prevailing in photosynthetic reaction centers, or from linked RPs that are bound to stay within a maximum separation, the influence of exchange interaction on MARY effects with freely diffusing RPs is generally considered as small. This assumption has obtained support from model calculations of MARY effects.[36] On the other hand, analysis of CIDNP data[57,159] suggested the necessity of explicitly taking into account the effect of exchange interaction in order to model the absolute CIDNP intensities quantitatively (cf. also Figure 8).

There has been recent evidence[119] that exchange interaction can cause distinct effects even in the case of freely diffusing RPs. In Figure 20 are shown MARY spectra of exciplex fluorescence obtained in the first derivative mode with Py/DMA exciplexes in MeOH. These spectra exhibit a low-field antiphase feature that corresponds to a MARY sign inversion from negative to positive at about 7 G. This feature disappears at higher donor concentrations which is attributed to the perturbation of spin evolution by electron hopping between different donor molecules during the lifetime of the RIP.

Two explanations of the low-field feature have been considered: a "J-resonance" occurring at a field where S/T$_-$ level crossing occurs ($B_{res} \triangleq 2J_{eff}$) and the effect of a breakdown of the rule of conservation of total (electronic + nuclear) spin, strictly valid only in zero-field and responsible for a somewhat reduced efficiency of S/T mixing in the RP. Recently the former explanation has obtained strong support from complimentary effects detected by magnetic resonance modulation of the fluorescence intensity[160] (cf. Section V).

The role of exchange energy need not be confined to interaction between the radicals constituting the geminate pair, but may also become effective when otherwise unreactive paramagnetic species are added to the reaction medium. There have been several reports of quenching of MARY effects by the trivalent lanthanide ions or neutral complexes of them. Such effects were first observed by Sakaguchi and Hayashi[161] and by Turro et al.[162] for triplet RPs in micelles, later by Basu et al.[163,164] for singlet RIPs from photoelectron transfer

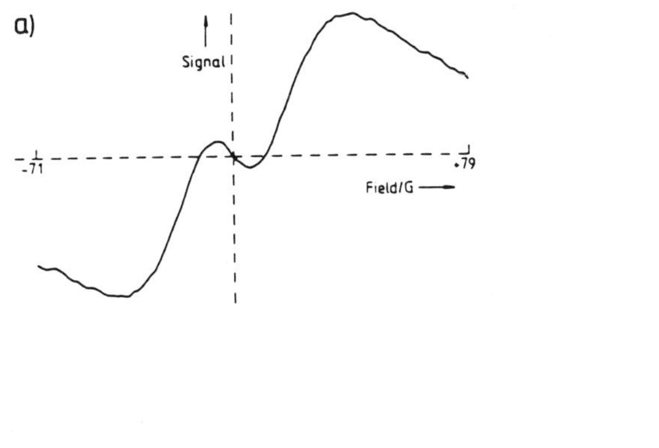
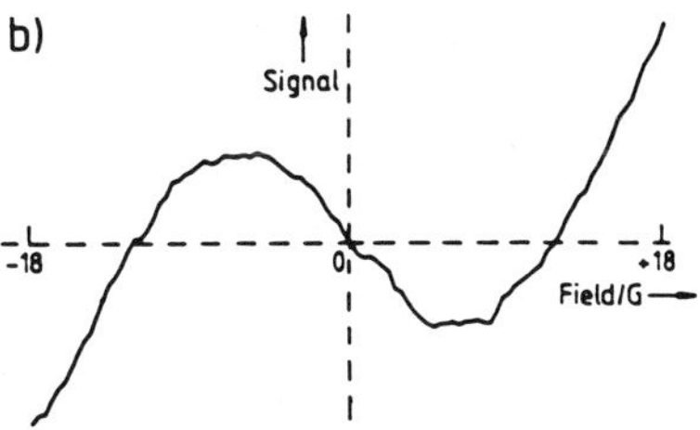
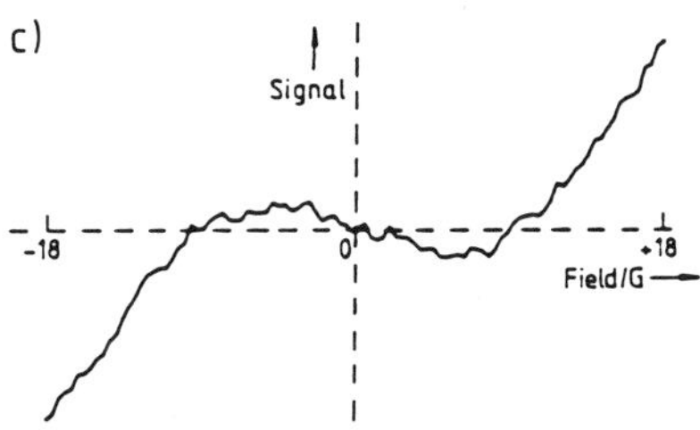
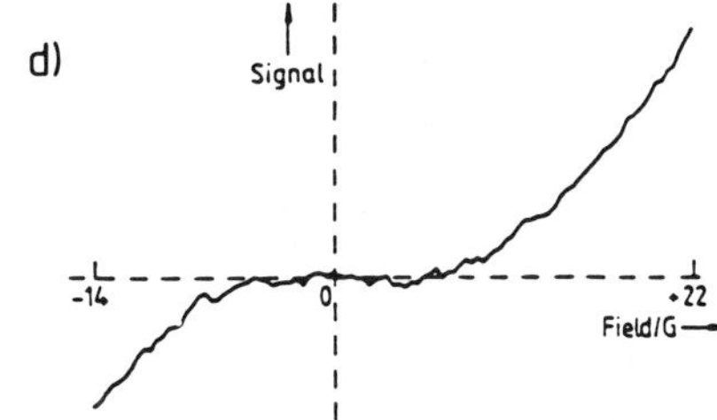

FIGURE 20.   Field-modulated MARY spectra from exciplex fluorescence in the pyrene(*1*)/dimethylaniline system in methanol. The field sweep starts from strongly reversed values and the spectra have inversion symmetry about the zero-field position. (a) Full spectrum with [DMA] = 10^{-2} *M*; (b-d) low-field feature only. [DMA] = 10^{-2} *M* (b), $6*10^{-2}$ *M* (c), 10^{-1} *M* (d). (Adapted from Hamilton et al.[119])

in homogeneous solution, and most recently by Tanimoto et al.[165] for chemically linked RPs. All these investigations have come out with rather similar relative effects of the lanthanide ions. As an example the results obtained by Basu et al.[163] for the quenching of the MARY effect with 9-cyanophenanthrene/*trans*-anethole (*2/3*) exciplex fluorescence by lanthanide acetylacetonates are shown in Figure 21. At first sight the quenching efficiency, formally determined as a Stern-Volmer constant, parallels the number of unpaired f-electrons in the lanthanide^{3+} ions with a maximum for Gd^{3+}. However, the case of Eu^{3+} represents a striking exception from this correlation. This peculiar behavior of Eu^{3+} together with the general trend of all the lanthanide^{3+} ions is nicely accounted for by attributing the quenching to the effect of Heisenberg exchange between the lanthanide^{3+} ion and any of the two radicals of the RIP. In the lanthanide^{3+} ions SOC is very strong and the electron spin orientation is not a constant of motion, but only the orientation of the total (orbital + spin) angular momentum. Following the approach of Elliott[166] the authors of Reference 163 suggested to use the stationary part of the electron spin, i.e., its projection onto the direction of total angular momentum, as a measure of efficiency for changing the RP spin multiplicity by exchange. This quantity (cf. Figure 22)

$$\langle JLS|S_{\parallel\,J}|JLS\rangle = (g - 1)\sqrt{J(J + 1)} \equiv \sqrt{G} \qquad (158)$$

corresponds to the square root of the de Gennes factor G. As is demonstrated in Figure 21 the correlation of the MARY quenching efficiency with $\sqrt{G}$ is quite striking.

In particular the peculiar behavior of Eu^{3+} that cannot be accounted for on the basis of its total spin (S = 3) or its magnetic moment (3.6β) is borne out by the de Gennes factor. However, a profound theoretical justification of this correlation derived from a dynamic theory of interaction processes including the role of spin relaxation at the lanthanide^{3+} ions has yet to come.

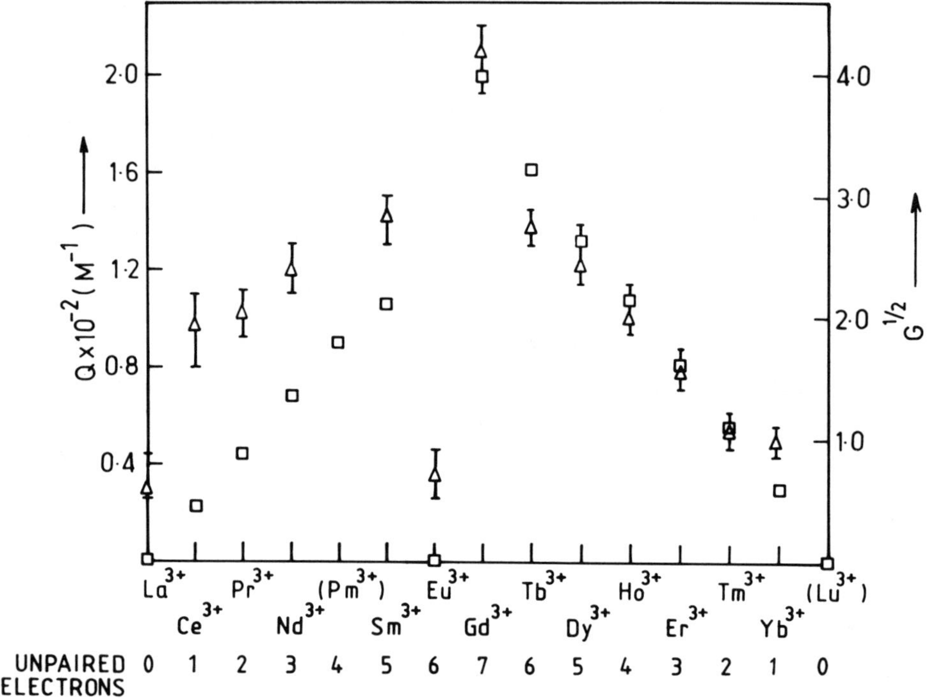

FIGURE 21. Quencing of MFE on 9-cyanophenanthrene/*trans*-anethole (*2/3*) exciplex fluorescence by lanthanide^{3+}-triacetylacetonate complexes. $\triangle$, magnetic quenching rate constants; $\square$, square root of de Gennes factor G (cf. text). (Reprinted from Basu, S., Nath, D., and Chowdhury, M., *J. Lumin.*, 40 and 41, 252, 1988. With kind permission of M. Chowdhury; copyright 1988, Elsevier Science Publishers B.V.)

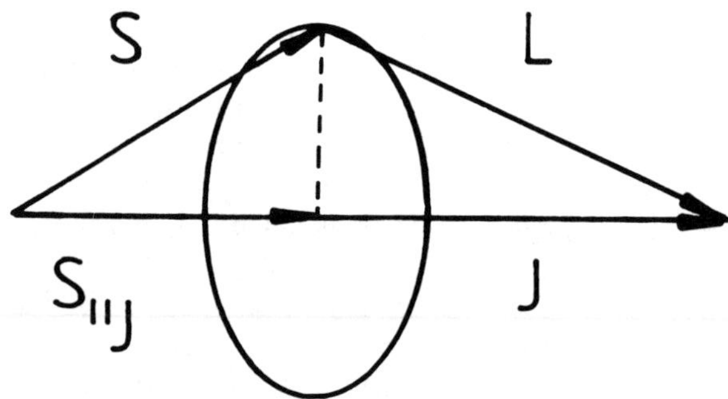

FIGURE 22.   The projection $S_{\parallel J}$ of S on the vector J as a constant of motion.

Exchange interaction is probably not sufficient for a quantitative account of the lanthanide^{3+} effect on MARY spectra. The importance of dipolar magnetic interactions between the lanthanide^{3+} complexes and the RP electron spins has been quantified by Tanimoto et al.[165] who analyzed absolute rate constants, directly determined in their experiments. Their conclusion was that the observed rate constants can be well acounted for by a relation for magnetic dipole interaction familiar from the theory of lanthanide^{3+} spin probes in ESR.[167] They did not comment on the Eu^{3+} case, however, for which they, too, observed a low quenching efficiency.

e. Chemically Linked Radical Ion Pairs

Magnetokinetic methods can provide an efficient approach to the study of distance dependence of exchange interaction and reactivity, and for the folding dynamics of polymer chains when applied to RPs linked, e.g., by polymethylene chains of variable length. Photo CIDNP investigations of cyclic ketones with triplet biradicals as magnetokinetically active intermediates represent the first studies of this kind.[45,77,168-170] Later the magnetic field-dependent kinetics of these systems has been observed directly by laser flash spectroscopy (cf. Section III. C). The most detailed magnetokinetic studies of biradical systems, S-RIPs generated by photoelectron transfer between dimethylaniline and singlet excited pyrene, linked by a polymethylene chain (Py[n]DMA) *(4)*, have been reported by Weller's group.[124,156,171-173]

$$(CH_2)_n - \bigcirc - N(CH_3)_2$$

4

The MARY effects investigated concerned the yield of pyrene triplets probed by time-selective absorption sampling with two laser pulses[115] or the yield of singlet recombination probed through the exciplex fluorescence.[124] A representative collection of results, MARY spectra of triplet recombination yields for linking chains with $n = 6$ to 13 (CH_2) groups, is depicted in Figure 23.

The various curves may be grouped into three types. For $n \leq 6$ the triplet yield is low and not magnetic field dependent; for $6 < n < 12$ the triplet yield increases with n and exhibits a characteristic magnetic field dependence: zero slope at $B_0 = 0$, followed by an increase to a maximum at $B = B_{max}(n)$, decreasing with increasing n. At higher fields Φ_T drops to a limiting value lower than $\Phi_T(B = 0)$. For $n \geq 12$ B_{max} is actually zero and the MARY curve is of the type typical for unlinked RIPs, although the effect of RIP linkage is still noticeable in a $B_{1/2}$ larger than in the case of free radicals but approaching it with still increasing n.

The n-dependent effects borne out in the MARY curves of Figure 23 reflect the distance dependence of exchange interaction. Transitions between the initially generated S-RIP and the T-RIP states, from which recombination to the locally excited triplet product occurs, are due to isotropic hfc that may be characterized by an effective value B_{hfc}. If even in the most extended conformation of the polymethylene chain $J_{ext}(n) > B_{hfc}$ (case $n \leq 6$) no hfc-induced ISC is possible and the triplet yield (due to SOC in the singlet exciplex) is not magnetic field-dependent. As soon as $J_{ext}(n) \leq B_{hfc}$ (case $n > 6$) magnetic field-dependent ISC in the RIP can ensue. The absolute values of Φ_T increase with increasing n, since S and T levels get closer to each other with decreasing J. Increasing the field for a given n will produce a T_-/S level crossing (cf. Figure 7) that leads to an increase of Φ_T with B_{max} reached at $B_0 = 2J_{eff}$, whereby J_{eff}, representing in a way to be discussed below the distribution of end-to-end distances of the polymethylene chain, decreases with increasing n. A further increase of the field disconnects both T_+ and T_- from S and causes the decrease of Φ_T with the saturating value corresponding to $T_0 \rightarrow S$ ISC only. The maximum at $B_0 = 2J_{eff}(n)$ will merge into zero-field if $2J_{eff} < B_{hfc}$.

A first-order interpretation of the B_{max} values might assign to $J_{eff} = B_{max}/2$ a separation r_{eff} of the linked RIP according to the exponential $J(r)$ dependence (Equation 55) and interpret r_{eff} as the average value of the end-to-end distance distribution of the polymethylene chain. The following empirical relation was found in Reference 171 for the $Py^-(n) DMA^+$ RIPs:

$$r_{eff}[\text{Å}] = 2.82\sqrt{n} \tag{159}$$

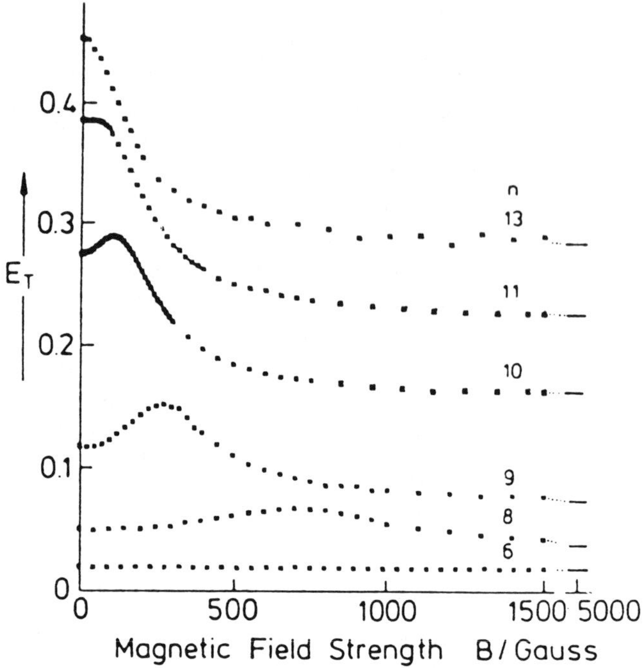

FIGURE 23.   Triplet absorbance observed as a function of magnetic field strength in linked EDA systems Py-$(CH_2)_n$-DMA (*4*) in acetonitrile. (Reprinted from Staerk, H., Treichel, R., and Weller, A., *Biophysical Effects of Steady Fields* (*Springer Proc. Phys.*, Vol. 11), Maret, G., Kiepenheuer, J., and Boccara, N., Eds., Springer-Verlag, Berlin, 1986, 85. With kind permission of A. Weller; copyright 1986, Springer-Verlag.)

The square root dependence on n is expected for a random walk-type vectorial addition of elementary lengths (of course, such an element must encompass several C–C bonds since due to the fixed C–C–C valence angle the orientation cannot be changed at random from one C–C link to the next).

In a more refined interpretation of the effects of exchange interaction on the MARY spectra of linked RIPs one has to consider that for dynamically variable RP separation the "spectrum" of exchange interaction does not, in general, coincide with that derived from the corresponding static distribution of RP separations. The effect of dynamic reorganization of the distribution has been discussed in Section II.A.1.c. Accordingly one should expect that for a fixed chainlength parameter n the MARY spectrum should be susceptible to variations of solvent viscosity. This has indeed been confirmed experimentally. Some data pertinent to solvent viscosity dependence of the J-resonance maximum in Φ_T MARY spectra of Py(n)DMA-linked systems are collected in Table 4. As the solvent is changed to higher viscosity the B_{max} values as well as the amplitudes R_{max} decrease.

The theory of MARY spectra for linked RPs as developed by Schulten and Bittl[46,49,64,81,174] has been applied to model the viscosity dependence of MARY spectra by Busmann et al.[51] In Figure 24 a comparison between experimental and theoretical results for Py(n)DMA is shown. The theoretical simulations are based on a numerical solution of the SLE entailing an exponentially decaying exchange interaction ($J_0 = 1.7*10^{17}$rad*s^{-1}, $r_J = 0.47$ Å).

The stationary distribution of end-to-end separations as obtained from a Monte Carlo calculation of chain conformations has been substituted by a smooth analytical function extending from 7 to 17 Å. For numerical calculation this interval has been divided into ten discrete segments. The rate constants k_d of hopping between adjoining segments used to

TABLE 4

Characteristic Parameters[a] of MFEs on Triplet Recombination Yield from Polymethylene-linked S-RIPs of Pyrene and Dimethylaniline (*4*) in Solvents of Different Viscosity[b]

	ACN		CEH		DEG	
n	B_{max}	R_{max}	B_{max}	R_{max}	B_{max}	R_{max}
7	1650	13	1120	1	985	3
8	750	33	515	7	380	8
9	280	29	210	14	150	12
10	110	5.5	74	7	50	5

[a] B_{max} in G, R_{max} in %.

[b] ACN, acetonitrile ($\eta = 0.39$ cP); CEH, *n*-octyltetraoxyethylene (η intermediate between ACN and DEG); DEG, diethylene glycol ($\eta = 38$ cP).

Data from Reference 156.

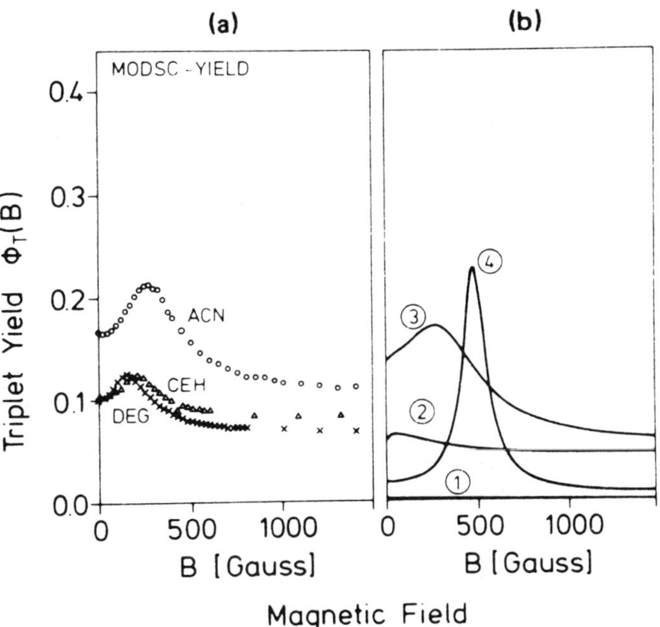

FIGURE 24.   Experimental (a) and theoretical (b) MARY spectra of triplet yields from linked RIPs Py$^{-\cdot}$-(9)-DMA$^{+\cdot}$ (cf. (*4*)) in solvents of different viscosity (cf. text for details of the theoretical calculation). (Reprinted from Busmann, H.-G., Staerk, H., and Weller, A., *J. Chem. Phys.*, 91, 4098, 1989. With kind permission of A. Weller; copyright 1989, American Institute of Physics.)

model the chain dynamics were $1*10^9$, $5*10^{10}$, $1*10^{12}$, and $2*10^{13}$ s^{-1} for curves 1 to 4, respectively. Actually for the individual segments these rate constants are multiplied by suitable weight factors accounting for their different equilibrium populations. The RIPs were assumed to be created and to undergo back electron transfer only in the first segment ($r = 7.5$ Å), with a rate constant of $k_{S,T} = 3*10^{12}$ s^{-1} for both singlet and triplet recombination. For the parameters of curve 3 the hopping rate corresponds to a sum of radical diffusion coefficients of about $4*10^{-5}$ cm^2 s^{-1} and the RIP lifetime calculated as 14.7 ns corresponds to the experimental one found for ACN. In fact, curve 3 in Figure 24b represents the MARY spectrum in ACN quite well. Decreasing the hopping rate by a factor of 200 qualitatively reproduces the observed change of the MARY spectrum when changing the solvent to the more viscous diethylene glycol. The model spectra 1 and 4 are close to the static and the motionally narrowed situation, respectively. These cases are far from what is observed experimentally.

It has been emphasized by the authors of Reference 51 that assuming a localized reaction at the distance of closest approach is essential for obtaining reasonable quantitative predictions of the MARY effects. This point has been investigated in more detail by Bittl and Schulten.[64] They showed that to reproduce the experimental RIP lifetimes in ACN for n = 8 to 10 (9, 14.5, 20.5 ns[175]) and to correctly model the observed MARY curves, a distance-dependent reaction rate $k_{S,T}(r)$ could be used described by an exponential law

$$k_{S,T}(r) = k_o \exp(-r/r_J) \tag{160}$$

with r_J the same as for exchange interaction and $k_0 = 1.3*10^{15}$ s^{-1} independent of n. The decay curves obtained were monoexponential. Furthermore, it was shown that using, instead of the universal (n-independent) $k_{S,T}(r)$, an r-independent $\bar{k}_{S,T}(n)$ corresponding to the effective exponential decay constant did not noticeably alter the decay curves nor the MARY spectra. Of course, the assumption of a universal (n- *and* r-independent) reactivity $k_{S,T}$ must lead to quantitatively unsatisfactory predictions.

In concluding this point we want to add that the absolute values characterizing any assumed r-dependence of $k_{S,T}(r)$ (the exponential function in Reference 64 and a step function in Reference 51) critically depend on the stationary distribution $p(r)$ of RIP separation used in the model. In fact, according to the exponential function with k_0 as in Reference 64 one would obtain $k_{S,T}$ (7 Å) $= 4*10^8$ s^{-1}, $k_{S,T}$ (8 Å) $= 0.5*10^8$ s^{-1}, values that are much smaller than the effective $4*10^{12}$ s^{-1} used for the interval 7 to 8 Å in Reference 51. Nevertheless, both types of $k_{T,S}(r)$ yield the correct experimental lifetime of the RIP. The reason is that the stationary distance distribution chosen in Reference 64 extends to closer separations (≈ 5 Å)[46] than that in Reference 51 (7 Å). The maxima of the distributions are at 10 and 14Å, respectively. Astonishingly enough, using the same $J(r)$ both calculations yield similar B_{max} values (≈ 300 G) corresponding to $r_{eff} \approx 8.2$ Å.

MFEs on exciplex fluorescence intensity from polymethylene-linked RIPs have also been investigated by Tanimoto et al.[151,152] using the Ph(n)DMA *(5)* system.

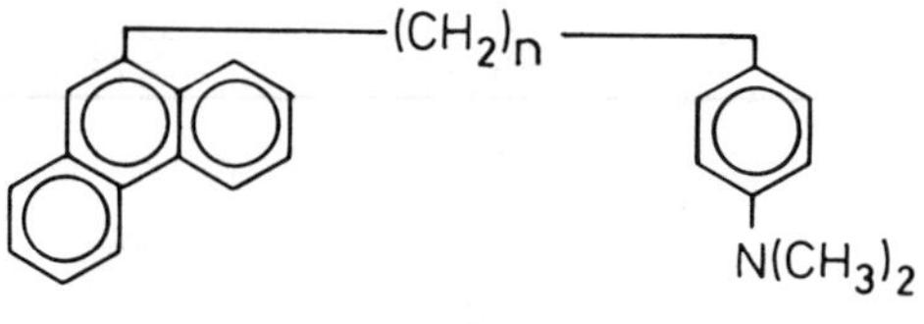

5

Only the cases n = 3, 6, 8, and 10 were studied. Qualitatively, the MARY spectra in ACN show a behavior similar to those reported by Weller and co-workers.[124] However, for n = 10 the saturating MFE ($R_\infty \approx + 50\%$) is alrady as high as for n = 16 with the Py(n)DMA system. Another interesting observation was that only exciplexes (I) generated from direct excitation of the pyrene moiety proved to be magnetic field-dependent, whereas exciplex emission (II) generated by excitation in the absorption band of DMA was not. Both emissions were also found largely different in decay time ($\tau_I = 50$ ns, $\tau_{II} = 10$ ns for n = 10 in ACN). The lifetime of type I emission was increased by 86% in a saturating magnetic field. As an explanation of the two types of exciplexes different conformations of the initially formed RIP have been suggested.

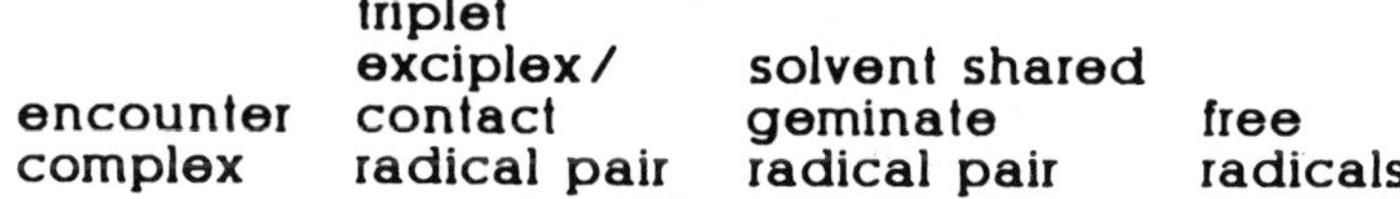

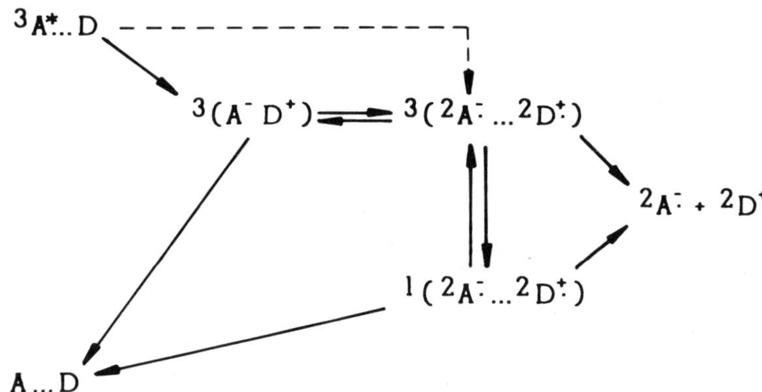

FIGURE 25.   General reaction scheme for triplet electron transfer reactions.

The situation of chemically linked S-RIPs has also been considered in an investigation of copolymers of monomeric olefines with pendant phenanthrene and DMA groups.[151] In this way saturated links of three, six, or nine atoms came to separate the two electron donor/acceptor moieties. Astonishingly enough, the magnetic field dependence of the exciplex emission observed from solutions of such polymers resembled that of homogeneous solutions of unlinked phenanthrene and DMA. The conclusion in Reference 151 was that the exciplex emission observed, for the main part, does not come from pairs of nearest neighbors on the same strand. Our suggestion is that one might also consider as an explanation the possibility of effective free mobility of bound radical ions due to hopping of both electrons and holes facilitated by the high local concentrations of donor and acceptor sites in the copolymers.

2. Triplet Electron Transfer Reactions

For triplet electron transfer reactions the reaction scheme of the singlet case is to be modified as shown in Figure 25. The rate constants of triplet electron transfer reactions show a similar dependence on the standard free enthalpy of RP formation, as in the singlet case,[176] i.e., they approach the diffusion-controlled limit as the exergonicity of the reaction increases. Due to the spin conservation rule the primary product after electron transfer originates with triplet spin correlation. The question whether the primary product is a contact RP (polar triplet exciplex) or a solvent-shared RP may be even more difficult to decide in the general case of a triplet reaction than for singlet reactions. Transition of the triplet exciplex to the ground state pair A . . . D should be conceived as an exergonic spin-inverted backward electron transfer and for the electronic coupling factor depends on spin-orbit interaction,[105,177-180] whereas for the Franck-Condon factor the energy gap law applies.[179,181] The conceivable reaction pathway

$$^3(A^-D^+) \rightarrow {}^1(A^-D^+) \rightarrow {}^1(AD)_o \tag{161}$$

involving spin conversion of the triplet exciplex to the singlet one faces the problem that SOC between electronic states of the same orbital configuration vanishes to first order.[102] Another point against it is that the singlet exciplex is higher in energy by about 0.1 eV.[182]

It must be mentioned, though, that in some cases the reverse ordering of energy (corresponding to a negative J, cf. Figure 6) has been advocated and that the pathway (161) has been claimed to apply in the case of oxygen-induced quenching of some triplet exciplexes.[181,183] We think that the arguments for such an energy ordering should be critically reexamined. In a RP where S and T correlate on adiabatic potential surfaces with S_0 and T_1, respectively, of the united system, as is the case when a σ-bond may be formed between them, this ordering (J < 0) is natural. However, in a RP with singly occupied LUMO of A and HOMO of D, which is the situation resulting from photoelectron transfer, T and S of the RP correlate adiabatically with T_1 and S_1 of the united system, where S_1 should always be above T_1.

In any case, whether J > 0 or < 0, in the contact RP (or polar exciplex) $|J| \gg B_{hfc}$ so that hfc-induced ISC is not efficient. This mechanism can take over only after dissociation of the triplet exciplex into a solvent-shared RP, where, on the other hand, the possibility of SOC-induced backward electron transfer is lost. (Note that in separate radicals SOC usually does not directly contribute to spin motion unless orbitals are available so close to the single occupied ones that they are energetically within reach of the SOC. Such cases require high symmetry as in linear molecules or in highly symmetric transition metal complexes, and strong SOC.)

Re-encounters of the RP after ISC from T to S are required to express the effect of hyperfine induced spin motion in the yield of geminate backward electron transfer. The switching of the ISC mechanism between SOC and hfc as a function of RP separation is an important concept in the interpretation of MFEs, heavy atom effects, and their mutual interference.

a. Triplet Radical Ion Pairs in Polar Solvents

In polar and low-viscous solvents dissociation of RIPs is usually very efficient and, unless extraordinarily high SOC is active, hfc-induced ISC in the solvent-shared RIP is not fast enough to compete. Hence, high yields of free radicals and only small, if any, MFEs on the free radical yield will ensue. The first MFE on the yield of free radicals from electron transfer reactions with excited triplets has been observed by Periasamy and Linschitz[184] in the electron transfer quenching of fluorenone (6) triplet by diazabicyclooctane (DABCO, (7)) in propylene carbonate.

6   7

The radical yield, detected by laser flash spectroscopy, was increased by a magnetic field, the effect saturating at about 270 G. This behavior clearly points to the influence of the hfc mechanism. Whereas at 343 k and $\eta = 1$ cP with a zero-field radical yield of 0.89 the (saturated) MFE was only 1.6%, at 223 K and $\eta = 45.3$ cP with a zero-field radical yield of 0.13 the MFE amounted to 24%. The η/T dependence of absolute radical yields Φ_r and saturated relative MFEs $R_\infty(\Phi_r)$ have been possible to correlate with each other on the basis of an exponential reaction model, equating the rate constant of recombination with k_{TS}, the effective rate constant of ISC, and setting $k_{TS}(B = 0)/k_{TS}(B_\infty) = 1.22$.

Similar observations have been reported by Kuz'min and co-workers[185] for the electron transfer reaction between 4-dimethylaminopyridine (*8*) and duroquinone (*9*) triplet investigated in glycerol/ethanol mixtures of various viscosities.

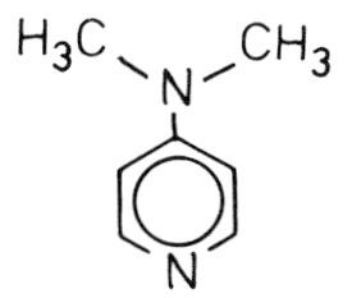

In contrast to the rapidly saturating field dependence with the system of Reference 184 the magnetic field dependence in this system showed a rather gentle decrease of slope up to 0.15 T, where the MFE for a 900 cP mixture amounted to a 80% increase of free radical yield. This type of magnetic field dependence points to a considerable participation of incoherent spin motion in the ISC process of the RP. As in the previous case[184] in solvents of viscosities below 5 cP the MFE dropped below 5%.

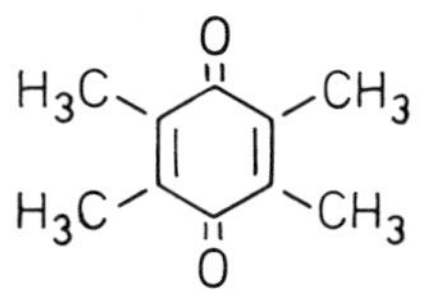

Another possibility of slowing down the complete separation of RIPs is decreasing the solvent polarity. In fact, in unpolar solvents the contact RIP (or polar exciplex) will not be able to dissociate at all. An illuminating study on the mechanistic features of spin-inverted backward electron transfer in the transition region between unpolar and polar solvents has recently been reported by Levin et al.[179] Applying ns-time-resolved laser flash spectroscopy they measured the recombination kinetics of T-RIPs generated by quenching a series of anthraquinones, naphthoquinones, and benzoquinones by three aromatic amines. The donor/acceptor systems covered a range of -2.2 eV $\leq \Delta G_{bet} \leq \sim 0.3$ eV for backward electron transfer to the donor/acceptor ground state. Solvents used were benzene ($\epsilon = 2.3$), $CHCl_3$ ($\epsilon = 4.7$), and $C_2H_2Cl_4$ ($\epsilon = 8.2$).

In benzene and $CHCl_3$ no free radical ions were formed. The RIP signals attributed to T contact RIPs or triplet exciplexes, in this case, decreased monoexponentially (cf. Figure 26a) with rate constants k_{bet} between $7.0*10^5$ and $3.7*10^7$ s^{-1}. In these cases the decay constants were insensitive to a magnetic field. Their order of magnitude may be considered as characteristic of SOC-induced ISC in exciplexes or contact RIPs in the absence of internal heavy atoms (cf. below). Furthermore, a characteristic "bell-shaped" correlation between lg k_{bet} and ΔG_{bet} was established for each of the solvents. These correlations have been possible to model with reasonable parameters by applying a theoretical relation derived from the semiclassical approximation of nonadiabatic electron transfer by Ulstrup and Jortner.[186] The small coupling matrix elements of 4 to $7*10^{-5}$ eV required by the fit are characteristic of the spin-forbidden nature of the process.

In $C_2H_2Cl_4$ the radical ion signals exhibited a two-stage decay (cf. Figure 26b) with the slowly decaying part corresponding to homogeneous recombination of free ions, and the fast one (k_{TE}) exhibiting a decay constant of the same order of magnitude as in less polar solvents. Using the yield of free radicals the rate constant k_{TE} of triplet exciplex decay could be decomposed into the sum $k_{bet} + k_{dis}$. Whereas k_{bet} again corresponded to the bell-shaped correlation with ΔG_{bet} the rate constants k_{dis} exhibited little variation (2 to $6*10^6$ s^{-1}). The latter have been shown to be consistent with diffusion-controlled escape from the Coulomb cage.

Associated with the appearence of free radical ions was the observation of a MFE (cf. Figure 26b) during the decay of the triplet exciplex/contact RIP signal. This MFE corresponds to a decrease of k_{bet} by 0.8 to $1.3*10^6$ s^{-1} in an almost saturating field of about 0.2 T. The

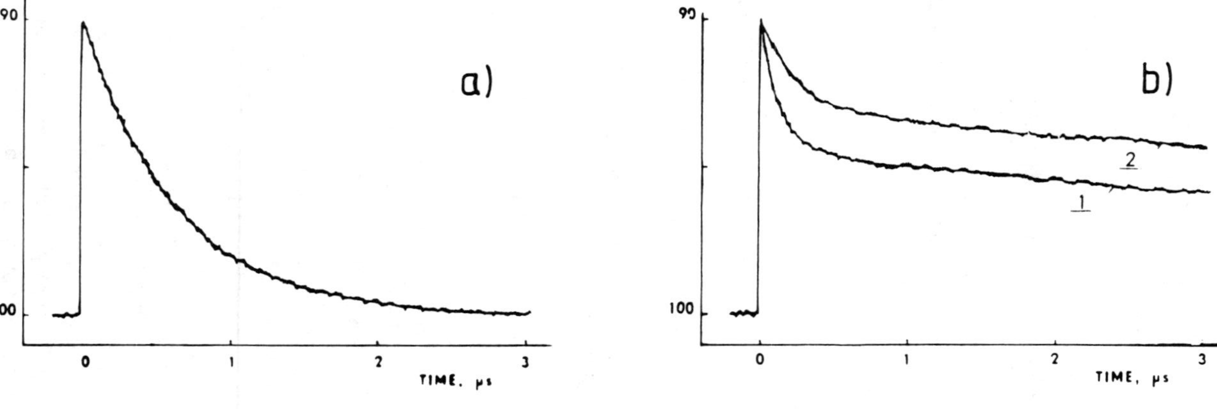

FIGURE 26.   Decay of radical ion absorption after triplet quenching of duroquinone (9) by triphenylamine. (a) In benzene (no magnetic field dependence detectable); (b) in $C_2H_2Cl_4$, 1: zero-field, 2: in a field of 0.24 T. (Reprinted from Levin, P. P., Pluzhnikov, P. F., and Kuz'min, V. A., *Chem. Phys.*, 137, 331, 1989. With kind permission of P. P. Levin and V. A. Kuz'min; copyright 1989, Elsevier Science Publishers B. V.)

only logical explanation is that in a solvent like $C_2H_2Cl_4$ where the radicals may separate, a contribution to spin-inverted backward electron transfer comes from the route

$$^3(A^-D^+) \rightarrow {}^3(^2A^-\ldots{}^2D^+) \rightarrow {}^1(^2A^-\ldots{}^2D^+) \rightarrow {}^1(A^-D^+) \rightarrow A\ldots D \qquad (162)$$

involving triplet exciplex dissociation, spin evolution in the separated RIP, and re-encounters with recombination. The magnetic field dependence of this contribution to k_{bet} exhibits a sharp drop between 0 and 100 G. The $B_{1/2}$ (for the interval between zero-field and a field of 2 kG) of 35 G is characteristic of the coherent hfc-induced spin motion. Since, however, a saturation value is not yet obtained at 2 kG incoherent contributions are also noticeable.

Similar effects as with the previous systems in $C_2H_2Cl_4$ have been found with anthraquinone and triphenylamine in $CHCl_3$ if its polarity is increased by the addition of 5 to 15% of ethanol[187] or in the polar but highly viscous solvent glycerol with the electron donor/acceptor pair *p*-phenylaniline/benzophenone-triplet.[188]

b. Systems with Enhanced SOC

As was mentioned above in Section II.C, SOC is of inherent anisotropic nature and may differentiate between different triplet spin orientations with respect to the molecular frame. However, in triplet exciplexes with normal SOC and $k_{bet}(SOC) = 10^6$ to 10^7 s^{-1}, these anisotropic processes are much slower than spin relaxation among the triplet substates and hence no MFEs according to the TM as described in Section II.C can ensue. This situation changes upon introducing heavy atom substituents into the constituents of the triplet exciplex. Of course, increasing $k_{bet}(SOC)$ by several orders of magnitude will make the triplet exciplex difficult to observe directly and will also drastically reduce the yield of free radicals that might serve as a probe. From the latter point of view it is an advantage to use reactant pairs of the type $A^+ \ldots D$ yielding on electron transfer only a charge shift to $A \ldots D^+$, so that the RP will not be trapped in a Coulomb cage.

A first systematic investigation of the internal heavy atom effect on $k_{bet}(SOC)$ as expressed in the yield of free radicals has been reported by Steiner and Winter[178] with a study of the electron transfer reaction between thionine (*10a*) triplet and the complete series of monohalogenated anilines in methanol.

	X	R	
a	S	H	thionine, TH$^+$
b	S	CH$_3$	methylene blue, MB$^+$
c	Se	H	selenine, SeH$^+$
d	O	H	oxonine, OxH$^+$

10

Here it was shown that the relative values of $k_{bet}(SOC)$ evaluated from the radical yields by assuming constant values of k_{dis}, the rate constant of triplet exciplex dissociation, were determined by the SOC contribution from the heavy atom (Br,I) substituent, following approximately the relation

$$k_{bet}(SOC) \propto C_x^2 \zeta_x^2 \qquad (163)$$

with ζ_x the atomic SOC constant and C_x^2 the spin density at the heavy atom substituent x in the singly occupied MO of the $D^{+\cdot}$ radical. The latter feature is borne out as a marked positional dependence of the internal heavy atom effect on $k_{bet}(SOC)$.

TABLE 5
Kinetic Parameters Evaluated from the Magnetic Field Dependence of Free Radical Yield Φ_r Due to Dissociation of Short-Lived Triplet Exciplexes Between Thionine[104] or Methylene Blue[191] and Heavy Atom-Substituted Anilines

Acceptor/donor[a]	Solvent[b]	Φ_r	$k_{bet}(SOC)$[c]	k_{dis}[c]	D_r[c]
TH$^+$/p-I-An	MeOH	0.135	68	5.0	8.2
TH$^+$/o-I-An	MeOH	0.225	42	6.9	12.0
TH$^+$/m-I-An or p-Br-An	MeOH	0.45 0.51	9.4	6.0	10.0
MB$^+$/p-I-An	MeOH	0.10	37.5	2.2	7.5
MB$^+$/p-I-An	10% egly	0.085	30.8	1.5	5.7
MB$^+$/p-I-An	20% egly	0.066	29.5	1.05	4.9
MB$^+$/p-I-An	40% egly	0.042	20.6	0.44	3.0

[a] TH$^+$: thionine (*10a*); MB$^+$: methylene blue (*10b*); I-An: iodoaniline; Br-An: bromoaniline.

[b] % egly: ethylene glycol fraction in mixtures with methanol. Viscosities at 20°C are MeOH (0.6 cP), 10% egly (0.828 cP), 20% egly (1.155 cP), 40% egly (2.143 cP).

[c] In units of 10^9 s^{-1}.

Similar internal heavy atom effects on the free radical yield from the decay of triplet exciplexes have been recently reported by Kikuchi et al.,[189] who also established a relation with the second-order rate constant of free radical recombination, indicating that T-RPs formed with a statistical probability of 3:4 may recombine directly under the effect of enhanced SOC. Thus the close correspondence of the two situations in the formation of a T contact RP through electron transfer between D and ^{3}A$^+$ and through encounters of free radicals A· and D$^+$· has been demonstrated.

MFEs with heavy atom substituted triplet exciplex systems of type ^{3}(A·D$^+$·) were first observed by Steiner[190] for thionine (*10a*) triplet reacting with bromo- or iodo-anilines. These effects that were first thought to be due to the Δg-mechanism in solvent-separated RPs have been investigated in greater detail and analyzed quantitatively according to the TM in Reference 104. The observed MFEs consist of a decrease of radical yield saturating slowly above 1 T with B$_{1/2}$ in the order of 0.2 T. An increase of the MFEs parallels the decrease of the radical yield caused by the heavy atom effect. In Table 5 are listed the pertinent kinetic parameters of the triplet exciplexes (k$_{bet}$(SOC) rate constant of SOC-induced backward electron transfer to singlet ground state, k$_{dis}$ rate constant of dissociation into a solvent-shared RP, D$_r$ constant of rotational diffusion) evaluated from the magnetic field dependence of the free radical yields. According to these data the ^{3}(A·D$^+$·)-type triplet exciplexes are very short-lived species with subnanosecond lifetimes. In fact, the observed growing-in of the MFE was synchronous with the decay of ^{3}A$^+$ and the rise of A· even on the shortest time scale (~20 ns) of triplet quenching investigated, indicating that the lifetime of the intermediate wherein the MFE develops must be considerably shorter.

The values of k$_{bet}$(SOC) exhibiting the position-dependent heavy atom effect are by three to four orders of magnitude larger than those noted above for triplet exciplexes with "normal" SOC. On the other hand, the rate constants of dissociation are practically independent of position and kind of the halogen substituent. It may be reasonably assumed that such a value also applies to the exciplex with unsubstituted aniline. On the basis of a normal k$_{bet}$(SOC) of 10^6 to 10^7 s^{-1} in this case the yield of free radicals would be expected to reach 100% within experimental accuracy, which has indeed been observed in a detailed quantitative reinvestigation of radical yields.[192] It is of interest to note that for a solvent of the viscosity of MeOH Weller's empirical relation for singlet exciplexes (cf. Equation 157) when specified

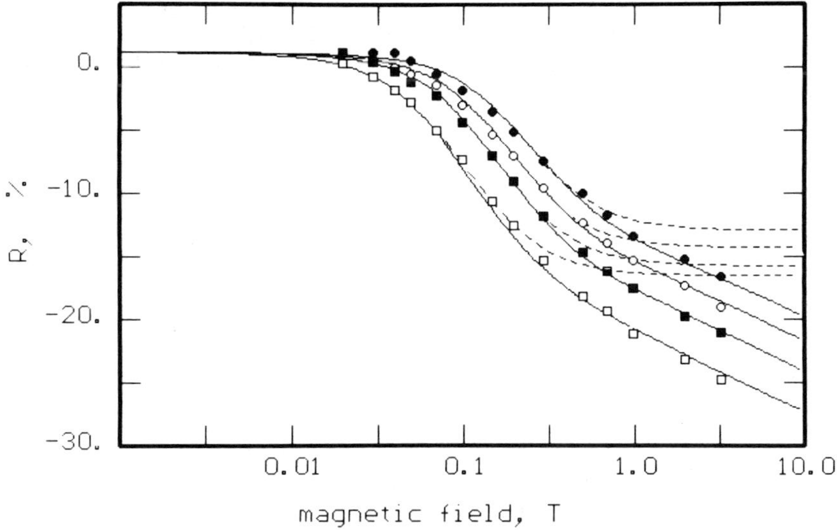

FIGURE 27.   Relative MFE R on free radical yield from electron transfer between *p*-I-aniline and methylene blue triplet (*10b*) in methanol/ethylene glycol mixtures (%MeOH: ●, 100; ○, 90; ■, 80; □, 60).[191] Solid lines: theoretical results from model calculations combining TM and Δg-RMP.[191] Parameters used to fit the Δg-RPM-dependent part of the curves are $\lambda_S = 1$, a = 7 Å, σ = 0.1Å, Δg = 0.015, $Q^x_{(P)}$,$J_0 = 0$, D = (2.15, 1.56, 1.12, 0.6)*10^{-5} cm² s^{-1} for 100, 90, 80, 60% MeOH, respectively. A Gaussian distribution of ω_n(hfc) (cf. Equation 34) corresponding to a width of 6 G was employed. Dashed lines: high-field behavior expected when employing the TM only.

to systems devoid of Coulomb interaction, affords a value of 4*10^9 s^{-1}, that is in good agreement with the values of the triplet exciplexes in Table 5.

The values obtained for the rotational diffusion constant D_r may be interpreted on the basis of Debye's equation

$$D_r = \frac{kT}{8\pi R^3 \eta} \qquad (164)$$

where R is the effective hydrodynamic radius of the triplet exciplex and η the viscosity of the solvent. According to Equation 164 R is about 3 Å. From a space-filling model of a sandwich-type exciplex conformation a value of 4.4 Å would be expected. The lower value from the experimental D_r might signalize that in the triplet exciplex the donor moiety, wherein the heavy atom center with its anisotropic SOC is located, might rotate rather independently, which is an argument for a rather loose structure of the triplet exciplex, corresponding to that of a RP in contact.

The role of rotational diffusion in the TM is to provide a mechanism of spin reorientation in the molecular frame of that moiety wherein the anisotropy of SOC is fixed. A slowing down of rotational diffusion must support the anisotropic SOC effects that are the source of MFEs according to the TM. This expectation has been confirmed experimentally.[191,193] Results of an investigation[191] employing methylene blue (*10b*) triplet as electron donor in various mixtures of MeOH and ethylene glycol are depicted in Figure 27. The experimental points display an increase of the MFE with increasing solvent viscosity. Also shown are theoretical curves representing best fits according to the TM. While these fits are excellent up to about 0.3 T it is clear that the saturation behavior characteristic for the TM is not borne out by the experimental results. The MFE increases, though slowly (note the lg B scale) up to the highest fields of about 3.3 T. The unsaturated high field part of the MARY curves is quantitatively accounted for by the Δg-type RPM in the solvent-shared RP originating from dissociation of the triplet exciplex/triplet contact RP. Due to the p-I substituent in the donor $\Delta g \approx 0.015$ is fairly high so that the effect of T_0/S mixing in the RP becomes appreciable at high fields and leads to a noticeable further decrease of the free radical yield through the reaction pathway denoted in Equation 162. The contribution of this mechanism, too, increases with solvent viscosity.

The viscosity dependence of the kinetic parameters evaluated for the triplet exciplex is given in Table 5. The rate constant k_{dis} varies inversely with η, with a slope of $2.6*10^9$ cP*s^{-1} in excellent agreement with Equation 157. The rate constant D_r assigned to rotational diffusion also varies linearly with $1/\eta$, however with an η-independent contribution of about $1.5*10^9$ s^{-1}. This points to viscosity-independent contributions of spin relaxation, possibly related to vibrations. The rate constant $k_{bet}(SOC)$ exhibits a weak viscosity dependence which is in line with the suggestion that the triplet exciplex conformation corresponds to that of a loose contact RP.

Observation of MFEs in triplet electron transfer reactions with erythrosine (*11c*), an iodine-substituted xanthene dye, as an electron-donating triplet has been reported by Levin et al.[107] Modified methylviologens (*12*) have been applied as electron acceptors, whereby charged substituents $(R,R' = -CH_2CH_2 -NR_3''^+, -CH_2 SO_3^-)$ have been used to vary the net charge of the electron acceptor between 0 and $+4$.

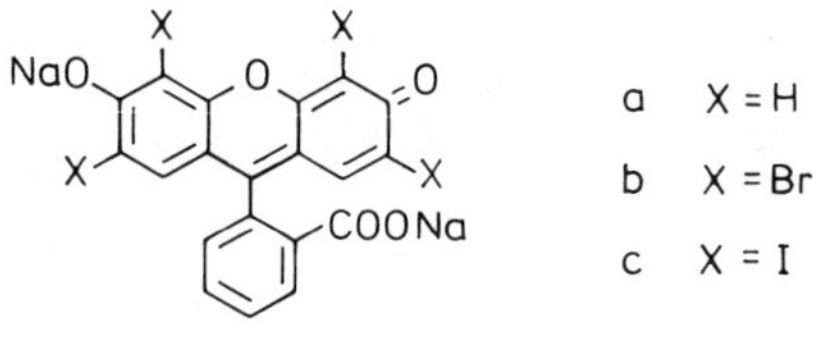

11

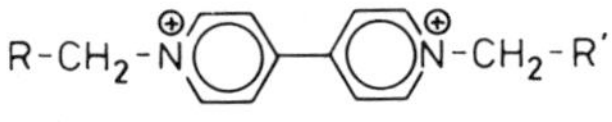

12

Since the dye was applied as a dianion the charge pairing type of the RP could be varied between attractive $(D^-\cdot A\cdot^{+++}, D^-\cdot A\cdot^{++}, D^-\cdot A^+\cdot)$, indifferent $(D^-\cdot A\cdot)$, and repulsive $(D^-\cdot A^-\cdot)$. The yield of free radicals increased in this order, as expected. In a magnetic field of 0.18 T a relative decrease of the radical yield by 6 to 8% was observed, that was attributed to the TM. However, the number of data (only one field value) is insufficient for a sensible evaluation of parameters.

c. MFEs in Photoelectron Transfer Reactions with Transition Metal Complexes

Particularly strong SOC effects are to be expected for reaction intermediates with unpaired spins located on transition metal ions. An example of such a case is the probably most often studied photochemical electron transfer reaction[194-196] between photoexcited Ru(bpy)$_3^{2+}$ (*13*) or its analogues with other diimine ligands and methylviologen (*12*, R = R' = H).

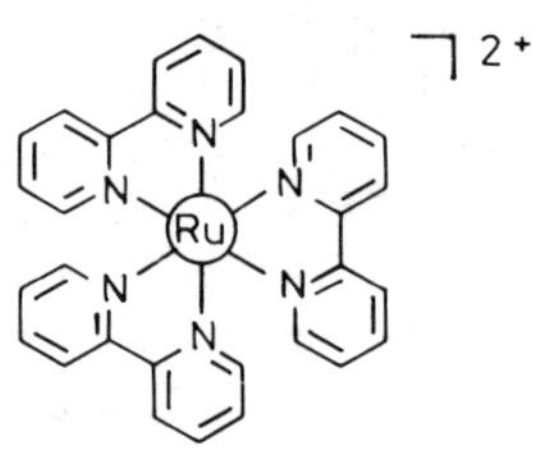

13

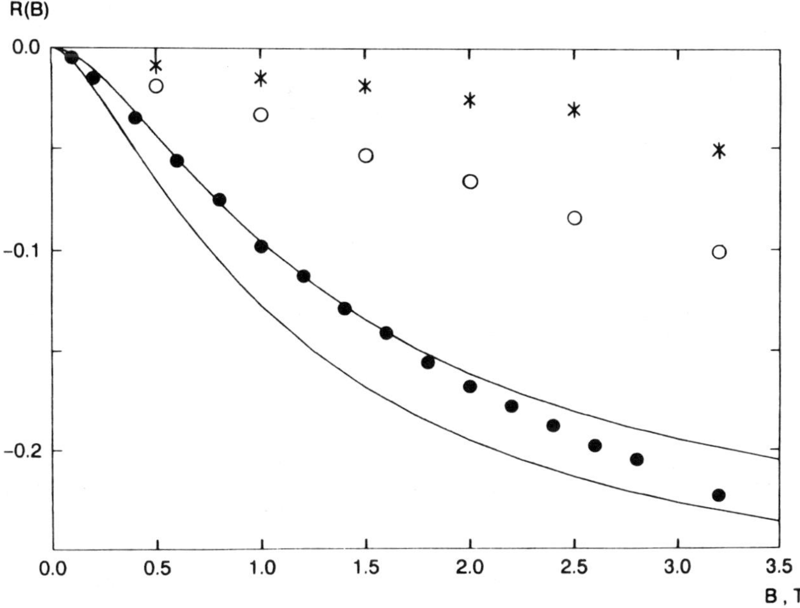

FIGURE 28.   Relative MFE on the yield of $MV^{+\cdot}$ radicals $[MV^{2+} = (12, R = R' = H)]$ in the electron transfer quenching of excited Ru-complexes in H_2O/ACN mixtures: •, $Ru(bpy)_3^{2+}$ (*13*); ○, $Ru(phen)_3^{2+}$; *, $Ru(dce)_3^{2+}$; phen = phenanthroline, dce = 4,4'-di(ethoxycarbonyl-2,2'-bipyridine. The solid lines are results of model calculations for $Ru(bpy)_3^{3+}/MV^{+\cdot}$ (cf. References 205 and 206) employing the parameter values $g_\perp = 2.71$, $g_\parallel = 1.12$,[207,208] $k_{dis} = 5$ ns^{-1}, $k_{bet} = 100$ ns^{-1}, and $w = 30$ ns^{-1} (lower curve), 60 ns^{-1} (upper curve), both sets yielding a free radical yield of about 25% consistent with the experiment.

The photoreactive state of the complex involved in the electron transfer reaction is the lowest CT triplet state (the question of its spin-purity is still a matter of controversy[197,198]). As may be judged from the quantum yield of the free $Ru(bpy)_3^{3+}/MV^{+\cdot}$ species that is typically in the order of 0.2,[199] fast backward electron transfer in the primary pair is quite efficient.

MFEs on the yield of $Ru^{III}/MV^{+\cdot}$ have been reported by Ferraudi and Argüello[200] and by Steiner et al.[201] An example of the magnetic field dependence of this effect for $Ru(bpy)_3^{2+}$ and several other derivatives as recently measured in our laboratory[202-204] is shown in Figure 28. The data in this figure have been obtained from stationary illumination experiments (cf. Reference 201) in solutions with EDTA added as a sacrificial electron donor to reduce the Ru^{III} species so that $MV^{+\cdot}$ radicals can be detected on a long time scale. These results are, however, in full agreement with ns time-resolved observations of the radicals produced in the absence of EDTA as investigated by laser flash spectroscopy.[201-204]

Whereas Ferraudi and Argüello attributed the MFE on the yield of electron transfer product to a magnetic reduction of the rate constant of forward electron transfer, in our work[201] it has been unequivocally shown that it is due to an increase of the rate of backward electron transfer in the primary $Ru(bpy)_3^{3+}/MV^{+\cdot}$ pair. The sign of the MFE is compatible with the TM and the Δg-type RPM, both relying on SOC effects. In fact, a closer inspection of the situation led us to conclude that, formally, the mechanism may be described as a mixture of both TM and Δg-type RPM (cf. Figure 29).

For an interpretation of the MFEs it is important to note that when using normal (i.e., without internal heavy atoms) electron donors or acceptors MFEs were detectable only in the oxidative quenching of $Ru(bpy)_3^{2+}$-type complexes, but not in the reductive mode.[201] The essential difference between them is that in the oxidative case an open-shell $(t_{2g}\text{-}d^5)$ electron configuration of the Ru ion is created, whereas in the reductive case the accepted

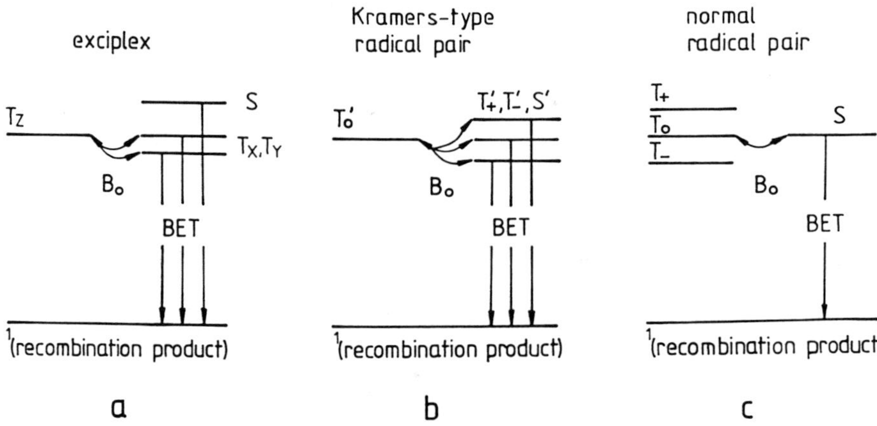

FIGURE 29.   Three types of SOC-induced magnetokinetic mechanisms for spin-forbidden backward electron transfer (BET). (a) Triplet mechanism (TM) in triplet exciplex; (c) Δg-type radical pair mechanism (RPM); (b) the Kramers-type RPM, comprising features of both TM and Δg-type RPM.

electron occupies a π^* MO of a ligand. Marked SOC effects can be expected only in the former case.

It has been mentioned above that in organic radicals SOC cannot cause a direct spin-reorientation. This situation is essentially different for open-shell d-electron configurations of heavy transition metals because spin flips can be combined with transitions among near-degenerate d orbitals. As a result in the latter case the α,β spin doublet used to describe the spin state of normal radicals has to be replaced with a spin-orbit-coupled Kramers doublet.

In the components (α',β') of the latter α and β spin states will always be coupled to some degree so that the actual spin orientation may not be fixed as sharply as in a normal organic radical. Combining a Kramers doublet species (as $Ru[bpy]_3^{3+}$) with a normal organic radical (as $MV^{+\cdot}$) yields a pair with four formal "spin" states. A convenient basis is

$$S' = \frac{1}{\sqrt{2}} (\alpha'_1\beta_2 - \beta'_1\alpha_2) \tag{165}$$

$$T'_o = \frac{1}{\sqrt{2}} (\alpha'_1\beta_2 + \beta'_1\alpha_2) \tag{166}$$

$$T'_+ = \alpha'_1\alpha_2 \tag{167}$$

$$T'_- = \beta'_1\beta_2 \tag{168}$$

These pair states are denoted by primed symbols in analogy to those of a normal RP. The difference is, however, that except for T'_0 which can be shown to correspond to a mixture of configurations with $|S| = 1$ only, the primed pair states are not pure $S = 0$ or $|S| = 1$ states. Singlet spin configurations ($S = 0$) are distributed over $T'_\pm$ and S', endowing them with the potential to undergo (partially spin-allowed) backward electron transfer to the singlet ground-state pair of reactants. The substate selectivity thus introduced tends to enrich the $Ru(bpy)_3^{3+}/MV^{+\cdot}$ ensemble in the T'_0 substate. (It should be noted that the axis of quantization in the primed basis of states is fixed in the molecular frame of the Kramers doublet species.) The effect of a magnetic field is to augment backward electron transfer by inducing T'_0-S' mixing (analogous to the normal Δg-mechanism) or T'_0-$T'_\pm$ mixing (analogous to the TM) (cf. Figure 29).

We have applied these principles in a dynamic treatment of the MFE on backward electron transfer in the $Ru(bpy)_3^{3+}/MV^{+\cdot}$ pair. In principle the experimental curves may be fitted quite well. A suitable set of parameters is given with Figure 28. The rate constant $k_{bet}(S)$ assigned to a hypothetical pair of pure singlet spin is of reasonable order of magnitude. The rate constant k_{dis} corresponds to the result obtained from the Eigen equation. The parameter w, corresponding to a rate constant of relaxation among the Kramers doublet states, provides interesting information on a process not otherwise accessible experimentally, since ESR spectra of these complexes have not been obtained in liquid solution at room temperature.

Another example of MFEs in photoelectron transfer reactions with transition metals has been reported by Horvath et al.[209] Investigating the yield of solvated electrons from photo ionization of Cu(I)cyanide complexes they observed that a magnetic field induces a significant decrease for $Cu(CN)_2^-$ but has no effect in case of the $Cu(CN)_3^{2-}$ complex. The effect has been attributed to a Δg-type mechanism in the primary $Cu^{II}(CN)_2/e^-$ pair, suggested to originate from a ligand field-excited state of $Cu^I(CN)_2^-$ with triplet spin.

d. Triplet Electron Transfer in Linked Donor/Acceptor Systems

Chemically linked triplet RPs produced by exergonic electron transfer with excited triplet states have been investigated by several groups.

Mataga and co-workers[210] studied a Zn-mesoporphyrin IX to which two MV^{2+} substituents were attached $[P(n)(MV^{2+})_2,\ n = 6,\ 11\ (14)]$.

14

In comparing their results, e.g., with the Py(n)DMA (4) system, one should note that the number of atomic links between the donor and acceptor π-systems is actually given by n + 4. Using laser flash spectroscopy the lifetime of the corresponding RPs was measured as a function of the magnetic field in two different solvents (cf. Figure 30). The lifetime increased with the magnetic field as to be expected for the case of T-RPs, where ISC according to the hfc mechanism is the rate determining step in backward electron transfer. The magnetic field-dependent curves exhibit weak but clearly noticeable minima at low fields (30 to 40 G for n = 6, $\leqslant$10 G for n = 11) that may be related to the effect of n-dependent exchange interaction. A pronounced solvent effect was observed on the magnetic field dependence of the RP lifetime for n = 6 (cf. Figure 30b). The strong reduction of the MFE in acetone/methanol indicates a predominance of SOC-induced backward electron transfer or perhaps decay by non-spinselective chemical processes.

Linked T-RIPs involving the Zn-tetraphenylporphyrin (TPP) chromophore as an electron donor with a pendant MV^{2+} acceptor [Zn-TPP(n)MV^{2+}, n = 4, 6, 8] have also been investigated by Nakamura et al.[211] Their results refer to linked RPs with 8 to 12 intervening atoms, including a phenyl ring as a member of the chain. They, too, observed a magnetic field-dependent increase of the RP lifetime. It is noteworthy that in the saturating limit of fields >0.5 T the lifetimes ($\sim$1.3 μs) were independent of n. These values, measured in acetonitrile as solvent, are by a factor of about 4 longer than those for the systems studied

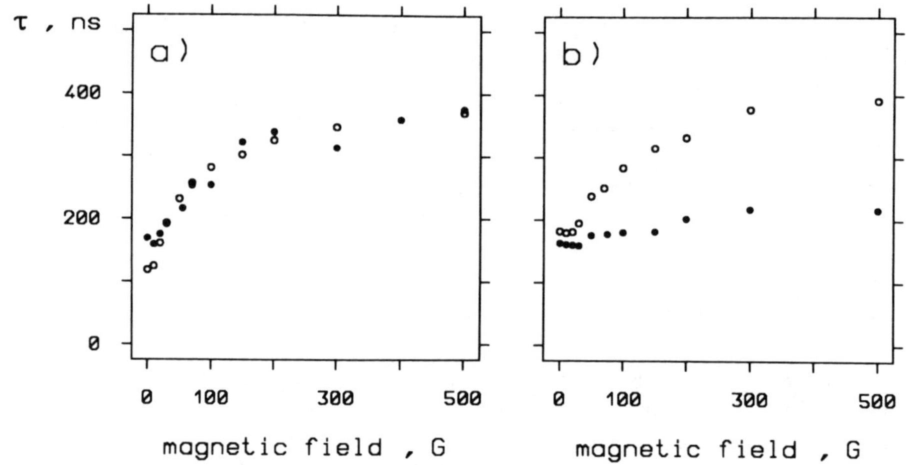

FIGURE 30.   Magnetic field dependence of the lifetime of $P^+(n)MV^{+\cdot}$ (cf. (*14*)); $\circ$, in DMSO; $\bullet$, in MeOH/acetone; (a) n = 11, (b) n = 6. (Adapted from Saito et al.[210])

by Mataga's group. When incorporated in dihexadecyltrimethylammonium chloride bilayers the $^3[Zn\text{-}TPP^+\cdot(n)MV^+\cdot]$ RPs showed similar MFEs as in ACN, however with a larger lifetime in the saturating field limit.

N-(ω-[p-Nitrophenoxy]alkyl)aniline[212] or corresponding compounds with phenothiazine or triphenylphosphine[213] replacing the anilino group represent another type of chemically linked systems where MFEs point to T-RPs as intermediates of photochemical conversions initiated by a primary photoelectron transfer. These reactions have been investigated by steady-state photolysis. Whereas in ACN electron transfer is the primary reaction, hydrogen transfer takes place instead in benzene (for more details on these systems cf. Section III.B below).

B. HYDROGEN ATOM ABSTRACTIONS

MFEs with RPs originating in H-atom abstractions have been studied using excited triplets of aromatic ketones,[188,214-217] quinones,[218,219] N-heterocycles,[220-222] nitroaromatics,[212,223] and xanthene dyes[224] as H-acceptors. Examples with triplet carbenes[225,226] as H-atom acceptors have also been reported. However, in this case MFEs have been studied in micellar solutions only (cf. Section IV).

Since no net charge is transferred with an H-atom the RPs originating in such reactions are not trapped in Coulomb cages. Hence the yield of escape from triplet RPs is high in low-viscous homogeneous solutions and the MFE on it is low. Therefore, the more detailed investigations with such types of RPs have utilized special means of enhancing the cage effect, as there is use of highly viscous homogeneous solvents, eventually at lower temperatures, application of chemical links, i.e., the study of biradicals, or of micellar supercages (cf. Section IV).

MFEs in highly viscous homogeneous solvents have been investigated in some detail by Kuzmin and co-workers.[185,214,215] Glycerol (η = 10.7 P at 293 K) was the solvent applied by them to study the recombination kinetics of radicals formed according to Equation 169.

$$\left(\begin{array}{c}^3\\ \text{Ph}_2\text{CO}\end{array}\right)^* + \begin{array}{c}\text{OH}\\ \\ \text{CH}_3\end{array} \longrightarrow \left(\begin{array}{c}^3\\ \text{Ph}_2\dot{C}\text{OH}\end{array} + \begin{array}{c}\cdot\text{O}\\ \\ \text{CH}_3\end{array}\right)$$

      15   16   17   18

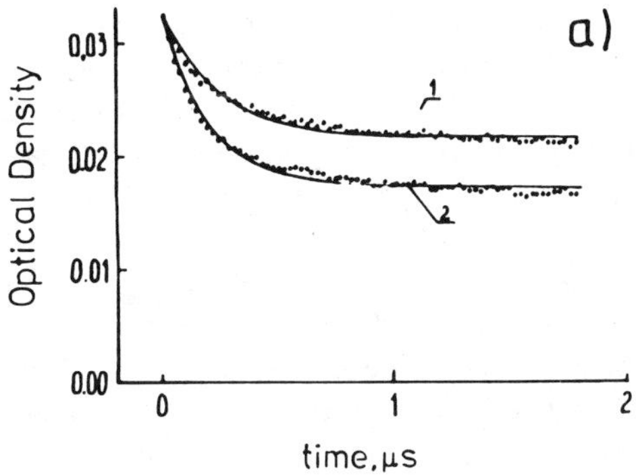
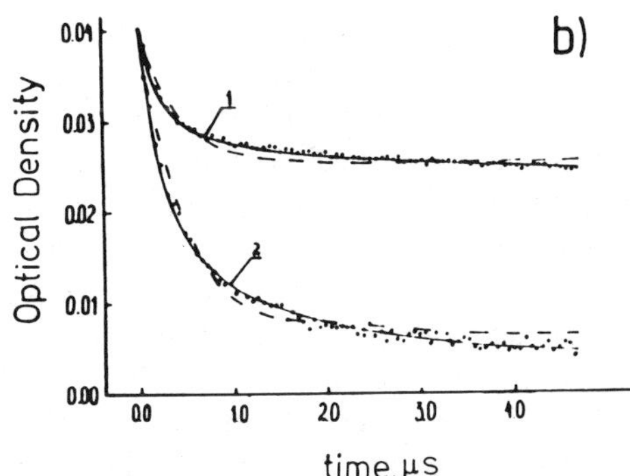

FIGURE 31.   Variation of the optical density of benzophenone ketyl radical (*17*) after laser excitation of benzophenone in a glycerol/cresol mixture. (a) At 239 K, in zero magnetic field (2) and in a field of 0.34 T (1); (b) in a magnetic field of 0.34 T at 293 K (1) and 253 K (2). Kinetic simulation by exponential law: solid lines in (a) and dashed lines in (b); by Noyes function: solid line in (b). (Reprinted from Levin, P. P., Khudyakov, I. V., and Kuz'min, V. A., *J. Phys. Chem.*, 93, 208, 1989. With kind permission of P. P. Levin and V. A. Kuz'min; copyright 1989, American Chemical Society.)

The recombination kinetics observed (cf. Figure 31) by laser flash spectroscopy does, in fact, show a fast geminate phase of recombination kinetics and resembles very much the typical behavior in micellar solutions. In a magnetic field the rate of geminate recombination is decreased and the yield of long-lived uncorrelated radicals is increased.

The geminate kinetics may be evaluated on the basis of a simple kinetic model leading to monoexponential decay and yielding two first-order rate constants k_{rec} and k_{esc}. However, a better fit is obtained when applying the re-encounter model of Noyes yielding a relative diffusion coefficient of $D^N = 1.4*10^{-9}$ cm^2 s^{-1}. This value is similar to the relative diffusion coefficient $D^E \approx 3.7*10^{-9}$ cm^2 s^{-1} evaluated from k_{dis} on the basis of Eigen's relation

$$k_{dis} = \frac{3D^E}{\sigma^2} \tag{170}$$

with σ the encounter distance. It is, however, much lower than the value of D^{SE} according to the Stokes-Einstein relation ($12*10^{-9}$ cm^2 s^{-1}). Since, according to general experience, the SE relation grossly *under*estimates experimental diffusion constants in highly viscous solvents, one may conclude that the relative diffusion constant appearing in the observed geminate kinetics is much lower than the sum $D_1 + D_2$ of the individual diffusion constants. Levin et al.[214] offer the explanation that the relative motions of the radicals in the pair may be correlated, or that an attractive potential, possibly by hydrogen bonding, could be present between the radicals. One might hesitate to overemphasize such consequences from an experiment conducted with high signal averaging ($\approx$ 128 to 1024 flashes) on a sample which is not flown, so that even low degrees of irreversibility may distort the observed kinetics. However, it should be pointed out that a similar discrepancy between the diffusion constant extracted from the geminate kinetics and the sum of the individual diffusion constants of the radicals has also been noted by Shushin[141] when analyzing the geminate RIP recombination kinetics observed by Werner et al.[114]

The observed MFE corresponds to a decrease of the effective rate constant of recombination, whereby the field dependence saturates at about 0.1 T when k_{rec} ($\approx 2*10^6$ s^{-1} at $B_0 = 0$, 293 K) reaches about 70% of its zero-field value. The sign of the MFE results from the triplet spin correlation with which the RP is formed and which is more effectively lost in zero-field than in high field. From the field dependence and the maximum effect one can draw the following qualitative conclusions: in zero-field the triplet-singlet spin evolution cannot be rate determining, since on the basis of the prevailing hyperfine coupling (B^{hfc}) this process should be by two orders of magnitude faster than the value observed for k_{rec}. At fields higher than B^{hfc} the coherent spin motion $T_\pm \rightarrow T_0,S$ is suppressed. However, incoherent relaxational processes still can allow for recombination during the geminate phase, although they, too, are slowed down in a magnetic field. This kinetic quenching of the relaxation processes seems to be complete at about 0.1 T. From the saturation value $k_{rec}(0)/k_{rec}$(high field) $\approx$ 1.3 being clearly smaller than 3 one has to conclude that even if $T_\pm \rightarrow$ S transitions due to independent spin motion in the individual radicals are suppressed RPs with triplet spin correlation still have a good chance to recombine in a re-encounter. This recombination channel is attributed to the effect of SOC effective in the contact of RPs and assumed to be independent of an external magnetic field. All these conclusions are similar to those reached earlier in connection with the interpretation of observations in micellar systems (*vide infra*).

Investigations employing triplets of the xanthene dyes fluorescein (*11a*), ethyleosin (*11b*, ethyl ester), and erythrosine (*11c*) as H-acceptors from *p*-cresol (*16*) in glycerol were undertaken to specify the role of SOC on spin-forbidden geminate recombination.[224] Astonishingly the heavy atom substituents had only moderate influence on the saturation MFE on the free radical yield. At 295 K and $\eta = 10$ P these effects amounted to +110% for fluorescein and to +50% for both ethyleosin and erythrosine. It is remarkable that the saturation MFE passed through a maximum as a function of η/T.

There are several other reports referring to H-atom abstractions in low-viscous solutions at room temperature where the mechanistic interpretation of the MFEs is unclear or rather speculative. Staerk and Rasi Naqvi[216] found a rather linear decrease with the magnetic field in the yield of ketyl radicals observed on laser excitation of benzophenone in *n*-hexane. The suggestion that this effect (-8% at 0.8 T) might be due to the Δg-mechanism assisting the geminate recombination of triplet RPs was not substantiated in a quantitative theoretical model calculation by Schulten and Epstein applying this mechanism.[227]

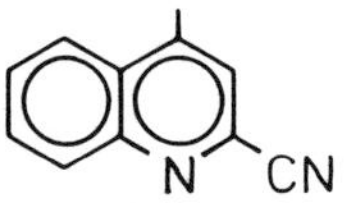

N   CN

19

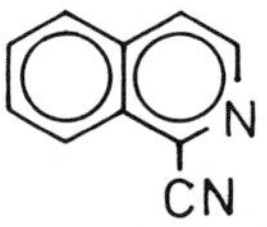

N
CN

20

Photosubstitution of the CN substituent in 4-methylquinoline-2-carbonitrile (*19*) or 1-isoquinolinecarbonitrile (*20*) by solvent radicals, when irradiating these compounds in ethanol or cyclohexane, is assumed to involve H-abstraction from the solvent by the ring nitrogen in the first step. MFEs on the yields of the final photosubstitution products have been reported by Hata and co-workers.[220-222] The magnetic field dependences investigated up to quite high fields of about 1.8 T are of different types and the explanations make use of various factors such as precursor multiplicities, hfc, and Δg-mechanism as well as level crossing effects due to exchange splitting and Zeeman interaction. Time-resolved investigations focusing directly on the MFEs of primary intermediates might be helpful for a better understanding of these interesting effects.

1. Biradicals

In compounds such as (*21*) and XO-(*n*)-XH (cf. Figure 33, below) photochemical H-abstraction by the triplet excited chromophore takes place as an intramolecular process leading to the formation of triplet biradicals. In the case of compounds XO-(*n*)-XH the carbonyl oxygen in the excited xanthone moiety accepts an H-atom from the CH_2-bridge in the xanthene moiety. The decay of the biradicals thus formed has been observed by laser flash spectroscopy by Tanimoto et al.[217] The biradicals decay almost completely by forming intramolecular

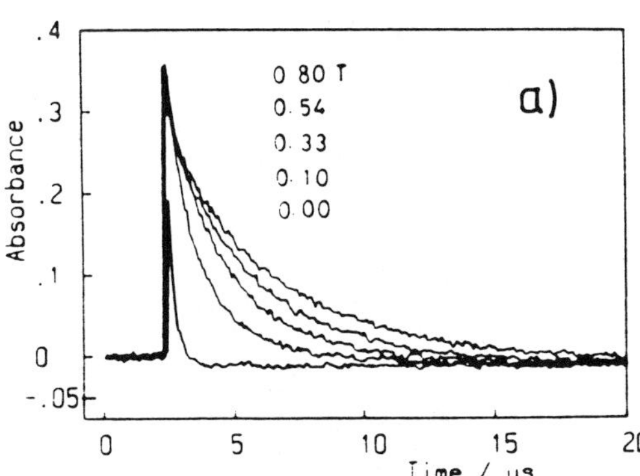
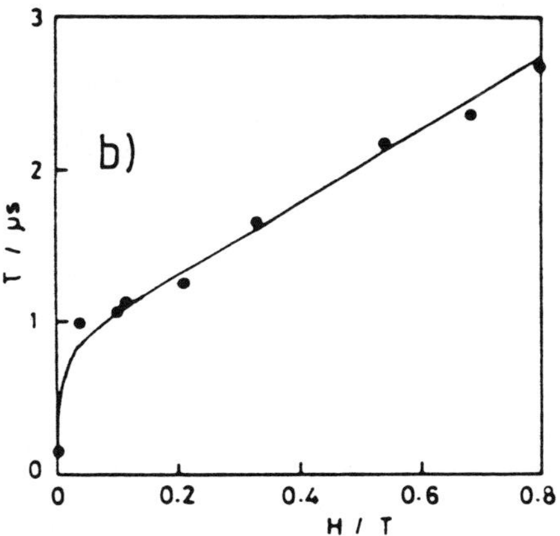

FIGURE 32.   MFE on the decay of xanthene radical absorption in biradical $\overset{.}{X}$OH-(12)-$\overset{.}{X}$ (cf. Figure 33). (a) Decay curves; (b) magnetic field dependence of decay times. (Reprinted from Tanimoto, Y., Takeshima, M., Hasegawa, K., and Itoh, M., *Chem. Phys. Lett.*, 137, 330, 1987. With kind permission of Y. Tanimoto; copyright 1987, Elsevier Science Publishers B. V.)

coupling products. As can be seen in Figure 32a the decay time of the biradical is strongly increased by a magnetic field. For n = 12 the field dependence of the decay time is shown in Figure 32b. After a large increase by a factor of about 7 between 0 and 20 mT the lifetime still continues to rise linearly, reaching an overall increase by a factor of 19 at 0.8 T. This large increase of the biradical lifetime demonstrates that ISC processes induced by interactions which are not strongly influenced by a magnetic field, e.g., SOC in the contact of the biradical ends, are fairly slow in this case. The two regions of magnetic field dependence apparent in Figure 32b may be attributed to the quenching of $T_{\pm} \rightarrow S$ transitions occurring by the coherent hfc mechanism (low-field effect) and by incoherent relaxation processes (high-field effect). It should be noted that the relaxational region was not observed with the biradical systems derived from electron transfer reactions described above. It is noteworthy, however, that a clear separation of the kinetics into a fast $T_0 \rightarrow S$ initial section and a magnetic field-dependent slow $T_{\pm} \rightarrow S$ section is not observed. At no field strength does the kinetics show significant deviations from monoexponential decay.

The chainlength dependence of the MFE with XO-(*n*)-XH has been studied for n = 2 to 12. (cf. Figure 33). The MFE $\tau(0.8\ T)/\tau(0\ T)$ decreases from 19 to about 3 as n is lowered from 12 to 2, whereby the most significant drop occurs below n = 6. This effect is attributed to the effect of exchange interaction impeding $T \rightarrow S$ transitions already at zero-field. However, contrary to what has been observed with the biradicals from electron transfer reactions, with these systems no extrema in the magnetic field dependence are found, which would be indicative of T_-/S level crossings due to the combined effects of exchange interaction and Zeeman splitting. It was not possible to explain the n-dependence of the maximum MFE in terms of a general distance dependence of J as Equation 55. Using the average distance between the radical centers ($\bar{r}$ = 12.4 Å for n = 2, $\bar{r}$ = 16.1 Å for n = 8) the exchange interaction to be expected (e.g., with the parameters given in the caption of Figure 8) is much too low to cause any effects. This argument applies also if the exchange interaction is averaged over the possible conformations from which a nonmonotonic behavior with n would be expected.

$$O_2N-\bigcirc-O(CH_2)_n-NH-\bigcirc$$

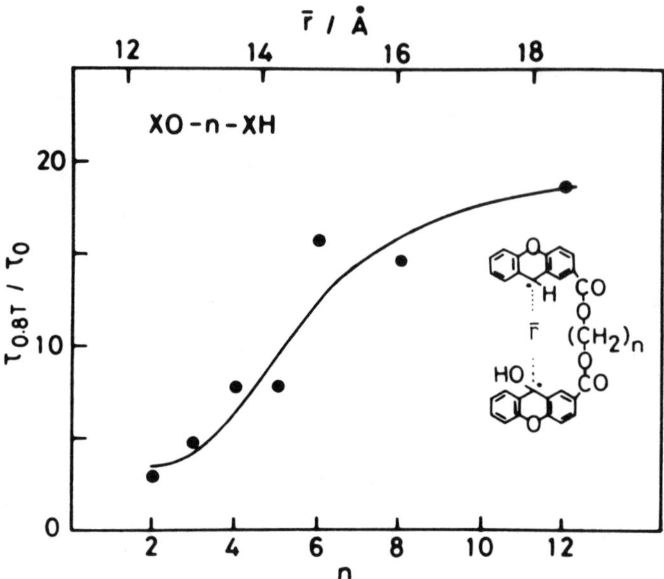

FIGURE 33. Chainlength dependence of MFE on lifetime of biradicals
$\dot{X}$OH-(n)-$\dot{X}$. $\bar{r}$ is the average interradical distance between XOH and X.
(Reprinted from Tanimoto, Y., Takeshima, M., Hasegawa, K., and Itoh,
M., *Chem. Phys. Lett.*, 137, 330, 1987. With kind permission of Y.
Tanimoto; copyright 1987, Elsevier Science Publishers B.V.)

The photochemistry of compounds (*21*) as well as of the corresponding nitronaphthoxyl analogues has been investigated in some detail by Nakagaki et al.[212,223,228,229] The behavior of both series of homologues is similar. However, only with the nitronaphthoxyl chromophore the absorption band is well separated from that of the anilino chromophore and selective excitation of the former is possible. The series of homologues of type (*21*) exhibit the interesting feature of reaction mechanism switching as a function of chain length n. Thus for $n \leq 6$ a photo-Smiles rearrangement is found, whereby, formally, the nitroaryl group migrates to the amino nitrogen, whereas for $n \geq 8$ an oxidative elimination of the anilino substituent is observed. A primary H-abstraction from the N-terminal CH_2 group by the nitro group to yield a biradical is the primary step in the latter case. Oxidative decoupling of the anilino group may occur then as an *intra*molecular (cage) or *inter*molecular (escape) process. Only in the former case the nitro group of the decoupled fragment is reduced to the nitroso group, so that the two reaction channels may be distinguished by product analysis. The product distribution was found to depend both on the chain length n and an external magnetic field. Application of a magnetic field favors formation of the escape product, as can be expected for a T-RP originating from the triplet excited nitroaromatic chromophore. For n = 8 it is even possible to reverse the ratio of cage to escape yields (61% cage product at zero-field, 36% at $B_0 = 0.64$ T), which represents a remarkable example of reaction control by a magnetic field. For $n \geq 11$ both chain length and MFE attain limiting values (about 83 and 60% cage product in zero-field and at 0.64 T, respectively).

C. HOMOLYTIC BOND CLEAVAGE

The first photochemical MFE with reactions of this type was reported in 1976 by Tanimoto et al.[16] They investigated the decomposition of benzoylperoxide, sensitized by singlet excited chrysene, in toluene for the yields of various reaction products formed in escape or cage reactions[230,231] from the primary RP originating in the process

$$PhCO_2\text{--}O_2CPh \xrightarrow{\ ^1Sens^*\ } {}^1\overline{PhCO_2\cdot\ \cdot O_2CPh} \tag{171}$$

The clearest effect was observed for phenyl benzoate produced as a cage product by the reaction sequence

$$^1\overline{PhCO_2\cdot\ \cdot O_2CPh} \to CO_2 + {}^1\overline{Ph\cdot\ \cdot O_2CPh} \to CO_2 + PhCO_2Ph \tag{172}$$

The yield of phenylbenzoate is decreased by the magnetic field. Rather high field values are necessary, however (at 4 T the yield decreases by 8%), and the field dependence follows a square-root relation. This behavior may be understood in terms of the Δg-type RPM. As the field is increased the S/T_0 transition frequency $\omega ST_0 = \Delta g\beta B_0/\hbar$ increases and the cage recombination probability decreases. The g-factor difference between the radicals $Ph\cdot$ and $PhCO_2\cdot$ amounts to about 0.01 so that at 4 T $\omega_{ST_0} \approx 3.5*10^9$ s^{-1}. Applying a rough estimate with competitive first-order rate processes the observed MFE would suggest a value of about 20 ps for the lifetime of the RP.

The square-root dependence observed is just what would be expected in lowest order from the diffusion model of the RPM (cf. Section II.A.2). The corresponding result for the case of an original S-RP recombining in the singlet spin state only is

$$^S p_s \approx \Lambda[1 - \frac{1}{2}(1 - \Lambda)\delta] \tag{173}$$

From the slope of 0.038 T and using $\Delta g = 0.01$ one obtains

$$(1 - \Lambda)\left(\frac{a^2}{D}\right)^{1/2} = 2.56*10^{-6}\ s^{-1/2} \tag{174}$$

Using reasonable values for a (≈ 3 Å) and D ($\approx 1*10^{-5}$ cm^2 s^{-1}) one obtains $\Lambda = 0.73$ which seems quite an acceptable value for the efficiency of S-RP recombination.

Another important type of reaction involving homolytic bond cleavage is the Norrish type I photoreaction of $n\pi^*$-excited ketones whereby α-CC-cleavage is the primary photochemical reaction. Examples of magnetokinetic effects have been investigated, e.g., with di-*tert*-butyl ketone[232] and dibenzylketone (DBK).[18,233-238] In homogeneous solution DBK has been mainly studied for ^{13}C-magnetic isotope effects. With di-*tert*-butyl ketone, however, the magnetic field dependence of the yield of cage and escape products has been determined in some detail by Fischer.[232] Photocleavage occurs in the excited triplet state yielding a triplet geminate RP (*22/23*)

$$^3\overline{\underset{22}{CH_3\text{-}\underset{CH_3}{\overset{CH_3}{\underset{|}{\overset{|}{C}}}}\text{-}\overset{O}{\underset{\cdot}{C}}} \qquad \underset{23}{\overset{CH_3}{\underset{CH_3}{\overset{|}{\underset{|}{\cdot C}}}}\text{-}CH_3}}$$

which, apart from the starting material, may form the cage recombination products pivaldehyde (*24*) and 2-methylpropene (*25*). Pivaloyl radicals (*22*) escaping from the geminate pair will undergo decarbonylation and upon homogeneous recombination of the *tert*-butyl radicals (*23*) hexamethylethane, *tert*-butane, and 2-methylpropene are formed.

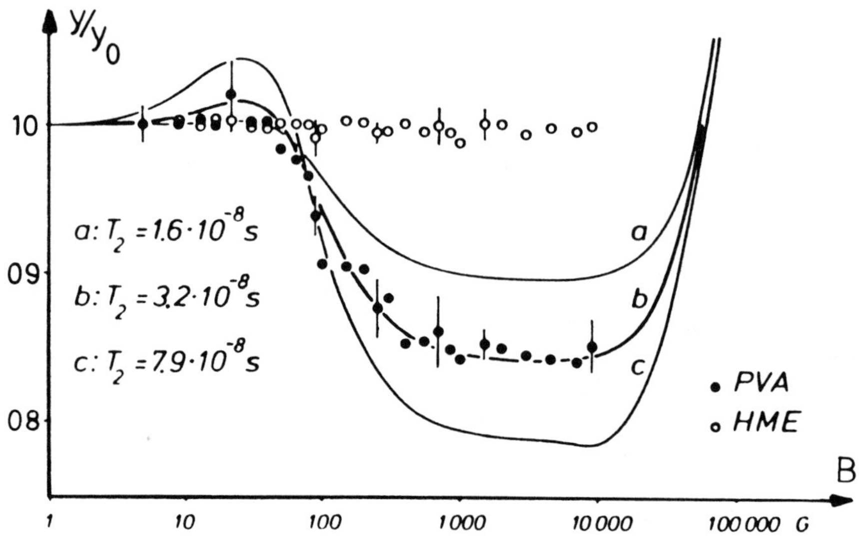

FIGURE 34. Magnetic field dependence of the yields of pivalaldehyde (PVA, [24]) and hexamethylethane (HME) during photolysis of di-*tert*-butyl ketone in *n*-decane at 68°C. The solid lines represent theoretical results from the RPM employing various values of the spin-rotational relaxation time constant T_2 for the pivaloyl radical (22). (Reprinted from Fischer, H., *Chem. Phys. Lett.*, 100, 255, 1983. With kind permission of H. Fischer; copyright 1983, Elsevier Science Publishers B.V.)

$$CH_3-\underset{\underset{CH_3}{|}}{\overset{\overset{CH_3}{|}}{C}}-CHO \qquad CH_2=C\overset{CH_3}{\underset{CH_3}{\diagdown}}$$

25   24

The MFE measured on the yields of pivalaldehyde (cage product) and hexamethylethane (escape product) in *n*-decane at 68°C are shown in Figure 34.

Whereas there is no detectable MFE on the yield of the escape product, the yield of cage product is decreased by about 16% in high magnetic field, whereby a saturation is achieved at about 0.1 T with a $B_{1/2}$ of about 11 mT. The field dependence could be reproduced theoretically by using the first-re-encounter version of the RPM.

The equivalence of the nine protons in the *tert*-butyl radical allows the spin motion in the geminate pair to be calculated exactly for all fields [A(*tert*-butyl) = 2.27 mT, A(pivaloyl) = 0.06 mT, the latter was neglected] if exchange interaction is neglected. The coherent spin motion was multiplied by an exponential damping factor accounting for the effects of electron spin relaxation in the pivaloyl radical. (The relaxation is attributed to the spin-rotational mechanism. In this case $T_1 = T_2$, both being independent of the field strength.) The effect of decreasing the spin relaxation time is to reduce the MFE which is a consequence of incompletely suppressed $T_\pm \rightarrow S$ spin transitions in the RP. The value of $T_1 = T_2$ yielding the best fit of the MFE compares favorably with independent experimental results from ESR line widths.

The $B_{1/2}$ value (field of half-saturation) is about 11 mT. This compares very well with a theoretical value of 11.5 mT obtained from Weller's formula (Equation 152) which is an indication that exchange effects are not borne out in the MFE.

At very low fields an initial increase and a subsequent sign inversion of the MFE are predicted by theory. The reason for this is the high degeneracy of spin energy eigenstates at zero-field, which does not allow for the most efficient T/S mixing. In a very weak magnetic field the degeneracies disappear and the T/S process becomes more efficient before, at somewhat higher fields, the Zeeman splitting starts to separate the $T_\pm$ and T_0,S levels to an extent that $T_{\pm} \rightarrow T_0,S$ transitions become slower. Experimental accuracy does not allow to prove or disprove this theoretically predicted effect in the present case.

The theoretical curves in Figure 34 turn upward again at fields >1 T. This behavior is due to the Δg-mechanism, enhancing $T_0 \rightarrow S$ transitions as the field increases. For the pivaloyl/*tert*-butyl RP Δg amounts to 0.0018 only. The high field region of increasing MFE was not reached in the experiments.

There was no MFE detectable for the yield of escape product. This observation is understandable in view of the high absolute yield of escape from the geminate T-RP ($\sim$97%). Thus, although the same absolute change affects both cage and escape channel the relative MFEs are of different order of magnitude.

We will briefly mention at this point that RPs produced by homolytic bond cleavage of triplet precursors have also been used as convenient molecular devices for magnetic isotope separation. Although, later, micellar media were found much more efficient for this purpose (*vide infra*) the first successful experiments of this kind were performed in homogeneous solution by Buchachenko et al.[18] In the photolysis of dibenzyl ketone an enrichment of the magnetic isotope ^{13}C was found in the nondecomposed educt, because cage recombination of the geminate T-RPs is assisted by hyperfine interaction with the magnetic ^{13}C nuclei. Reviews on the investigation of magnetic isotope effects may be found in References 1, 7, and 239 to 241.

The photo-Fries rearrangement of phenol esters to acyl phenols is another type of reaction starting with formation of geminate RPs by homolytic bond cleavage. As was derived from CIDNP studies (cf. Reference 57 and references given therein) the precursor multiplicity is singlet.

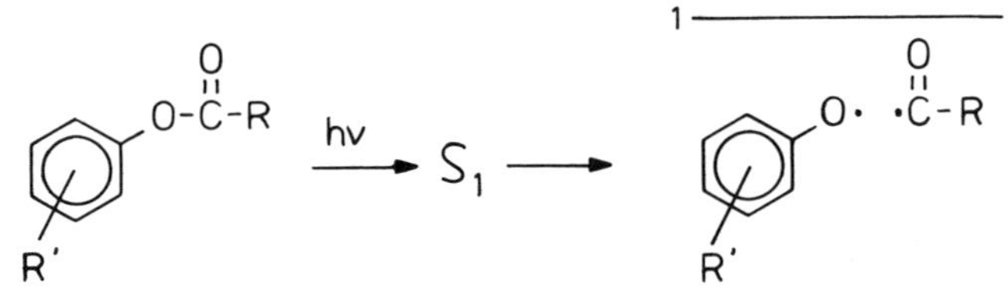

Homolytic cleavage takes place at the esteric C–O bond yielding a phenoxyl and an acyl radical. Cage recombination regenerates in part the initial ester in competition with C–C coupling leading to various isomers of acyl phenols, the typical products of the Fries rearrangement. Escape products are phenols, the diacyl recombination product, and, possibly, polymerization products.

In spite of a number of CINDP investigations direct detection of MFEs on product yields has been reported only once. Nakagaki et al.[242] investigated the photo-Fries reaction of 1-naphthyl acetate (*28*) in acetonitrille and some other solvents. The main products detectable were the acetyl naphthols (*29*) and (*30*) (cage products) and naphthol (escape product).

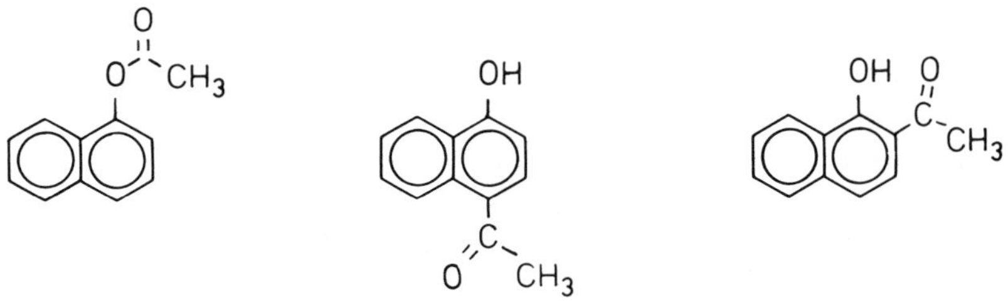

The quantum yield of formation of (*30*) was particularly convenient to detect from the change of the UV spectrum during photolysis. Whereas no MFE was detectable with natural isotope constitution such an effect became apparent when substituting ^{13}C for the carbonyl C atom. For this isotope-labeled compound the quantum yield of formation of (*30*) was increased by $3 \pm 1\%$ in a magnetic field of 0.64 T. The $^{13}C/^{12}C$ magnetic isotope effect in zero-field was reported to be $4 \pm 2\%$.

The signs of both effects indicate the singlet multiplicity of the original RP corroborating the CIDNP results of analogous reactions. The smallness of the effects, too, is a consequence of the initial spin correlation. Since the RP is immediately ready to recombine, its chemical lifetime (estimated as about 30 ps in Reference 242) is very short and does not allow for extensive spin evolution. In fact, only considerable increase of the hfc strength in the acyl radical by isotopic substitution $^{13}C/^{12}C$ can render the S $\rightarrow$ T process in the RP sufficiently fast to make it detectable as a reduction of cage recombination yield. In a magnetic field this zero-field magnetic isotope effect is largely suppressed so that the cage recombination yield increases in comparison with zero-field. The quantitative applicability of RP theory to achieve good agreement with experimental results in the analysis of net CIDNP intensities observed during photo-Fries rearrangement of *p*-methylphenyl-*p*-chlorobenzoate has been demonstrated by Vollenweider and Fischer.[57]

A homolytic N–C bond cleavage is known to occur on photolysis of hexaphenylbiimidazolyl (HPBI, [*31*]):

$$\text{(176)}$$

31   *32*

The reaction is thermally reversible and since the radicals (*32*) show a violet color Equation 176 represents an interesting photochromic system, particularly in a polymer matrix where the back reaction is slow.

The magnetic field dependence of the bleaching reaction of (*31*) was investigated in poly(methylmethacrylate) by Sato et al.[243] Kinetic traces at zero-field and at 0.44 T are shown in Figure 35. In the beginning the absorbance decay shows a faster kinetic component which after about 5 s turns into the slower one of second order. Both components are retarded in a magnetic field. Whereas no definite interpretation was given of the fast kinetic component the slower one is attributed to homogeneous second-order recombination of the radicals formed. The rate constant (k_2) of this process shows approximately linear decrease with the magnetic field. The effect amounts to about -30% at 0.4 T.

Homogeneous recombination of uncorrelated radicals involves statistical combination of spin states in the encounter pairs, i.e., the rate constant k_2 may be written as

$$k_2 = \left(\frac{1}{4} \Phi_{rec}^S + \frac{3}{4} \Phi_{rec}^T \right) k_{diff} \tag{177}$$

where Φ_{rec}^S and Φ_{rec}^T are the specific geminate recombination yields of RPs originating with singlet and triplet spin, respectively. By reference to the scheme in Figure 1 one can derive that by a deceleration of k_{TS} and k_{ST} through the magnetic field Φ_{rec}^S will increase and Φ_{rec}^T will decrease, if only singlet pair recombination is possible as to be assumed in the present case. Thus the kinetic effect observed here must be due to the MFE on the statistically formed *triplet* RPs. A necessary condition that this will occur is

$$k_{TS} \lesssim k_{esc} \tag{178}$$

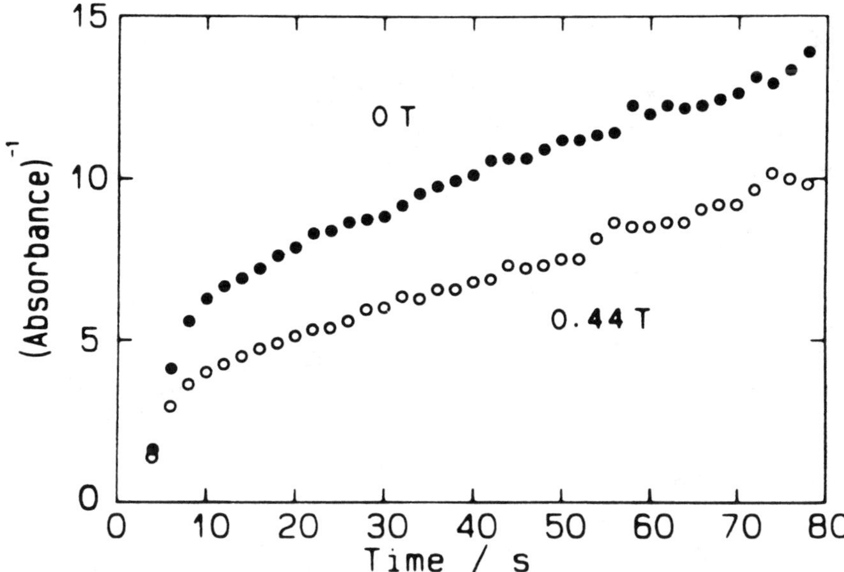

FIGURE 35.   Second-order plot of thermal fading in zero-field and in a magnetic field of 0.44 T of photochromic absorbance developed after irradiation of hexaphenylbiimidazolyl (*31*) in a poly-methylmethacrylate matrix at 40°C. (Reprinted from Sato, H., Kasatani, K., and Murakami, S., *Chem. Phys. Lett.*, 151, 97, 1988. With kind permission of H. Sato; copyright 1988, Elsevier Science Publishers B. V.)

because for $k_{TS} \gg k_{esc}$ either spin equilibrium would be attained during the cage lifetime (if $k_S \ll k_{esc}$) and hence $\Phi_{rec}^S = \Phi_{rec}^T$, independent of the field, or (if $k_S \gg k_{esc}$) both Φ_{rec}^S and Φ_{rec}^T would approach unity and, again, would be independent of the magnetic field. The continuously increasing MFE up to 0.4 T is not compatible with the coherent hyperfine mechanism but rather points to the effect of the relaxation mechanism. Since in this case k_{TS} corresponds to k_{rel} it would be of great interest to know the order of magnitude of k_{esc} to which the value of k_{rel} must be related according to the condition in Equation 178. Experiments investigating the instationary recombination kinetics after pulse photolysis would be desirable.

1. Biradicals from Homolytic Bond Cleavage Reactions

A most instructive series of investigations on the behavior of biradicals produced and reacting according to the pathways represented in Figures 36 and 37 has been performed by Turro and co-workers.[244-250] For a summary of this work cf. also Reference 251.

Two types of biradicals exhibiting distinct characteristics in their kinetic behavior arise from the photocleavage. Whereas the acyl-benzylic biradicals originating from monophenyl cycloalkanones (cf. Figure 36) are stable with respect to decarbonylation, such a process efficiently converts the primary biradical into a dibenzylic biradical in case of 2,*n*-diphenyl cycloalkanones (cf. Figure 37). The photochemical precursor of the biradicals is a triplet $n\pi^*$ state. Hence regeneration of the educt or rearrangements to other intramolecular coupling products requires an ISC process to occur before or simultaneously (*vide infra*) with formation of the new bond.

The lifetime of the T-RP is by one to three orders of magnitude longer than that of the singlet biradicals studied, e.g., by Weller's group, with the linked electron donor/acceptor systems. Thus, in case of the triplet biradicals dealt with in this section convenient observation of their decay kinetics was possible by nanosecond laser flash spectroscopy.

The essential aspects characterizing the dynamic behavior of the σ-type triplet biradicals are similar to those of the π-type singlet biradicals described above. Chain dynamics varies

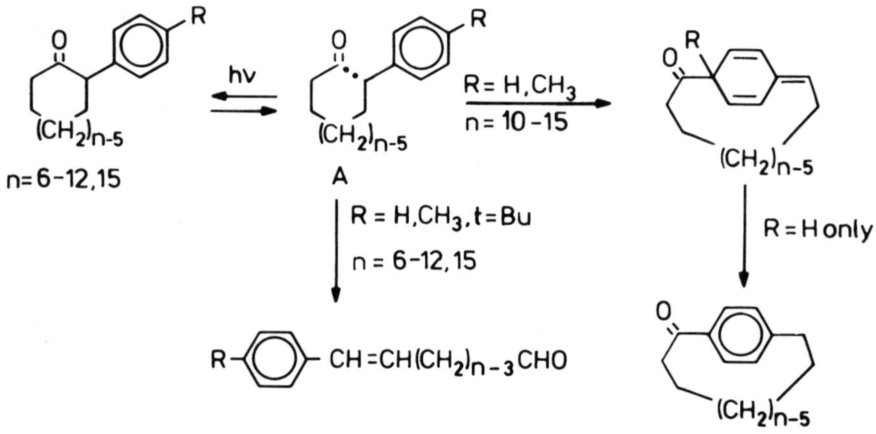

FIGURE 36.   Reaction scheme with biradicals of type A. (Adapted from Doubleday et al.[251])

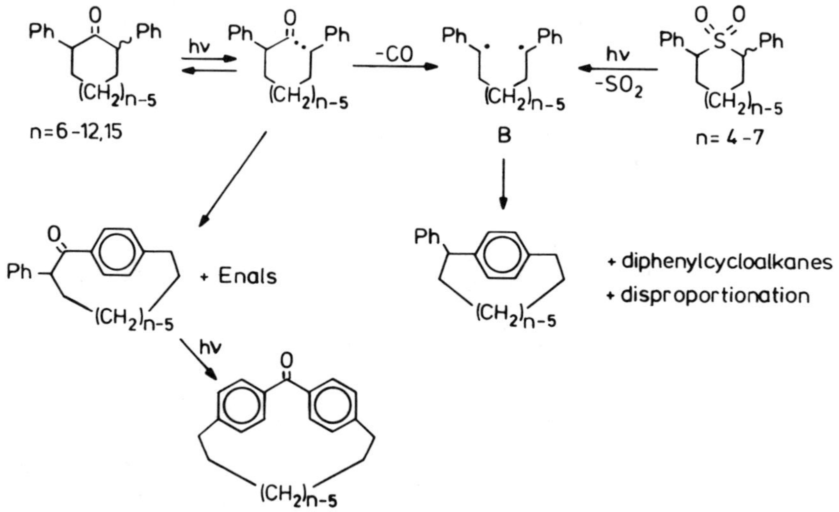

FIGURE 37.   Reaction scheme with biradicals of type B. (Adapted from Doubleday et al.[251])

the distance between the localized radical sites. This, in turn, modulates the exchange interaction responsible for the T/S energy gap. In the extended conformation this energy gap is small and allows for hfc-induced ISC processes. In addition, in the work of Turro's group it has been unambiguously shown that appreciable ISC does also occur at close distances due to the effect of SOC. The elucidation of the role of SOC in combination with chain dynamics and hfc-induced ISC is a particular merit of the work reviewed in this section.

To understand the effect of SOC in biradicals it is important to recall that for this type of interaction the angular momentum change associated with a spin flip must be compensated by a change of (direction of) orbital momentum. In molecular systems with purely real orbitals this can be only achieved if an orbital transition takes place simultaneously with the spin flip. As a consequence, SOC cannot couple the lowest singlet and triplet states of a biradical if they are of identical orbital nature, as is the case if the biradical ends are far apart. (This restriction does not apply to hfc when the electron spin flip is compensated by a nuclear spin flip.) If the radical ends of the biradical with their half-filled orbitals approach each other and the orbitals acquire some overlap, S and T functions adopt increasingly

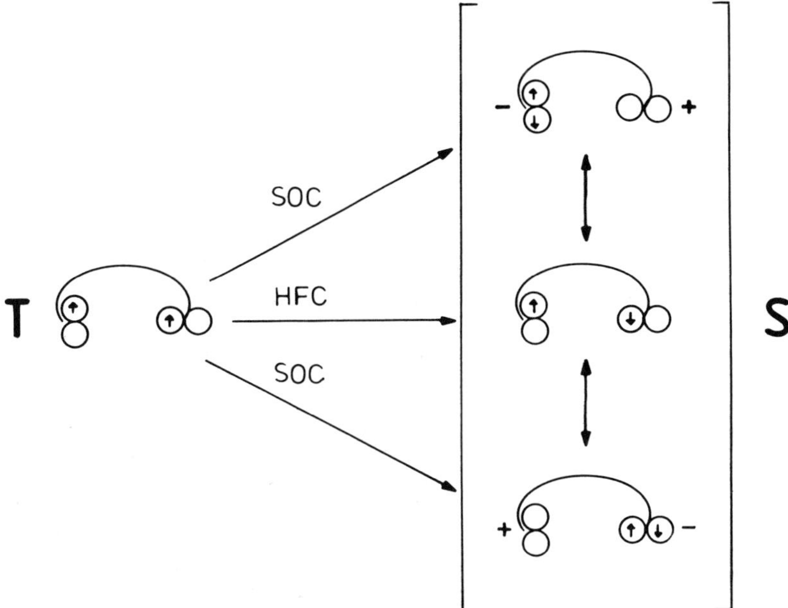

FIGURE 38.   Electronic mechanism of S/T transitions in a 1,n-alkanediyl-type biradical. At close contact of the biradical ends the zwitterionic mesomeric structures become increasingly important in a realistic representation of the singlet state. HFC and SOC connect the triplet state with different mesomeric components of the singlet state.

different orbital character which is also expressed energetically as a lowering of the singlet energy due to bonding interaction, whereby, in terms of VB theory, zwitterionic mesomeric structures become involved, and as an increase in triplet energy due to the antibonding interaction. Now SOC of T and S becomes feasible in that the spin flip of a radical electron may be accompanied by an orbital transition (cf. Figure 38). A semiempirical relation[252] has been suggested for the SOC matrix element coupling S and T states of a pair of two radicals or two biradical ends on an aliphatic carbon chain with the unpaired spins localized in atomic p-orbitals:

$$\beta_{SOC} = Be^{-br} \sin\Phi \tag{179}$$

Here r denotes the separation of the active centers carrying the two half-filled p-AOs, and Φ is the angle between the p-orbital axes, the sine term indicating optimum condition for perpendicular arrangement. For carbon 2 p orbitals b $\approx$ 3.03 Å^{-1} and B $\approx$ 15 cm^{-1}.

As a consequence of the r and Ψ dependence of β_{SOC}, which goes along with a related change of the singlet-triplet splitting energy ΔE_{ST}, the ISC transitions brought about by SOC cannot be adequately treated without due consideration of dynamic changes of geometry, e.g., in terms of the Landau-Zener-Stueckelberg approach (cf. Reference 253). A rigorous handling of this problem is an even more difficult task than treating hfc-induced ISC under conditions of dynamic modulation of ΔE_{ST}. Without going into details, however, one can note two important implications:

1.   In the case of the SOC mechanism ISC and chemical reaction can no longer be considered as separate processes. Rather, SOC-induced ISC should be conceived as a nonspin-conserving chemical reaction. That such reactions may occur even without thermal access to a S/T energy crossing point has been demonstrated by Fisher and Michl[254] through the observation of spin-forbidden proton tunneling in the triplet ground-state biradical 1,3-perinaphthadiyl.

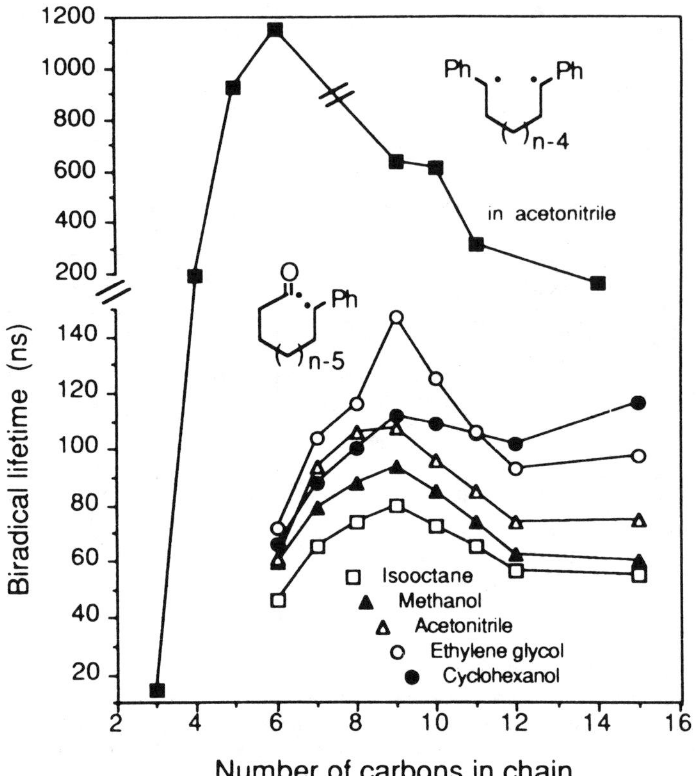

FIGURE 39.   Biradical lifetime vs. chainlength n for 1,n-biradicals of type A (bottom) and type B (top) (cf. text and Figures 36 and 37) in different solvents at room temperature. (Reprinted from Doubleday, C., Jr., Turro, N. J., and Wang, J.-F., *Acc. Chem. Res.*, 22, 199, 1989. With kind permission of N. J. Turro; copyright 1989, American Chemical Society.)

2.   The rate constant $k_{isc}(SOC) = k_{rec}$, if ISC by hfc is absent, will strongly *decrease* with *increasing* r.

This behavior contrasts with the case of hfc-induced ISC. Here β_{hfc} is independent of r, but since ΔE_{ST} with its exponential r-dependence prohibits hfc-induced ISC processes, $k_{isc}(hfc)$ *decreases* sharply with *decreasing* r. On the other hand, one should bear in mind that $k_{isc}(hfc) \leq k_{rec}$ since ISC and chemical reaction are distinct processes in this case, the equality sign holding only for r-averaged quantities in the case of rapid chain dynamics.

Comparing the behavior of type A and type B biradicals (cf. Figures 36 and 37) it has been demonstrated that the significance of SOC and HFC contributions to triplet biradical kinetics may be strongly controlled by differences in the chemical structure. In Figure 39 taken from Reference 251, the observed biradical lifetimes are shown as a function of chain length for both types of biradicals. It is seen that for type B biradicals with n $\geq$ 6 (top) the lifetime is by about one order of magnitude larger than for type A biradicals (bottom), the latter being somewhat solvent dependent. Turro and co-workers[244,251] have argued that the lifetime of type A biradicals is largely determined by $k_{isc}(SOC)$ whereby the n-dependence reflects the ease of formation of conformers with short end-to-end distance r. Support of this interpretation is provided by a smilar n-dependence that has been reported for the efficiency of cyclization reactions.[255,256] The n-dependence is supposed to reflect the strain energy of the cyclic compounds formed.

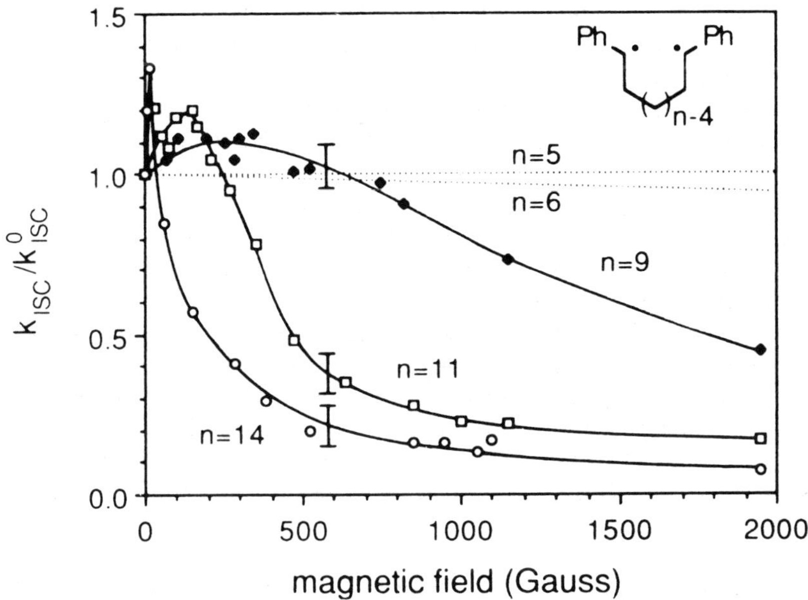

magnetic field (Gauss)

FIGURE 40.   Magnetic field dependence of relative value of biradical decay constant $1/\tau$ $\equiv k_{isc}$ for type B biradicals with various chainlengths. (Reprinted from Wang, J., Doubleday, C., Jr., and Turro, N. J., *J. Phys. Chem.*, 93, 4780, 1989. With kind permission of N. J. Turro; copyright 1989, American Chemical Society.)

For type B biradicals the SOC-induced ISC process is assumed to govern k_{rec} only for $n < 6$. For the same n (>6) k_{isc}(SOC) is much smaller for B-type biradicals than for the A-type ones possessing an acyl radical end. This difference is attributed to an increase of SOC in the acyl end due to the higher nuclear charge of the O as compared to C (note that for elements in the same period SOC increases with the fourth power of nuclear charge). Another explanation could be, however, that the unpaired electron on the acyl end has some probability to be located in the oxygen n-orbital and that a one-center n-π contribution of SOC at the oxygen atom is involved. One may recall that such a situation is also responsible for the preference of n-π ISC transitions over π-π ones in electronically excited molecular states (El-Sayed rule, Reference 257).

As another possibility to change the influence of SOC the effect of heavy atom substituents has been investigated.[247] In biradicals of type B an increase of τ^{-1} was observed on introduction of p-Br substituents at the phenyl rings. Thus, with n = 5, τ^{-1} amounted to $1.1*10^6$, $3.6*10^6$, and $5.3*10^6$ s^{-1} for the unbrominated, the monobrominated, and the dibrominated compound, respectively. Substitution by Cl did not have a similar effect so that explanations by other electronic factors than SOC could be excluded.

Figure 39 shows that for n >6 the lifetime of B-type triplet biradicals decreases again. This effect is attributed to the increase of k_{isc}(hfc) which becomes more and more efficient as ΔE_{ST} decreases. The suggested difference in the dominant ISC mechanism for n >6, i.e., SOC in the case of A-type biradicals and hfc for B-type biradicals, is also borne out by the quantitative characteristics of the magnetic field dependence of τ^{-1} in these cases. The results for the B-type biradicals are shown in Figure 40. They exhibit a behavior analogous to that observed for the EDA-type singlet biradicals, namely, an initial increase of τ^{-1} (corresponding to k_{isc}), passage through a maximum at a field B_{max} that increases with decreasing chainlength, and a decrease of τ^{-1} at higher fields leading to a limiting high field value below the zero-field value. Qualitatively this behavior is explained by the T_{-1}/S level crossing phenomenon that has been described in detail in Section II.A.1.c. It is also observed with biradicals of type A. However, whereas in the latter case $k_{isc}(B = 0.2$ T$)/k_{isc}(B = 0$ T$) = 0.91$ (n $= 11$), for B-type biradicals this ratio is as low as 0.17 (n $= 11$) and even smaller for larger n (0.07 for n $= 14$).[250]

If τ^{-1} is determined by k_{isc} the extent to which it may be modified by a magnetic field depends on the relative contributions of SOC and hfc mechanisms. Since it is only the hfc part of k_{isc} that is decreased in high magnetic fields the limiting τ^{-1} value may be used to assess the SOC part of k_{isc} and the value of $f^{\infty} = k_{isc}^{\infty}/k_{isc}^{\circ}$ reflects the relative contribution of SOC at zero-field. It is then clear that type B biradicals where (for n >6) k_{isc} is assumed to be essentially based on hfc must exhibit a much larger ratio of f^{∞} than type A biradicals where the SOC contribution dominates.

The implications of f^{∞} values significantly higher than 3 have been discussed in some detail in Reference 250, particularly in relation with the observation that in a magnetic field the decay kinetics of the type B biradicals (n >6) was purely monoexponential. Since the T_0-S ISC process is independent of a magnetic field, a biexponential decay is to be expected at fields where $T_{\pm}$ levels become sufficiently separated from T_0 and S, so that hfc-induced transitions are suppressed. If this is not observed there seem to be only two possible explanations:

1.   Spin relaxation among the triplet sublevels could be rapid. In this case k_{isc} would correspond to an average value of all triplet substates and could not fall below $\frac{1}{3}$ of the zero-field value. The observation of f^{∞} values of about 14 (n $= 14$) and 6 (n $= 11$) is at variance with this interpretation.
2.   It is conceivable that in a magnetic field the initial population of the triplet biradical substates is not uniform but, due to spin polarization in the carbonyl triplet precursor (cf. the so-called triplet mechanism of CIDEP[258,259]), the population of one of the $T_{\pm}$ sublevels might be dominant to such an extent that its decay (via spin relaxation to T_0 and S) determines the observed kinetics. This is the explanation favored by the authors of Reference 250. They estimated that for spin population of 80% or higher in T_+ or T_- the kinetic contribution of T_0 might go unnoticed.

For a given biradical the relative importance of ISC and chain dynamics as rate-determining steps may be changed by variation of temperature and viscosity. Thus, for both type A and B biradicals with n >6 the Arrhenius plots are curved displaying a steeper slope at low temperature.[249,251] Typical activation parameters evaluated below $-40°$C were A $= 10^{10}$ to 10^{11} s^{-1} and $E_a = 3$ to 4 kcal, whereas in the region 0 to 100°C typical values were A $= 10^8$ s^{-1} and $E_a = 1$ kcal. In two cases with B-type biradicals (n $= 9, 11$) even slightly negative values of E_a were found.

The particular temperature dependence points to a change in the rate-determining step of biradical recombination. At low temperature chain dynamics is rate determining, whereas at higher temperature ISC takes over the role of the rate-determining step. For the cases with a negative activation energy in the region of rate-determining ISC an explanation was offered in terms of the following expression separating the $<r>$ and T dependence of k_{isc}:[249]

$$\frac{d}{dT} k_{isc} = \left(\frac{\partial k_{isc}}{\partial T}\right)_{\langle r \rangle} + \left(\frac{\partial k_{isc}}{\partial \langle r \rangle}\right)_T \frac{\partial \langle r \rangle}{\partial T} \tag{180}$$

At a constant mean end-to-end separation $<r>$, k_{isc} may be assumed to be temperature independent. Thus the first term on the right-hand side of Equation 180 may be neglected. Experimental data from polymers[260-262] and flexible nitroxide biradicals[263] indicate that the mean end-to-end distance tends to decrease with higher temperature, i.e., d$<r>$/dT < 0. As pointed out above, the term $(\partial k_{isc}/\partial <r>)_T$ is positive for the hfc mechanism but negative

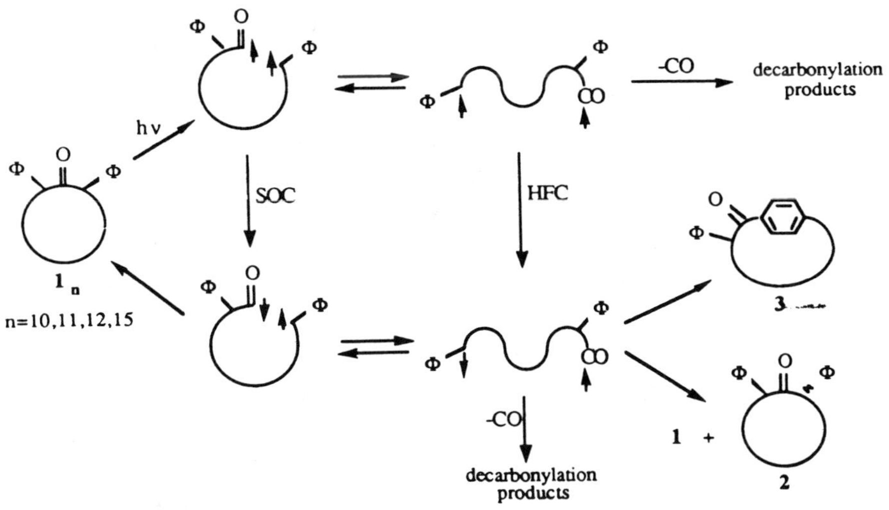

FIGURE 41.   Reaction scheme explaining the tendency of hyperfine coupling (HFC) and spin-orbit coupling (SOC) to produce different products from biradical intermediates. (Reprinted from Turro, N. J., Doubleday, Ch., Jr., Hwang, K.-C., Cheng, C.-C., and Fehlner, J. R., *Tetrahedron Lett.*, p.2929, 1987. With kind permission of N. J. Turro; copyright 1987, Pergamon Press.)

for the SOC mechanism. This would give the correct prediction of observed positive E_a for type A biradicals with dominant SOC and negative E_a for type B biradicals if n is large enough that hfc is predominant.

Variations of solvent viscosity are expected to affect the biradical lifetime only in case that chain dynamics is the rate-determining step of recombination. Such a behavior was, in fact, observed[246] with type A biradicals measured in solvents of different viscosities at -81 and at $+15°C$. A pronounced viscosity dependence was found at $-81°C$ where chain dynamics had been established as rate determining but not at 15°C where ISC is rate determining.

Finally, an interesting magnetic isotope effect with these biradical systems must be mentioned. A general requirement of magnetic isotope separation in RP reactions is the kinetic competition of one or more reaction channels that depend on spin with one or more that do not. The competition between these channels acts as a sorting device for radicals with different magnetic isotopes. For RPs cage reactions usually represent the spin dependent reactions and escape reactions the spin-independent ones. True escape reactions are, of course, not possible in biradicals. However, chemical transformations as decarbonylations are also spin independent and may provide the required second type of channels necessary for magnetic isotope separation. From these considerations it follows that magnetic isotope separation should be observable in recombination products of type A biradicals if they can undergo transformation to type B biradicals. Corresponding experiments with 2,n-diphenylalkanones (n = 10 to 15) have, in fact, afforded very high factors of ^{13}C isotope enrichment (up to 180%)[248] in some isomerization products **2** and **3** obtained from the starting material **1** (cf. Figure 41). Magnetic isotope enrichment in the starting material itself was, however, fairly low.

From these results Turro and co-workers concluded that the starting material is mainly regenerated by SOC-induced ISC from the initial biradical conformation, whereas formation of isomers **2** and **3** in Figure 41 requires a conformational change passing through situations with longer end-to-end distances so that hfc-induced ISC with a differentiation between biradicals with different magnetic isotopes can ensue. Note that the 2,n-diphenyl *trans* isomer is the enriched one when the starting material is *cis* and vice versa. The difference in the

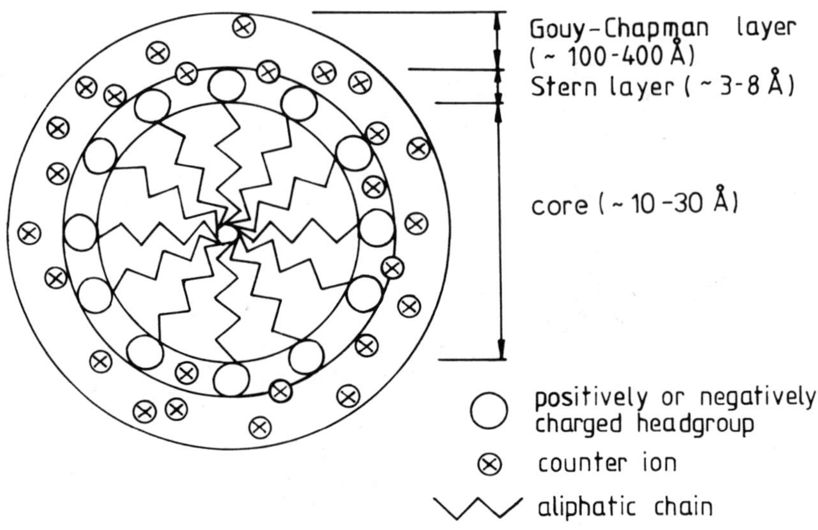

FIGURE 42.   Schematics of structure of a normal micelle of an ionic surfactant. (Adapted from Pfüller.[267])

enrichment factors of the possible coupling products is taken as evidence of "the inherent tendency of hfc and SOC to produce different products" and the effect is suggested as a probe to "assign the dynamic pathways through which each product is formed".

To corroborate this interpretation it should be quantitatively assessed to what extent the analyzed starting material has been regenerated from the biradical or, else, is simply unreacted. This requires a detailed analysis involving quantum yield measurements and determination of the single step conversion factor (cf. Section IV.C.1).

IV. REACTIONS IN MICELLAR SOLUTION

A. MICELLAR AGGREGATES AS SUPERCAGES FOR RADICAL PAIRS

Micellar solutions as specific microstructured reaction media have attracted the interest of chemists and, in particular, photochemists[264] for several decades. The intriguing potential of micelles as "supercage" microreactors for RPs was first described by Turro and co-workers,[234,265,266] who detected an enormous increase of cage recombination for geminate RPs when produced in micelles as compared to homogeneous solutions. Along with the enhancement of the cage effect a strong sensitivity to external magnetic fields was observed. This discovery marked the starting point of a very fruitful section of magnetokinetic research.

Micellar aggregates (micelles, reverse micelles, swollen micelles, microemulsion nanodroplets) are formed at critical concentrations of surfactants in various solvents. The surfactants may be anionic (e.g., sodium dodecylsulfate = SDS, or sodium-bis-[2-ethylhexyl]-sulfosuccinate = Aerosol OT [AOT]), cationic (e.g., cetyltrimethylammoniumchloride = CTAC, -bromide = CTAB, dodecyltrimethylammoniumbromide = DTAB), amphoteric (so far not used in relation with magnetokinetic studies), or nonionic (e.g. {p-[1,1,3,3-tetramethylbutyl]phenyl}polyethyleneglycol-[9/10]-ether = Triton X 100, dodecyl-polyethyleneglycol-[23]-ether = Brij 35). In polar solvents (usually water) the surfactants form normal micelles (cf. Figure 42) with a hydrophobic unpolar core and a hydrophilic polar surface, whereas reverse micelles are formed in unpolar solvents. In the latter case the polar head groups, generally associated with some hydration water, are crowded together in the center and the surface of the aggregate is lipophilic. Sizable amounts of hydrophobic or hydrophilic co-solutes may be incorporated into the cores of the respective types of micelles,

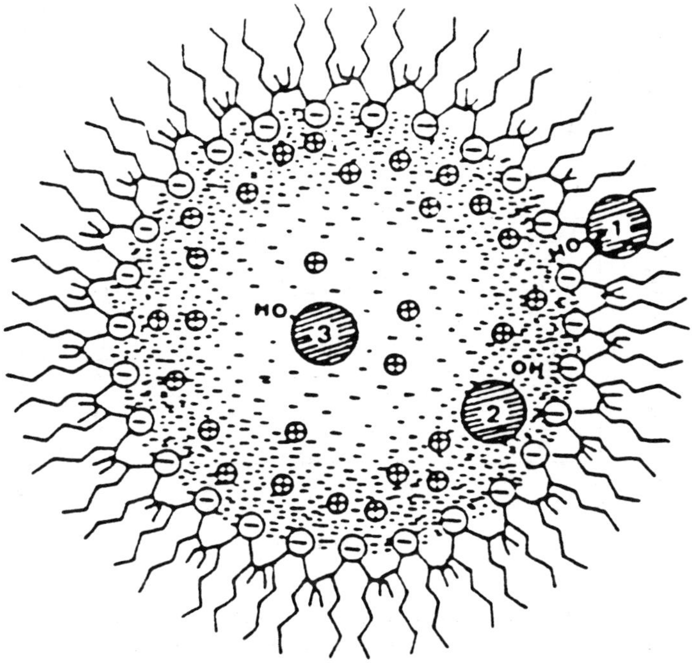

FIGURE 43.   Schematics of a water-in-oil microemulsion nanodroplet with AOT as surfactant. The probe molecules depicted reside in the surfactant layer (1), the ionic double layer (2), and the interior of the water pool (3). (Reprinted from Valeur, B. and Bardez, E., in *Structure and Reactivity in Reverse Micelles*, Pileni, M. P., Ed., Elsevier, Amsterdam, 1989, 103. With kind permission of B. Valeur; copyright 1989, Elsevier Science Publishers B.V.)

producing swollen micelles and finally so-called "nanodroplets"[268] of microemulsions (cf. Figure 43) which, in the case of reverse micelles with an aqueous interior, may be conceived as water droplets enclosed by a monolayer of surfactant molecules.[270]

RPs in micellar aggregates can be conveniently produced by photoreactions of substrates solvated somewhere in the micellar aggregates. The initial situation at the birth of a RP may be more uniform for monomolecular production processes as Norrish type I cleavage of ketones, than for bimolecular reactions as hydrogen abstraction or electron transfer, which both require diffusional encounter of a photoexcited molecule with a substrate that may be already present in the micelle at the instant of photoexcitation or may have to intrude from the bulk solvent. RPs produced in a micellar aggregate may undergo cage and escape reactions as in the case of homogeneous solutions, however, with the diffusion-dependent part of these processes characteristically modified by the microheterogeneous environment. Thus "escape" means that one of the radicals will have to leave the micellar supercage, a process which may take several orders of magnitude longer time than to leave a region of corresponding volume in a homogeneous solvent. Thus the cage time available for spin evolution processes is, as a rule, much longer than in homogeneous solvents and different mechanisms of spin motion with different magnetic field dependence may become apparent. Concerning the diffusional motion and the sequence of reencounters during the cage period of a RP in a micellar aggregate one has to consider that the specific microheterogeneous structure of the aggregate and the specific affinity of the radicals may confine the freedom of diffusional motion to restricted regions, e.g., the interface or the interior, of the aggregate. Furthermore, the size of the aggregate will determine the stochastic distribution of RP separation distances during the periods between re-encounters in the same supercage.

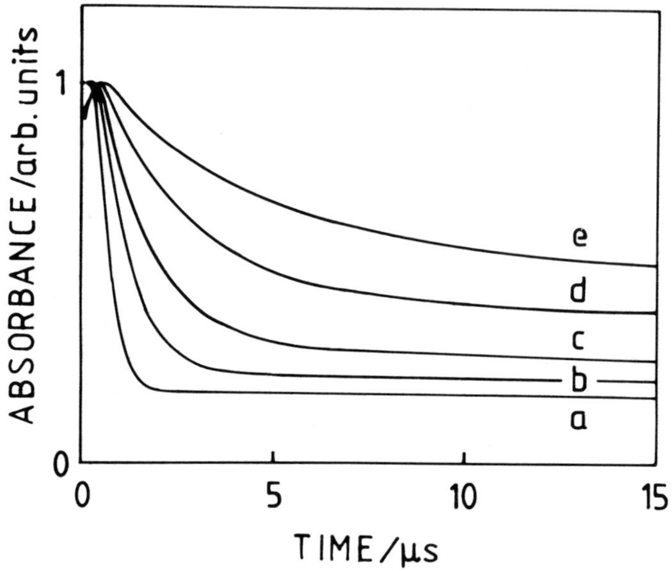

FIGURE 44.   Radical decay kinetics observed after laser flashing of p,p′-difluorobenzophenone in SDS micellar solution in various magnetic fields (in T): (a) 0, (b) 0.1, (c) 0.2, (d) 0.5, (e) 1.34. (Reproduced from Sakaguchi, Y. and Hayashi, H., *Chem. Phys. Lett.*, 87, 539, 1982. With kind permission of H. Hayashi; copyright 1982, Elsevier Science Publishers B.V.)

In the following we will attempt to illustrate, by a selection of representative examples rather than by a comprehensive account, the most significant results from magnetokinetic investigations of photoinduced RP reactions in micellar media. In this selection the literature up to the end of 1989 has been considered. For a more comprehensive compilation of work published before 1987 the reader may consult Reference 1.

B. RADICAL PAIRS FROM H-ATOM ABSTRACTION
1. Reactions of Triplet Excited Carbonyl Compounds with Surfactants

Reactions of this type have been studied by the groups of Scaiano,[271,272] Hayashi,[273-276] and Tanimoto.[277-282] The first investigations employed benzophenone, mostly in aqueous SDS micellar solution, where the ketone solubilized in the micelle abstracts an H-atom from the surfactant alkane chain. Time-resolved experiments[272,274] applying nanosecond laser flash photolysis showed that the carbonyl triplet is partially converted to the ketyl radical. Some small fraction of long-lived radicals appeared that was increased by a magnetic field and was attributed to radicals that had become isolated by escape of the ketyl radical from the micelle of its origin. This MFE was attributed to the retarding effect of the external magnetic field on the rate of spin-forbidden recombination of the geminate triplet RPs. Initially the rate of geminate recombination was thought to be determined by the triplet-to-singlet spin conversion of the RP even at zero-field. Since in the reaction of benzophenone triplet with surfactants the rate constant of triplet decay is similar to that of intramicellar RP recombination it is difficult to analyze the MFE on the intramicellar RP kinetics in reliable detail.

A very pronounced effect was observed with 4,4′-difluorobenzophenone in SDS micelles[276] (cf. Figure 44). The decay of the curves shown in Figure 44 is attributed to intramicellar RP decay. The rate constants (in s^{-1}) determined at various magnetic fields are $2.9*10^6$ (0.0 T), $1.3*10^6$ (0.1 T), $0.75*10^6$ (0.2 T), $0.45*10^6$ (0.5 T), and $0.3*10^6$ (1.34 T). Such a large MFE on a rate constant had never been observed before. The values reported indicate that the MFE is not yet saturated at 1.34 T and $B^{1/2}$ must be larger than 0.2 T. It was noted that such a field dependence could not be explained by a suppression of *coherent* $T_{\pm} \leftrightarrow S$ transitions in the RP. The possibility of a magnetic field-dependent relaxational $T_{\pm} \leftrightarrow S, T_0$ process as a mechanism underlying the observed magnetic field dependence of intramicellar RP kinetics was suggested here for the first time. It should be noted, however, that the curves in Figure 44 show no indication of a distinct kinetic component due to recombination of T_0-RPs which should be expected to be discernible from the magnetic field-dependent recombination of $T_{\pm}$-RPs, at least in high fields.

A number of reactions involving H-atom abstractions from surfactants in micelles by triplet excited quinones have been investigated by the groups of Hayashi[41,275] and Tanimoto.[277-283] With quinones the triplet reaction appears to be faster so that the initial decay kinetics of intramicellar RPs can be followed more directly. Investigating naphthoquinone and 2-methylnaphthoquinone in SDS micelles Sakaguchi and Hayashi[275] provided evidence that the intramicellar RP recombination in magnetic fields becomes biexponential. The fast, magnetic field-independent component ($4.7*10^6$ s^{-1}) was tentatively attributed to T_0-RP decay, although it could not be definitely excluded that it was partly due to the observation of decaying quinone triplets. The slow kinetic component of intramicellar recombination strongly decreased in a magnetic field (to $0.5*10^6$ s^{-1} at 1 T). The $B^{1/2}$-value ($\sim$0.05 T for the smaller of the recombination constants and $\sim$0.1 T for the escape yield)* clearly points to the contribution of field-dependent relaxation processes to the overall $T \rightarrow S$ ISC process in the RP.

O

R

R = H, COOH, SO₃Na

O

33

In a study of RP kinetics following photoexcitation of the anthraquinones (*33*) Tanimoto et al.[279] reported a magnetic field dependence of RP decay that exhibited monoexponential kinetics at any field. It is noteworthy that, while the effective rate constants of intramicellar recombination were fairly independent of substituents R, the rate of escape of the anthrasemiquinone radical was three times as high for the SO_3^--substituted derivative than for the others, which may be attributed to the different solubilities of the corresponding radicals in the aqueous phase.

In time-resolved studies of RP kinetics in micelles laser flash photolysis is usually combined with probing of transients by absorption spectroscopy. By this type of probing it is often not feasible to monitor product formation because of unspecific product absorption or of a lack of sensitivity. In the case of the anthraquinone-SDS adduct resulting from geminate coupling of corresponding RPs Tanimoto et al.[282] successfully applied a two-step laser excitation of product fluorescence to monitor its formation in a time-resolved fashion. Hereby a second (probing) laser pulse follows the photolyzing pulse at an adjustable delay time. It excites the fluorescence of product present at the instant of the delayed probe pulse. The intensity of the fluorescence pulse emitted is a measure of the concentration of product. The rise time observed in this way for the product and its magnetic field dependence was in accord with previous transient absorption measurements of the radicals.[279] It should be noted that a similar technique based on time-resolved fluorescence excitation had also been applied by Staerk and Razi Naqvi[216] for monitoring the formation of ketyl radicals in ho-

---

* It should be noted that although the magnetic field dependence of $\Phi_{esc} \equiv k_{esc}/(k_{esc} + k_{rec})$ is completely determined by that of k_{rec}, the respective $B_{1/2}$ values are generally different. From the kinetic expression for Φ_{esc} one can derive that $B_{1/2}(\Phi_{esc})$ must be always higher than $B_{1/2}(k_{rec})$.

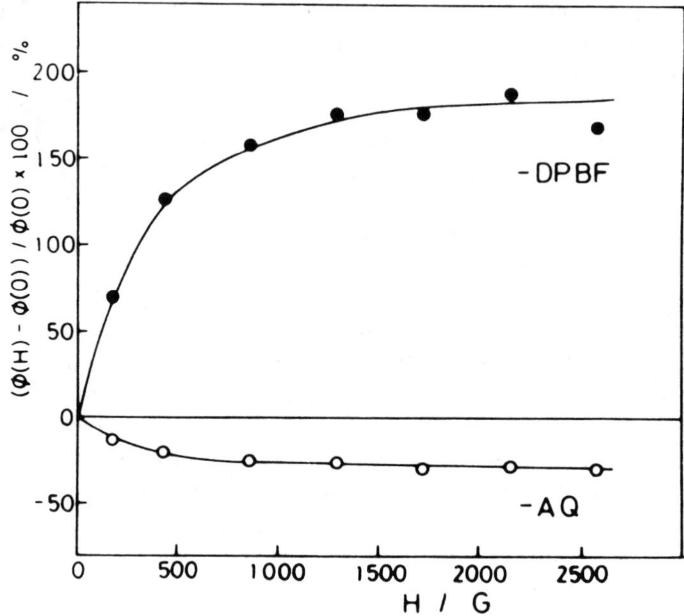

FIGURE 45.   Magnetic field dependence of the relative change of quantum yield, Φ, of disappearance of DPBF (*34*) and AQ (*33*, R = H) in aerated SDS micellar solution. (Reprinted from Tanimoto, Y., Shimizu, K., and Itoh, M., *J. Am. Chem. Soc.*, 106, 7257, 1984. With kind permission of Y. Tanimoto; copyright 1984, American Chemical Society.)

mogeneous solution (cf. also the investigation of McLauchlan and Nattrass[284] described below).

An interesting photochemical application of magnetic field control of RP kinetics in micellar solution has been described in Reference 281 for a photosensitized oxidation reaction of 1,3-diphenylisobenzofuran (DPBF, (*34*)).

34

It is proposed that in oxygen containing SDS micellar solutions sensitization by anthraquinone (AQ, [*33*, R = H]) occurs through the formation of hydroperoxy radicals in the reaction

$$AQH\cdot + O_2 \rightarrow AQ + HO_2\cdot \tag{181}$$

Since in a magnetic field the lifetime of $AQH\cdot$ is increased the yield of $HO_2\cdot$ is expected to increase as well. In fact, the quantum yield of DBPF disappearance is increased by 200% at saturating fields of *circa* 0.2 T with $B^{1/2} \approx 35$ mT (cf. Figure 45). This MFE is much larger than the one on the quantum yield of disappearance of anthraquinone, which is negative and saturates at about -30%.

2. Reactions of Triplet Excited Carbonyl Compounds with H-Donor Additives

By adding suitable H-donors that are solubilized in the micelles together with the photoreactive carbonyl compound the rate of triplet quenching can be increased so that the kinetic distinction between the latter process and the initial fast part of the intramicellar RP decay becomes clearer. Such experiments have been reported by Scaiano et al.[271,272,285] using cyclohexadiene as a H-donor, by Tanimoto[277,283] using xanthene or 9,10-dihydroanthracene, and more recently by Levin, Kuz'min, and co-workers[286-293] using 4-phenylphenol or 4-phenylaniline. In the latter case, however, an electron transfer must be assumed as the primary process, followed by a protonation of the ketylanion if the medium is slightly acidified.

Whereas most of these studies were carried out in SDS micelles a variation of the type of surfactant was employed in a recent investigation by Scaiano and co-workers.[294] Here a series of five phenols (*35a* to *e*) was used as H-donors to react with triplet excited butyrophenone in micelles of the anionic SDS, the cationic CTAC, CTAB, and DTAB, and the neutral Brij 35.

35a

35b

35c

35d

35e

The RP kinetics was monitored through the transient absorption due to the H-donor-derived phenoxyl radicals. In general the kinetics and MFEs with these RPs involving O-centered radicals were similar to those observed with C-centered radicals. An example of the kinetics is given in Figure 46. The observed signals were analyzed in terms of two first-order rate processes, for recombination and escape. Exit rates varied between $0.5*10^6$ and $2.7*10^6$ s^{-1}. However, the variations exhibited were due to structural differences of the phenols rather than to different types of surfactant. On the other hand, the values of k_{rec} at zero-field showed a clear dependence on the micellar size. For example, for the RPs involving phenol (*35e*) $k_{rec}(B_0 = 0)$ was found to be $0.94*10^6$ s^{-1} in CTAB (C_{16}-chain) and $3.3*10^6$ s^{-1} in DTAB (C_{12}-chain), i.e., a decrease of the recombination rate is observed on increasing the micellar size. More quantitative examples of this relation will be given below, and for the case of microemulsions cf. Section IV.D. Effects possibly associated with the charge type of the surfactant did not become apparent as one would anticipate for the case of neutral reactants forming pairs of neutral radicals.

In the systems studied by Levin, Kuz'min, and co-workers various quinones and benzophenones were reacted with 4-phenyl substituted phenol (*36*) or aniline (*37*).

36

37

In SDS micelles the efficiency of radical escape is rather low with these reactants, i.e., geminate recombination of the initially triplet spin-correlated RPs is the dominating process and pronounced kinetic effects of an external magnetic field can be observed. In these systems the RP recombination kinetics was found monoexponential (down to a low level of

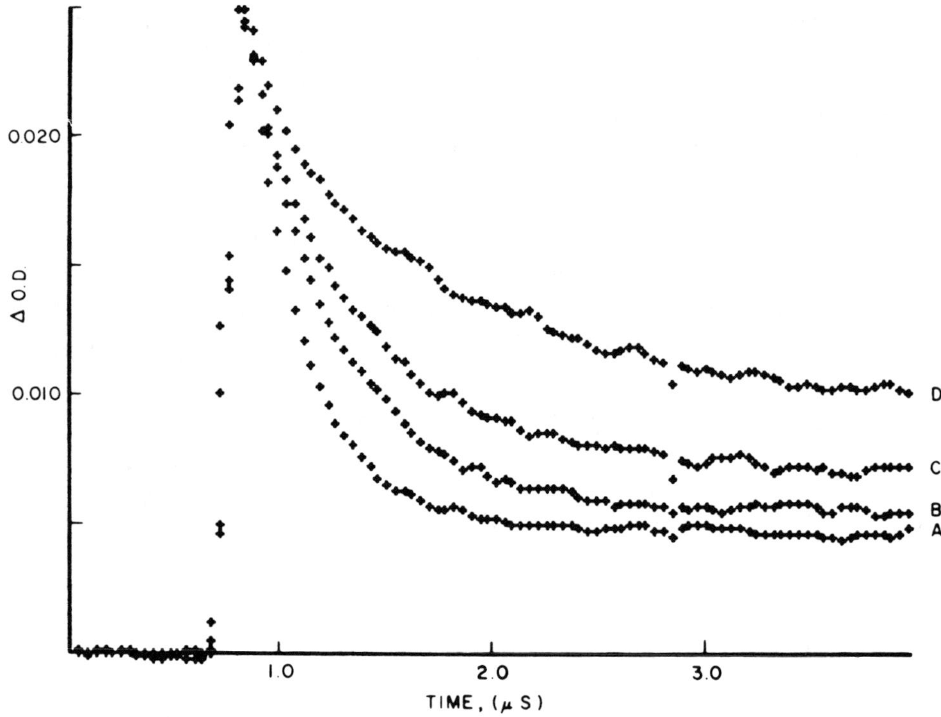

FIGURE 46. MFE on the recombination kinetics of the vitamin E phenoxyl (*35e*)/butyrophenone ketyl radical pair in 0.2 *M* SDS: (A) 0 G; (B) ~100 G (stirring magnet); (C) 500 G (horeshoe magnet); (D) 1400 G. (Reprinted from Evans, C., Ingold, K. U., and Scaiano, J. C., *J. Phys. Chem.*, 92, 1257, 1988. With kind permission of J. C. Scaiano; copyright 1988, American Chemical Society.)

long-lived escape radicals) at low fields but displayed biexponential decay behavior at higher fields where a rapid magnetic field-independent component, attributed to T_0-RPs, is separable from the slower magnetic field-dependent component attributed to $T_\pm$-RPs.[286,289]

In Reference 290 the influence of micellar size variation achieved by changing the chain length of surfactants (sodium $[CH_2]_n$ sulfonates with n = 10, 12, 13) or by increasing the ionic strength by addition of NaCl was studied on the magnetokinetic behavior of RPs produced in the reaction of triplet excited quinone (*38*) with (*36*) or (*37*).

38

With increasing size of the micelle (C_{10} to C_{13}) the recombination constant at zero-field decreased by about a factor of 3, whereas the corresponding high field value (a maximum field of 0.34 T was applied) decreased only by a factor of 2 or less. This feature is borne out even more clearly when the concentration of NaCl was increased, which is known to have an increasing effect on the size of the micelles (cf. Reference 295).

The observed effects on k_{rec} at zero-field and at a field of 0.34 T are shown in Figure 47a. Whereas in zero-field k_{rec} decreases linearly with the concentration of NaCl up to 0.5 *M*, the high-field value is found constant up to about 0.25 *M* implying that different and independent processes are rate determining in zero-field and in high fields. An obvious

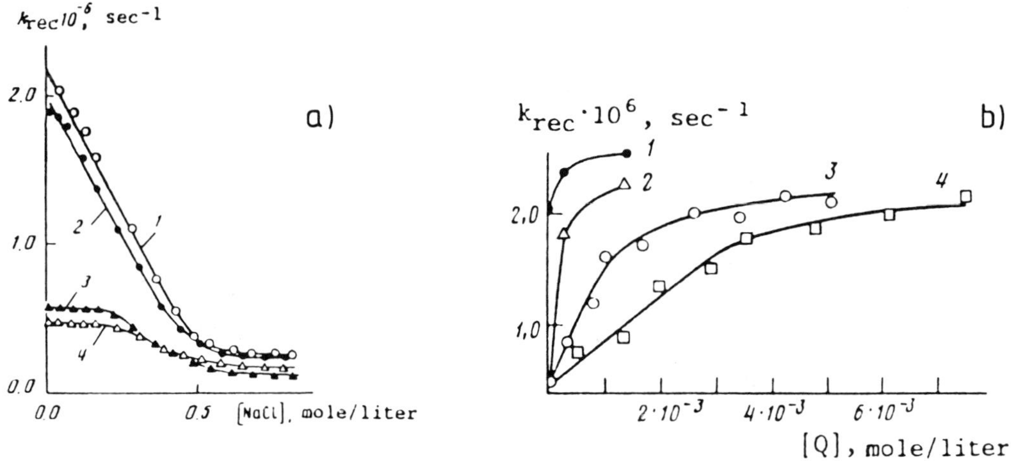

FIGURE 47.   Dependence of the rate constants of recombination of RPs in SDS micelles. X = (*38*), POH = (*36*), PNH$_2$ = (*37*). (a) On the concentration of NaCl. RP XH·/PNH$_2^+$· at zero-field (1) and at 0.34 T (3); RP XH·/PO· at zero-field (2) and at 0.34 T (4). (b) On the concentration of paramagnetic quenchers. RP XH·/PO· with O$_2$ at zero-field (1) and at 0.34 T (2), with complex CoCl$_2$Pz$_2$ (Pz = [*41*]) at 0.34 T (3), and with radical (*39*) at 0.34 T. (Reprinted from Kuz'min, V. A. and Levin, P. P., *Bull. Acad. Sci. U.S.S.R. Div. Chem. Sci.*, 35, 1291, 1986; 36, 691, 1987. With kind permission of P. P. Levin and V. A. Kuz'min; copyright 1986 and 1987, Plenum Publishing Corporation.)

interpretation is that in zero-field the rate of RP encounters, i.e., diffusion, is the limiting parameter, whereas in high fields it is the rate constant of T → S conversion processes that adopts this role. Whereas the former is related to the micellar size the latter is obviously not. Such a behavior is expected for T → S RP processes depending on the coherent hyperfine coupling mechanism and, more important at high fields, on individual electron spin relaxation in both radicals.

The effect of paramagnetic additives to quench the magnetic field dependence of spin-forbidden RP recombination processes has been mentioned in Section III.A.1 with the example of Lu^{3+} ions in homogeneous[163,164] and in micellar systems.[161,162] A study implying the paramagnetic species O$_2$, the nitroxyl radicals (*39*) and (*40*), and the paramagnetic complex CoCl$_2$Pz$_2$ (Pz = (*41*)) has been reported by Kuzmin and Levin.[287]

It was found (cf. Figure 47b) that under high-field conditions the rate constant k$_{rec}$ of RPs derived from triplet excited quinone (*38*) in the reaction with H-donor (*36*) in SDS micelles was increased by adding the paramagnetics, approaching the zero-field value at higher concentrations of the additives. In the case of oxygen even the zero-field value is somewhat increased. From the initial slope of the plots of k$_{rec}$ vs. concentration of paramagnetic additive effective rate constants for the paramagnetic interaction were derived. For O$_2$ which is distributed rather homogeneously over the whole system, a value of 5*10^9 M^{-1} s^{-1} was obtained, indicating that in this case the underlying interaction is apparently diffusion controlled. For the hydrophilic nitroxyl radical (*40*) that does not penetrate into the micelle no effect was found up to concentrations of 0.05 M. For the functionalized nitroxyl radical (*39*)

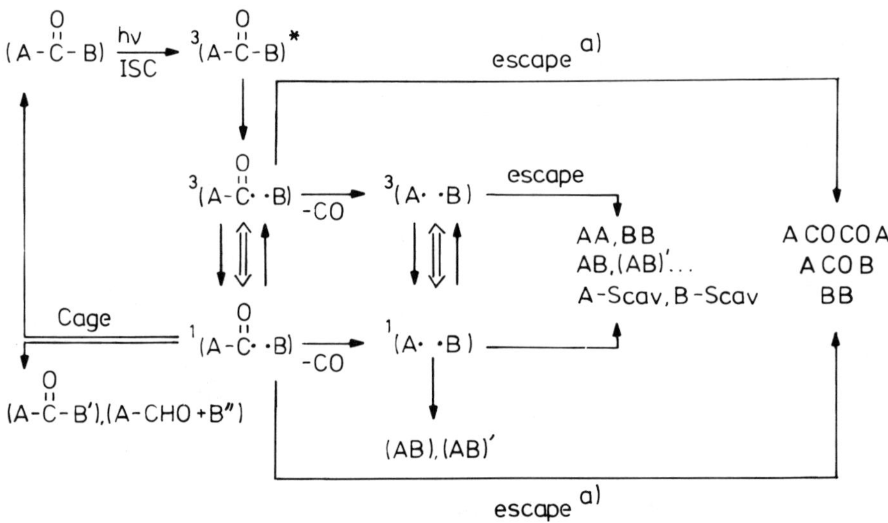

FIGURE 48.   General reaction scheme for ketones undergoing Norrish type-I photocleavage and investigated for magnetokinetic effects in micellar solution. ( ) denotes micellar supercage; A: benzyl, α-R,R'-benzyl, phenyl, or alkyl; B: benzyl or α-R,R'-benzyl; B' is an isomer of B, B'' a dehydrogenated product. A-Scav and B-Scav denote products formed upon reaction with radical scavengers. [a]Where CO elimination occurs it is usually faster than the escape process.

and the Co complex, both being essentially localized in the micelles, effective quenching constants (per one quencher molecule in the micelle) of $4*10^5$ and $1.1*10^6$ s^{-1}, respectively, were obtained from Figure 47b. These values are about one order of magnitude smaller than to be expected for a diffusion-controlled process in the micelle. As a mechanistic explanation the authors suggested that the effect of the paramagnetic was due to electron spin dipole-dipole and spin exchange interaction accelerating electron spin relaxation in the RPs. The vanishing or (in the case of O_2, cf. also References 286 and 288) small effect of paramagnetics on k_{rec} in zero-field indicates that spin processes are not rate determining for intramicellar recombination of triplet RPs under this condition.

The internal heavy atom effect on intramicellar RP magnetokinetics has been studied for H-abstracting systems by introducing Br substituents[288,289,293] in the H-acceptor benzophenone or in the H-donor phenol. The results are very similar to those obtained for the electron transfer reaction between phenylaniline and the various substituted benzophenone triplets.[44,291] We will deal with these results in more detail below (cf. Section IV.D).

C. NORRISH TYPE-I CLEAVAGE OF KETONES

This type of reaction has been most fruitful in providing examples of magnetic field-dependent reaction kinetics or the closely related magnetic isotope effects in micellar media. The majority of contributions has come from Turro's laboratory where the "micellar supercage effect" was also discovered.[265,266] SDS or CTAC (= HDTC) were the surfactants usually applied in these studies.

A general reaction scheme applying to the photolysis of ketones in micellar solution is depicted in Figure 48. Some analogous sulfones have also been studied and have been found to behave similarly to the ketones (cf. below). As a rule, the ketones exhibit very efficient ISC to populate the lowest triplet state which is of nπ* character and readily undergoes α-CC cleavage at a rate that has not been time-resolved in the nanosecond laser flash experiments applied so far.

Triplet RPs are thus produced whereby cleavage appears to occur selectively at the bond connecting the most stable pair of radicals.[296,297]

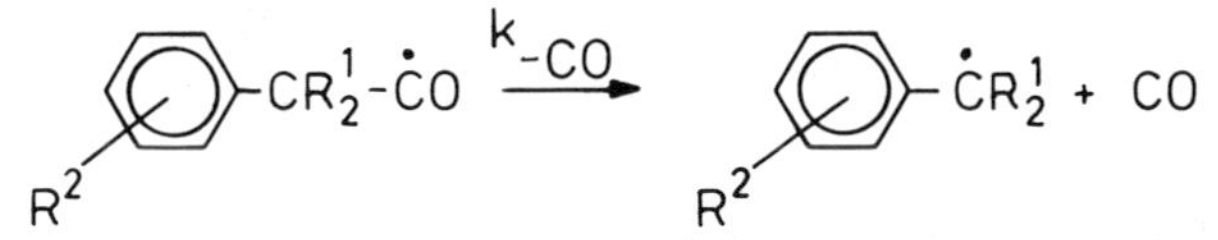

$$\text{(182)}$$

42   *43*

The acyl radical, if it is of the 2-phenylacetyl type (*42*), undergoes decarbonylation with a rate constant between $6*10^6$ s^{-1} (R^1 = R^2 = H) and $1.5*10^8$ s^{-1} (R^2 = H, R^1 = CH$_3$) at room temperature. Alkyl or phenyl substituents have a much stronger effect on k_{-CO} at the α-benzylic position (R^1) than as ring substituents (R^2).

Decarbonylation, if it does occur, is usually faster than radical escape from the micelles. The latter process has been separately monitored in a time-resolved experiment employing Fremy's salt (*45*) as a water phase resident scavenger of benzyl radicals (*44*):[298]

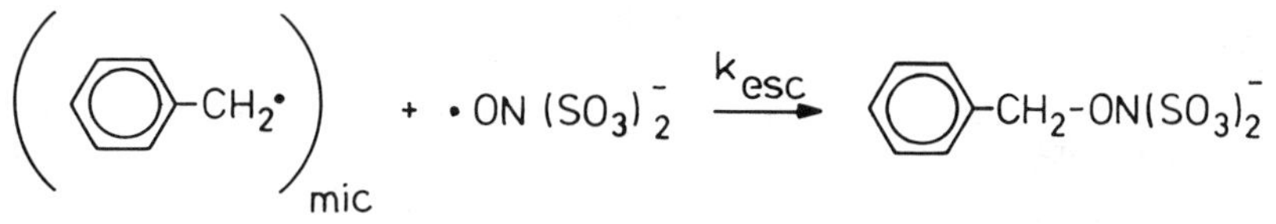

$$\text{(183)}$$

44   *45*   *46*

Probing by transient absorption and by time-resolved CIDNP yielded consistent results for the micellar exit rate of the benzyl radical of about 10^6 s^{-1}. This value was decreased to about 2 to $3*10^5$ s^{-1} for the methylated (ring or α-C) benzyl radicals reflecting their increased hydrophilic character. As is shown by the extreme cage effects in the photolysis of p,p'-di*tert*butyldibenzylketone this effect is still more pronounced with *tert*butyl substituents.[299]

Processes occurring in competition with decarbonylation or micellar exit are RP ISC and intramicellar coupling reactions of RPs with singlet spin correlation. Hereby the starting material ACOB may be recovered either unchanged or in isomerized form ACOB'. A recombination product formed, e.g., in the photolysis of dibenzylketone, is 1-phenyl-*p*-methylacetophenone (PMAP, [*47*]) obtained through a coupling of the 2-phenylacetyl radical

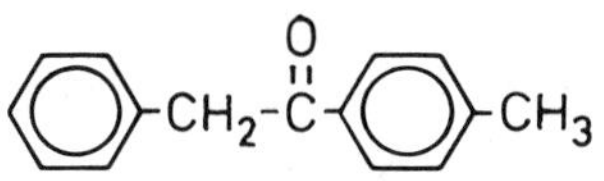

47

to the phenyl ring of the benzyl radical, followed by a hydrogen shift.[300] If the benzylic carbon is a center of chirality, reformation of the original C–C bond will yield a statistical mixture of stereoisomeres, a result that can be used to assess the amount of cage recombination if the starting material is enriched in, or is a pure stereoisomer.[98,301]

Another reaction possibility, particularly important for RPs originating in the photolysis of deoxybenzoins, is a disproportionation:[302]

$$\text{Ph}\dot{\text{C}}\text{O} + \cdot\text{C(CH}_3)_2\text{Ph} \rightarrow \text{PhCHO} + \text{H}_2\text{C=C(CH}_3)\text{Ph} \qquad \text{(184)}$$

yielding benzaldehyde and methylstyrene.

Coupling products A-B may be formed between radicals from the same or from different micelles, i.e., before or after micellar escape, whereas A-A- and B-B-type products can be formed only after micellar exit. Whereas in homogeneous solution the ratio of AB/AA/BB will be approximately statistical, i.e., 2:1:1, the supercage effect in micellar systems will increase the fraction of A-B coupling products. A quantitative measure of the cage effect may be defined as:[303]

$$\text{cage effect} = \frac{[AB] - ([AA] + [BB])}{[AB] + [AA] + [BB]} \tag{185}$$

Another possibility to determine the efficiency of intramicellar recombination is by scavenging the radicals in the aqueous phase so that AB will be only formed by the intramicellar reaction and its yield provides a measure of the cage effect. As a convenient scavenger $CuCl_2$ has been used.[302,304] The C-radicals are oxidized by the Cu^{2+} ions and the carbonium ions thus produced may add OH^- or Cl^-.

1. Spin-Dependent Effects in Continuous Photolysis
a. Magnetic Isotope Effects

A purely intrinsic magnetic effect that has been intensively studied in connection with the photolysis of benzyl ketones is the phenomenon of magnetic isotope enrichment. As far as the increased hyperfine coupling due to magnetic isotopes can increase the rate of RP ISC it will increase the efficiency of cage recombination of RPs generated with triplet spin correlation and thus lead to an enrichment of magnetic isotopes in the cage products. Spin evolution and spin-selective reaction thus combine to act as a sorting mechanism for isotopes differing in their magnetic moments.

The first application of this principle has been reported by Buchachenko et al.[18] utilizing the photolysis of dibenzylketone in homogeneous solution for obtaining ^{13}C enrichment of the recovered starting material. So far, magnetic isotope effects have been studied with ^2H, ^{13}C, ^{15}N, ^{17}O, and ^{73}Ge. For detailed information on this topic the reader is referred to several reviews that have appeared during the last years.[1,7,240,305,306] Here we will focus on several general aspects that are of interest in relation to the understanding of spin-selective reactions of RPs in micellar systems. The examples refer to the enrichment of cage products in ^{13}C.

Considering the case of recovered starting material, e.g., dibenzylketone (DBK), the quantity of interest is

$$\alpha = \frac{\Phi^{12}_{-DBK}}{\Phi^{13}_{-DBK}} \tag{186}$$

where Φ^{12}_{-DBK} and Φ^{13}_{-DBK} denote the photochemical quantum yields of disappearance of DBK with ^{12}C and ^{13}C, respectively. Actually for each ^{13}C substituent position a separate α can be defined. If the positions of ^{13}C are not distinguished in the final analysis the α-value determined corresponds to a weighted sum of the individual positions.

The quantity α is called a single-stage enrichment factor and may be determined experimentally by measuring the overall enrichment factor

$$S_f = \frac{[^{13}C]_f / [^{12}C]_f}{[^{13}C]_o / [^{12}C]_o} \tag{187}$$

as a function of f, the fraction of the starting material that has been converted to products. The following relation,[234,307] first derived by Bernstein,[308] relates S_f to f:

$$\log S_f = [(1 - \alpha)/\alpha]\log(1 - f) \tag{188}$$

Strictly speaking, in order that Equation 188 be exact, f must be referred to the conversion of the nonmagnetic fraction of the starting material. However, for samples with low ^{13}C content a distinction between f and the overall conversion factor is not necessary. (It should be noted that the definition of α, Equation 186, as applied here corresponds to the usage in

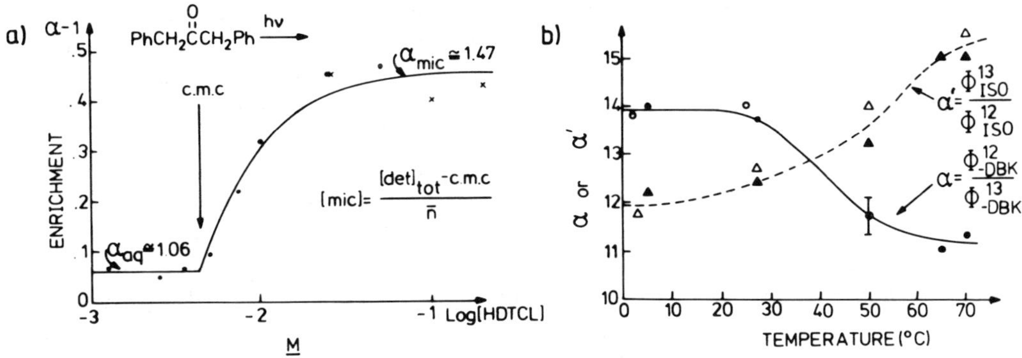

FIGURE 49. ^{13}C enrichment of recovered dibenzylketone (DBK) and isomeric ketone (PMAP, [47]) from DBK photolysis in HDTC (CTAC) surfactant solution. (a) Enrichment factor ($\alpha - 1$) for DBK vs. concentration of HDTC (log-scale); (b) temperature dependence of enrichment factors α for DBK and α' for PMAP as determined from ^{13}C enrichment studies (●,▲) and quantum yield values (○,△). (Reprinted from Kräutler, B. and Turro, N. J., *Chem. Phys. Lett.*, 70, 270, 1980; Turro, N. J., Chow, M.-F., and Kräutler, B., *Chem. Phys. Lett.*, 73, 545, 1980. With kind permission of N. J. Turro; copyright 1980, Elsevier Science Publishers B.V.)

Turro's work. The same quantity is denoted as α_G in Buchachenko's terminology,[306] where α is defined as $[\Phi^{13}_{-DBK} - \Phi^{12}_{-DBK}]/\Phi^{12}_{-DBK} = [1 - \alpha_G]/\alpha_G$.

According to Equation 188 the enrichment factor α may be evaluated from the slope of a straight line obtained when plotting log S_f vs. log$(1 - f)$. Typical values in low viscous homogeneous solvents are about 1.05,[306] whereas in micellar solution α is in the order of 1.4,[305] corresponding to about an eightfold increase of isotope-enrichment efficiency. With homogeneous solvents viscosities in the order of 20 P are required to reach comparable efficiencies.[238]

The correspondence of the enrichment factor α as determined from the dependence of enrichment vs. conversion according to Equation 188 to the ratio of quantum yields defined in Equation 186 has been verified experimentally by Turro et al.[304] The quantum yields Φ^{12}_{-DBK} and Φ^{13}_{-DBK} were measured with two samples of DBK that were of natural ^{13}C abundance (99% ^{12}C) and 90% ^{13}C-enriched in the carbonyl carbon, respectively. The quantum yields for disappearance of DBK, found identical to that of diphenylethane formation, amounted to $\Phi^{12}_{-DBK} = 0.24 \pm 0.03$ and $\Phi^{13}_{-DBK} = 0.17 \pm 0.02$ yielding $\Phi^{12}_{-DBK}/\Phi^{13}_{-DBK} = 1.41$, the same value that was obtained with an enrichment experiment evaluated according to Equation 188.

An instructive study of the change in α, when passing from homogeneous to micellar solution, has been reported by Kräutler and Turro.[300] Photolyzing DBK in solutions of variable concentration of HDTC (CTAC) they measured α for ^{13}C enrichment in DBK (cf. Figure 49a) as well as the yield γ of cage product PMAP (47). A marked increase in both quantities was observed with a sharp onset at the surfactant concentration corresponding to the c.m.c., the point where micelles start to appear in equilibrium with surfactant monomers. In this way the micellar cage effect has been directly demonstrated.

Whereas in homogeneous solution the time span τ^{hom}_{cage} available for geminate recombination is much shorter than the characteristic time required for spin evolution

$$\tau^{hom}_{cage} \ll \tau_{isc} \lesssim \tau_{-CO} \tag{189}$$

the situation for micellar solubilization of RPs corresponds to

$$\tau_{isc} \lesssim \tau_{-CO} \ll \tau^{mic}_{cage} \tag{190}$$

i.e., the chemical lifetime τ_{-CO} limits the time span during which magnetic isotope-dependent differential spin evolution can occur. In systems such as deoxybenzoins where no decarbonylation takes place after α-cleavage it is the micellar exit time τ_{cage}^{mic} that adopts this limiting role. Under such conditions, too, remarkable magnetic isotope effects on cage recombination yields in zero-field have been observed.[302]

In the cases where decarbonylation occurs, a variation of temperature may be applied as a convenient means to vary the time span of spin evolution. For example, in the case of the 2-phenylacetyl radical formed in the photolysis of DBK τ_{-CO} changes from about 185 ns at 27°C to 37 ns at 70°C.[309] Such changes have different effects on the single-stage magnetic isotope enrichment factors α and α' for recovered DBK and the cage product PMAP, respectively. Let us denote by Φ_{isc}^{12}, Φ_{isc}^{13}, Φ_{-CO}, Φ_{ir} the efficiencies of the corresponding process in relation to their directly competing processes according to the reaction scheme in Figure 48. The subscript "ir" refers to formation of the isomerized cage product ACOB′ (i.e., PMAP for DBK photolysis). Then the enrichment factors α and α' may be expressed as

$$\alpha \equiv \frac{\Phi_{-DBK}^{12}}{\Phi_{-DBK}^{13}} = \frac{\Phi_{-CO} + \Phi_{isc}^{13} * \Phi_{ir}}{\Phi_{-CO} + \Phi_{isc}^{12} * \Phi_{ir}} \qquad (191)$$

and

$$\alpha' \equiv \frac{\Phi_{PMAP}^{13}}{\Phi_{PMAP}^{12}} = \frac{\Phi_{isc}^{13}}{\Phi_{isc}^{12}} \qquad (192)$$

In the case of the isomerized cage product the enrichment factor α' directly reflects the ratio of ISC efficiencies with the respective magnetic isotopes. At long times τ_{-CO} the efficiencies Φ_{isc}^{13} and ϕ_{isc}^{12} approach values characteristic of spin equilibrium. Only at times τ_{-CO} shorter than τ_{isc}^{12} and τ_{isc}^{13} the ratio of Φ_{isc}^{13} and Φ_{isc}^{12} reflects the magnetic isotope effect on the ISC rate. Hence α' will be at maximum at short τ_{-CO}, implying an increase with temperature. In the case of the recovered ketone the extent at which the actual magnetic isotope effect $\Phi_{isc}^{13}/\Phi_{isc}^{12}$ is expressed in α depends on Φ_{-CO}. The absolute value of $\Phi_{isc}^{13} * \Phi_{ir}$ must become larger or at least comparable to Φ_{-CO} in order that the magentic isotope effect in α be borne out most clearly. An increase in Φ_{-CO} as occurring with increasing temperature will correspondingly reduce the value of α. Exactly this behavior has been observed and analyzed in Reference 309 (cf. Figure 49b).

As stated above, due to their different hfc constants magnetic nuclei in different positions are expected to exhibit different effects on RP ISC and thus must also be differentiated in magnetic isotope enrichment experiments. In 2-phenylacetyl (*48*) and benzyl radicals (*49*) originating in DBK photolysis ^{13}C nuclei have the highest hyperfine coupling in the carbonyl group, position 1 ($a_{13_C} = 124$ G) and at the methylene groups, position 2 ($a_{13_C} = 51$ G) and 3 ($a_{13_C} = 24$ G).

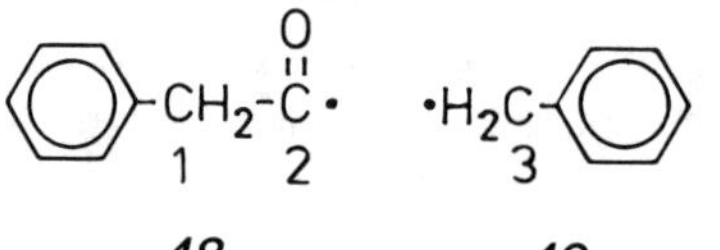

$$\text{C}_6\text{H}_5-\text{CH}_2-\overset{\overset{\text{O}}{\|}}{\text{C}}\cdot \qquad \cdot\text{H}_2\text{C}-\text{C}_6\text{H}_5$$

$$\underset{1}{} \quad \underset{2}{} \qquad \underset{3}{}$$

48   *49*

Their individual effects have been distinguished by ^{13}C-NMR analysis of the product PMAP (*47*) obtained from micellar photolysis of DBK with natural isotopic composition. In fact, only enrichment in positions corresponding to 1 to 3 were found to be significant [S(1) = 1.23, S(2) = 1.17, S(3) = 1.06] and their relative values reflect the order of the respective ^{13}C hfc constants.

Other experiments[233] employing DBK samples specifically enriched in ^{13}C at position C(1) or C(2) and C(3) and analyzing by MS and by ^1H-NMR ^{13}C-satellite intensities also corroborated the above findings. More indirect evidence of the positional selectivity between C(1) and C(2,3) came from a comparison of ^{13}C enrichment of DBK and the escape product 1,2-diphenylethane,[238] carried out in viscous homogeneous solution.

It has been mentioned that the magnetic isotope effects are much larger than any mass isotope effects to be expected, e.g., between ^{12}C and ^{13}C. One obvious proof that the observed enrichments cannot be due to mass isotope effects was provided in a study where ^{17}O isotope enrichment during DBK photolysis was compared for samples of natural isotopic abundance and others enriched in ^{17}O or ^{18}O. The magnetic ^{17}O isotope was found to be enriched in the recovered DBK.[310,311] For a mass isotope effect a monotonic change of the enrichment factor from ^{17}O to ^{18}O would have been expected. It shall be noted that aqueous micellar solutions were not found ideal for the O-isotope enrichment experiments, since the magnetokinetic effect was obscured by a superposition of thermal ^{16}O-exchange with the solvent water. Better results were obtained when the reaction was performed on porous silica.

Other evidence that the ^{13}C isotope effects occurring in the photolysis of ketones are, in fact, due to the magnetic properties of the nuclei comes from the sensitivity of these effects to the influence of external magnetic fields: in high fields the separation effects are largely quenched.[304,312] Thus, e.g., for ^{13}C enrichment in DBK photolyzed in HDTC micellar solution α was found to be 1.45 at zero-field, 1.15 at 1.45 T and 1.02 at 10 T. The first two values correspond to situations where hfc-induced T $\rightarrow$ S conversion in the RP occurs from three sublevels or from one sublevel (T_0), respectively. In very high fields T_0-S ISC becomes dominated by the $\Delta g^* B_o$ contribution which does not depend on magnetic isotope constitution.

b. Magnetic Field-Dependent Effects

More detailed magnetic field dependences of cage effects investigated by continuous photolysis have been reported in References 302 and 313. As an example, the magnetic field dependence of the percentage of diphenylethane formed as a cage product in the photolysis of DBK in HDTC micellar solution is shown in Figure 50. The cage product yield was obtained from the corresponding product yields in the presence and absence of $CuCl_2$ as a scavenger of escaped radicals. The cage effects show a main decrease with magnetic fields below 0.1 T. They are sensitive to magnetic isotope substitution of the benzylic C or H atoms but not of the carbonylic carbon. Obviously RPs undergoing T $\rightarrow$ S ISC before decarbonylation recombine rapidly to DBK or PMAP so that the benzyl/benzyl RPs originate with almost pure triplet spin correlation, i.e., with no information on the possible presence of ^{13}CO in their precursor. For the samples with ^{13}C at the benzylic positions (a), with natural isotope abundance (b) and with d_4- deuteration at the benzylic positions (c) the cage effects are in the order of a > b > c at all magnetic field values, reflecting the order of effective hyperfine coupling strength. This order is also borne out by the corresponding $B_{1/2}$ values (*circa* 550, 200, 120 G), though it must be noted that they are by at least a factor of 2 larger than the corresponding $B_{1/2}^{hfc}$ values (cf. Equation 152). Nevertheless, this relation documents that ISC is achieved through hyperfine coupling and that the efficiency of the underlying mechanism is decreased with increasing Zeeman splitting.

Due to the isotope-dependent difference in magnetic field sensitivity the efficiency of ^{13}C magnetic isotope enrichment passes through a pronounced maximum at about 300 G for the h_4-benzylic DBK. It is strange, however, that such a maximum is not observed for the d_4-benzylic DBK.

All methods of assessing cage effects in continuous photolysis experiments must be based either on quantum yield measurements or on the determination of relative yields of cage products. As far as the starting material itself may be regenerated as a cage product it

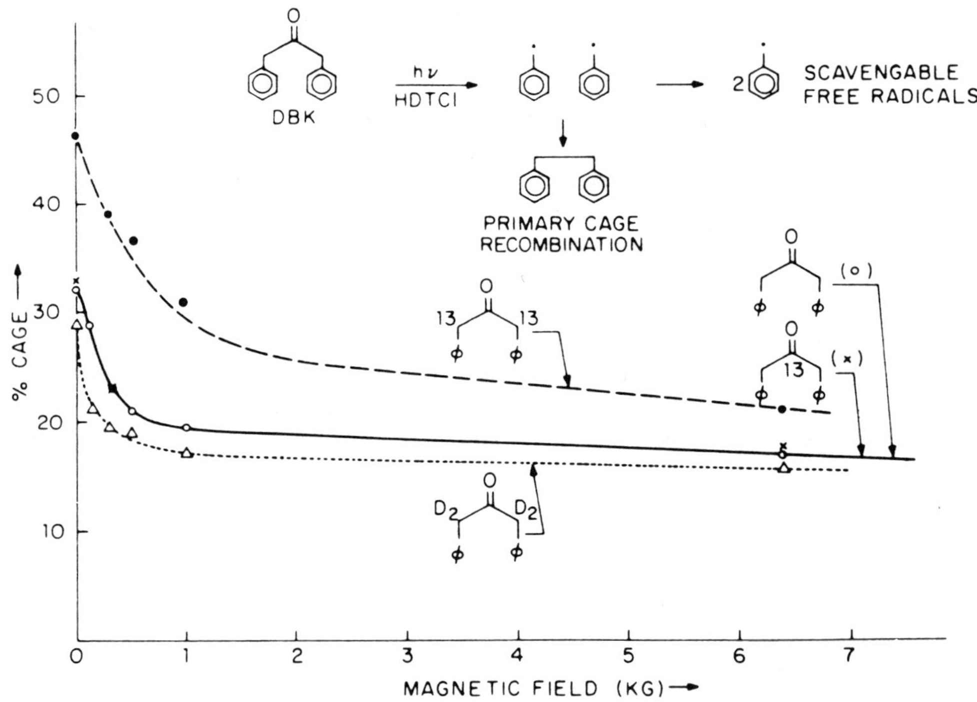

FIGURE 50.   Cage effect in the formation of diphenylethane as a function of magnetic field for DBK and several labeled DBKs. (Reprinted from Turro, N. J., Chow, M.-F., Chung, C.-J., Weed, G. C., and Kräutler, B., *J. Am. Chem. Soc.*, 102, 4843, 1980. With kind permission of N. J. Turro; copyright 1980, American Chemical Society.)

must differ in some detectable way from the constitution of the initial compound in order that the latter method of cage effect determination can be applied. For a distinction such subtle differences as isotopic constitution or steric configuration will suffice, however.

A recent example of the latter case has been reported by Buchachenko and co-workers.[98] Here the starting keton, the optically active (S)-(+)-α-methyldeoxybenzoin (*50*), was photolyzed in SDS micellar solution.

50

A variety of products is formed in cage and escape reactions from the initial T-RP of a benzoyl and a *sec*-phenethyl radical. Regeneration of the starting material is a cage process. During the RP lifetime stereochemical information on the configuration of the chiral precursor is lost and the recombination product is obtained as a racemate. The progress of racemization of the starting material during photolysis may be analyzed in close analogy to the kinetic formalism applied to magnetic isotope enrichment. Denoting by p the optical purity of ketone (*50*) and by f the degree of decomposition, the relation

$$\ln(p/p_o) = \beta * \ln(1 - f) \tag{193}$$

could be derived. Plots of ln (p/p_o) vs. ln $(1-f)$ yielded straight lines with a slope β related to the cage recombination probability P_r by

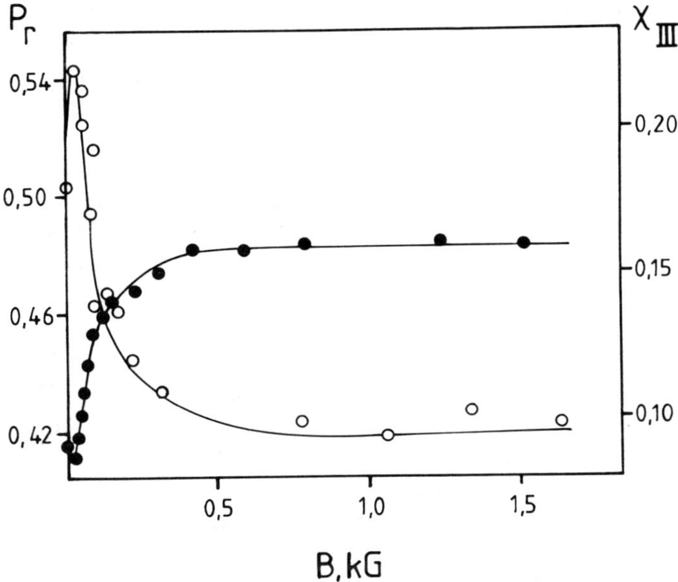

FIGURE 51. Magnetic field dependence of benzoyl/*sec*-phenethyl RP recombination probability P_r (o) and the yield χ_{III} of *d,l*- and meso-diphenylbutane (●) in continuous photolysis of α-methyldeoxybenzoin (*50*) in SDS micellar solution. (Adapted from Tarasov et al.[98])

$$P_r = \frac{\beta}{1 + \beta} \tag{194}$$

The magnetic field dependence observed for P_r is shown in Figure 51. For the probability of disproportionation of the RP (formation of benzaldehyde and styrene), another cage reaction, the same magnetic field dependence was found. The yield of escape products, e.g., 2,3-diphenylbutane, showed a complementary magnetic field dependence (cf. Figure 51).

A characteristic feature of the magnetic field dependence of P_r is that it passes through a pronounced maximum at small fields (<50 G) and then sharply falls approaching a nearly constant value of -14% below the zero-field at $B_0 > 600$ G. The field $B_{1/2}$ where $P_r = \frac{1}{2}(P_r^{zf} + P_r^{hf})$ is at 100 ± 20 G, which clearly exceeds the value of $B_{1/2}^{hfc} \approx 60$ G ($B_{sec\text{-}phenethyl}^{hfc} = 31$ G, $B_{benzoyl}^{hfc} = 1.5$ G) characteristic of the effective isotropic hyperfine coupling in the RP.

In order to account for the observed MFE quantitatively, the authors of Reference 98 performed model calculations employing ISC by isotropic hyperfine coupling and by spin relaxation due to electronic dipolar spin-spin coupling. The latter could be fitted to the experimental data only when adopting rather short orientational correlation times of $\tau_c \sim 8$ ps and an average distance of the RP of 6 Å, but still it was not as satisfactory as when applying the isotropic hyperfine coupling mechanism only. With the latter model the observed maximum could be reproduced (though it was overestimated in its absolute height) and the calculated $B_{1/2}$ (108 G) was close to the observed one.

The maximum in the $P_r(B)$ curve was rationalized by the breakdown, in low magnetic field, of the zero-field selection rule of conservation of total (nuclear + electronic) spin (cf. also Section III.C). Furthermore, it was pointed out that the $B_{1/2}$ value obtained when defining it as the field where $P_r = \frac{1}{2}(P_r^{max} + P_r^{hf})$ is close to the effective hyperfine coupling strength of the RP $[\approx(\{B_1^{hfc}\}^2 + \{B_2^{hfc}\}^2)^{1/2}]$.

From the limiting high field effect on cage (-14%) and escape products ($+80\%$) it was inferred that the total probability for the RP to undergo cage processes is 0.85 in zero-field and 0.73 in high field. The latter value is much higher than $^1/_3$ of the zero-field value, a result that would be expected for high fields if hfc only is responsible for the $T \rightarrow S$ process. In the model calculations this discrepancy was resolved by assuming that the original RP is not formed in a pure triplet state. Another possibility to account for it, namely, ISC contributions from a magnetic field-independent relaxation mechanism, was not considered.

In concluding this section on magnetokinetic effects observed with ketones under conditions of continuous photolysis we mention an interesting application of potential practical importance. As was shown by Turro and co-workers,[314-316] emulsion polymerization of styrene and methyl methacrylate photosensitized by dibenzyl ketones was markedly influenced by an external magnetic field. Since the initial stage of such polymerizations starts with monomeres solubilized in micelles, it is favorable that initiator RPs produced there undergo efficient escape in order to avoid fast intramicellar recombination with termination of the chain reaction. Since the amount of escape for T-RPs increases in a magnetic field, higher yield and higher molecular weights can be obtained under such conditions. ^{13}C magnetic isotope effects, which have also been observed, can be understood by the same reasoning.

The importance of spin correlation effects for these phenomena has been demonstrated by using diphenylazoethane [Ph(CH$_3$)CH–N=N–CH(CH$_3$)Ph] as an initiator of polymerization. Three modes of radical generation have been compared, viz., thermal, direct photochemical, and triplet-sensitized photochemical.[315]

Only in the last case was a MFE observed, indicating that only the kinetics of initially triplet spin-correlated pairs is significantly influenced by a magnetic field.

2. Time-Resolved Investigations

Relatively few time-resolved studies have been reported with micellar RPs from ketone photocleavage. Early investigations with RPs from DBK photolysis were of rather poor quality and no quantitative magnetic field dependence was obtained for the fast geminate recombination rate constant,[274,276,299,317,318] determined to be on the order of $5*10^6$ s^{-1}. It was found, though, that the yield of long-lived radicals increased from about 0.5 in zero-field to about 0.65 at fields above 700 G with $B_{1/2} \approx 100$ G.[317] Concerning the decay of long-lived radicals occuring with a $t_{1/2}$ of 5 to 10 μs,[299,317] it was noticed that it was not (with HDTC micelles) or only weakly quenched (in SDS micellar solution) by CuCl$_2$ scavenger. Thus, obviously, no or very little micellar escape of radicals took place. The nature of the slow decay process was not definitely assessed. Attribution of the corresponding kinetics to spin-relaxation[299] in the geminate RP seems doubtful in view of its low rate even in zero-field and its small[317] or vanishing[299] magnetic field dependence.

More detailed kinetic information was obtained with RPs from laser photolysis of ketones (*51a* to *f*).[319]

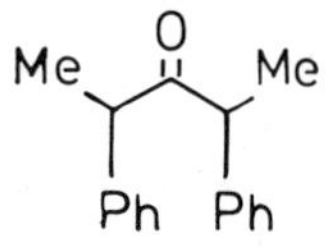

51a 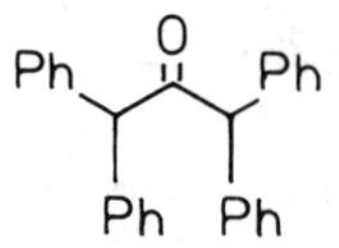   *51b* 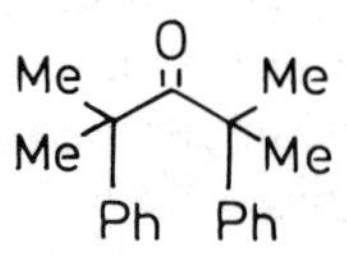   *51c*

TABLE 6

Rate Parameters and MFEs Pertaining to Fast Decay of RPs from Ketone Cleavage in Micellar Solution[319,320]

Radical pair	Micelle	k_{rec}^{zf} [a]	k_{esc} [a]	$R(\Phi_{esc})$ [b]	$B_{1/2}^{hfc}$ [c]
sec-Phenethyl/*sec*-phenethyl	SDS	7.2	4.8	57	62
	HDTC	3.3	1.6	100	62
Diphenylmethyl/diphenylmethyl	SDS	5.2	1.3	280	30
	HDTC	2.6	0.2	900	30
Diphenylmethyl/benzoyl	SDS	5.4	3.6	45	27
	HDTC	4.0	2.4	24	27
Cumyl/cumyl	HDTC	4.4	1.1	265	72
Cumyl/benzoyl	SDS	7.0	5.0	28	69

[a] In units of 10^6 s^{-1}.
[b] Relative increase of Φ_{esc} between zero-field and 2.5 kG.
[c] Calculated according to Equation 152.

51d

51e

51f

51a-d$_6$

51a-d$_8$

51b-d$_2$

Here the symmetric ones (*51a* to *c*) after α-cleavage and fast decarbonylation of the 2-phenyl-substituted acetyl radical yield two identical radicals, whereas (*51d* to *f*) yield unsymmetric ones, since the benzoyl and acetyl radicals do not undergo decarbonylation. The radical signals observed exhibited the typical fast (submicrosecond) and slow ($\geqslant$ 10 µs) decay attributed to intramicellar (geminate) and intermicellar (random) recombination. From the observed fast decay rate constant and the ratios of radical absorbance at the end and the beginning of the fast slope (corresponding to Φ_{esc}), the decomposition of $k_{obs} = k_{rec} + k_{esc}$ was possible by setting $k_{esc} = \Phi_{esc}*k_{obs}$. The pertinent results obtained for various RPs in SDS and HDTC micelles[319] are shown in Table 6.

There is an obvious influence of the nature of the surfactant on k_{rec} and k_{esc}, both being larger for a given RP in SDS than in HDTC. This difference is more pronounced for k_{esc} than for k_{rec}. In the case of the *sec*-phenethyl/acetyl RP from ketone (*51f*) escape (obviously of the acetyl radical) seems to be 100% efficient. In Table 6 are also given the theoretical $B_{1/2}^{hfc}$ values characterizing the effective isotropic hfc in the RPs. There is no correlation between k_{rec}^{zf} and $B_{1/2}^{hfc}$.

The MFEs observed between zero-field and 2.5 kG for the yield of escaping radicals vary between +24 and +900%, indicating magnetic field-dependent reductions of k_{rec} by a factor of 1.45 (diphenylmethyl/benzoyl RP in HDTC) to 38 (diphenylmethyl/diphenylmethyl RP in HDTC). Based on the mechanistic concept that for a RP produced with equal probabilities in its triplet substates T_0, $T_{\pm}$, which cannot be kinetically distinguished at zero-field but will behave differently in a high field where $T_{\pm} \rightarrow$ S ISC is suppressed, the yield of escape should not be able to increase by more than a factor of 3 (R = +200%) between zero-field and high field. There are several cases in Table 6 where R is significantly higher. To account for this observation it was suggested[319] that spin polarization from the $S_1 \rightarrow T_1$ ISC process in the parent ketone might lead to a nonequilibrium population of the RP triplet substates (actually this is the so-called triplet mechanism of CIDEP, cf. References 103, 258, and 259).

Other valuable magnetokinetic information on the role of hfc in the recombination of the symmetric *sec*-phenethyl and diphenylmethyl RPs has been reported in Reference 320, where ketones (*51a*) and (*51b*) have been investigated with various degrees of deuteration ([*51a*-d$_6$], [*51a*-d$_8$], [*51b*-d$_2$]). The fast initial part of the RP decay signals observed by laser flash photolysis in HDTC micellar solution was reported to be monoexponential in fields below 1 kG and to separate into a fast, field-independent component and a slower one (but still distinct from the much slower decay of escape radicals), the rate constant of which falling to very low values in high fields. The field-dependent rate constant (k_{obs}) is plotted as a function of the magnetic field strength in Figure 52a for ketone (*51a*) and its deuterated analogues. The zero-field value for (*51a*-h$_8$) is somewhat larger than given in Table 6. At low fields there appears a little maximum that is not present for the deuterated compounds. The effect of deuteration on k_{obs} is rather small at zero-field but increases at intermediate fields up to 400 G. At high fields the differences due to deuteration disappear again. The ratio between k_{obs} in zero-field and in high field amounts to about 5, implying that after transforming $k_{obs} = k_{rec} + k_{esc}$ to k_{rec} this ratio will be even higher. From Figure 52a the $B_{1/2}$ values are 100, 50, and 40 G for the RPs from (*51a*-h$_8$), (*51a*-d$_6$), and (*51a*-d$_8$), respectively. These values are by a factor of two or more larger than the theoretical $B_{1/2}^{hfc}$ values.

Due to the different field dependencies for different isotope substitution the relative magnetic isotope effect k_{rec}^d/k_{rec}^h is a function of the field (cf. Figure 52b). It approaches a minimum value at about 100 G, keeping constant up to about 500 G, and disappears (i.e., $k_{rec}^d/k_{rec}^h \rightarrow 1$) in the limit of high fields. This kind of magnetic field dependence bears important information concerning the mechanism of RP ISC. While in the field region 0 to 100 G where the suppression of coherent $T_\pm \rightarrow S$ transitions is to be expected, k_{rec} exhibits different field dependence for different degrees of isotopic substitution; at fields >100 G k_{rec}, though still strongly magnetic field dependent, changes in the same way for all variants of isotopic substitution. It is plausible to attribute the different field dependence below 100 G to the magnetic field dependence of coherent $T_\pm \rightarrow S$ processes (the small but definite magnetic isotope effect at zero-field indicates that even for degenerate $T_0, T_\pm, S$ hyperfine-induced spin evolution is not so fast as to establish complete spin equilibrium during micellar RP decay), whereas at fields >100 G the field dependence is due to $T_\pm \rightarrow T_0, S$ relaxation processes. As follows from Equations 40 and 41 for $\omega_o \tau_c \gg 1$, k_{rel} will vary as k_{rel}^o/ω_o^2 whereby the specific constant k_{rel}^o, depending on the effective anisotropic couplings that will differ among various isotopes, is just a constant of proportionality. Thus the field dependence of k_{rec} as well as the constancy of relative magnetic isotope effects on k_{rec} may be explained.

Results obtained with symmetric diphenylmethyl RPs from ketones (*51b*-h$_2$) and (*51b*-d$_2$) led to similar conclusions. For these RPs the ratio $k_{obs}^{zf}/k_{obs}^{hf}$ reached a remarkably high value of about 12. The kinetic magnetic isotope effect between fields of 100 and 500 G amounted to $k_{obs}^d/k_{obs}^h = 0.7$.

3. Radical Pairs from Photolysis of Sulfones

Photolysis of sulfones (*52a* to *d*) has been investigated for MFEs in micellar solution.[321,322]

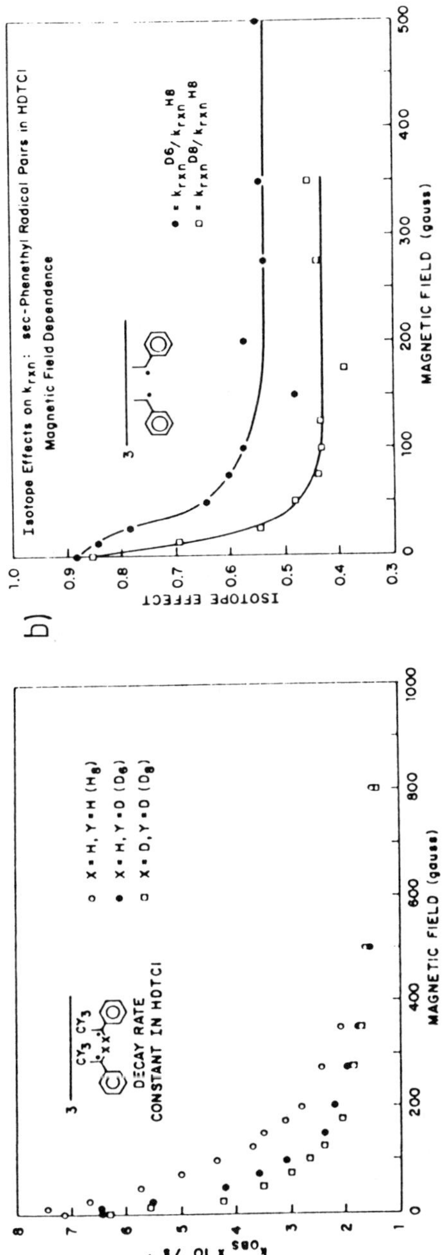

FIGURE 52.   Decay rate constants $k_{obs} = k_{rec} + k_{esc}$ (a) and relative magnetic isotope effects (b) on the rate constants k_{rec} (k_{rxn}) of *sec*-phenethyl RPs, generated with triplet spin in micelles of HDTC, as a function of magnetic field strength and isotopic substitution. (Reprinted from Turro, N. J., Zimmt, M. B., and Gould, I. R., *J. Phys. Chem.*, 92, 433, 1988. With kind permission of N. J. Turro; copyright 1988, American Chemical Society.)

$R,R' = H$ or CH_3

Sulfones (*52a*) and (*52b*) undergo photochemical S–C cleavage in their excited triplet state with subsequent rapid SO_2 elimination.[321] Under conditions of continuous photolysis they exhibited the cage effects and MFEs typical of RPs originating from triplet precursors. Sulfones (*52c*) with a β-naphthyl substituent show practically no RP formation from the excited triplet. In the case of compound (*52c*) with $R' = R = H$ which shows 100% cage product formation and no MFE on direct photolysis, the cage effect decreases to 90% if the reaction is triplet sensitized and a relative MFE of -40% on it appears.[321]

Due to rapid SO_2 elimination the radicals recombining in the photolysis of compounds (*52a* to *c*) are carbon centered as for the case of the corresponding ketones. In contrast to these cases photolysis of (*52d*) produces sulfur-centered radicals[322] that do not eliminate SO_2. The transient absorption of the benzenesulfonyl radical was detected. Its decay consisted of a fast ($\sim 2*10^6$ s^{-1}) and a slow ($\sim 5*10^4$ s^{-1}) component. Although the corresponding rate constants did not reveal a MFE, the amplitude of the slow component, attributed to the yield of escape radicals, increased in a magnetic field as to be expected for the case of T-RPs. The field dependence of the escape yield exhibited a fairly rapid increase of 20% up to 0.1 T followed by a gradual increase of another 30% up to 1 T. The observed MFE is the first one involving RPs with S-centered radicals. Its existence documents that SOC in such radicals would not be strong enough as to obscure kinetic effects due to hyperfine induced ISC. Thus magnetic isotope separation of S-isotopes should, in fact, be feasible on the basis of the RPM.

D. ELECTRON TRANSFER REACTIONS

In normal micellar solution few studies with RPs from photoelectron transfer reactions have been reported, employing excited carbonyl triplets as electron acceptors and primary or secondary aromatic amines as donors in SDS micelles.

In the first publication on such a system (duroquinone with diphenylamine) by Tanimoto et al.[323] only a very weak MFE of about $+5\%$ increase in the yield of escape radicals has been reported. A slow rate of recombination of RPs, even after ISC from triplet to singlet, was suggested as an explanation.

More prominent effects have been found, however, by this group[324] with Reaction 195.

$$\left(\text{*53*}\right) + (Ph_2)NH \xrightarrow{H^+} \text{*54*} + (Ph)_2NH^{+\bullet} \qquad (195)$$

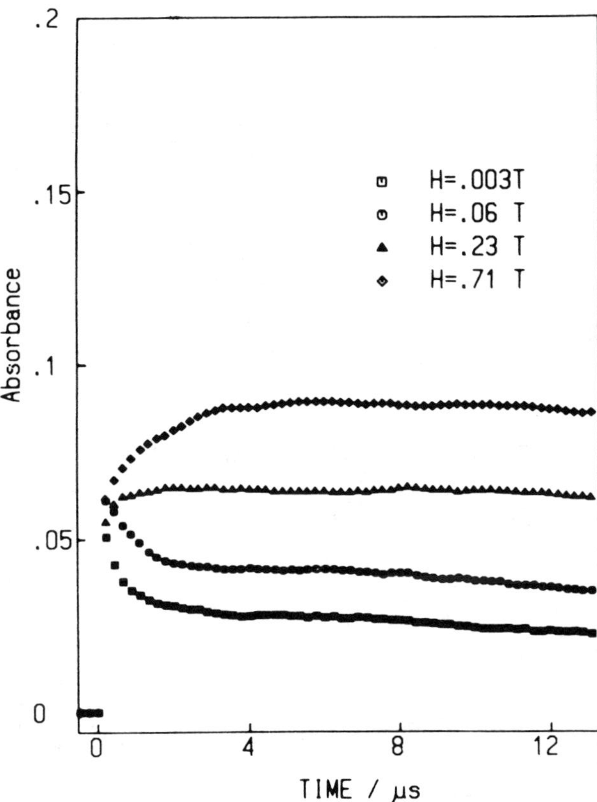

FIGURE 53.   MFE on the evolution of the absorbance of diphenylamine radical cations produced by electron transfer reaction with 1-acetonaphtone triplet in SDS micellar solution. (Reprinted from Tanimoto, Y., Takayama, M., Itoh, M., Nakagaki, R., and Nagakura, S., *Chem. Phys. Lett.*, 129, 414, 1986. With kind permission of Y. Tanimoto; copyright 1986, Elsevier Science Publishers B. V.)

The MFE exhibited by the transient absorption signals due to the diphenylamine cation radical is shown in Figure 53. It is remarkable that in the presence of a magnetic field not only the radical decay becomes slower, which is due to the retardation of the intramicellar recombination of triplet RPs, but that at high fields (>0.23 T) the amount of radicals formed still increases during a time interval (4 μs) much larger than the intramicellar RP decay time in zero-field. The explanation is that with the donor concentrations applied the radicals produced arise from both fast intramicellar ($k = 8*10^6$ s^{-1}) and slow intermicellar ($k \approx 8*10^7$ M^{-1} s^{-1} corresponding to an effective first-order rate constant of $2.4*10^5$ s^{-1}) triplet quenching. The contribution of the latter process can only appear as a rise in radical absorption when the intramicellar recombination rate of the triplet RPs is sufficiently slowed down. Kinetic deconvolution has been applied to evaluate k_{rec} as $7*10^6$ s^{-1} at zero field and $0.7*10^6$ s^{-1} at a field of 0.56 T, with a B$^{1/2}$ of about 500 G.

Another system employing photoelectron transfer between *p*-phenylaniline (*37*) and various triplet-excited benzophenone derivatives including halogen-substituted ones has been studied in SDS micellar solution by Levin and Kuzmin.[44,291] These reactions were investigated in acidified solutions so that the ketyl anion radical produced in the electron transfer steps was rapidly protonated. The magnetic field dependence of the recombination kinetics of the RPs has been measured and analyzed in detail (cf. Figure 54 and Table 7). The decay was almost completely due to intramicellar recombination with only little escape and formation

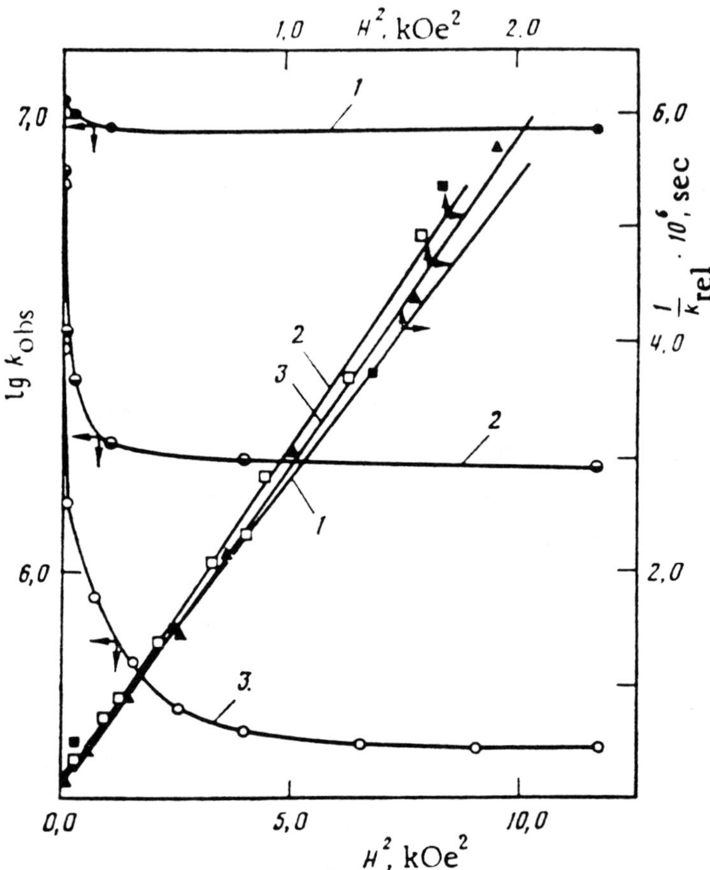

FIGURE 54. Dependence of RP decay constant k_{obs} and spin relaxation rate constant k_{rel} on square of magnetic field strength H^2 for $PNH_2^{+\cdot}/XH\cdot$ RPs with X = 4-bromobenzophenone (1), 4,4′-dichlorobenzophenone (2), and 4,4′-dimethylbenzophenone (3). PNH_2 = (*37*). (Reprinted from Levin, P. P. and Kuz′min, V. A., *Bull. Acad. Sci. U.S.S.R. Div. Chem. Sci.*, 37, 224, 1988. With kind permission of P. P. Levin and V. A. Kuz′min; copyright 1988. Plenum Publishing Corporation.)

TABLE 7
Kinetic Characteristics of RPs Entailing the Ketyl Radical of Substituted Benzophenones and the Radical Cation of 4-Phenylaniline in Micelles of SDS at 25°C

Ketone	k_{obs}^{a} (B = 0 T)	k_{obs}^{a} (B = 0.34 T)	$k_{S}^{a,b}$	$k_{T}^{a,c}$	$k_{rel}^{o\ a}$	τ_{c}^{d}
4,4′-Dimethoxybenzophenone	7.9	0.35	30	0.26	3.7	1.8
4,4′-Dimethylbenzophenone	7.3	0.41	28	0.35	5.0	2.1
Benzophenone	7.1	0.50	27	0.35	4.0	1.8
4-Benzoylbenzoic acid	5.8	0.63	21	0.53	3.3	1.7
4,4′-Dichlorobenzophenone	7.6	1.7	25	1.5	4.3	2.0
2-Bromobenzophenone	8.2	4.6	19	4.4	3.7	1.7
4-Bromobenzophenone	11.2	9.2	18	9.0	3.6	1.7

[a]   In units of 10^6 s^{-1}
[b]   Corresponding to k_{chem} in Reference 291.
[c]   Corresponding to k_{SOI} in Reference 291.
[d]   In units of 10^{-10} s.

Data from Reference 291.

of long-lived radicals. At zero-field the intramicellar decay was monoexponential; at higher fields it became biexponential. The lower, magnetic field-dependent one of the decay constants (k_{obs}) was used for a quantitative analysis. A heavy atom effect of Cl and Br substituents is borne out as a slight increase of $k_{obs}(0)$ but as a strong increase of $k_{obs}(0.34\ T)$, such that the MFE on k_{obs} is strongly reduced by halogen substituents. For B >200 G the magnetic field dependence of k_{obs} could be consistently modeled on the basis of the scheme depicted in Figure 5, by fitting four individual parameters for each donor/acceptor system (the notations used by Levin and Kuz'min are given in parentheses): k_S (k_{chem}) the rate constant of spin-allowed recombination, k_T (k_{SOI}) the rate constant of recombination of triplet pairs due to the effect of spin-orbit interaction during ^3RP contacts, $k_{r,1} + k'_{r,1} = k_{rel}$ characterized by k^o_{rel} and τ_c, the zero-field spin relaxation rate constant, and the effective orientational correlation time determining the magnetic field dependence of relaxational $T_\pm \rightarrow S,T_0$ processes according to the relation

$$k_{rel} = k^o_{rel}/[1 + (\bar{g}\beta B_o\tau_c/\hbar)^2] \tag{196}$$

Since the yield of escape was small, consideration of a fifth parameter, k_{esc}, the rate constant of radical escape represented only a minor correction. The fit was based on the following equations:

$$k_{obs}(B = 0) = \frac{1}{4}k_S + \frac{3}{4}k_T + k_{esc} \tag{197}$$

$$k_{obs}(B = 0.34\ T) = k_T + k_{esc} \tag{198}$$

$$k_{obs}(B > 0.02\ T) = k_-(k_S,k_T,k_{rel},k_{esc}) \tag{199}$$

where k_- is given by Equation 49.

The parameter values obtained from the experiments are given in Table 7. From these it can be seen that the heavy atom effect on k_{obs} is largely represented in k_T (k_{SOI}), whereas k_S (k_{chem}) displays only a minor "heavy atom effect", meaning that the assumption of SOC effecting direct recombination of T-RPs is essentially correct, i.e., that a magnetic field-independent SOC contribution to k_{rel} is of minor importance (cf. below). It is gratifying to note that the relaxation parameters k^o_{rel} and τ_c are very close for all the systems represented in Table 7. The value of about $2*10^{10}$ s for τ_c is on the order of what is expected from independent ESR information.[325] The relaxation mechanism may be attributed to the hyperfine tensor anisotrophy.

1. Radical Pairs in Microemulsion Nanodroplets

In our laboratory we have studied in some detail the magnetokinetic behavior of RPs produced by electron transfer reactions between aromatic amines and photoexcited dyes ($^3TH^+$ [*10a*], $^3SeH^+$ [*10c*], $^1OxH^+$ [*10d*]) solubilized within nanodroplets of water dispersed in unpolar solvents[326] (cf. Figure 43).

Investigations with triplet RPs have been reported within the surfactant CDBA (cetyldimethylbenzylammonium chloride) in benzene as bulk solvent. Various amounts of water covering a water/surfactant molar ratio (w) from 8 to 32 were added. In this way the radius (R_{nd}) of the nanodroplets could be varied in a well-defined way [$R_{nd} \approx w*(1.9\ Å)$], providing an ideal means to study supercage effects on RP kinetics.

In Figure 55a, as an example, is shown the radical decay kinetics after quenching $^3TH^+$ by aniline. With the quencher concentration (0.05 M) used the triplet lifetime was about 15 ns, so that the radical decay kinetics observed is not superimposed by radical formation

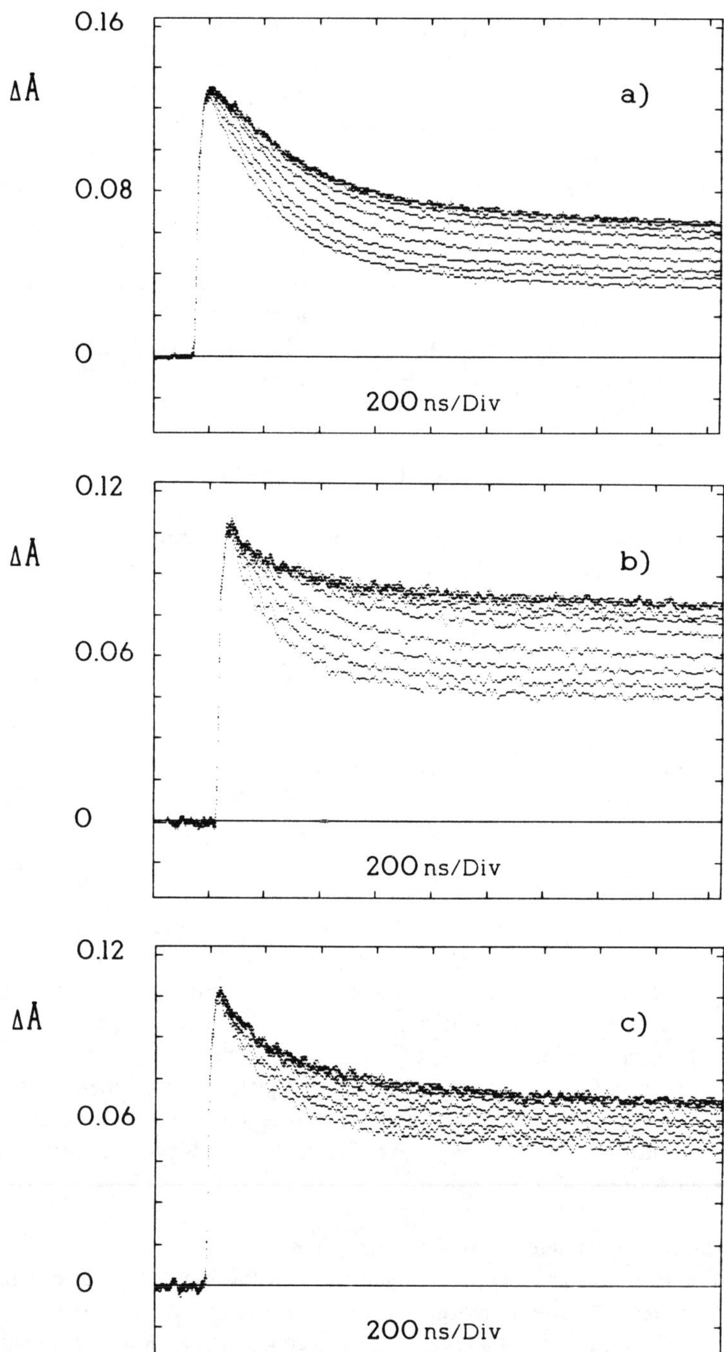

FIGURE 55. Magnetic field dependence of T-RP recombination in H_2O/CDBA/ benzene microemulsion nanodroplets (w = 15).[327,328] (a) TH·/An$^{+·}$, (b) C_{16}-TH·/ An$^{+·}$, (c) C_{16}-TH·/p-C_{16}-An$^{+·}$. TH$^+$ = (*10a*), C_{16}-TH$^+$ = (*55a*), p-C_{16}-An = (*56*). In each diagram the signal traces from bottom to top correspond to magnetic fields of 0, 0.005, 0.01, 0.02, 0.05, 0.1, 0.2, 0.5, 1.0, 3.3 T.

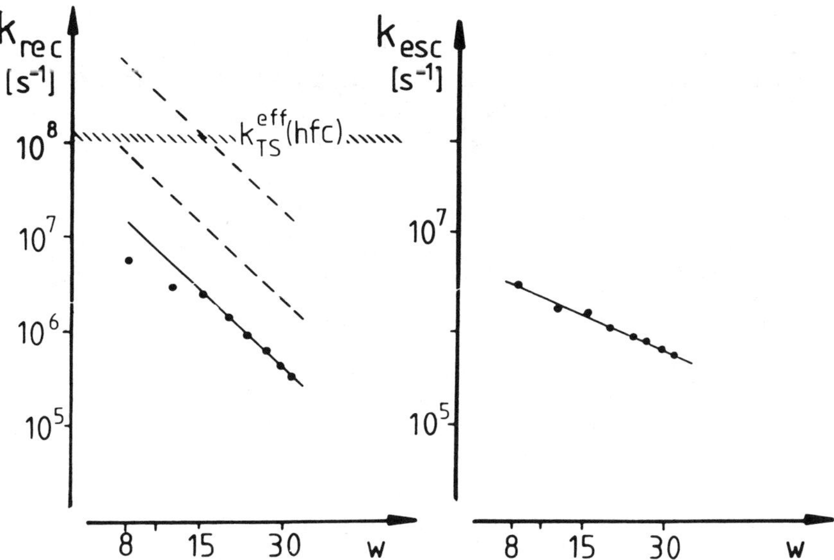

FIGURE 56.   Variation of rate constants k_{rec} and k_{esc} (zero-field values) for T-RPs TH·/An$^{+·}$ in H$_2$O/CDBA/benzene microemulsion nanodroplets as a function of water/surfactant molar ratio w. The dashed lines (left) mark theoretical expectations for diffusion-controlled recombination, assuming η = 1 cP as in bulk water (upper line) or η = 10 cP (lower line). TH$^+$ = (*10a*). (Reprinted with permission from Ulrich, T. and Steiner, U. E., *Chem. Phys. Lett.*, 112, 365, 1984. © 1984 Elsevier Science Publishers B.V.)

kinetics. The RP decay consists of a fast (intramicellar) recombination with k_{rec} = 3*10^6 s^{-1} at zero-field and a slow intermicellar one in the millisecond time regime. The rate constant k_{esc} of micellar exit was determined as 1.75*10^6 s^{-1}. On applying magnetic fields the fast intramicellar recombination is slowed down and the yield of escape is increased.

Before discussing the details of the magnetic field dependence of k_{rec} some experimental information pertaining to the preferred location of the radicals inside the microemulsion aggregates and to their diffusional mobility may be of interest. When the radius of the nanodroplets is increased the rate constants k_{rec} and k_{esc} are systematically changed as depicted in Figure 56. For w >15 the values of k_{rec} vary approximately as w^{-3} implying an inverse proportionality to the volume of nanodroplets, as would be expected for a process where the rate is proportional to the frequency of encounters in such a volume. In fact, in zero-field the rate of T $\rightarrow$ S conversion as induced by isotropic hyperfine coupling is expected to be much faster (~2*10^8 s^{-1})[42] than found for k_{rec} and hence will not be rate determining for triplet RPs in zero-field. With spin equilibrium rapidly established the recombination rate constant should approach a value of ¼ of the diffusion-controlled one as is indeed found in homogeneous solution. However, in the microemulsion the observed rate constants are by a factor of about 50 smaller than to be expected for a diffusion-controlled reaction in nanodroplets of corresponding size (dashed line in Figure 56). This is an argument against free diffusion through the volume of the nanodroplets. It receives support from investigations with modified reactants[327,328] C$_{16}$-TH$^+$ (*55a*), di-C$_{16}$-TH$^+$ (*55b*), and p-C$_{16}$-An (*56*) carrying n-hexadecyl substituents in order to anchor them in the surfactant layers and to restrict the mobility of the hydrophilic reacting pendants to the waterpool/surfactant interface.

TABLE 8
Rate Constants of Intramicellar Recombination (k_{rec}, 10^6 s^{-1}) and Escape (k_{esc}, 10^6 s^{-1}) for Triplet RPs TH·/An$^+$· and n-C$_{16}$-Alkyl-Substituted Derivatives in H$_2$O/CDBA/Benzene Microemulsions[327,328] (cf. Figure 55)

| | Electron Acceptor | | | | | |
| | TH$^+$ | | C$_{16}$-TH$^+$ | | di-C$_{16}$-TH$^+$ | |
Electron donor	k_{rec}	k_{esc}	k_{rec}	k_{esc}	k_{rec}	k_{esc}
An	2.8	1.2	2.7	1.9	1.0	3.4
p-C$_{16}$-An	1.8	1.2	2.3	1.9	1.9	3.8
N-C$_{16}$-An	2.1	1.0	2.1	2.0	2.0	3.6

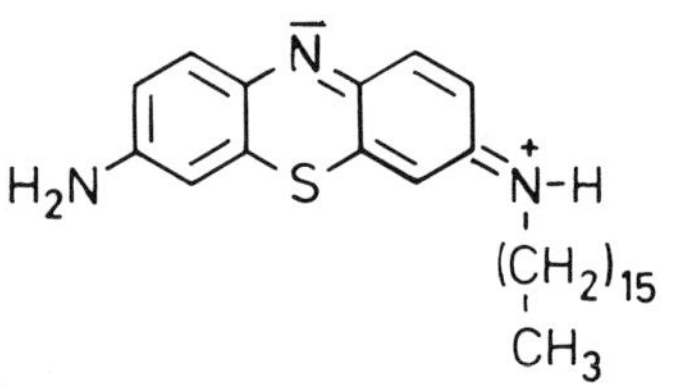

C$_{16}$TH$^+$

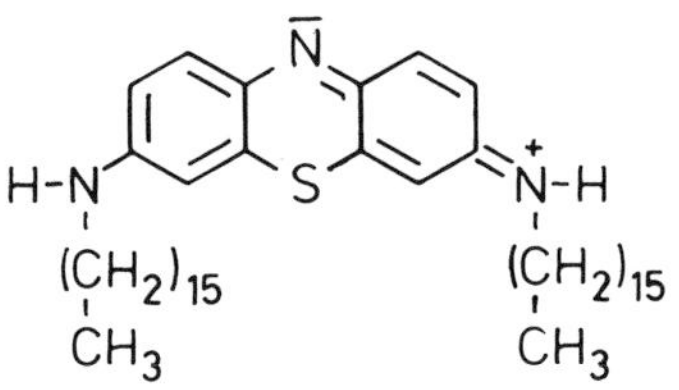

di-C$_{16}$TH$^+$

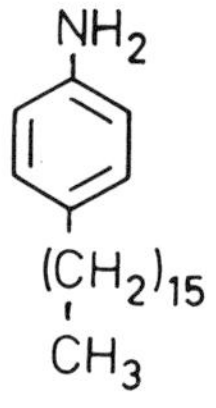

p-C$_{16}$An

55a

55b

56

Observed values of k_{rec} and k_{esc} for the various donor/acceptor combinations are given in Table 8.

The recombination rate constant does not exhibit marked changes when anchoring one or both of the radicals in the surfactant layer (cf. Figure 55a to c). From this result one has to conclude that, as for the anchored system, for the unsubstituted radicals, too, diffusion and recombination take place in the approximately two-dimensional space of the Stern layer. From geometric reasons for this case an inverse square law should be expected for k_{rec} as a function of nanodroplet size parameter w. The higher power law found might be explained by structural changes in surfactant packing, facilitating diffusion as the interface becomes less curved with increasing w.

For TH·/An$^+$· the rate of escape increases approximately as the inverse of the nanodroplet radius. Obviously, as would be expected from intuition, penetration of the escaping radical through the surfactant layer is impeded if it is less curved. There is a systematic increase of k_{esc} in the series of radicals TH·, C$_{16}$-TH·, di-C$_{16}$TH·, independent of the counter radical. This is a clear indication that, as to be expected, it is the uncharged dye semiquinone radical that escapes into the unpolar bulk phase, a process that is favored if the radical becomes more lipophilic.

The magnetic field dependence k_{rec}, the rate constant of intramicellar recombination, has been studied for the RP TH·/An$^+$· as a function of the waterpool size. The results are presented as a double log plot in Figure 57. Although the absolute values of k_{rec} vary by a factor of about 20 over the range of w-variation, the shapes of the curves representing the magnetic field dependence are essentially unchanged. The field dependence seems to approach saturation at fields of about 1 T, where k_{rec} is reduced by a factor of 2 to 3 with respect to its corresponding zero-field values. On each curve is marked the interpolated position where

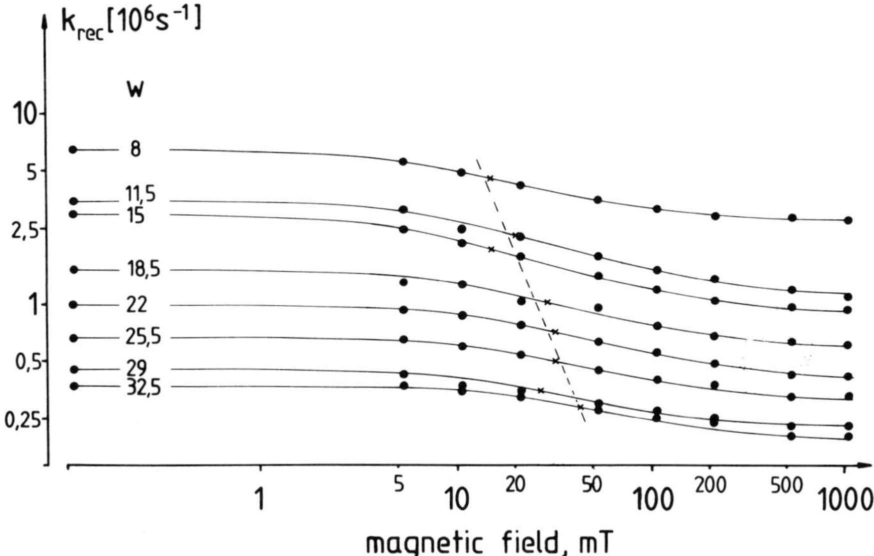

FIGURE 57.   Magnetic field dependence of rate constants k_{rec} of geminate recombination of $^3(TH\cdot/$ $An^{+\cdot})$ RPs in H_2O /CDBA/benzene microemulsions for various w values corresponding to nanodroplet radii between 15 and 60 Å. $TH^+ = (10a)$. The solid lines have been drawn to fit the experimental points. (Reprinted with permission from Ulrich, T. and Steiner, U. E., *Chem. Phys. Lett.*, 112, 365, 1984. © 1984 Elsevier Science Publishers B.V.)

$$k_{rec}(B_{1/2}) = \frac{1}{2} (k_{rec}^{zf} + k_{rec}^{hf}) \tag{200}$$

which defines a corresponding $B^{1/2}$ value for each w. Despite the scatter displayed by these $B^{1/2}$ values, there is a clear trend for $B^{1/2}$ to increase with the nanodroplet radius ($B^{1/2} \approx 14$ mT for w = 8 to $B^{1/2} \approx 40$ mT for w = 32.5). The slope indicated corresponds to a value of -2, implying a relation:

$$k_{rec}(B_{1/2}) \sim B_{1/2}^{-2} \tag{201}$$

The remarkable constancy of the ratio f

$$f \equiv k_{rec}^{zf}/k_{rec}^{hf} \tag{202}$$

of about 2.5 implies that k_{rec}^{hf} depends in the same way on the nanodroplet size as k_{rec}^{zf}. Thus it must be also assigned to a mechanism involving RP encounters. Furthermore, since the field dependence of k_{rec} seems to be saturated at 1 T the limiting high-field value of k_{rec}^{hf} must be considered as characteristic of a magnetic field-independent process, whereby triplet RPs can recombine if they come into contact. SOC in contact RPs, as suggested by Levin and Kuzmin,[44] is such a mechanism. As in their work, the analysis of the field dependence can be based on the scheme in Figure 5 and on Equation 49 expressing $k_{rec}(B) = k_- - k_{esc}$ as a function of k_S, k_T (both field-independent, but w-dependent), and $k_{rel}(B)$, an effective rate constant of $T_\pm \rightarrow (T_0,S)$ transitions.

Using Equation 49, and setting $k_{obs} = k_{rec} + k_{esc}$ in Equations 197 and 198, the value $k_{rel}(B^{1/2}$ can be expressed as a function of the limiting values of k_{rec} at zero-field (k_{rec}^{zf}) and at high field (k_{rec}^{hf}):

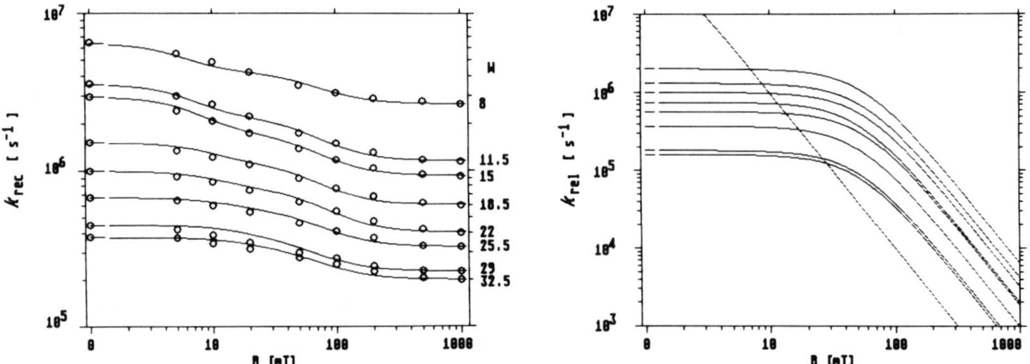

FIGURE 58.   Theoretical simulation (left-hand side) of the magnetic field dependence of geminate recombination constant k_{rec} of T-RPs TH·/An^{+}· in H_2O/CDBA/benzene microemulsion nanodroplets with various water/surfactant ratios w (cf. Figure 57). The magnetic field dependence of the expression $k^0_{rel}/(1 + \omega^2_0\tau^2_c)$ with parameters k^0_{rel} and τ_c as used for the fit is represented by the set of curves in the right-hand-side diagram (higher values of k_{rel} for smaller w). The straight line represents the phenomenological term A/ω^2_0 employed in the fit. (From Steiner and Wu[329] and Wu.[330])

$$k_{rel}(B_{1/2}) = \frac{3}{4}(k^{zf}_{rec} - k^{hf}_{rec}) = \frac{3}{4}\frac{f-1}{f}k^{zf}_{rec} \tag{203}$$

This relation indicates that, as k^{zf}_{rec} decreases with increasing size of the nanodroplet, the value of k_{rel} where the half-field value of k_{rec} is reached must be proportionally decreased. This implies higher values of $B_{1/2}$ with increasing w as has been observed (cf. Figure 57).

Following the procedure suggested by Levin and Kuzmin,[44] i.e., using Equations 197 to 199, we formally evaluated "$k_{rel}(B)$" from the magnetic field dependence of the observed RP decay constant $k_{obs} = k_{rec} + k_{esc}$.[329] This phenomenological quantity "$k_{rel}(B)$" could be satisfactorily fitted over the complete range of B (including fields <200 G) by setting

$$\text{``}k_{rel}(B)\text{''} = \frac{A}{\omega^2_o} + \frac{k^o_{rel}}{1 + \omega^2_o\tau^2_c} \tag{204}$$

wherein the first term provides a phenomenological description of the effect of spin motion due to isotropic hyperfine coupling. The curves shown on the left-hand side in Figure 58 were calculated using $A = 3*10^{24}$ rad*s^{-1} and $\tau_c = 1.1*10^{-10}$ s, independent of w. The relaxation parameter k^o_{rel} turned out to depend on the nanodroplet radius (proportional to w) in a similar way as k_{rec} (cf. Figure 58, right-hand side). This finding strongly suggests that the relevant spin relaxation mechanism requires RP encounters, which in turn allows the conclusion that the mechanism involves interactions between the radicals. The relaxation mechanism due to dipolar spin-spin interaction may account for this observation.

Among the T_1 relaxation mechanisms anisotropic hfc belongs to the most important ones in organic radicals. In this respect it is of interest that the *largest anisotropic* couplings are to be expected for radicals with high π-orbital spin densities on ^{19}F substituents ($a_\parallel - a_\perp = 300$ G for a π-spin density of 1[331]). This particular role of fluorine is due to its high nuclear magnetic moment and the smallness of its 2p orbital. Investigating the magnetic field-dependent recombination rate constants of triplet RPs TH·/p-F-An^{+}·, o-F-An^{+}·, and m-F-An^{+}· in microemulsions as specified for Figure 55, we found significant effects on the $B_{1/2}(k_{rec})$ values (160, 125, and 580 G for o-, m-, and p-F-An, respectively) but not on the f-values. This indicates that there is a substituent effect of fluorine on k_{rel} but not on $k^{hf}_{rec}(\triangleq k_T)$ as to be expected. The order of the $B_{1/2}$ values is in line with the order of π-spin densities

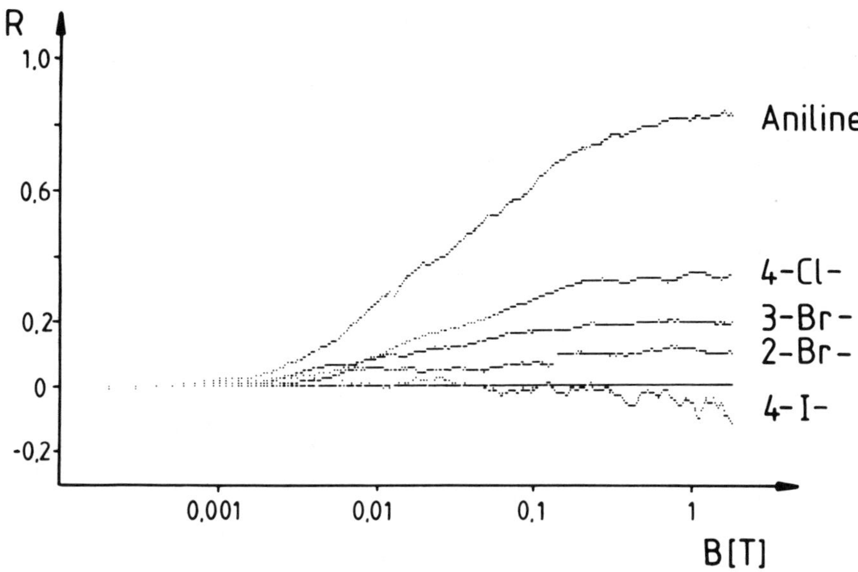

FIGURE 59.   Relative MFE of the yield of radicals escaping geminate recombination in T-RPs of TH·/X-An^{+}· in H_2O/CDBA/benzene microemulsion nanodroplets. (Reprinted with permission from Ulrich, T., Steiner, U. E., and Schlenker, W., *Tetrahedron*, 42, 6131, 1986. © 1986 Pergamon Press PLC.)

on the F-atoms. For *m*-F-An with the lowest F-spin density $B_{1/2}$ is not significantly different from the aniline case.

It has been emphasized above that SOC can contribute magnetic field-independent pathways to the recombination of T-RPs. A demonstration of such effects with T-RPs of thionine and several monohalogenated anilines in microemulsion nanodroplets is given in Figure 59. Here is shown the relative MFE on the yield of long-lived (escape) radicals. The magnetic field-dependent increase of the yield of escape is due to a corresponding decrease of k_{rec}. The f values determined were 3.2, 1.9, 1.5, and 1.0 for aniline, and the 4-Cl, 3-Br, 4-Br derivatives, respectively, reflecting the order of magnitude of position-dependent SOC[178] of the halogen substituent.

Two possible explanations may be discussed: a magnetic field-independent contribution k_{rel}^{SOC} to spin relaxation[43] or an increase of k_T.[44] Since the SOC effect was only weakly expressed in k_{rec}^{zf} we suggested the first of these possibilities, invoking the spin-rotational relaxation as the relevant mechanism. As a consequence of this explanation the limiting high field value k_{rec}^{hf} should have been weakly dependent on the size of the nanodroplet. Recently we measured the magnetic field dependence of k_{rec} with 4-Cl-An and 3-Br-An for various nanodroplet radii. Actually the factor f $-$ $k_{rec}^{af}/k_{rec}^{hf}$ was independent of w, i.e., the SOC-enhanced value of k_{rec}^{hf} shows the same w-dependence as k_{rec}^{zf} and thus must be due to SOC in contact T-RPs. We have observed corresponding effects when increasing the SOC on the side of the electron acceptor. Thus with SeH·(*10c*)/An^{+}· the observed f-value of 1.6 is clearly reduced with respect to the TH·/An^{+}· pairs.[330]

It has also been noted[43] that the initial amount of RPs detected in the nanodroplets directly after triplet quenching was decreased in parallel to the SOC effect previously detected with these systems in homogeneous solutions[178] and explained in terms of SOC-induced deactivation of triplet exciplexes. Thus, as has also been pointed out by other workers,[44,189] it may be concluded that both heavy atom effects—increase of the deactivation rate of exciplexes and of the recombination rate of T-RPs—are intimately related and that triplet exciplexes and triplet RPs in contact may be actually considered as the same type of species (cf. also Section III.A.2).

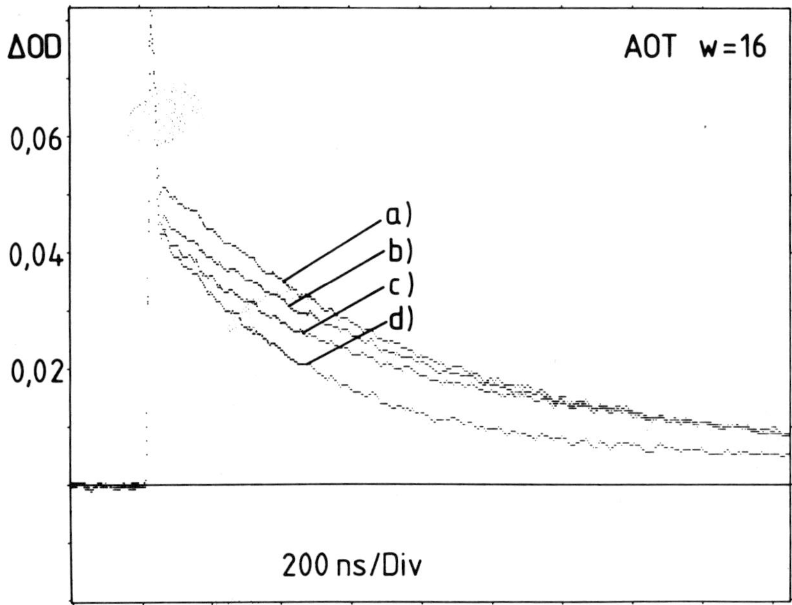

FIGURE 60.   Magnetic field dependence of OxH·/TMPDA$^{+·}$ S-RP recombination kinetics in H$_2$O/AOT/iso-octane microemulsion nanodroplets monitored through OxH· radical absorption. OxH$^+$ = (*10d*). Field values are (in T): (a) 0, (b) 0.0075, (c) 0.05, (d) 1.0. (Reprinted with permission from Baumann, D., Ulrich, T., and Steiner, U. E., *Chem. Phys. Lett.*, 137, 113, 1987. © 1984 Elsevier Science Publishers B.V.)

If this is true, then k_T is expected to be susceptible to MFEs according to the triplet mechanism or the Δg-mechanism. Since these effects use to be comparatively small, however, or require rather extreme SOC, they are probably difficult to detect. In the case of the TH$^+$/*p*-I-An reactant pair there is an indication of decreasing yield of free radicals at high fields (cf. Figure 59). Since we did not see a corresponding effect on k_{rec} it was attributed to the efficiency of RP formation and could be due to either the TM or the Δg-type RPM, whereby the field dependence is rather in favor of the latter (cf. the corresponding investigation in homogeneous solution, Section III.A.2, and Reference 191).

Whereas there are many time-resolved studies of magnetokinetic effects with singlet RPs or singlet biradicals in homogeneous solution or in photosynthetic reaction centers, there is only one example of such a case for micellar systems, studied in our laboratory.[332] Here single RPs were produced in nanometer-sized water droplets through electron transfer quenching of singlet excited oxonine (OxH$^+$(*10d*) by *N*,*N*,*N'*,*N'*-tetramethylparaphenylenediamine (TMPDA). In the microemulsion the yield of radicals detected right after the laser pulse is about half that of a homogeneous solution, implying that there is a considerable amount of very fast geminate recombination. After the laser pulse a radical decay with a first-order rate constant of $1.1*10^6$ s^{-1} is observed (cf. Figure 60). It must also be attributed to intramicellar recombination. There is a strong absorption decaying during the laser pulse. It is mainly due to S$_1$-S$_n$ absorption of the excited dye but it may contain some contribution from rapidly recombining singlet RPs.

When applying magnetic fields two effects become apparent:

1.   The yield of radicals detectable right after the laser pulse decreases with a field dependence characterized by $B_{1/2} \approx 80$ G and a saturating field of about 500 G.
2.   The effective rate constant of the time resolved recombination process remains fairly constant up to fields of about 500 G, but then increases by a factor of 1.8 when the field is raised to 1 T.

Obviously the S-RP recombination occurs in two stages: the first one within a few nanoseconds during the laser pulse and the second one within about 1 μs. Though with different characteristics, in principle both processes are accelerated by a magnetic field. This is in accord with the singlet multiplicity of the originating RP. For such a case any spin equilibration, being due to coherent or incoherent ISC processes, will decrease the singlet probability and thus the recombination probability in an encounter. The different magnetic field dependences of both recombination stages are related to their different time scales. During the fast recombination stage of a few nanoseconds only coherent hyperfine-induced ISC can produce efficient S $\rightarrow$ T transitions. Rather low fields are sufficient to suppress this type of ISC process. During the second stage of recombination of about 1 μs spin relaxation can take over the role of the S $\rightarrow$ T ISC mechanism even at fields where the coherent process is suppressed. Higher fields are required to sufficiently slow down S $\rightarrow$ T$_{\pm}$ relaxation. Whereas in zero-field the recombination rate constant should correspond to 25% singlet character (S,T$_0$,T$_{\pm}$ mixture), in high fields it should correspond to 50% singlet character (S,T$_0$ mixture), implying a maximum magnetokinetic effect of 2 which is close to the observed one of 1.8.

In order to account for the two different domains of recombination we discussed two possibilities for assigning the first one: (1) "free" diffusion of the radicals in the interior of the water pool and (2) persistence of local correlation of the initial RP before it distributes randomly over the polar interface of the nanodroplet. Evidence against the first one came from an investigation with varying nanodroplet size, when it was found that the fraction of radicals involved in the fast recombination stage did not depend on the nanodroplet size, as should have been the case if interpretation (1) is correct.

Thus we are left with interpretation (2) which is actually a two-dimensional analogue to geminate and homogeneous recombination in both solvents. However, the present two-dimensional "homogeneous" recombination is still recombination of the original pair, characterized by a quasi-stationary distribution of distances over the sphere represented by the polar interface within one nanodroplet.

V. MAGNETOKINETICS AND MAGNETIC RESONANCE

So far our treatise of MFEs in photochemistry has dealt with the application of static external magentic fields only. A review of this kind would, however, fall short without touching on the effects of resonant magnetic fields, too. In fact, the methods of magnetokinetics span a continuous range between conventional chemical kinetics and traditional magnetic resonance (cf. Figure 61).

The essential requirement of magnetically sensitive reactions is the occurrence of intermediates with more than one unpaired electron spin. Since this is the usual property of electronically excited states, too, photochemical reactions are particularly disposed to exhibit magnetokinetic effects. These are usually borne out as MAgnetic effects on Reaction Yields, wherefore the acronym MARY has been coined.[333] Correspondingly, the function of R(B$_0$), R denoting the relative change of the yield, may be called a "MARY spectrum". As a consequence of the interaction of electronic and nuclear spins and of spin-selective reactivity in RPs, the population of electronic-nuclear spin substates may deviate strongly from thermal equilibrium, which causes exceptional line intensities (polarization) in the ESR spectra of the radicals (CIDEP) and in the NMR spectra of their diamagnetic products (CIDNP)[10,32] if the reactions are carried out in a magnetic field. In CIDEP and CIDNP the function of the resonant magnetic fields is only to probe the magnetokinetic effects leading to the unusual spin level populations. If, however, strong microwave[334] or radiofrequency fields[335,336] are applied, they may have a significant influence on the population (and coherence) of the magnetic sublevels during the lifetime of the magnetokinetically active intermediates and

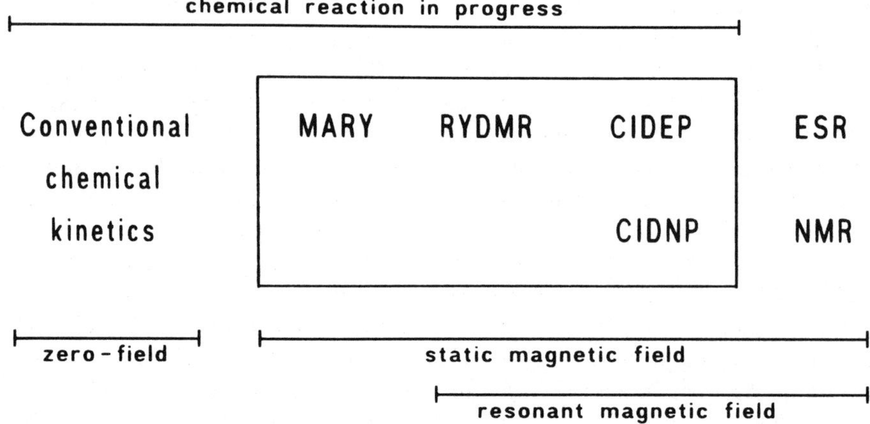

FIGURE 61.   Scope of magnetokinetics. The "magnetokinetic" methods are listed in the box.

thus exhibit effects on their reactivity. Under such conditions the magnetic resonance spectrum may be detected by observing reaction yields as a function of resonant field frequency, or as usual in magnetic resonance, by scanning B_0 and keeping the frequency of the B_1-field fixed. This technique is called Reaction Yield Detected Magnetic Resonance (RYDMR;[337] 219for other acronyms cf. Reference 1).

To complete the list of magnetokinetic effects one should finally mention that population inversions of spin levels created by magnetokinetically active reactions can be used to drive the coherent emission of radiofrequency power, i.e., to realize a chemical RASER.[338]

The principle of detecting magnetic resonance by indirect but more sensitive methods has been known since long from Optically Detected Magnetic Resonance (ODMR). However, in these cases the species wherein the magnetic resonance transition occurred and the species detected by luminescence are the same. By definition, in RYDMR methods a chemical transformation has to occur between magnetic resonance transition and probing, which can be done by optical (emission, absorption) or other methods (e.g., photoconductivity or spin trapping; cf. below).

The principle of RYDMR spectroscopy may be explained with reference to Figure 62 pertaining to the case of a RP. First one has to recall that application of a static magnetic field causes a MARY effect. If, for example, we assume that the RP is produced in the singlet state and that the products of the singlet reaction channel shall be monitored, then the MARY effect will correspond to a signal enhancement (positive MARY effect), since the channels of $T_\pm$ reactions are decoupled from the initial state as a consequence of the Zeeman splitting. This destructive effect of the static magnetic field may be partially compensated by tuning a resonant microwave field to the $T_\pm \rightarrow T_0$ transitions. Thus the RYDMR spectrum (measured as usual in magnetic resonance by applying a fixed microwave frequency and scanning the B_0-field through the resonance, therefore it will be denoted as a RYDMR-B_0 spectrum) may be conceived as a resonant compensation of the MARY signal (cf. the example given at the end of this section). From the informational point of view a RYDMR-B_0 spectrum represents the ESR spectrum of the magnetokinetically active species, i.e., RPs, TT-pairs, or TD-pairs. The method is highly selective and discriminates the ESR contribution of reactive pairs from a possibly huge background of single paramagnetic particles. It is also highly sensitive. For example, in the case of recombining RPs produced by high energy radiation stationary concentrations of only 20 RPs in the sample have been detected[339,340] by monitoring the radioluminescence.

Since the lifetime of magnetokinetically active transients as RPs may be as short as a few nanoseconds, the microwave powers, conveniently expressed in terms of B_1-field strength,

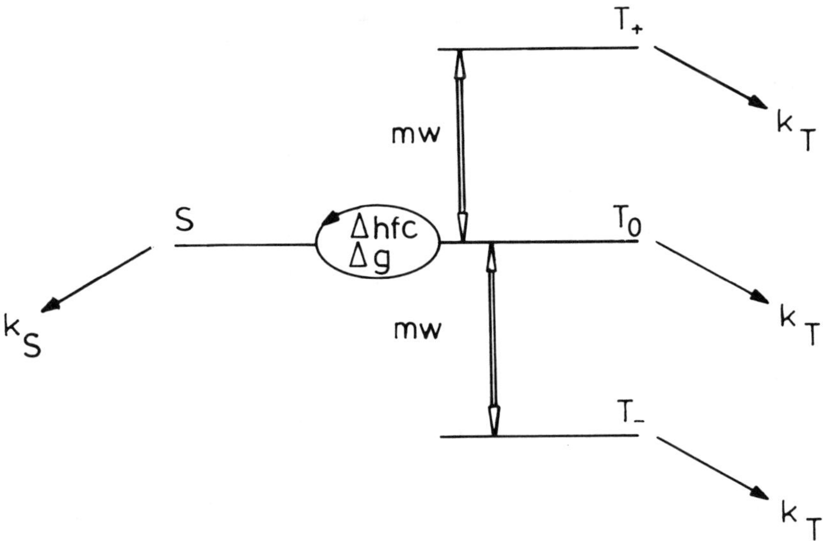

FIGURE 62.   Reaction scheme of RYDMR in a radical pair. Transitions denoted by mw
are due to resonant interaction with a microwave field.

required to achieve detectable kinetic effects are considerably higher than in conventional
ESR. Typical B_1 values are on the order of several Gauss (note that for $B_1 = 1$ G the Rabi
frequency characterizing the time for an electron spin flip is $17.6*10^6$ rad*s^{-1}), whereas in
conventional cw ESR B_1 is usually below 0.05 G in order to avoid saturation effects.

Very recently a most illuminating experiment has been reported by the Novosibirsk
group[341] who succeeded in directly observing coherent $T_0 \rightarrow T_{\pm}$ transitions induced by a
resonant B_1-field. The process giving rise to the observed signal was fluorescent recombination of p-terphenyl cation/anion RPs produced as singlet pairs in squalane solution excited
with β^- radiation from a ^{90}Sr source. The basic idea of the experiment is sketched in Figure
63. The RPs are shown in a $\beta_1\alpha_2$ constellation with the transverse components oscillating
at a frequency $\Delta\omega$ between out-of-phase and in-phase representing periodic S/T_0 transitions.
(For a critical remark on this representation cf. Section II.A). The difference $\Delta\omega$ of precession
frequencies is due to different actual hyperfine fields at the two radical spins. If a microwave
field of frequency ω_0 is applied which is in resonance with the radical 1, for example, then
the transverse component of spin 1 becomes stationary in the rotating frame where the
microwave B_1-vector is at rest, whereas spin 2 precesses with a frequency of $\Delta\omega$. On the
other hand, the transversal B_1-field induces a precession of spin 1 with frequency $\omega_1 =
g\beta B_1/\hbar$, causing periodic transitions of the RP between $\beta_1\alpha_2$ (T_0/S) and $\alpha_1\alpha_2$ (T_+). As long
as the coherence is maintained the fluorescence intensity which is a measure of the instant
singlet probability should exhibit quantum beats at the frequency ω_1. Such a behavior has,
in fact, been detected (Figure 63d), with the expected proportionality between beat frequency
and B_1. The RYDMR-B_1 quantum beat is, in fact, the analogue to transient nutations (Torrey
oscillations) well known from nuclear magnetic resonance. Its observability requires that
$\Delta\omega > \omega_1$. If the B_1-field becomes too strong, i.e., $\omega_1 > \Delta\omega$, no separate resonance of the
two radicals is possible. *A fortiori*, in the latter case both spins precess coherently around
B_1. They become locked (in T or S) and hyperfine-induced $T_0 \leftrightarrow S$ transitions can no longer
occur. If such conditions prevail the RYDMR effect acts to enhance the MARY effect.
Therefore as B_1 is systematically increased (RYDMR-B_1 spectrum) from low to high values
the RYDMR intensity will first increase with opposite sign of the MARY effect, then
decrease, change its sign, and increase again with the same sign as the MARY effect.

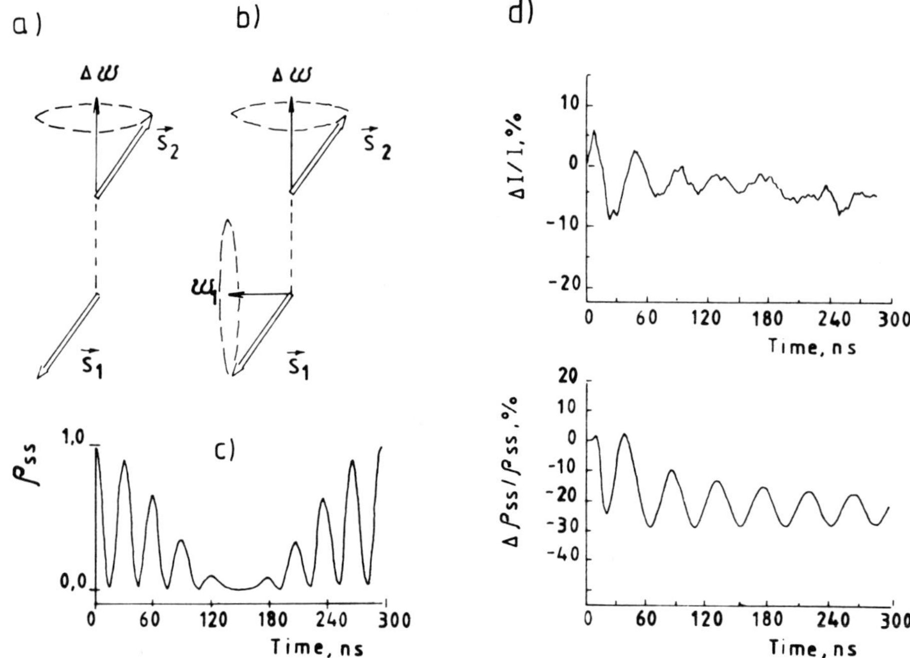

FIGURE 63.   Observation of RYDMR quantum beats in a resonant B_1-field. (a) Rephasing mechanism of S/T_0 mixing in a static magnetic field; (b) mechanism of coherent $S/T_0 \rightarrow T_{\pm}$ transitions due to Rabi precession in resonant B_1-field; (c) time dependence of singlet probability under precessional motions according to (b); (d) top: experimental observation of rf-field-induced quantum beats in radioluminescence from p-terphenyl-h_{14} solution in squalane. (B_1 = 8.0 ± 1 G); bottom: theoretical simulation of beats corresponding to Rabi precession. (Reprinted from Saik, V. O., Anisimov, O. A., Koptyug, A. V., and Molin, Yu. N., *Chem. Phys. Lett.*, 165, 142, 1990. With kind permission of Yu. N. Molin; copyright 1990, Elsevier Science Publishers, B. V.)

RYDMR-B_1 spectra are particularly valuable for investigating RPs with a fixed exchange interaction. The RYDMR effects exhibited by such systems are comparable to the MARY effects of biradicals observed at low fields where one of the Zeeman levels (usually T_-) crosses the S level, split away from T_0 by 2 J (cf. Figure 7). In a resonant RYDMR-B_1 field the T_0 state is distributed over two states t_+ and t_-, quantized parallel and antiparallel to B_1 in the rotating frame, that are increasingly split up as B_1 increases. A level crossing of t_- with S occurs at some B_1 corresponding to a maximum (anti-MARY) RYDMR effect. At higher B_1 the RYDMR intensity decreases with final sign inversion and enhancement of the MARY effect. For obtaining a maximum of information the two-dimensional (B_0,B_1) RYDMR spectrum should be considered.[333]

The feasibility of RYDMR experiments was first investigated theoretically by Kubarev and Pshenichnov.[342] The first experimental observation of a RYDMR-B_0 spectrum was reported by Frankevich and Pristupa[343] for the charge transfer fluorescence of rubrene crystals doped with rubrene peroxide as an electron acceptor. Since then the technique has been applied to various magnetokinetic phenomena in the solid state (mainly in connection with the luminescence of organic molecular crystals; cf. References 337, and 344 to 347), and in liquid solutions with applications to radioluminescence (cf. References 340, and 348 to 351) or laser flash photolysis.[352,353] We do not intend to give a survey here (for systematic reference one may consult previous reviews),[1,334] but will only touch upon some recent reports that have direct relations to certain aspects emphasized elsewhere in this contribution.

In the high resolution MARY spectra obtained with the magnetic field modulation method[119] (cf. Section III.A) a particular low-field feature reminiscent of a level crossing phenomenon has been noted when observing the exciplex fluorescence intensity from geminate pyrene$^-\cdot$/dimethylaniline$^+\cdot$ RP recombination. Two explanations have been considered: (1) T_--S level crossing based on a 2J exchange energy splitting of T_0 and S or (2) increase of $T \to S$ ISC at low fields due to the breakdown of the conservation rule for total (electronic + nuclear) spin, valid in zero-field. Very recently the RYDMR-B_1 spectra have been reported for the same system and also for the RPs Py$^+\cdot$/1,2-DCB$^-\cdot$, 1,3-DCB$^-\cdot$, and 1,4-DCB$^-\cdot$ (Py = pyrene (*1*), DCB = dicyanobenzene).[160] The spectra (cf. Figure 64) showed a corresponding low field feature with the initial RYDMR intensity in anti-MARY direction (indicating enhancement of ISC by the RYDMR transition), followed by a sign inversion. For Py$^-\cdot$/DMA$^+\cdot$ (DMA = dimethylaniline) the B_1 at the RYDMR-B_1 spectral minimum ($B_{1,min}$) was in good correspondence with the field B_{inv} where the sign inversion occurred in the modulated MARY spectrum (note that this point corresponds to a minimum in the normal MARY spectrum). For the Py$^+\cdot$/DCB$^-\cdot$ system the $B_{1,min}$ values were quite small (3, 6, and 3 G for 1,2-, 1,3- and 1,4-DCB, respectively). Corresponding low field features had not been possible to resolve with the modulated MARY technique.

The observed B_1 sign inversion features have been suggested to represent "J-resonances", i.e., to reflect a B_1-induced $t_-(T_0)$-S level crossing. (The possibility [2] discussed above cannot apply at the B_0-fields where the RYDMR spectra were observed.) From the $B_{1,min}$ values an effective exchange interaction in the RPs can be tentatively assigned. An alternative explanation for the RYDMR-B_1 spectral zero crossing would have been to relate it to an effective hfc strength. As noted above the point $\omega_1 \approx \Delta\overline{\omega}$ will be also characteristic of a sign inversion in the RYDMR effect. The characteristic hfc constants (B^{hfc} = 6.3, 10.4, and 4.6 G for 1,2-, 1,3-, and 1,4-DCB, respectively) of the DCB$^-\cdot$ radicals, although several times larger than the corresponding $B_{1,min}$ values, are in the same relative order. The main argument provided against an interpretation in terms of hfc was derived from an observed solvent polarity effect on $B_{1,min}$. Whereas the hfc constants under consideration should not really be sensitive to solvent polarity effects, such an effect is rather plausible for the exchange interaction, in that the average distance of the RIP can be expected to depend on the kind of solvation.

Independent evidence of a small but finite effective J detectable in RPs came from CIDEP experiments. As was first noted by Murai and co-workers,[354] the hyperfine lines of the CIDEP spectra observed shortly after a laser pulse, when photolyzing bezophenone in SDS micellar solution, showed unusual splittings into emission/absorption doublets. This feature disappeared within a few microseconds. The experiments have been repeated and reproduced by Closs et al.[355] (cf. Figure 65) who established it as an effect peculiar to micelles. At the same time, however, Buckley and co-workers[356] reported this type of phenomenon also for isopropyl ketyl T-RPs generated by acetone photolysis in isopropanol solution at 199 K. Independently, the latter two groups arrived at the same interpretation. The line splitting into an emission/absorption doublet was assigned to the effect of exchange interaction observed directly in the ESR spectrum of the T-RP and decaying in parallel with the diffusional separation of the latter. In Figure 66 we give a simplified schematic representation of the underlying principle. The energy represented corresponds to the Zeeman plus hyperfine energy of spin 1 (unprimed) and the exchange energy with spin 2 (primed). Such a simplified picture may be applied to the case where the exchange energy is small compared with the spacing between hyperfine transitions in different radicals. By exchange coupling to the second radical the two spin levels considered of the first radical are split. Since the RP is produced in its triplet state the populations of electron spin states $\alpha\alpha'$, $\alpha\beta'$, $\beta\beta'$, and $\beta\alpha'$ are $^1/_3$, $^1/_6$, $^1/_3$, and $^1/_6$. Consequently, each hyperfine transition in the ESR spectrum of radical 1 will be split by 2 J, one component appearing in emission and one in absorption, whereby a negative sign of J (singlet below triplet, this is the situation represented

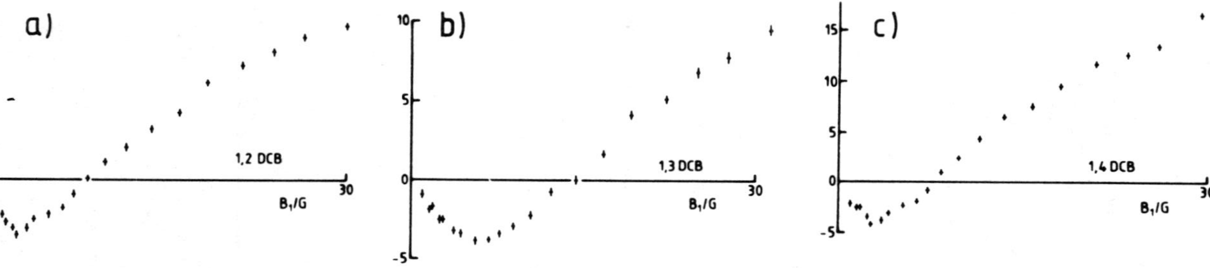

FIGURE 64.   RYDMR-B_1 spectra observed by probing exciplex fluorescence intensity from pyrene (*1*)/dicyanobenzene (DCB) solutions in cyclohexanol/acetonitrile. (a) 1,2-DCB; (b) 1,3-DCB; (c) 1,4-DCB. (Reprinted from Hamilton, C. A., McLauchlan, K. A., and Peterson, K. R., *Chem. Phys. Lett.*, 162, 145, 1989. With kind permission of K. A. McLauchlan; copyright 1989, Elsevier Science Publishers B. V.)

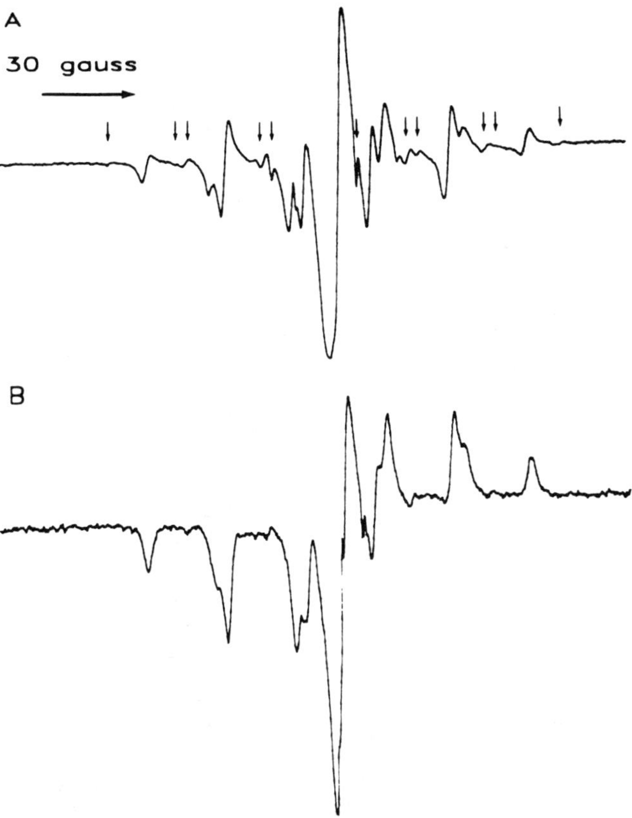

FIGURE 65. ESR spectra taken 0.8 μs (A) and 2.8 μs (B) after excitation of benzophenone-d_{10} in SDS micellar solution. The broad signal in the center is assigned to the benzophenone ketyl radical. Lines attributed to the minor component $H_3C-\overset{\cdot}{C}H-CH_2-$ are marked by arrows. The major signals are due to radicals of the type $-CH_2-\overset{\cdot}{C}H-CH_2-$. Note that signals above the baseline are in absorption and below in emission. (Reprinted from Closs, G. L., Forbes, M. D. E., and Norris, J. R., Jr., *J. Phys. Chem.*, 91, 3592, 1987. With kind permission of G. L. Closs; copyright 1987, American Chemical Society.)

in Figure 66) corresponds to low-field emission and high-field absorption as was observed experimentally. The observed splittings correspond to exchange energies J in the order of 1 G.

In References 355 and 356 the observed CIDEP spectra were simulated by employing model spin Hamiltonians with time-independent J, either for a fixed effective value[355] or for the particular distance distribution that has developed at the instant of probing. On the other hand, it is clear, however, that, as with the well-known examples of biradicals, the diffusional motion of the RP can be adequately accounted for by a dynamic modulation of exchange interaction only. A theoretical analysis of the problem, recently presented by Shushin,[357] did not remove this shortcoming of the initial treatments, but reasoned that only for J fixed within rather narrow bounds the J-splitting feature in the CIDEP spectrum can be accounted for. From this analyis, Shushin obtained that for the experimental conditions where the J-splitting feature is observed the RP resides at a fixed distance for several hundreds of nanoseconds. Such a slow hopping dynamics seems difficult to accept as a reasonable picture of RP motion.

The above-mentioned conclusions about effective exchange interaction in RPs should

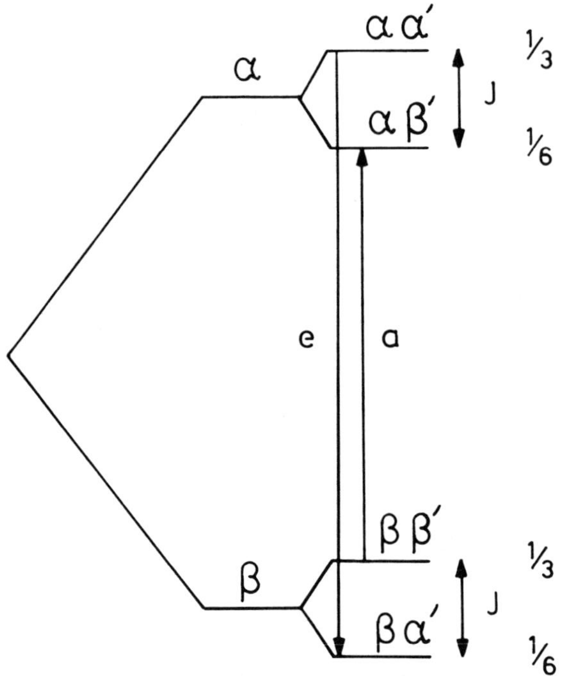

FIGURE 66.   Level scheme explaining the emission/absorption splitting of hyperfine lines in a triplet RP with weak exchange interaction J. A single hyperfine transition in the unprimed radical is considered. The numbers denote the level population probabilities in a RP with uniform triplet sublevel population. Transition e is observed in emission, a in absorption.

be compared with other information coming from magnetic field-dependent ^{13}C-CIDNP investigations of recombination products from micellar RPs.[358] These spectra exhibited B_{max} values of CIDNP intensity, that vary inversely with the size of the micelles and are comparable to the effective 2J values determined for chemically linked RPs. They are by two orders of magnitude larger than the CIDEP J-splittings. In conclusion, it seems that a complete understanding of the ''J-features'' observed in CIDEP and RYDMR of unlinked RPs has not been reached at present.

As a final, most remarkable example of a RYDMR observation we refer to a recent investigation by Okazaki et al.[359,360] These authors obtained RYDMR-B_0 spectra of intramicellar triplet RPs produced by continuous photolysis of menadione (*57*) or anthraquinone in SDS micellar solution. The yield of escape radicals (SDS·) was determined with the help of spin traps (PBN = phenyl-*tert*-butyl nitrone, or DMNS = perdeutero-2,4-dimethyl-3-nitrosobenzene sulfonate) by which the SDS radicals were transformed to stable nitroxyl radicals (PBN-SDS (*58*), DMNS-SDS (*59*)).

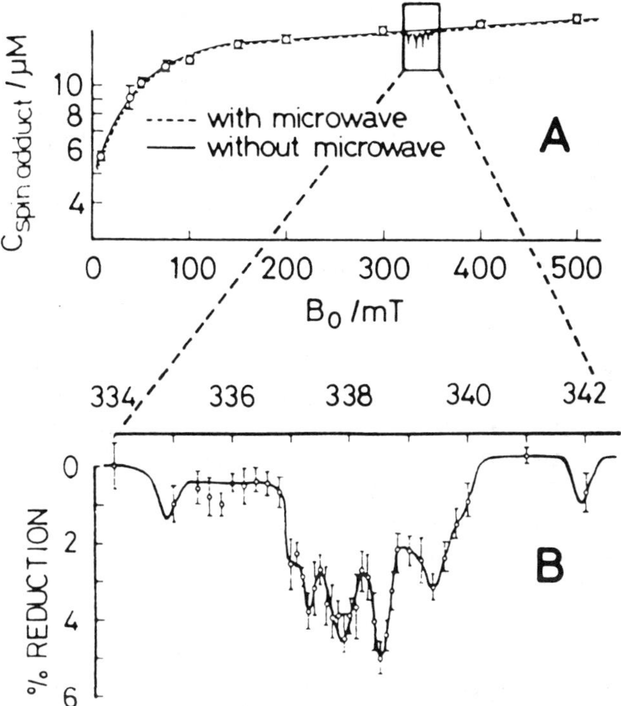

FIGURE 67. (A) Magnetic field dependence of spin adduct yield from menadion (*57*) photoreduction in SDS micellar solution in the presence of spin trap DMNS (*59*) with an X-band microwave field off or on. (B) Shows the relative effect of the microwave field on an enlarged scale. It corresponds to the RYDMR spectrum of the menadione-semiquinone/SDS·RP. (Reprinted from Okazaki, M, Sakata, S., Konaka, R., and Shiga, T., *J. Chem. Phys.*, 86, 6792, 1987. With kind permission of T. Shiga; copyright 1987, American Institute of Physics.)

The samples were photolyzed for 30 s in the cavity of the ESR spectrophotometer with the field B_0 set to variable values and with the B_1-field on or off. After another 60 s in the dark the ESR spectrum of the spin-trapped radicals was recorded, its amplitude providing a measure of the yield of escape radicals. In Figure 67 the result for the case of menadione photolysis and usage of the DMNS spin trap is illustrated. In part A of the figure the MARY spectrum indicates the increase of escape yield in a magnetic field. Superimposed on it is the MARY spectrum measured with B_1 (at X-band frequency) switched on. The difference between both, notable only around 0.34 T corresponding to an X-band resonance field with $g \approx 2$, represents the RYDMR-B_0 spectrum, shown in enlarged form in part B of the figure. From this representation it is clear that the RYDMR effect observed corresponds to a resonant anti-MARY effect. The structure of the RYDMR spectrum corresponds to the superposition of the ESR spectra of an SDS and a semiquinone radical. As has been worked out theoretically[360] the situation determining shape and sign of the RYDMR spectra observed is well described by the limits $J < \delta\omega_{hfc}$ (the hyperfine linewidth) and $B_1 \approx 1$ G $< \Delta\Delta\omega_{hfc}$ (the spacing between hyperfine transitions), since there is no indication of J-splitting of the hyperfine lines and the microwave power does not wipe out the hyperfine structure.

Another report of RYDMR-B_0 and -B_1 spectra from micellar RPs obtained with time-resolved optical probing has been given by McLauchlan and co-workers.[284] In this case the hyperfine structure of the radicals was only weakly resolved. A sign inversion of RYDMR intensity has not been observed even with the highest B_1 (≈ 13 G) applied.

VI. CONCLUSION

What is most intriguing in MFEs on chemical kinetics is that very weak perturbations do, in fact, control the course and rate of conversions of much higher chemical energy. Such effects become possible due to the spin-selective nature of chemical reactions and the finite rate at which different spin states, even with energy separations much less than kT, are converted into each other, so that spin motion, controllable by external or internal magnetic fields, may become a rate-determining step in the overall chemical process. Since, on the other hand, the details of spin motion depend also in a specific way on the molecular and chemical dynamics of the reaction intermediates involved (radical pairs, excited triplet states), information on the characteristics of this dynamics is borne out in the magnetic field dependence of magnetokinetic effects. Hence such effects represent a unique diagnostic tool to look deeper into the details of chemical reactions involving these types of intermediates.

The research work done so far in the field of magnetokinetics has been mainly carried out in such a diagnostic sense, whereby aiming for the development of systems where the MFEs are optimized and expressed most clearly and informatively and trying to extract the inherent dynamic information by developing and applying suitable theoretical models has had, and still has, predominance over routine mechanistic applications.

On the other hand, although the large effects in micellar systems and in chemically linked biradicals are rather suggestive, practical applications of magnetic fields are, so far, rare. The control of emulsion photopolymerization[314-316] by magnetic fields is probably the most prominent example. The possibility of isotope enrichments by taking advantage of the magnetic isotope effect with or without an external field is also noteworthy. There ought to be a challenge to exploit this mechanism for magnetic isotopes other than ^{13}C for which it has been studied most extensively. Applications up to elements including the third period of the periodic system should be possible.

Photochemical MFEs might also be involved in biomagnetic phenomena. Thus evidence has been provided for the involvement of the visual system in mediating MFEs on pineal melatonin* synthesis in the rat. It is remarkable that very small changes (≤ 1.0 G) in the order of the earth magnetic field are responsible for these effects.[361] Influences of low magnetic field-strength variations on the retina and pineal gland of quail and humans have also been reported.[362]

The most important intermediates exhibiting magnetokinetic sensitivity are RPs that may be generated photochemically by electron transfer, H-atom transfer, or homolytic bond cleavage reactions. Exciplexes that may be considered as some form of contact RPs are magnetokinetically active only in the triplet state, provided they have strong internal SOC. A special case of RPs are those involving Kramers doublet species with strong SOC.

The most important interactions controlling the spin dynamics are isotropic hyperfine coupling, exchange interaction, and SOC. Since hyperfine coupling data of the radicals involved can be obtained from independent sources, in mechanistic interpretations of kinetic MFEs spin motion due to hyperfine coupling usually plays the role of an internal clock providing an absolute time scale to calibrate the rates of other dynamic processes in the mechanism. There is, however, also the possibility to determine the hyperfine structure by the magnetokinetic RYDMR technique and assign thereby the nature of the radicals involved in RP intermediates.

Exchange interaction causes an energy splitting of spin states belonging to sets of different multiplicity and, in combination with a magnetic field, may give rise to level crossing effects (J-resonances). Due to the stochastic variation of RP separation and the exponential distance dependence of exchange interaction, its effect is a dynamic one. Dynamic exchange effects have been studied in great detail with chemically linked RPs providing information on the

---

* Melatonin is a hormone thought to be involved in the control of the mammalian circadian system.

parameters of the exponential distance dependence of J and the folding dynamics of short polymer chains. On the other hand, with unlinked RPs in viscous homogeneous solution or in micelles distinct J-resonances have been detected at very low fields. The implications of these latter effects with respect to the time dependence of RP separation are not yet fully understood. In a normal RP the effects of SOC, too, decrease strongly with RP separation, because SOC can only induce multiplicity changes if they are combined with a change in the orbital state. For this to occur the RP must be in close contact. Only in RPs involving strongly spin-orbit mixed Kramers doublet species can this type of interaction couple states with different total spin in the separated RP.

Spin-orbit coupling-induced reactions in RPs are usually not sensitive to magnetic fields. Therefore the amplitude of MFEs may be utilized to assess the relative contributions of SOC-induced processes and of those involving spin motion in the separated RP. In case of strong SOC specific MFEs may ensue in triplet exciplexes due to the triplet substate selectivity of SOC (triplet mechanism) or in separated RPs due to the Δg-type radical pair mechanism.

In highly viscous homogeneous solutions and, particularly, in micellar media providing for long cage times of geminate RPs their spin correlation can be probed over fairly long periods of time (up to several microseconds). Under such conditions incoherent spin motion, i.e., spin relaxation, becomes part of the overall reaction kinetics. Although considerable progress has been made in the analysis of the magnetic field dependence of magnetokinetic effects under such conditions, their information potential has not yet been fully exploited.

In concluding we note that, today, the kit of magnetokinetic methods represents a continuous range of techniques linking the conventional fields of chemical kinetics and magnetic resonance spectroscopy. Thereby, the relation between chemical kinetics and spectroscopy has become more symmetric. In magnetokinetics either of the two approaches may take the role of the probe technique for investigating problems traditionally considered as typical for the other field. Thus, whereas formerly the use of magnetic resonance has served chemical kinetics as a means to probe the time dependence of concentrations, in the magnetokinetic RYDMR technique chemical kinetics is used to probe spectroscopic transitions. The development of magnetokinetics has certainly helped to deepen our understanding of many elementary aspects of chemical reactions, particularly photochemical ones. It may be anticipated that it will continue to do so during the near future.

ACKNOWLEDGMENTS

We wish to thank Jeannette Dreher for her qualified help with the figures. The financial support of our research work by the Deutsche Forschungsgemeinschaft and the Fonds der Chemischen Industrie is gratefully acknowledged.

REFERENCES

1. **Steiner, U. E. and Ulrich, T.,** *Chem. Rev.,* 89, 51, 1989.
2. **Bhatagnar, S. S., Mathur, R. N., and Kapur, R. N.,** *Philos. Mag.,* 8, 457, 1929.
3. **Sokolik, I. A. and Frankevich, E. L.,** *Usp. Fiz. Nauk,* 111, 261, 1973.
4. **Swenberg, Ch. E. and Geacintov, N. E,.** *Organic Molecular Photophysics,* Vol. 1, Birks, J. B., Ed., John Wiley & Sons, New York, 1973, 489.
5. **El-Sayed, M. A.,** *Adv. Photochem.,* 9, 311, 1974.
6. **Hausser, K. H. and Wolff, H. C.,** *Adv. Magn. Reson.,* 8, 85, 1976.
7. **Zel'dovich, Ya. B., Buchachenko, A. L., and Frankevich, E. L.,** *Sov. Phys. Usp.,* 31, 385, 1988.
8. **Ward, H. R.,** *Acc. Chem. Res.,* 5, 18, 1972.
9. **Lawler, R. G.,** *Acc. Chem. Res.,* 5, 25, 1972.

10. **Lepley, A. R. and Closs, G. L.,** *Chemically Induced Magnetic Polarization,* John Wiley & Sons, London, 1978.
11. **Sagdeev, R. Z., Molin, Yu. N., Salikhov, K. M., Leshina, T. V., Kamha, M. A., and Shein, S. M.,** *Org. Magn. Reson.,* 5, 603, 1973.
12. **Brocklehurst, B.,** *Chem. Phys. Lett.,* 28, 357, 1974.
13. **Brocklehurst, B., Dixon, R. S., Gardy, E. M., Lopata, V. J., Quinn, M. J., Singh, A., and Sargent, F. P.,** *Chem. Phys. Lett.,* 28, 361, 1974.
14. **Schulten, K., Staerk, H., Weller, A., Werner, H.-J., and Nickel, B.,** *Z. Phys. Chem. (Frankfurt/Main),* 101, 371, 1976.
15. **Michel-Beyerle, M. E., Haberkorn, R., Bube, W., Steffens, E., Schröder, H., Neusser, H. J., and Schlag, E. W.,** *Chem. Phys.,* 17, 139, 1976.
16. **Tanimoto, Y., Hayashi, H., Nagakura, S., Sakuragi, H., and Tokomaru, K.,** *Chem. Phys. Lett.,* 41, 267, 1976.
17. **Hata, N.,** *Chem. Lett.,* p.547, 1976.
18. **Buchachenko, A. L., Galimov, E. M., and Ershov, V. V.,** *Dokl. Akad. Nauk SSSR,* 228, 379, 1976.
19. **Sagdeev, R. Z., Leshina, T. V., Kamkha, M. A., Belchenko, O. L., Molin, Yu. N., and Rezvukhin, A. J.,** *Chem. Phys. Lett.,* 48, 89, 1977.
20. **Steubing, W.,** *Verh. Dtsch. Phys. Ges.,* 15, 1181, 1913.
21. **Franck, J. and Grotrian, W.,** *Z. Phys.,* 6, 35, 1921.
22. **Van Vleck, J. H.,** *Phys. Rev.,* 40, 544, 1932.
23. **Weast, R. C., Ed.,** *CRC Handbook of Chemistry and Physics,* 58th ed., CRC Pres. Boca Raton, FL, 1978.
24. **Boxer, S. G., Goldstein, R. A., and Franzen, S.,** *Photoinduced Electron Transfer,* Part B, Fox, M. A. and Chanon, M., Eds., Elsevier, Amsterdam, 1988, 163.
25. **Noyes, R. M.,** *J. Chem. Phys.,* 22, 1349, 1954.
26. **Kaptein, R.,** *J. Am. Chem. Soc.,* 94, 6251, 1972.
27. **Schulten, Z. and Schulten, K.,** *J. Chem. Phys.,* 66, 4616, 1977.
28. **Turro, N. J. and Kräutler, B.,** *Diradicals,* Borden, W. T., Ed., John Wiley & Sons, New York, 1982, 259.
29. **Einstein, A., Podgolsky, B., and Rosen, N.,** *Phys. Rev.,* 47, 777, 1935.
30. **Schulten, K. and Wolynes, P.,** *J. Chem. Phys.,* 68, 3292, 1978.
31. **Blum, K.,** *Density Matrix Theory and Applications,* Plenum Press, New York, 1981.
32. **Salikhov, K. M., Molin, Yu. N., Sagdeev, R. Z., and Buchachenko, A. L.,** *Spin Polarization and Magnetic Effects in Radical Reactions,* Elsevier, Amsterdam, 1984.
33. **Knapp, E.-W. and Schulten, K.,** *J. Chem. Phys.,* 71, 1878, 1979.
34. **Schlenker, W., Ulrich, T., and Steiner, U. E.,** *Chem. Phys. Lett.,* 103, 118, 1983.
35. **Haberkorn, R.,** *Chem. Phys.,* 24, 111, 1977.
36. **Werner, H.-J., Schulten, Z., and Schulten, K.,** *J. Chem. Phys.,* 67, 646, 1977.
37. **Lüders, K. and Salikhov, K. M.,** *Chem. Phys.,* 117, 113, 1987.
38. **Carrington, A. and McLachlan, A. D.,** *Introduction to Magnetic Resonance,* Chapman and Hall, London, 1979.
39. **Atkins, P. W. and Kivelson, D.,** *J. Chem. Phys.,* 44, 169, 1966.
40. **Hubbard, P. S.,** *Phys. Rev.,* 131, 1155, 1963.
41. **Hayashi, H. and Nagakura, S.,** *Bull. Chem. Soc. Jpn.,* 57, 322, 1984.
42. **Ulrich, T. and Steiner, U. E.,** *Chemical Phys. Lett.,* 112, 365, 1984.
43. **Ulrich, T., Steiner, U. E., and Schlenker, W.,** *Tetrahedron,* 42, 6131, 1986.
44. **Levin, P. P. and Kuzmin, V. A.,** *Chem. Phys. Lett.,* 165, 302, 1990.
45. **De Kanter, F. J. J. and Kaptein, R.,** *J. Am. Chem. Soc.,* 104, 4759, 1982.
46. **Schulten, K. and Bittl, R.,** *J. Chem. Phys.,* 84, 5155, 1986.
47. **Burri, J. and Fischer, H.,** *Chem. Phys. Lett.,* 139, 497, 1989.
48. **Salikhov, K. M., Molin, Yu. N., Sagdeev, R. Z., and Buchachenko, A. L.,** *Spin Polarization and Magnetic Effects in Radical Reactions,* Elsevier, Amsterdam, 1984, 71.
49. **Bittl, R. and Schulten, K.,** *J. Chem. Phys.,* 90, 1794, 1989.
50. **Kubo, R.,** *Fluctuation, Relaxation and Resonance in Magnetic Systems,* ter Haar, D., Ed., Oliver and Boyd, Edinburgh, 1962, 23.
51. **Busmann, H.-G., Staerk, H., and Weller, A.,** *J. Chem. Phys.,* 91, 4098, 1989.
52. **Adrian, F. J.,** *J. Chem. Phys.,* 53, 3374, 1970.
53. **Adrian, F. J.,** *J. Chem. Phys.,* 54, 3912, 1971.
54. **Adrian, F. J.,** *J. Chem. Phys.,* 54, 3918, 1971.
55. **Pedersen, J. B.,** *J. Chem. Phys.,* 67, 4097, 1977.
56. **Salikhov, K. M.,** *Theor. Exp. Chem.,* 13, 551, 1977.
57. **Vollenweider, J.-K. and Fischer, H.,** *Chem. Phys.,* 124, 333, 1988.

58. **Haberkorn, R.,** *Mol. Phys.,* 32, 1491, 1976.
59. **Salikhov, K. M., Sarvarov, F. S., Sagdeev, R. Z., and Molin, Yu. N.,** *Kinet. Katal.,* 16, 279, 1975.
60. **Tang, J. and Norris, J. R.,** *Chem. Phys. Lett.,* 92, 136, 1982.
61. **Boxer, S. G., Chidsey, C. E. D., and Roelofs, M. G.** *Proc. Natl. Acad. Sci. U.S.A.,* 79, 4632, 1982.
62. **Mints, R. G. and Pukhov, A. A.,** *Chem. Phys.,* 87, 467, 1984.
63. **Pedersen, J. B. and Freed, J. H.,** *J. Chem. Phys.,* 59, 2869, 1973.
64. **Bittl, R. and Schulten, K.,** *Chem. Phys. Lett.,* 173, 387, 1990.
65. **Pedersen, J. B. and Freed, J. H.,** *J. Chem. Phys.,* 61, 1517, 1974.
66. **Schulten, K. and Windemuth, A.,** *Biophysical Effects of Steady Fields (Springer Proc. Phys.,* Vol. 11), Maret, G., Kiepenheuer, J., and Boccara, N., Eds., Springer-Verlag, Berlin, 1986, 99.
67. **Schulten, K., Swenberg, Ch. E., and Weller, A.,** *Z. Phys. Chem. (Frankfurt/Main),* 111, 1, 1978.
68. **Zientara, G. P. and Freed, J. H.,** *J. Phys. Chem.,* 83, 3333, 1979.
69. **Shushin, A. I.,** *Chem. Phys. Lett.,* 130, 452, 1986.
70. **Shushin, A. I.,** *Chem. Phys. Lett.,* 140, 111, 1987.
71. **Shushin, A. I.,** *Mol. Phys.,* 64, 65, 1988.
72. **Salikov, K. M.,** *Teor. Eksp. Khim.,* 13, 732, 1977.
73. **Pedersen, J. B. and Freed, J. H.,** *J. Chem. Phys.,* 57, 1004, 1972.
74. **Pedersen, J. B. and Freed, J. H.,** *J. Chem. Phys.,* 58, 2746, 1973.
75. **Pedersen, J. B. and Freed, J. H.,** *J. Chem. Phys.,* 62, 1790, 1975.
76. **Freed, J. H.,** *Chemically Induced Magnetic Polarization,* Muus, L. T., Ed., Reidel, Dordrecht, 1977, 309.
77. **De Kanter, F. J. J., den Hollander, J. A., Huizer, A. H., and Kaptein, R.,** *Mol. Phys.,* 34, 857, 1977.
78. **de Kanter, F. J. J., Kaptein, R., and van Santen, R. A.,** *Chem. Phys. Lett.,* 45, 575, 1977.
79. **de Kanter, F. J. J. and Kaptein, R.,** *Chem. Phys. Lett.,* 58, 340, 1978.
80. **de Kanter, F. J. J., Sagdeev, R. Z., and Kaptein, R.,** *Chem. Phys. Lett.,* 58, 334, 1978.
81. **Bittl, R. and Schulten, K.,** *Chem. Phys. Lett.,* 146, 58, 1988.
82. **Evans, G. T., Fleming, P. D., III, and Lawler, R. G.,** *J. Chem. Phys.,* 58, 2071, 1973.
83. **Shushin, A. I.,** *Chem. Phys. Lett.,* 85, 562, 1982.
84. **Evans, G. T.,** *Mol. Phys.,* 31, 777, 1976.
85. **Sarvarov, F. S. and Salikhov, K. M.,** *React. Kinet. Catal. Lett.,* 4, 33, 1976.
86. **Minaev, B. F.,** *Opt. Spectrosc. (U.S.S.R.),* 44, 148, 1978.
87. **Haberkorn, R.,** *Chem. Phys.,* 26, 35, 1977.
88. **Haberkorn, R. and Michel-Beyerle, M. E.,** *Biophys. J.,* 26, 489, 1979.
89. **Werner, H. J., Schulten, K., and Weller, A.,** *Biochim. Biophys. Acta,* 502, 255, 1978.
90. **Haberkorn, R., Michel-Beyerle, M. E., and Marcus, R. A.,** *Proc. Natl. Acad. Sci. U.S.A.,* 76, 4185, 1979.
91. **Hoff, A. J. and Hore, P. J.,** *Chem. Phys. Lett.,* 108, 104, 1984.
92. **Chidsey, C. E. D., Takiff, L., Goldstein, R. A., and Boxer, S. G.,** *Proc. Natl. Acad. Sci. U.S.A.,* 82, 6850, 1985.
93. **Goldstein, R. A. and Boxer, S. G.,** *Biophys. J.,* 51, 937, 1987.
94. **Salikhov, K. M. and Mikhailov, S. A.,** *Theor. Exp. Chem.,* 19, 508, 1983.
95. **Geacintov, N. E. and Swenberg, C. E.,** *Luminescence Spectroscopy,* Lumb, M. D., Ed., Academic Press, London, 1978, 239.
96. **Merrifield, R. E.,** *J. Chem. Phys.,* 48, 4318, 1968.
97. **Bittl, R., Schulten, K., and Turro, N. J.,** *J. Chem. Phys.,* 93, 8260, 1990.
98. **Tarasov, V. F., Shkrob, I. A., Step, E. N., and Buchachenko, A. L.,** *Chem. Phys.,* 135, 391, 1989.
99. **Schwoerer, M. and Wolf, H. C.,** *Proc. Colloq. Ampere,* 14, 87, 1966.
100. **De Groot, M. S., Hesselmann, I. A. M., and Van der Waals, J. H.,** *Mol. Phys.,* 12, 259, 1967.
101. **Hall, L., Armstrong, A., and El-Sayed, M. A.,** *J. Chem. Phys.,* 48, 1395, 1968.
102. **McGlynn, S. P., Azumi, T., and Kinoshita, M.,** *Molecular Spectroscopy of the Triplet State,* Prentice Hall, Englewood Cliffs, NJ, 1969.
103. **Atkins, P. W. and Evans, G. T.,** *Mol. Phys.,* 27, 1633, 1974.
104. **Ulrich, T., Steiner, U. E., and Föll, R. E.,** *J. Phys. Chem.,* 87, 1873, 1983.
105. **Steiner, U. E.,** *Ber. Bunsenges. Phys. Chem.,* 85, 228, 1981.
106. **Serebrennikov, Yu. A. and Minaev, B. F.,** *Chem. Phys.,* 114, 359, 1987.
107. **Levin, P. P., Shafirovich, V. Ya., Klimchuk, E. S., Kuzmin, V. A., Maier, V. E., and Yablonskaya, E. E.,** *Bull. Acad. Sci. U.S.S.R. Div. Chem. Sci.,* 37, 1270, 1988.
108. **Pedersen, J. B. and Freed, J. H.,** *J. Chem. Phys.,* 62, 1706, 1975.
109. **Brocklehurst, B.,** *Nature (London),* 221, 921, 1969.
110. **Anisimov, O. A. and Molin, Yu. N.,** *High Energy Chem.,* 14, 225, 1980.
111. **Brocklehurst, B.,** *Int. Rev. Phys. Chem.,* 4, 279, 1985.

112. **Weller, A.,** *Z. Phys. Chem. (Frankfurt/Main),* 130, 129, 1982.

113. **Rehm, D. and Weller, A.,** *Ber. Bunsenges. Phys. Chem.,* 73, 834, 1969.

114. **Werner, H.-J., Staerk, H., and Weller, A.,** *J. Chem. Phys.,* 68, 2419, 1978.

115. **Treichel, R., Staerk, H., and Weller, A.,** *Appl. Phys.,* B31, 15, 1983.

116. **Wasielewski, M. R., Bock, Ch. H., Bowman, M. K., and Norris, J. R.,** *J. Am. Chem. Soc.,* 105, 2903, 1983.

117. **Fedotova, E. Ya., Sokolik, I. A., and Frankevich, E. L.,** *High Energy Chem.,* 14, 261, 1980.

118. **Bube, W., Haberkorn, R., and Michel-Beyerle, M. E.,** *J. Am. Chem. Soc.,* 100, 5993, 1978.

119. **Hamilton, C. A., Hewitt, J. P., McLauchlan, K. A., and Steiner, U. E.,** *Mol. Phys.,* 65, 423, 1988.

120. **Tanimoto, Y., Watanabe, T., Nakagaki, R., Hiramatsu, M., and Nagakura, S.,** *Chem. Phys. Lett.,* 116, 341, 1985.

121. **Boxer, S. G., Chidsey, C. E. D., and Roelofs, M. G.,** *J. Am. Chem. Soc.,* 104, 1452, 1982.

122. **Weller, A., Nolting, F., and Staerk, H.,** *Chem. Phys. Lett.,* 96, 24, 1983.

123. **Leshina, T. V., Salikhov, K. M., Sagdeev, R. Z., Belyaeva, S. G., Maryasova, V. I., Purtov, P. A., and Molin, Yu. N.,** *Chem. Phys. Lett.,* 70, 228, 1980.

124. **Staerk, H., Kühnle, W., Treichel, R., and Weller, A.,** *Chem. Phys. Lett.,* 118, 19, 1985.

125. **Nath, D. N. and Chowdhury, M.,** *Indian J. Pure Appl. Phys.,* 22, 687, 1984.

126. **Basu, S., Nath, D. N., and Chowdhury, M.,** *Chem. Phys. Lett.,* 161, 449, 1989.

127. **Hoff, A. J.,** *Photochem. Photobiol.,* 43, 727, 1986.

128. **Hoff, A. J.,** *Q. Rev. Biophys.,* 17, 153, 1984.

129. **Schulten, K.,** *J. Chem. Soc.,* 82, 1312, 1985.

130. **Tribel, M. M., Morozov, A. K., Leksin, A. N., and Frankevich, E. L.,** *Dokl. Akad. Nauk SSSR,* 284, 1170, 1985.

131. **Tribel, M. M., Morozov, A. K., and Frankevich, E. L.,** *Dokl. Phys. Chem.,* p.455, 1987.

132. **Basu, S., Nath, D. N., and Chowdhury, M.,** *Proc. Indian Natl. Sci. Acad.,* 54A, 830, 1988.

133. **Nath, D. N. and Chowdhury, M.,** *Chem. Phys. Lett.,* 109, 13, 1984.

134. **Nath, D. N., Basu, S., and Chowdhury, M.,** *J. Indian Chem. Soc.,* 63, 56, 1986.

135. **Staerk, H., Treichel, R., and Weller, A.,** *Chem. Phys. Lett.,* 96, 28, 1983.

136. **Michel-Beyerle, M. E., Krüger, H. W., Haberkorn, R., and Seidlitz, H.,** *Chem. Phys.,* 42, 441, 1979.

137. **Krüger, H. W., Michel-Beyerle, M. E., and Seidlitz, H.,** *Chem. Phys. Lett.,* 87, 79, 1982.

138. **Krüger, H. W., Michel-Beyerle, M. E., and Knapp, E. W.,** *Chem. Phys.,* 74, 205, 1983.

139. **Nolting, F. Staerk, H. and Weller, A.,** *Chem. Phys. Lett.,* 88, 523, 1982.

140. **Lavrik, N. L. and Nechaev, O. V.,** *Chem. Phys.,* 124, 273, 1988.

141. **Shushin, A. I.,** *Chem. Phys. Lett.,* 118, 197, 1985.

142. **Mauzerall, D. and Ballard, S. G.,** *Annu. Rev. Phys. Chem.,* 33, 377, 1982.

143. **Brocklehurst, B.,** *Faraday Discuss. Chem. Soc.,* 63, 96, 1977.

144. **Brocklehurst, B.,** *Chem. Phys. Lett.,* 44, 245, 1976.

145. **Klein, J. and Voltz, R.,** *Can. J. Chem.,* 55, 2102, 1977.

146. **Anisimov, O. A., Bizyaev, V. L., Lukzen, N. N., Grigoryants, V. M., and Molin, Yu. N.,** *Chem. Phys. Lett.,* 101, 131, 1983.

147. **Veselov, A. V., Melekhov, V. I., Anisimov, O. A., and Molin, Yu. N.,** *Chem. Phys. Lett.,* 136, 263, 1987.

148. **Basu, S., Nath, D., and Chowdhury, M.,** *J. Chem. Soc. Faraday Trans. 2,* 83, 1325, 1987.

149. **Petrov, N. Kh., Fedotova, E. Ya., and Frankevich, E. L.,** *High Energy Chem.,* 14, 257, 1980.

150. **Petrov, N. Kh., Shushin, A. I., and Frankevich, E. L.,** *Chem. Phys. Lett.,* 82, 339, 1981.

151. **Tanimoto, Y., Hasegawa, K., Okada, N., Itoh, M., Iwai, K., Sugioka, K., Takemura, F., Nakagaki, R., and Nagakura, S.,** *J. Phys. Chem.,* 93, 3586, 1989.

152. **Tanimoto, Y., Okada, N., Itoh, M., Iwai, K., Sugioka, K., Takemura, F., Nakagaki, R., and Nagakura, S.,** *Chem. Phys. Lett.,* 136, 42, 1987.

153. **Nath, D. N. and Chowdhury, M.,** *Pramana,* in press.

154. **Nath, D. N., Basu, S., and Chowdhury, M.,** *J. Chem. Phys.,* 91, 5857, 1989.

155. **Gordon, M. and Ware, W. R., Eds.,** *The Exciplex,* Academic Press, New York, 1975.

156. **Staerk, H., Busmann, H.-G., Kühnle, W., and Weller, A.,** *Chem. Phys. Lett.,* 155, 603, 1989.

157. **Frankevich, E. L. and Fedotova, E. Ya.,** *High Energy Chem.,* 14, 355, 1980.

158. **Fedotova, E. Ya., Stolovitskii, Yu. M., and Frankevich, E. L.,** *Dokl. Phys. Chem.,* 254, 779, 1980.

159. **Burri, J. and Fischer, H.,** *Chem. Phys.,* 139, 497, 1989.

160. **Hamilton, C. A., McLauchlan, K. A., and Peterson, K. R.,** *Chem. Phys. Lett.,* 162, 145, 1989.

161. **Sakaguchi, Y. and Hayashi, H.,** *Chem. Phys. Lett.,* 106, 420, 1984.

162. **Turro, N. J., Lei, X., Gould, I. R., and Zimmt, M. B.,** *Chem. Phys. Lett.,* 120, 397, 1985.

163. **Basu, S., Nath, D., and Chowdhury, M.,** *J. Lumin.,* 40 and 41, 252, 1988.

164. **Basu, S., Kundu, L., and Chowdhury, M.,** *Chem. Phys. Lett.,* 141, 115, 1987.

165. **Tanimoto, Y., Kita, A., Itoh, M., Okazaki, M., and Nakagaki, R.,** *Chem. Phys. Lett.,* 165, 184, 1990.

166. **Elliot, R. J.**, *Magnetic Properties of Rare Earth Metals*, Elliott, R. J., Ed., Plenum Press, New York, 1972, 13.
167. **Hyde, J. S., Swarts, H. M., and Antholine, W. E.**, *Spin Labeling*, Vol. 2, Berliner, L. J., Ed., Academic Press, New York, 1979.
168. **Closs, G. L. and Doubleday, C. E.**, *J. Am. Chem. Soc.*, 95, 2735, 1973.
169. **Doubleday, Ch., Jr.**, *Chem. Phys. Lett.*, 77, 131, 1981.
170. **Doubleday, Ch., Jr.**, *Chem. Phys. Lett.*, 81, 164, 1981.
171. **Weller, A., Staerk, H., and Treichel, R.**, *Faraday Discuss. Chem. Soc.*, 78, 271 and 332, 1984.
172. **Staerk, H., Treichel, R., and Weller, A.**, *Biophysical Effects of Steady Fields (Springer Proc. Phys.*, Vol. 11), Maret, G., Kiepenheuer, J., and Boccara, N., Eds., Springer-Verlag, Berlin, 1986, 85.
173. **Weller, A.**, *Supramolecular Photochemistry*, Balzani, V., Ed., Reidel, Dordrecht, 1987, 343.
174. **Bittl, R. and Schulten, K.**, *Biophysical Effects of Steady Fields (Springer Proc. Phys.*, Vol. 11), Maret, G., Kiepenheuer, J., and Boccara, N., Eds., Springer-Verlag, Berlin, 1986, 90.
175. **Treichel, R.**, Ph.D. thesis, Universität Göttingen, 1985.
176. **Vogelmann, E., Schreiner, S., Rauscher, W., and Kramer, H. E. A.**, *Z. Phys. Chem. N. F.*, 101, 321, 1976.
177. **Steiner, U. E., Winter, G., and Kramer, H. E. A.**, *J. Phys. Chem.*, 81, 1104, 1977.
178. **Steiner, U. E. and Winter, G.**, *Chem. Phys. Lett.*, 55, 364, 1978.
179. **Levin, P. P., Pluzhnikov, P. F., and Kuz'min, V. A.**, *Chem. Phys.*, 137, 331, 1989.
180. **Levin, P. P. and Kuz'min, V. A.**, *Russ. Chem. Rev.*, 56, 307, 1987.
181. **Levin, P. P., Pluzhnikov, P. F., and Kuz'min, V. A.**, *Chem. Phys. Lett.*, 152, 409, 1988.
182. **Weller, A.**, *The Exciplex*, Gordon, M. and Ware, W. R., Eds., Academic Press, New York, 1975, 23.
183. **Levin, P. P., Pluzhnikov, P. F., and Kuz'min, V. A.**, *Chem. Phys. Lett.*, 147, 283, 1988.
184. **Periasamy, N. and Linschitz, H.**, *Chem. Phys. Lett.*, 64, 281, 1979.
185. **Margulis, L. A., Khudyakov, I. V., and Kuz'min, V. A.**, *Chem. Phys. Lett.*, 119, 244, 1985.
186. **Ulstrup, J. and Jortner, J.**, *J. Chem. Phys.*, 63, 4358, 1975.
187. **Levin, P. P. and Kuz'min, V. A.**, *Bull. Acad. Sci. U.S.S.R. Div. Chem. Sci.*, 36, 2426, 1987.
188. **Levin, P. P., Khudyakov, I. V., and Kuz'min, V. A.**, *Bull. Acad. Sci. U.S.S.R. Div. Chem. Sci.*, 37, 432, 1988.
189. **Kikuchi, K., Hoshi, M., Abe, E., and Kokubun, H.**, *J. Photochem.*, 45, 1, 1988.
190. **Steiner, U. E.**, *Z. Naturforsch.*, 34a, 1093, 1979.
191. **Steiner, U. E. and Haas, W.**, *J. Phys. Chem.*, 95, 1991, in press.
192. **Waschi, H.-P.**, Ph.D. thesis, Universität Stuttgart, 1983.
193. **Schlenker, W.**, Ph.D. thesis, Universität Stuttgart, 1986.
194. **Kalyanasundaram, K.**, *Coord. Chem. Rev.*, 46, 159, 1982.
195. **Kalyanasundaram, K. and Grätzel, M.**, *Coord. Chem. Rev.*, 69, 57, 1986.
196. **Juris, A., Balzani, V., Barigelletti, F., Campagna, S., Belser, P. and von Zelewsky, A.**, *Coord. Chem. Rev.*, 84, 85, 1988.
197. **DeArmond, M. K. and Myrick, M. L.**, *Acc. Chem. Res.*, 22, 364, 1989.
198. **Krausz, E. and Ferguson, J.**, *Prog. Inorg. Chem.*, 37, 293, 1989.
199. **Hoffman, M. Z.**, *J. Phys. Chem.*, 92, 3458, 1988.
200. **Ferraudi, G. and Argüello, G. A.**, *J. Phys. Chem.*, 92, 1846, 1988.
201. **Steiner, U. E., Wolff, H.-J., Ulrich, T., and Ohno, T.**, *J. Phys. Chem.*, 93, 5147, 1989.
202. **Wolff, H.-J.**, Diploma thesis, Universität Konstanz, 1988.
203. **Wolff, H.-J.**, Ph.D. thesis, Universität Konstanz, in preparation.
204. **Wiedorn, M.**, Diploma thesis, Universität Konstanz, 1988.
205. **Steiner, U. E. and Bürßner, D.**, *Z. Physik. Chem.*, Frankfurt am Main, 1991, in press.
206. **Bürßner, D.**, Diploma thesis, Universität Konstanz, 1989.
207. **Kober, E. M. and Meyer, T. J.**, *Inorg. Chem.*, 22, 1614, 1983.
208. **DeSimone, R. and Drago, R. S.**, *J. Am. Chem. Soc.*, 92, 2343, 1970.
209. **Horvath, A., Zsilak, Z., and Papp, S.**, *J. Photochem.*, 50, 129, 1989.
210. **Saito, T., Hirata, Y., Sato, H., Yoshida, T., and Mataga, N.**, *Bull. Chem. Soc. Jpn.*, 61, 1925, 1988.
211. **Nakamura, H., Uehata, A., Motonaga, A., Ogata, T., and Matsuo, T.**, *Chem. Lett.*, p.543, 1987.
212. **Nakagaki, R., Hiramatsu, M., Mutai, K., Tanimoto, Y., and Nagakura, S.**, *Chem. Phys. Lett.*, 134, 171, 1987.
213. **Hiramatsu, M., Nakagaki, R., Tanimoto, Y., Mutai, K., Tukada, H., and Nagakura, S.**, *Chem. Phys. Lett.*, 142, 413, 1987.
214. **Levin, P. P., Khudyakov, I. V., and Kuzmin, V. A.**, *J. Phys. Chem.*, 93, 208, 1989.
215. **Khudyakov, I. V., Prokof'ev, A. I., Margulis, L. A., and Kuzmin, V. A.**, *Chem. Phys. Lett.*, 104, 409, 1984.
216. **Staerk, H. and Razi Naqvi, K.**, *Chem. Phys. Lett.*, 50, 386, 1977.
217. **Tanimoto, Y., Takeshima, M., Hasegawa, K., and Itoh, M.**, *Chem. Phys. Lett.*, 137, 330, 1987.

218. **Tanimoto, Y., Takashima, M., and Itoh, M.,** *J. Phys. Chem.,* 88, 6053, 1984.
219. **Tanimoto, Y., Takashima, M., Uehara, M., Itoh, M., Hiramatsu, M., Nakagaki, R., Watanabe, T., and Nagakura, S.,** *Chem. Lett.,* p.15, 1985.
220. **Hata, N. and Yamada, Y.,** *Chem. Lett.,* p.989, 1980.
221. **Hata, N. and Hokawa, M.,** *Chem. Lett.,* p.507, 1981.
222. **Hata, N. and Nishida, N.,** *Chem. Lett.,* p.1043, 1983.
223. **Nakagaki, R., Mutai, K., and Nagakura, S.,** *Chem. Phys. Lett.,* 154, 581, 1989.
224. **Klimchuk, E. S., Khudyakov, I. V., Margulis, L. A., and Kuz'min, V. A.,** *Bull. Acad. Sci. U.S.S.R. Div. Chem. Sci.,* 37, 2462, 1988.
225. **Tanimoto, Y., Jinda, C., Fujiwara, Y., Itoh, M., Hirai, K., Tomioka, H., Nakagaki, R., and Nagakura, S.,** *J. Photochem.,* 47, 269, 1989.
226. **Scaiano, J. C. and Lougnot, D.-J.,** *Chem. Phys. Lett.,* 105, 535, 1984.
227. **Schulten, K. and Epstein, I. R.,** *J. Chem. Phys.,* 71, 309, 1979.
228. **Nakagaki, R., Hiramatsu, M., Mutai, K., and Nagakura, S.,** *Chem. Phys. Lett.,* 121, 262, 1985.
229. **Nakagaki, R., Mutai, K., Hiramatsu, M., Tukada, H., and Nagakura, S.,** *Can. J. Chem.,* 66, 1989, 1988.
230. **Sakaguchi, Y., Hayashi, H., and Nagakura, S.,** *Bull. Chem. Soc. Jpn.,* 53, 39, 1980.
231. **Sakaguchi, Y., Hayashi, H., and Nagakura, S.,** *Bull. Chem. Soc. Jpn.,* 53, 3059, 1980.
232. **Fischer, H.,** *Chem. Phys. Lett.,* 100, 255, 1983.
233. **Turro, N. J., Chow, M.-F., Chung, C.-J., and Kräutler, B.,** *J. Am. Chem. Soc.,* 103, 3886, 1981.
234. **Turro, N. J. and Kräutler, B.,** *J. Am. Chem. Soc.,* 100, 7432, 1978.
235. **Tarasov, V. F., Pershin, A. D., and Buchachenko, A. L.,** *Izv. Akad. Nauk SSSR Ser. Khim.,* 8, 1927, 1980.
236. **Sterna, L., Ronis, D., Wolfe, S., and Pines, A.,** *J. Chem. Phys.,* 73, 5493, 1980.
237. **Turro, N. J., Anderson, D. R., and Kräutler, B.,** *Tetrahedron Lett.,* 21, 3, 1980.
238. **Tarasov, V. F., Askerov, D. B., and Buchachenko, A. L.,** *Bull. Acad. Sci. U.S.S.R. Div. Chem. Sci.,* 31, 1786, 1982.
239. **Buchachenko, A. L.,** *Prog. React. Kinet.,* 13, 163, 1984.
240. **Gould, I. R., Turro, N. J., and Zimmt, M. B.,** *Adv. Phys. Org. Chem.,* 20, 1, 1984.
241. **Turro, N. J. and Kräutler, B.,** *Isot. Org. Chem.,* 6, 107, 1984.
242. **Nakagaki, R., Hiramatsu, M., Watanabe, T., Tanimoto, Y., and Nagakura, S.,** *J. Phys. Chem.,* 89, 3222, 1985.
243. **Sato, H., Kasatani, K., and Murakami, S.,** *Chem. Phys. Lett.,* 151, 97, 1988.
244. **Zimmt, M. B., Doubleday, Ch., Jr., Gould, I. R., and Turro, N. J.,** *J. Am. Chem. Soc.,* 107, 6724, 1985.
245. **Zimmt, M. B., Doubleday, Ch., Jr., and Turro, N. J.,** *J. Am. Chem. Soc.,* 107, 6726, 1985.
246. **Zimmt, M. B., Doubleday, Ch., Jr., and Turro, N. J.,** *J. Am. Chem. Soc.,* 108, 3618, 1986.
247. **Zimmt, M. B., Doubleday, Ch., Jr., and Turro, N. J.,** *Chem. Phys. Lett.,* 134, 549, 1987.
248. **Turro, N. J., Doubleday, Ch., Jr., Hwang, K.-C., Cheng, C.-C., and Fehlner, J. R.,** *Tetrahedron Lett.,* p.2929, 1987.
249. **Wang, J., Doubleday, C., Jr., and Turro, N. J.,** *J. Am. Chem. Soc.,* 111, 3962, 1989.
250. **Wang, J., Doubleday, C., Jr., and Turro, N. J.,** *J. Phys. Chem.,* 93, 4780, 1989.
251. **Doubleday, C., Jr., Turro, N. J., and Wang, J.-F.,** *Acc. Chem. Res.,* 22, 199, 1989.
252. **Carlacci, L., Doubleday, Ch., Jr., Furlani, T. R., King, H. F., and McIver, J. W., Jr.,** *J. Am. Chem. Soc.,* 109, 5323, 1987.
253. **Eyring, H., Walter, J., and Kimball, G.,** *Quantum Chemistry,* John Wiley & Sons, New York, 1944, 326.
254. **Fisher, J. J. and Michl, J.,** *J. Am. Chem. Soc.,* 109, 583, 1987.
255. **Winnik, M. A.,** *Chem. Rev.,* 81, 491, 1981.
256. **Winnik, M. A.,** *Acc. Chem. Res.,* 18, 73, 1985.
257. **El-Sayed, M. A.,** *J. Chem. Phys.,* 38, 2834, 1963.
258. **Atkins, P. W. and McLauchlan, K. A.,** *Chemically Induced Magnetic Polarization,* Lepley, A. R. and Closs, G. L., Eds., John Wiley & Sons, London, 1978, 41.
259. **Wong, K. S., Hutchinson, D. A., and Wan, J. K. S.,** *J. Chem. Phys.,* 58, 985, 1973.
260. **Ciferri, A., Hoeve, C., and Flory, P.,** *J. Am. Chem. Soc.,* 83, 1015, 1961.
261. **Flory, P., Ciferri, A., and Chiang, R.,** *J. Am. Chem. Soc.,* 83, 1023, 1961.
262. **Abe, A., Jernigan, R., and Flory, P.,** *J. Am. Chem. Soc.,* 88, 631, 1966.
263. **Dikanov, S. A., Shchukin, G. I., Grigor'ev, I. A., Rukin, S. I., and Volodarski, L. B.,** *Bull. Acad. Sci. U.S.S.R. Div. Chem. Sci.,* 34, 515, 1985.
264. **Turro, N. J. and Grätzel, M.,** *Angew. Chem.,* 92, 675, 1980.
265. **Turro, N. J. and Cherry, W. R.,** *J. Am. Chem. Soc.,* 100, 7431, 1978.
266. **Turro, N. J. and Weed, G. C.,** *J. Am. Chem. Soc.,* 105, 1861, 1983.

267. **Pfüller, U.**, *Mizellen-Vesikel-Mikroemulsionen, Tensidassoziate und ihre Anwendung in Analytik und Biochemie*, Springer-Verlag, Berlin, 1986.
268. **Eicke, H.-F., Geiger, S., Sauer, F. A., and Thomas, H.**, *Ber. Bunsenges. Phys. Chem.*, 90, 872, 1986.
269. **Valeur, B. and Bardez, E.**, *Structure and Reactivity in Reverse Micelles*, Pileni, M. P., Ed., Elsevier, Amsterdam, 1989, 103.
270. **Pileni, M. P., Ed.**, *Structure and Reactivity in Reverse Micelles*, Elsevier, Amsterdam, 1989.
271. **Scaiano, J. C. and Abuin, E. B.**, *Chem. Phys. Lett.*, 81, 209, 1981.
272. **Scaiano, J. C., Abuin, E. B., and Stewart, L. C.**, *J. Am. Chem. Soc.*, 104, 5673, 1982.
273. **Sakaguchi, Y., Nagakura, S., and Hayashi, H.**, *Chem. Phys. Lett.*, 72, 420, 1980.
274. **Sakaguchi, Y., Hayashi, H., and Nagakura, S.**, *J. Phys. Chem.*, 86, 3177, 1982.
275. **Sakaguchi, Y. and Hayashi, H.**, *J. Phys. Chem.*, 88, 1437, 1984.
276. **Sakaguchi, Y. and Hayashi, H.**, *Chem. Phys. Lett.*, 87, 539, 1982.
277. **Tanimoto, Y., Takashima, M., and Itoh, M.**, *Bull. Chem. Soc. Jpn.*, 57, 1747, 1984.
278. **Tanimoto, Y. and Itoh, M.**, *Chem. Phys. Lett.*, 83, 626, 1981.
279. **Tanimoto, Y., Udagawa, H., and Itoh, M.**, *J. Phys. Chem.*, 87, 724, 1983.
280. **Tanimoto, Y., Udagawa, H., Katsuda, Y., and Itoh, M.**, *J. Phys. Chem.*, 87, 3976, 1983.
281. **Tanimoto, Y., Shimizu, K., and Itoh, M.**, *J. Am. Chem. Soc.*, 106, 7257, 1984.
282. **Tanimoto, Y., Shimizu, K., and Itoh, M.**, *Chem. Phys. Lett.*, 112, 2217, 1984.
283. **Tanimoto, Y., Takashima, M., and Itoh, M.**, *Chem. Lett.*, p.1981, 1984.
284. **McLauchlan, K. A. and Nattrass, S. R.**, *Mol. Phys.*, 65, 1483, 1988.
285. **Scaiano, J. C. and Lougnot, D. J.**, *J. Phys. Chem.*, 88, 3379, 1984.
286. **Levin, P. P. and Kuz'min, V. A.**, *Izv. Akad. Nauk SSSR Ser. Khim.*, 35, 464, 1986.
287. **Kuz'min, V. A. and Levin, P. P.**, *Bull. Acad. Sci. U.S.S.R. Div. Chem. Sci.*, 35, 1291, 1986.
288. **Levin, P. P. and Kuz'min, V. A.**, *Bull. Acad. Sci. U.S.S.R. Div. Chem. Sci.*, 36, 923, 1987.
289. **Levin, P. P. and Kuz'min, V. A.**, *Dokl. Phys. Chem.*, p.26, 1987.
290. **Levin, P. P. and Kuz'min, V. A.**, *Bull. Acad. Sci. U.S.S.R. Div. Chem. Sci.*, 36, 691, 1987.
291. **Levin, P. P. and Kuz'min, V. A.**, *Bull. Acad. Sci. U.S.S.R. Div. Chem. Sci.*, 37, 224, 1988.
292. **Efremkin, A. F., Levin, P. P., and Ivanov, V. B.**, *Dokl. Phys. Chem.*, p.476, 1988.
293. **Yankelevich, A. Z., Khudyakov, I. V., Levin, P. P., and Serebrennikov, Yu. A.**, *Bull. Acad. Sci. U.S.S.R. Div. Chem. Sci.*, 38, 18, 1989.
294. **Evans, C., Ingold, K. U., and Scaiano, J. C.**, *J. Phys. Chem.*, 92, 1257, 1988.
295. **Turro, N. J. and Okubo, T.**, *J. Am. Chem. Soc.*, 103, 7224, 1981.
296. **Gould, I. R., Baretz, B. H., and Turro, N. J.**, *J. Phys. Chem.*, 91, 925, 1987.
297. **Johnston, L. J., de Mayo, P., and Wong, S. K.**, *J. Am. Chem. Soc.*, 104, 307, 1982.
298. **Turro, N. J., Zimmt, M. B., and Gould, I. R.**, *J. Am. Chem. Soc.*, 105, 6347, 1983.
299. **Turro, N. J.**, *Tetrahedron*, 38, 809, 1982.
300. **Kräutler, B. and Turro, N. J.**, *Chem. Phys. Lett.*, 70, 270, 1980.
301. **Baretz, B. H. and Turro, N. J.**, *J. Am. Chem. Soc.*, 105, 1309, 1983.
302. **Turro, N. J. and Mattay, J.**, *J. Am. Chem. Soc.*, 103, 4200, 1981.
303. **Turro, N. J., Chung, Ch.-J., Jones, G., II, and Becker, W. G.**, *J. Phys. Chem.*, 86, 3677, 1982.
304. **Turro, N. J., Kräutler, B., and Anderson, D. R.**, *J. Am. Chem. Soc.*, 101, 7435, 1979.
305. **Turro, N. J. and Kräutler, B.**, *Isot. Org. Chem.*, 6, 107, 1984.
306. **Buchachenko, A. L.**, *Prog. React. Kinet.*, 13, 163, 1984.
307. **Buchachenko, A. L.**, *Russ. J. Phys. Chem.*, 55, 936, 1981 (English transl.).
308. **Bernstein, R.**, *J. Phys. Chem.*, 56, 893, 1952.
309. **Turro, N. J., Chow, M.-F., and Kräutler, B.**, *Chem. Phys. Lett.*, 73, 545, 1980.
310. **Turro, N. J., Paczkowski, M. A., and Wan, P.**, *J. Org. Chem.*, 50, 1399, 1985.
311. **Turro, N. J., Cheng, C.-C., Wan, P., Chung, C.-J., and Mahler, W.**, *J. Phys. Chem.*, 89, 1567, 1985.
312. **Kräutler, B. and Turro, N. J.**, *Chem. Phys. Lett.*, 70, 266, 1980.
313. **Turro, N. J., Chow, M.-F., Chung, C.-J., Weed, G. C., and Kräutler, B.**, *J. Am. Chem. Soc.*, 102, 4843, 1980.
314. **Turro, N. J., Chow, M.-F., Chung, C.-J., and Tung, C.-H.**, *J. Am. Chem. Soc.*, 102, 7391, 1980.
315. **Turro, N. J., Chow, M.-F., Chung, Ch.-J., and Tung, Ch.-H**, *J. Am. Chem. Soc.*, 105, 1572, 1983.
316. **Turro, N. J.**, U.S. Patent 4,448,657 (Cl. 204—155; CO8F2/50), 1984.
317. **Hayashi, H., Sakaguchi, Y., and Nagakura, S.**, *Chem. Lett.*, p.1149, 1980.
318. **Turro, N. J., Chow, M.-F., Chung, Ch.-J., Tanimoto, Y., and Weed, G. C.**, *J. Am. Chem. Soc.*, 103, 4574, 1981.
319. **Gould, I. R., Zimmt, M. B., Turro, N. J., Baretz, G. H., and Lehr, G. F.**, *J. Am. Chem. Soc.*, 107, 4607, 1985.
320. **Turro, N. J., Zimmt, M. B., and Gould, I. R.**, *J. Phys. Chem.*, 92, 433, 1988.
321. **Gould, I. R., Tung, C.-H., Turro, N. J., Givens, R. S., and Matuszewski, B.**, *J. Am. Chem. Soc.*, 106, 1789, 1984.

322. **Hayashi, H., Sakaguchi, Y., Tsunooka, M., Yanagi, H., and Tanaka, M.,** *Chem. Phys. Lett.,* 136, 436, 1987.
323. **Tanimoto, Y., Shimizu, K., Udagawa, H., and Itoh, M.,** *Chem. Lett.,* p.353, 1983.
324. **Tanimoto, Y., Takayama, M., Itoh, M., Nakagaki, R., and Nagakura, S.,** *Chem. Phys. Lett.,* 129, 414, 1986.
325. **Atherton, N. M. and Strack, S. J.,** *J. Chem. Soc. Faraday Trans. 2,* 68, 374, 1972.
326. **Steiner, U. E.,** *Structure and Reactivity in Reverse Micelles,* Pileni, M. P., Ed., Elsevier, Amsterdam, 1989, 156.
327. **Steiner, U. E. and Haas, W.,** to be published.
328. **Haas, W.,** Ph.D. thesis Universität Konstanz, in preparation.
329. **Steiner, U. E. and Wu, J. Q.,** to be published.
330. **Wu, J. Q.,** Ph.D. thesis, Universiät Konstanz, in preparation.
331. **Morton, J. R. and Preston, K. F.,** *J. Magn. Res.,* 30, 577, 1978.
332. **Baumann, D., Ulrich, T., and Steiner, U. E.,** *Chem. Phys. Lett.,* 137, 113, 1987.
333. **Lersch, W. and Michel-Beyerle, M. E.,** *Chem. Phys.,* 78, 115, 1983.
334. **Frankevich, E. L. and Kubarev, S. I.,** *Triplet State ODMR Spectroscopy,* Clark, R. H., Ed., John Wiley & Sons, New York, 1982, 138.
335. **Sagdeev, R. Z., Grishin, Yu. A., Gogolev, A. Z., Dooshkin, A. V., Semenov, A. G., and Molin, Yu. N.,** *Zh. Strukt. Khim.,* 20, 1132, 1979.
336. **Trifunac, A. D. and Evanochko, W. T.,** *J. Am. Chem. Soc.,* 102, 4598, 1980.
337. **Frankevich, E. L., Pristupa, A. I., and Lesin, V. I.,** *Chem. Phys. Lett.,* 47, 304, 1977.
338. **Buchachenko, A. L. and Berdinskii, V. L.,** *Russ. Chem. Rev.,* 52, 1, 1983.
339. **Anisimov, O. A., Grigoryants, V. M., Molchanov, V. K., and Molin, Yu. N.,** *Chem. Phys. Lett.,* 66, 265, 1979.
340. **Molin, Yu. N., Anisimov, O. A., Grigoryants, V. M., Molchanov, V. K., and Salikhov, K. M.,** *J. Phys. Chem.,* 84, 1853, 1980.
341. **Saik, V. O., Anisimov, O. A., Koptyug, A. V., and Molin, Yu. N.,** *Chem. Phys. Lett.,* 165, 142, 1990.
342. **Kubarev, S. I. and Pshenichnov, E. A.,** *Chem. Phys. Lett.,* 28, 66, 1974.
343. **Frankevich, E. L. and Pristupa, A. I.,** *Pisma Zh. Eskp. Teor. Fiz.,* 24, 397, 1976.
344. **Frankevich, E. L., Lesin, V. I., and Pristupa, A. I.,** *Chem. Phys. Lett.,* 58, 127, 1978.
345. **Frankevich, E. L., Tribel, M. M., Sokolik, I. A., and Pristupa, A. I.,** *Phys. Status Solidi B,* B87, 373, 1978.
346. **Frankevich, E. L., Lesin, V. I., and Pristupa, A. I.,** *Magnetic Resonance and Related Phenomena,* Kundla, E., Lippmaa, E., and Saluvere, T., Eds., Springer-Verlag, Berlin, 1979, 45.
347. **Steudle, W. and von Schütz, J. U.,** *J. Lumin.,* 18/19, 191, 1979.
348. **Smith, J. P. and Trifunac, A. D.,** *J. Phys. Chem.,* 85, 1645, 1981.
349. **Anisimov, O. A., Grigoryants, V. I., and Molin, Yu. N.,** *Chem. Phys. Lett.,* 74, 15, 1980.
350. **Molin, Yu. N., Anisimov, O. A., Melekhov, V. I., and Smirnov, S. N.,** *Faraday Discuss. Chem. Soc.,* 78, 289, 1984.
351. **Lefkowitz, S. M. and Trifunac, A. D.,** *J. Phys. Chem.,* 88, 77, 1984.
352. **Wasielewski, M. R., Norris, J. R., and Bowman, M. K.,** *Faraday Discuss. Chem. Soc.,* 78, 279, 1984.
353. **Grant, A. I., McLauchlan, K. A., and Nattrass, S. R.,** *Mol. Phys.,* 55, 557, 1985.
354. **Murai, H., Sakaguchi, Y., Hayashi, H., and I'Haya, Y.,** *J. Phys. Chem.,* 90, 113, 1986.
355. **Closs, G. L., Forbes, M. D. E., and Norris, J. R., Jr.,** *J. Phys. Chem.,* 91, 3592, 1987.
356. **Buckley, C. D., Hunter, D. A., Hore, P. J., and McLauchlan, K. A.,** *Chem. Phys. Lett.,* 135, 307, 1987.
357. **Shushin, A. I.,** *Chem. Phys. Lett.,* 162, 409, 1989.
358. **Zimmt, M. B., Doubleday, Ch., Jr., and Turro, N. J.,** *J. Am. Chem. Soc.,* 106, 3363, 1984.
359. **Okazaki, M., Sakata, S., Konaka, R., and Shiga, T.,** *J. Am. Chem. Soc.,* 107, 7214, 1985.
360. **Okazaki, M., Sakata, S., Konaka, R., and Shiga, T.,** *J. Chem. Phys.,* 86, 6792, 1987.
361. **Olcese, J., Reuss, S., and Vollrath, L.,** *Brain Res.,* 333, 382, 1985.
362. **Cremer-Bartels, G., Krause, K., and Küchle, H. J.,** *Graefe's Arch. Clin. Exp. Ophthalmol.,* 220, 248, 1983.
363. **Molin, Yu. N., Salikhov, K. M., and Zamaraev, K. I.,** *Spin Exchange,* Springer-Verlag, Berlin, 1980.

Chapter 2

PHOTOCATALYSIS BY SEMICONDUCTOR POWDERS — PREPARATIVE AND MECHANISTIC ASPECTS

Horst Kisch and Ronald Künneth

TABLE OF CONTENTS

I. SCOPE AND INTRODUCTION

Photochemistry, and especially *photocatalysis,* has experienced a powerful revival within the last two decades. This originated in the first oil crisis which stimulated worldwide research programs for the photochemical conversion and storage of solar energy. In addition to the generation of electricity with the help of photovoltaic cells using single crystal semiconductor electrodes, the formation of hydrogen and oxygen from water by irradiation of metallized semiconductor powders or colloids seemed to be the most promising approach. Despite some initial encouraging progress no practicable system is available at the present.

Besides the relation to solar energy conversion, photocatalysis by semiconductors per se is an important new field of photochemistry. It opens the possibility to perform reductive and oxidative reactions simultaneously at a single catalyst which is not possible with a homogeneous photocatalyst (photosensitizer). In addition to the cleavage of water, the reduction of dinitrogen and carbon dioxide, cycloadditions, and oxidative fragmentations were the main reactions investigated as summarized in recent books and review articles.[1-3] Some of these transformations are related to prebiotic chemistry like the TiO_2^{4-} or CdS^5-assisted photoreduction of N_2, the SiO_2-assisted formation of aldehydes and acids from CO and water,[6] and the generation of amino acids from CH_4, N_2, and H_2O in the presence of platinized TiO_2.[7] The fading of paints[8] and the detoxification of water[9] illustrate important technical and environmental aspects of semiconductor photocatalysis.

This new field is still a domain of physical chemistry and applications in preparative chemistry are very limited. In this article the authors therefore first give a short introduction to the underlying principles and then summarize reactions of synthetic interest. Since zinc and cadmium sulfide photocatalysts were employed in these transformations, the authors also include the photoreduction of water, dinitrogen, and carbon dioxide in the presence of these semiconductor powders.

II. GENERAL ASPECTS OF PHOTOCATALYSIS

The term *photocatalysis* consists of the combination of *photo*-chemistry and *catalysis* and implies that light and a catalyst are necessary to bring about or accelerate a chemical transformation. Reactions which fulfill these simple requirements have been named by an amazing collection of terms in the literature. Examples include *catalyzed, assisted, induced, accelerated, promoted,* and *stimulated.* Part of the confusion stems from the fact that historically almost any photoreaction was regarded as photocatalytic since the accelerating action of light on an otherwise slow thermal reaction was considered as a catalytic effect. In the following we make an attempt to deduce a general acceptable definition of photocatalysis by discussing the differences between a stoichiometric and catalytic thermal reaction, between a stoichiometric thermal and a stoichiometric photochemical reaction, and between a stoichiometric and catalytic photoreaction.

A. THERMAL CATALYSIS

The word *catalysis* probably was used the first time in the 16th century by the chemist A. Libavius in his book *Alchymia*. It originates from the Greek and means "decomposition" or "dissolution". Berzelius applied this term in the 19th century to reactions occurring in the presence of compounds which are not directly involved in the chemical transformation. Shortly later, Ostwald defined catalysis as a kinetic phenomenon in the sense that a catalyst changes the rate of a reaction without appearing in the products and without changing the chemical equilibrium. This means that product formation can be catalyzed only for exergonic or weakly endergonic reversible reactions. In the case of photocatalysis this may be different, since part of the absorbed light energy can also be used to drive strongly endergonic reactions.

With the advent of reaction rate theory Ostwald's definition was modified into "a catalyst is a compound which lowers the free activation enthalpy of the reaction". Thus, the difference between a stoichiometric (Equation 1) and catalytic thermal reaction (Equations 2 and 3) can be schematically described by the potential energy curves 1 and 3, respectively (Figure 1). By the presence of a catalyst the activation energy is lowered (E_3 vs. E_1) by opening a new reaction path which may include formation of a substrate-catalyst complex (Equation 2) as indicated by the shallow minimum (I) of curve 3.

$$S \rightarrow P \tag{1}$$

$$S + C \rightleftharpoons [S{\cdots}C] \tag{2}$$

$$[S{\cdots}C] \rightarrow P + C \tag{3}$$

From these definitions it follows that one catalyst molecule should induce the formation of many product molecules, i.e., the turnover number, TON, usually expressed as number of moles of product, n(p), formed per mole of catalyst, n(c), of a

$$TON = \frac{n(p)}{n(c)} \tag{4}$$

catalytic reaction has to be larger than one. When this number is divided by the reaction time, usually expressed in seconds, one obtains the turnover rate, TOR, which may be smaller than

$$TOR = \frac{n(p)}{n(c) \cdot s} \tag{5}$$

one but may exceed several thousands in the case of very active biological or synthetic catalysts.

When the turnover number is equal to or smaller than one, the pseudocatalyst A (assistor) is converted to A' and the reaction becomes stoichiometric (Equation 6).

$$S + A \rightarrow P + A' \tag{6}$$

Different to a common bimolecular reaction no parts of A are incorporated into the product P, nor is A a simple oxidizing or reducing agent. These transformations are therefore usually called *assisted, promoted,* or *induced reactions*. Examples of "metal ion promotion" of organic reactions are well documented[10] and are common in organometallic chemistry.[11] These reactions are often bimolecular versions of Equation 6 and involve complexation of one or both substrates to a transition metal complex (A). The latter is transformed to an

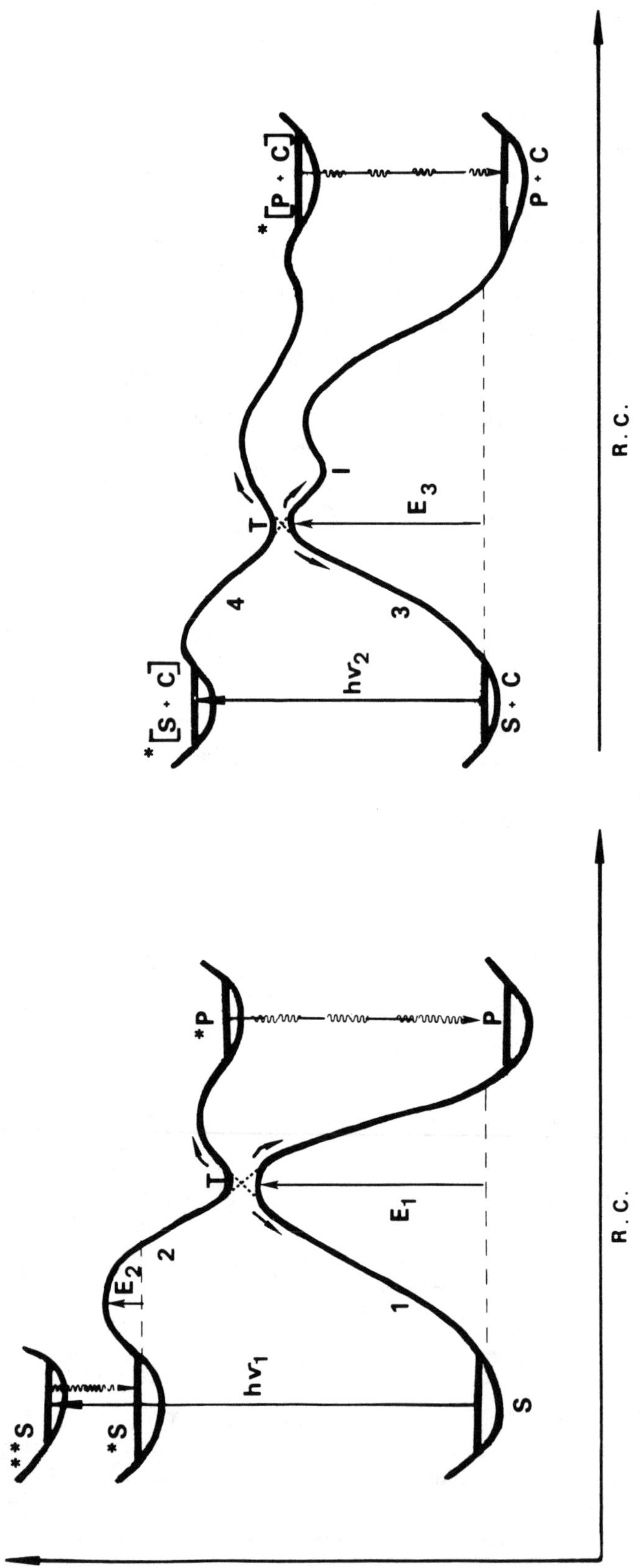

FIGURE 1.   Schematic description of the differences in potential energy changes as function of the reaction coordinate for a (1) thermal, (2) photochemical, (3) catalytic reaction, and (4) catalyzed photoreaction.

inactive compound (A′) during or after product formation. Typical examples are the cobalt-assisted synthesis of indoles from azobenzenes and alkynes,[12] or the manganese-assisted alkylation of molecular nitrogen to 1,2-diazenes.[13]

B. HOMOGENEOUS PHOTOCATALYSIS

With respect to the generally accepted definition of thermal catalysis, *photocatalysis* can be defined as ''acceleration of a photoreaction by the presence of a catalyst''.[14-22] This definition includes *photosensitization,* ''a process by which a photochemical alteration occurs in one molecular entity as a result of initial absorption of radiation by another molecular entity called photosensitizer,''[23] but it excludes the photoacceleration of a stoichiometric thermal reaction, exergonic or endergonic, and irrespective of whether it occurs in homogeneous solution[24] or at the surface of an illuminated electrode;[25] otherwise, any photoreaction would then be catalytic. The catalyst may accelerate the photoreaction by interaction with the substrate in its ground or excited state and/or with a primary photoproduct, depending on the detailed mechanism of the photoreaction.

Similar to thermal catalysis, turnover numbers and turnover rates can be used to experimentally proof the catalytic nature of the photoreaction. When only the requirements of thermal ''stoichiometric catalysis'' are fulfilled (Equation 6), the reaction is called ''*assisted photoreaction.*'' Accordingly, only those photoreactions which meet also the requirements for thermal catalysis are called *catalytic photoreactions.*

Since the efficiency of a photoreaction is best expressed as the product quantum yield Φ_p, i.e., moles of product formed divided by moles of light quanta absorbed by the homogeneous catalyst (sensitizer), one may define ''photocatalytic'' TON and TOR which is obtained through dividing Φ_p by the number of moles of catalyst employed.[26]

To illustrate the basic difference between *thermal catalysis* and *photocatalysis* we first compare a thermal and photochemical stoichiometric reaction. The two corresponding pathways of the reaction of S to P may be represented by the hypothetical potential energy curves 1 and 2, respectively (Figure 1). It is assumed, that a reactive state of different multiplicity, *S, is reached by intersystem crossing from a higher state **S which is populated by direct light absorption. From *S the system may have to overcome a small activation barrier (E_2) before product formation can proceed by two different reaction paths. If the distance between the two curves at the point T is large, transformation to P occurs along the excited-state curve 2 via *P followed by radiative or radiationless deactivation to the product ground state (adiabatic photoreaction). However, when the distance is small — the more common case — the system makes a radiationless transition (. . .) at point T from where it can return to S or form the product directly in its ground state (diabatic photoreaction).[27]

At the final point we are now able to discuss the difference between a *stoichiometric* and *catalytic* photoreaction by comparing the potential energy curves 2 and 4 (Figure 1), respectively. When the light is absorbed by the catalyst C (Equation 7), the system represents a *sensitized photoreaction* which may occur either via energy transfer (Equations 8 and 9) or electron transfer (Equations 10 to 12).

$$C \xrightarrow{h\nu} {}^*C \tag{7}$$

$$^*C + S \longrightarrow {}^*S + C \tag{8}$$

$$^*S \longrightarrow P \tag{9}$$

$$^*C + S \longrightarrow S^- + C^+ \tag{10}$$

$$S^- \longrightarrow P^- \tag{11}$$

$$C^+ + P^- \longrightarrow P + C \tag{12}$$

In the case of energy transfer, *S is converted to the product along the potential energy curve 2 while a new reaction path (e.g., curve 4) is opened when the *photosensitizer* transfers an electron to the substrate. Examples of photosensitization by electron transfer have been recently reviewed.[28]

The action of the catalyst on the substrate may lead to an excited substrate-catalyst complex located somewhere along curve 4. When the corresponding shallow minimum occurs around the point T, the system may change onto the potential energy curve of the thermal catalytic reaction, similar as discussed for the adiabatic and diabatic stoichiometric photoreaction; in this case the catalyst can accelerate product formation, therefore, also by favoring generation of the ground-state intermediate I over returning to the substrate. Note, that due to the high energy of the excited catalyst-substrate complex *[S··C], conversion to the end product is also thermodynamically feasible for endergonic reactions.

Different to *dynamic sensitization* where the excited sensitizer and the substrate collide by diffusion, *static sensitization,* where the sensitizer forms with the substrate a stable ground-state complex prior to light absorption, may be considered as a typical case of photocatalysis. Examples are the photodimerization of olefins to cyclobutanes[16] and the valence isomerization of norbornadiene to quadricyclane, both sensitized by copper(I)chloride.[29] Light absorption occurs by the initially formed olefin copper complex, like in static sensitization. Since the copper compound acts both like a sensitizer and like a thermal catalyst, by complexing the olefin, it is obvious why the copper chloride is usually considered as a *photocatalyst* and the reaction as a *copper-catalyzed* photocycloaddition. In many other cases, as recently summarized,[17] substrate complexation occurs via ligand substitution and differs even more from simple static sensitization.

It was recently proposed that the term *catalyzed photoreaction* should be used only when light is absorbed by the substrate and the catalyst thereafter interacts with *S.[18] Due to the short life time of excited states it seems more likely that the interaction occurs with a primary photoproduct. Accordingly only very few reactions follow this mechanism and the kinetic criterion proposed, linear increase of the inverse reaction quantum yield with the inverse catalyst concentration, is not an unambiguous proof.[30] This definition of a catalyzed photoreaction, although justified from a theoretical point of view,[20] seems not very practical since the required mechanistic details, especially in heterogeneous photocatalysis (*vide infra*), are difficult to obtain and may change too much with increasing mechanistic knowledge about the system. It is therefore proposed to adhere to the use of *photocatalysis* in all systems where intermediate catalyst-substrate complexes are most likely involved, as is usually the case for transition metal sensitizers and heterogeneous systems. Although this difference between a *sensitized* and *catalyzed* photoreaction is arbitrary, it is of practical advantage since it immediately signals the reader whether the action of the "sensitizer" occurs by classical sensitization or by a more complicated mechanism.

The special case of *photoinduced (photogenerated) catalysis* (Equations 13 and 14) is then included in the general field of photocatalysis. In this reaction a thermally active catalyst C is photochemically produced from the precursor C'.

$$C' \overset{h\nu}{\longrightarrow} C \tag{13}$$

$$S \overset{C}{\longrightarrow} P \tag{14}$$

According to Equations 13 and 14, photogenerated catalytic reactions are expected to have an induction period during which the catalyst concentration reaches an optimum value.

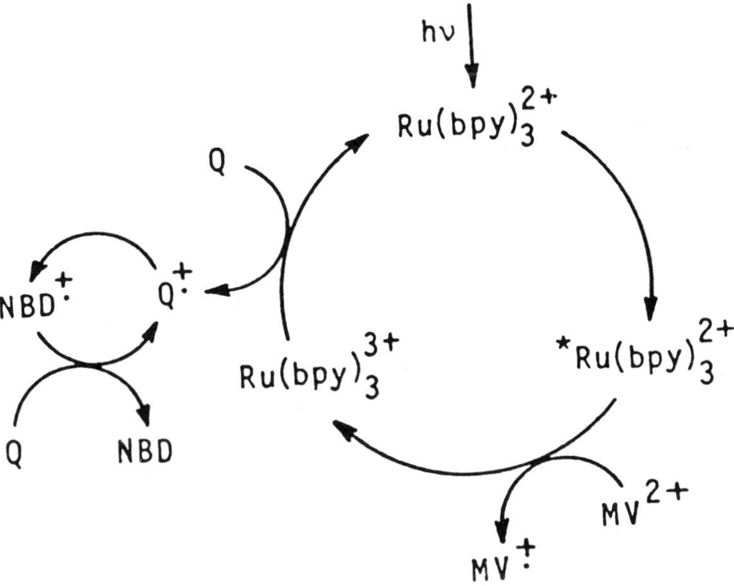

FIGURE 2. Photogenerated catalysis of the valence isomerization of quadricyclane (Q) to norbornadiene. (Adapted from Kutal et al.[32])

Thereafter, the reaction should continue with the same rate when the light source is turned off. In many systems, however, the rate is somewhat lower, since C is partially converted to C′ or some other inactive compounds. Would the system behave like a *catalyzed* or *sensitized photoreaction,* the rate should become zero when the irradiation is stopped.[16,17]

Quantum yields of product formation may exceed one as observed for the iron carbonyl photoinduced isomerization of 1-pentene to 2-pentene.[31] Photoinduced catalysis thus amplifies the chemical consequences of light absorption and offers an efficient way for the conversion of light into chemical energy.

The photoinduced catalysis of the valence isomerization of quadricyclane (Q) to norbornadiene (NBD) is summarized in Figure 2 to illustrate a reaction sequence which contains a bimolecular electron transfer step.[32] In the presence of methyl viologen (MV^{2+}) the excited ruthenium(II) complex is oxidized to the ruthenium(III) compound which in turn oxidizes quadricyclane to the radical cation. The latter then catalyzes the thermal valence isomerization.

C. HETEROGENEOUS PHOTOCATALYSIS

In heterogeneous photocatalysis the situation is more complicated, since the substrate is adsorbed onto the catalyst surface forming a surface catalyst-substrate complex at the solid-liquid interface. This can be considered as a type of supermolecule which is expected to change its structure even when light absorption occurs only by the photocatalyst. In a few cases, like in the CdS-catalyzed photodealkylation of Rhodamine B, it was shown that the same reaction proceeds whether the semiconductor or the substrate absorbs the light.[33-35] In addition to this direct effect of light, the original surface structure may be converted into a catalytic one during an induction period, as it is well known from thermal heterogeneous catalysis.

When a metallized semiconductor (SC/Pt), like platinized titanium dioxide, is used in the catalytic photoreduction of water to hydrogen by some reducing agent (Red), it is again more appropriate to consider the catalyst rather as a *photocatalyst* than as a *photosensitizer.* In this case the catalyst is at least bifunctional. The TiO_2 part absorbs light like a sensitizer

(Equation 15) and the platinum islands catalyze the transfer of electrons to water (Equation 16). Capture of an electron from the reducing agent regenerates the catalyst (Equation 17). Since basic aspects of electrocatalysis are involved in elementary reaction steps of this type of photocatalysis, the term *photoelectrocatalysis* is sometimes used in the literature.[36] Thus, the reaction should be called "titanium dioxide-catalyzed photoreduction of water".

$$SC/Pt \xrightarrow{h\nu} SC/Pt(h^+, e^-) \tag{15}$$

$$SC/Pt(h^+, e^-) + H_2O \longrightarrow SC/Pt(h^+) + \frac{1}{2} H_2 + OH^- \tag{16}$$

$$SC/Pt(h^+) + Red \longrightarrow SC/Pt + OX \tag{17}$$

In the field of homogeneous photocatalysis it is usually no problem to differentiate between assisted and catalytic photoreactions on the basis of the easy accessible turnover numbers (*vide supra*) which have to be <1 and >1, respectively. However, in heterogeneous photocatalysis this is often difficult since TON (Equation 4) is now defined by dividing the number of product molecules produced by the number of active sites, a value unknown for most photocatalysts. It was proposed to use the total surface area as an approximate value to allow estimation of a "photocatalytic TON" by dividing the quantum yield by this total surface area.[26] However, quantum yields are difficult to measure in these heterogeneous systems due to problems like reflectance, influence of reactor geometry, and particle size.[26,37-39]

Another possibility is to express the catalytic activity of the semiconductor powder just as in *thermal* heterogeneous catalysis by the turnover rate, $(TOR)_{het}$, according to Equation 18 where $N(p)$ is the number of product molecules formed

$$(TOR)_{het} = \frac{N(p)}{N_{(mono)} \cdot s} \tag{18}$$

and $N_{(mono)}$ is the number of substrate molecules contained in one monolayer of the powder employed. An estimation, based on the assumption that one monolayer consists of about 10^{14} molecules/cm^2, gave maximal TOR around 0.1 monolayers per second.[26,40-42] Excluded multilayer adsorption occurs; the correspondingly calculated turnover number signals a catalytic or assisted photoreaction whether its value is above or below unity. Without knowledge of the surface area of the semiconductor powder it is therefore not possible to make this decision.

Many of the photoreactions conducted in the presence of semiconductor particles, claimed to be "photocatalytic", in reality are *semiconductor-assisted photoreactions*.[26] An important example is the titanium dioxide-assisted photoreduction of dinitrogen.[4] Accordingly, the term *photocatalytic reaction*, although extensively appearing in the literature for all kinds of stoichiometric and catalytic photoreactions, should not be used further since it implies for the nonphotochemist that the reaction is catalytic with respect to photons, in which case it would belong to the field of *photogenerated catalysis*.

III. GENERAL ASPECTS OF HETEROGENEOUS PHOTOCATALYSIS AT SEMICONDUCTOR POWDERS

The mechanistic principles which enable a semiconductor powder, suspended in a solution of substrates, to catalyze a photoreaction originate in the field of photoelectrochemistry. From investigations with single crystal electrodes in photoelectrochemical cells the basic features of thermodynamic and kinetic details of interfacial electron transfer to dissolved

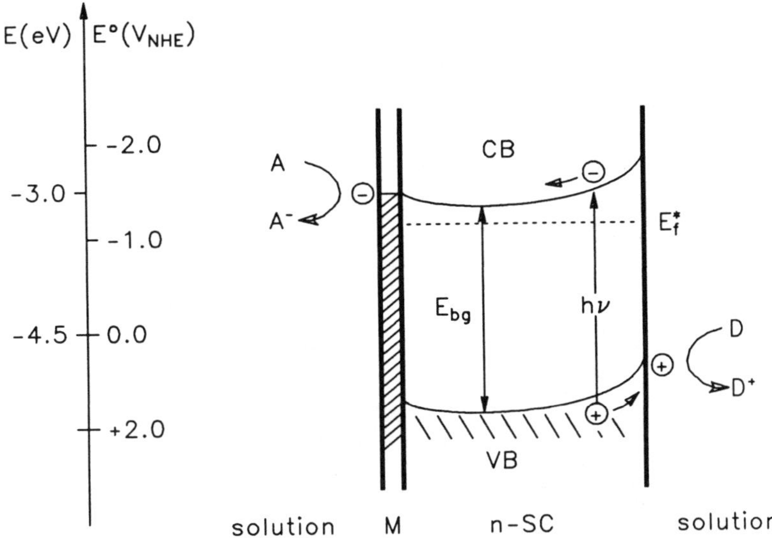

FIGURE 3.   Simplified description of the energetics of an irradiated metallized *n*-semiconductor particle in contact with a solution of D and A; E_F^* = Fermi level under irradiation; CB, VB: conduction and valence band.

molecules were established. For a critical discussion of the basic assumptions made in the deduction of the resulting theories and their relevance to catalysis the reader is referred to a recent article by Tributsch.[36] The use of a metallized semiconductor particle as a kind of "short-circuit" microphotoelectrochemical cell was introduced through the work of Bard on the Photo-Kolbe reaction.[43] Differences to the macroelectrodes employed in more conventional photoelectrochemistry have been discussed.[44-48] In the following we therefore give only a short summary restricted to *n*-type semiconductors.

A. THERMODYNAMIC ASPECTS

Imagine a semiconductor particle which is surface-loaded with a few mol% of a metal like platinum. When the particle diameter is larger than a few hundred nanometers[49] the energy bands are bent upward as drawn in Figure 3. This is a consequence of migration of electrons (majority carriers) from the bulk of the semiconductor to surface states or to a dissolved electron acceptor. The remaining positively charged depletion layer is responsible for the band-bending. Note that on this energy scale, as drawn in the left part of Figure 3, the electron energy increases upward. Note further that the energy barrier for conduction band electrons is larger at the free surface than at the SC/M contact. At the latter it can be initially as high as 0.4 and 1.6 eV for CdS/RuO$_2$ and CdS/Pt, respectively, but it is strongly decreased during early reaction stages. This converts the initial Schottky contact into an ohmic one.[45] From the variation of catalytic activities of CdS powders loaded with different metals it was proposed that the degree of band-bending is proportional to the difference between the metal's work function and the energy of the conduction band edge.[50]

When this modified semiconductor particle absorbs a quantum of ultraband-gap light, the initially formed electron-hole pair partially recombines and partially is separated by the inhomogeneous electrical field of the depletion layer. Both charges are driven toward the direction of lower energy, i.e., the electron into the bulk and the hole to the surface. There it may be trapped (h_{tr}^+) before it oxidizes an adsorbed donor D if its redox potential is negative of or approximately at the valence band edge at the surface. In contrast to the hole, the electron has to pass an energy barrier to reach the surface. Since this is usually much smaller at the SC/M contact than at the nonmetallized surface, the electron will arrive at

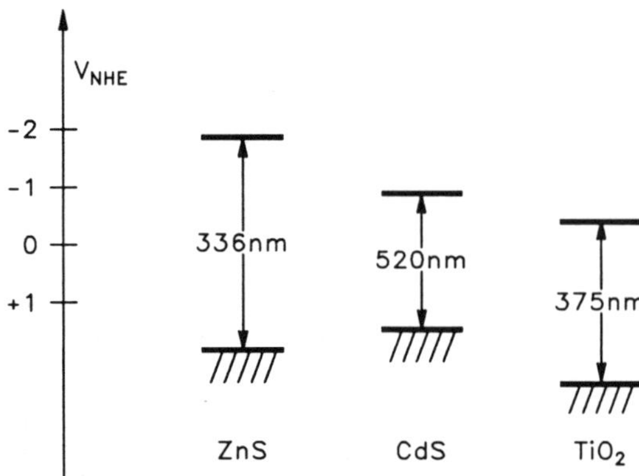

FIGURE 4.   Energetic positions of band edges at pH 7 of some typical *n*-semiconductors.

the former and thus the compartmentation of the photogenerated charges is completed. Once trapped at the surface (e_{tr}^-) the electron may reduce an adsorbed acceptor A if the redox potential A/A^- is more positive or approximately at the energy of the conduction band edge at the surface. In competition with this desired reaction of the trapped charges, they may also react with the semiconductor material itself inducing undesired photocorrosion processes (see Section III.C).

Thus, to catalyze the endergonic redox reaction $A + D \rightarrow A^- + D^+$, the positions of conduction and valence band edges at the surface must lie slightly negative and positive, respectively, of the corresponding half-cell potentials $E^0(A/A^-)$ and $E^0(D^+/D)$. From these criteria it is therefore possible to predict the thermodynamic feasibility of a semiconductor-photocatalyzed redox reaction.[51]

While the required redox potentials of the two half-cell reactions are usually easily accessible,[52] the band energies are that only at a first glance. In aqueous systems they usually shift cathodically with higher pH value by approximately 0.06 V per pH unit as demonstrated for ZnS,[53] CdS,[54] and TiO₂.[55] In neutral aqueous solution the values for the conduction and valence band of ZnS[53,56] and CdS[54,57] are $-1.84/+1.84$ and $-0.9/+1.5$ V, respectively (Figure 4). Apparent differences in band positions of single crystals and powders are usually due to the presence of surface states in the latter.[58] In Figure 4 titanium dioxide, the most commonly employed photocatalyst is included for comparison.

In addition to this pH dependence, surface impurities, adsorbed compounds, and the change to organic solvents may induce strong band shifts. Some results obtained with CdS single crystal electrodes may illustrate these effects.

Thus, the removal of traces of elemental sulfur and cadmium from the surface results in a cathodic shift of 1 V![58] Neutral RSH shows no influence, while adsorption of the charged RS^- shifts the band positions by 0.4 V cathodically.[59] The same shift of 0.4 V is observed when neutral water is replaced by acetonitrile as the solvent.[2]

From the previous discussion it follows that for a given semiconductor photocatalyst, variation of catalyst preparation, kind of solvent, and nature of substrate adsorption may induce a shift of band positions and therefore change the driving force of the electron and hole reaction with the acceptor and donor, respectively. This may represent one way to introduce chemoselectivity when the solution contains several donor/acceptor pairs. At fixed band positions the same may be achieved by proper variation of the potentials of one or both substrate half-cell reactions. As an example for the latter it was reported that introduction of two methoxy groups in the para-positions of benzophenone shifts the reduction potential to the negative of the conduction band edge of highly pure cadmium sulfide and thus prevents the reduction to alcohols as observed with other aromatic ketones.[60]

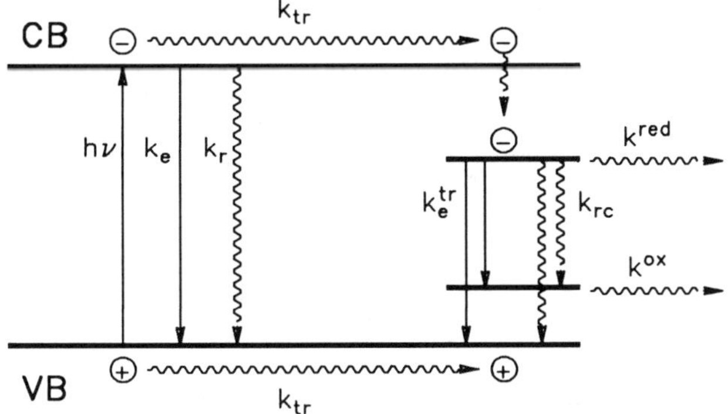

FIGURE 5. Basic photophysical processes of a photoexcited semi-conductor.

B. KINETIC ASPECTS

The kinetics of semiconductor-catalyzed photoreactions combines the basic features of heterogeneous catalysis and homogeneous stoichiometric photoreactions. Both aspects have been treated recently[9,61] and we therefore will focus only on some basic problems.

1. Basic Kinetic Parameters

A simplified energetic and kinetic scheme (Figure 5, Equations 19 to 23) where adsorption and desorption steps are omitted summarizes photophysical and primary photochemical processes.

We first consider the reaction $A + D \rightarrow B + C$ which is photocatalyzed by the semiconductor SC. Light absorption by SC (Equation 19a) occurs with the rate I_a (intensity of absorbed light) and affords a delocalized electron-hole pair which can recombine by radiative and radiationless processes (Equations 19b and 19c), or can be localized through trapping by surface states (Equations 19d and 19e).[62] The corresponding rate constants k_{tr} are assumed to be identical for holes and electrons.[63] The trapped electron-hole pair may undergo the reductive and oxidative primary reaction with the adsorbed substrates A and D (Equations 20 and 21) or recombine according to Equations 22 and 23. For simplicity, photocorrosion and consecutive reactions of D^+ and A^- to the end products are considered to be included in the radiationless processes according to Equations 19c and 23, and 20 and 21, respectively.

$$SC \xrightarrow{h\nu} SC^* \qquad\qquad I_a \qquad\qquad\qquad (19a)$$

$$SC^* \longrightarrow SC + h\nu' \qquad k_e[SC^*] \qquad\qquad (19b)$$

$$SC^* \longrightarrow SC + heat \qquad k_{re}[SC^*] \qquad\qquad (19c)$$

$$SC^* \longrightarrow e_{tr}^- \qquad\qquad k_{tr}[SC^*] \qquad\qquad (19d)$$

$$SC^* \longrightarrow h_{tr}^+ \qquad\qquad k_{tr}[SC^*] \qquad\qquad (19e)$$

$$e_{tr}^- + A_{ad} \longrightarrow A_{ad}^- \qquad k_r^{red}[e_{tr}^-][A_{ad}] \qquad\qquad (20)$$

$$h_{tr}^+ + D_{ad} \longrightarrow D_{ad}^+ \qquad k_r^{ox}[h_{tr}^+][D_{ad}] \qquad\qquad (21)$$

$$e_{tr}^- + h_{tr}^+ \longrightarrow h\nu'' \qquad\qquad k_e^{tr}[e_{tr}^-][h_{tr}^+] \qquad\qquad (22)$$

$$e_{tr}^- + h_{tr}^+ \longrightarrow \text{heat} \qquad\qquad k_{rc}^{tr}[e_{tr}^-][h_{tr}^+] \qquad\qquad (23)$$

Assuming stationary states for SC* and e_{tr}^-, h_{tr}^+, one obtains first $[SC^*] = I_a/k_e + k_{re} + k_{tr}$ and with $k_{tr}[SC]^* = e_{tr}^- I_a$ finally

$$\eta_{tr} = \frac{k_{tr}}{k_e + k_{re} + k_{tr}} \qquad\qquad (24)$$

When it is further assumed that $[e_{tr}^-] \approx [h_{tr}^+]$ then the steady-state approximation for e_{tr}^- leads to

$$\eta_{tr} I_a = [e_{tr}^-]k_r^{red}[A_{ad}] + [e_{tr}^-]^2(k_e^{tr} + k_{rc}^{tr}) \qquad\qquad (25)$$

This expression can be simplified for two extreme cases of light intensity.[64]

a. Light Intensity — Case 1

At high intensity, when approximately $I_a \geq 5{\cdot}10^{15}$ photons/s,[65] the quadratic term of Equation 25 is much larger than the first-power term and the latter may therefore be neglected. This means that the photophysical processes of charge recombination (Equations 22 and 23) prevail over the chemical reaction (Equations 20 and 21). In this case the steady-state concentration of the trapped charges is given by Equation 26.

$$[e_{tr}^-] = \sqrt{\frac{\eta_{tr}}{k_e^{tr} + k_{rc}^{tr}} \cdot I_a} \qquad\qquad (26)$$

When the rate of product formation corresponds to that of the primary reactions (Equations 20 or 21), substitution of $[e_{tr}^-]$ in Equation 20 by Equation 26 leads to the rate expression (Equation 27):

$$\text{rate} = k_r^{red}[A_{ad}] \sqrt{\frac{\eta_{tr}}{k_e^{tr} + k_{rc}^{tr}} I_a} \qquad\qquad (27)$$

Since, in general, the rate of a photoreaction is the product of light intensity absorbed and product quantum yield Φ_p, the latter is given by Equation 28 for the case discussed here.

$$\Phi_p = k_r^{red}[A_{ad}] \sqrt{\frac{\eta_{tr}}{k_e^{tr} + k_{rc}^{tr}} \frac{1}{I_a}} \qquad\qquad (28)$$

The linear dependence of reaction rate on the square root of I_a is typical for homogeneous and heterogeneous photoreactions where recombination of primary products is predominant. It has been observed for many semiconductor-catalyzed reactions like the TiO_2-catalyzed photooxidation of isopropanol,[64] the TiO_2/Pt-catalyzed photo-Kolbe reaction,[66] and the CdS-catalyzed photoisomerization of cis-stilbene.[67]

b. Light Intensity — Case 2

At low light intensity, approximately $I_a \leq 5{\cdot}10^{15}$ photons/s,[65] the quadratic term of Equation 25 can be neglected and the concentration of trapped electrons becomes now

$$[e_{tr}^-] = \eta_{tr} I_a/k_r^{red}[A_{ad}] \qquad\qquad (29)$$

Substitution of Equation 29 into the rate equation for product formation (Equation 20) affords Equation 30.

$$\Phi_p = \eta_{tr} \tag{30}$$

Thus, the product quantum yield is now equal to the efficiency of formation of the trapped electron-hole pair. Note that in this case the reaction rate increases linearly with the first power of I_a. This is observed when the above-mentioned reactions[64,67] are conducted at low light intensities.

An analogue consideration[68] for reactions involving two electrons or two holes (for example, $2\,e_{tr}^- + 2\,H_2O_{ad} \rightarrow H_2 + 2\,OH^-$) demonstrates that in this case the reaction rate should increase linearly with I_a over the whole intensity range. Examples are the zinc sulfide-catalyzed photodehydrodimerizations and the oxidation of formate to carbon dioxide (Chapters 4 and 5).

From the above discussion it is obvious that Φ_p may differ widely upon minor changes in catalyst preparation and solution composition. Higher light intensities may increase surface recombination,[69] and photophysical processes like emission are strongly influenced by adsorbed impurities.[70,71] It is therefore not surprising that comparison of quantum yields and selectivities reported by different research groups are not unambiguous. Even when for a series of substrates all the measurements of Φ_p were performed with one unique catalyst sample, care must be taken in the interpretation of the results since observed trends may not only reflect changes in k_r^{red} or k_r^{ox} (Equations 20 and 21), but also in η_{tr}. Furthermore, accurate determination of Φ_p is difficult in these heterogeneous systems[72] due to problems of light reflection and overlapping absorption spectra of substrate and photocatalyst. An example for the latter is the CdS-catalyzed photodealkylation of Rhodamin B where light absorption occurs by the dye and not by the semiconductor.[33-35]

Increasing the temperature may influence the reaction rate slightly by changing η_{tr}, more pronounced by increasing the rate of the slowest secondary reaction step, and also by shifting adsorption equilibria of educts and products. Depending on the relation of the two latter steps and on the temperature range investigated, the measured activation energy, E_a, may be zero, and larger or smaller than zero.[73-77] Since these factors depend on the detailed properties of the photocatalyst, minor modifications may induce variations of E_a. Accordingly, values of 1.7 and 4.2 kcal/mol were reported for two different CdS samples used in the photodimerization of *N*-vinylcarbazole.[78] No activation energy was observed for the ZnS-catalyzed photodehydrodimerization of cyclic ethers (Chapter 5).

2. Influence of Surface Area and Particle Size

For the influence of the specific surface area of the semiconductor powder on the rate of product formation two opposite effects are of major importance.[8] One is the rate of electron-hole recombination (Equations 22 and 23) which increases linearly with surface area and accordingly the reaction rate should decrease. The other is a linear increase of the reaction of the electron-hole pair with the adsorbed substrates (Equations 20 and 21) which should increase the overall reaction. It is therfore expected that, depending on the nature of semiconductor and substrates, the reaction rate, or Φ_p, may be constant, increase, or decrease with increasing surface area. This is nicely reflected by the CdS/Pt-catalyzed photoreduction of water by a mixture of sodium sulfide and sulfite. Highest Φ_p are observed with small surface areas and are constant up to 2 m²/g. From there a linear decrease to almost zero at a specific surface area of 6 m²/g takes place. Upon further increase to 100 m²/g this low quantum yield stays constant.[79]

Besides the surface area, the particle size[80] may govern the type of redox reaction which can be catalyzed and therefore determines the chemoselectivity of the photocatalyst. Especially the possibility whether one or more electrons are transferred from and to adsorbed substrates may be decided by the particle size. For an incident light intensity of 10^{17} photons/s and particles of 1000 and 10 nm diameter, time intervals of 10 and 100 ns, respectively, between the successive absorption of two photons can be estimated.[81] Thus, the chance of the one-electron reduction product to accept a second electron within its resident time at the surface should be much greater at a larger particle. However, until now there is no experimental evidence in the literature to support these arguments.

3. Inhibition by Hole and Electron Scavengers

One basic test for the presence of semiconductor photocatalysis in these heterogeneous sytems is to study the influence of hole and electron scavengers. A major problem is that not only the electron-hole pair, but also the primary products (A^-, D^+) may be quenched. In some cases a decision can be made on the basis of varying the redox potential of the scavenger. The inhibiting effect of 1,2,4,5-tetramethoxybenzene on the CdS-catalyzed dimerization of N-vinylcarbazole most likely occurs by quenching of the corresponding radical cation produced in the primary oxidative step (Equation 21).[78] Depending on the concentrations employed, nitrate may scavenge e_{tr}^- or the HCO_2^- radical which is the primary oxidation product of formate at illuminated ZnS (see Section IV.A).

C. PHOTOCORROSION

The industrial use of a mixture of zinc sulfide and barium sulfate as a white pigment (lithopone) at the beginning of our century already stimulated investigations to avoid the darkening occurring upon prolonged exposition to sunlight. In 1922 Lenard assumed that this photocorrosion process leads to a cleavage into sulfur and metallic zinc.[82] Shortly later it was shown that UV light and the presence of water are necessary to observe this photocorrosion.[83] The first quantitative investigation by Platz and Schenk[84] revealed the additional formation of hydrogen, zinc sulfate, and zinc hydroxide.

The decomposition of photoexcited ZnS into its elements seems to be a consequence of the quasi-element electronic configuration, "Zn(0)S(0)", of the lowest lying electron-hole pair.[85] Due to the highly covalent character of the ZnS bond, the effective charges of zinc and sulfur can be assumed as $+1$ and -1, respectively.[85-87] Since the valence band has sulfur 3p and the conduction band zinc 4s character,[88,89] the calculated quasi-element nature of excited ZnS becomes rationable.

The mechanism of photocorrosion of powders suspended in aqueous solutions seems to be the same for both zinc and cadmium sulfide. We focus on the latter since it was investigated in greater detail mainly because of its important role in the attempted cleavage of water to hydrogen and oxygen.[90] Unless otherwise cited the following summary is based on results from the groups of Memming[58] and Henglein.[3]

In the absence of air anodic photocorrosion (Equation 31)

$$CdS + 2h^+ \rightarrow Cd^{2+} + S \tag{31}$$

affords elemental sulfur and cadmium ions which are reduced to cadmium metal upon prolonged irradiation.

$$Cd^{2+} + 2e^- \rightarrow Cd(0) \tag{32}$$

This latter reaction competes with the reduction of lattice metal ions (Equation 33)

$$CdS + 2e^- \rightarrow Cd(0) + S^{2-} \tag{33}$$

In the case of ZnS or metallized CdS powder the reduction of water is a further competitive process (Equation 34)

$$2H_2O + 2e^- \rightarrow H_2 + 2OH^- \tag{34}$$

In the presence of oxygen, at short illumination times the overall reaction is

$$CdS + 2O_2 \xrightarrow{h\nu} Cd^{2+} + SO_4^{2-} \tag{35}$$

with the cathodic and anodic parts according to Equations 36 and 37, respectively.

$$O_2 + 4e^- + 2H^+ \rightarrow 2OH^- \tag{36}$$

$$CdS + 4h^+ + 2H_2O + O_2 \rightarrow Cd^{2+} + SO_4^{2-} + 4H^+ \tag{37}$$

The anodic process consists of Reactions 38 to 41 which are based on detailed comparative investigations of the processes occurring at single crystal electrodes and at powder suspensions.

$$CdS + h^+ \rightarrow Cd^{2+} + S_{ad}^{-\cdot} \tag{38}$$

$$S_{ad}^{-\cdot} + O_2 \rightarrow SO_{2ad}^{-\cdot} \tag{39}$$

$$SO_{2ad}^{-\cdot} + h^+ \rightarrow SO_{2ad} \tag{40}$$

$$SO_{2ad} + H_2O + 2h^+ \rightarrow SO_4^{2-} + 2H^+ \tag{41}$$

The fist step consists of the oxidation of sulfide to the $S^{-\cdot}$ radical (Equation 38) which then reduces oxygen via formation of the sulfur dioxide radical anion. The latter is finally oxidized via Equations 40 and 41 to the sulfate ion.

Note the surprising feature that oxygen is reduced via a preceding *primary oxidation* (Equation 39). The intermediate $S^{-\cdot}$ radical has been detected at ZnS powder by ESR[91,92] and at both colloidal sulfides by pulse radiolysis with UV-VIS detection (absorption maximum around 500 nm).[3,93] However, $S^{-\cdot}$ can also be formed via the *primary reduction* $S + e^- = S^{-\cdot}$ when elemental sulfur is present as impurity.

At prolonged irradiation time in the presence of oxygen the reduction of cadmium ions according to Equation 32 becomes more important and the overall corrosion reaction follows Equation 42.

$$2CdS + 2H_2O + 3O_2 \rightarrow 2Cd(0) + 2SO_4^{2-} + 4H^+ \tag{42}$$

All these photocorrosion processes are, of course, undesired and it is obvious that their relative importance strongly depends on the presence of surface states which may facilitate other reactions of the trapped electron-hole pair like recombination or redox reactions with adsorbed substrates. It is well known from ESR[69,70,92] and emission spectra[92] that most of these metal sulfide powders contain surface states. They are introduced during preparation of the powder as a result of lattice defects,[94,95] trapped holes,[92] surface impurities,[68] and metallization,[45] and during the catual catalytic reaction as a consequence of irradiation and substrate adsorption. The stabilizing effect of platinization is exemplified by Figure 6 for the ZnS-catalyzed reduction of water in the presence of formate.[96] Note that the presence

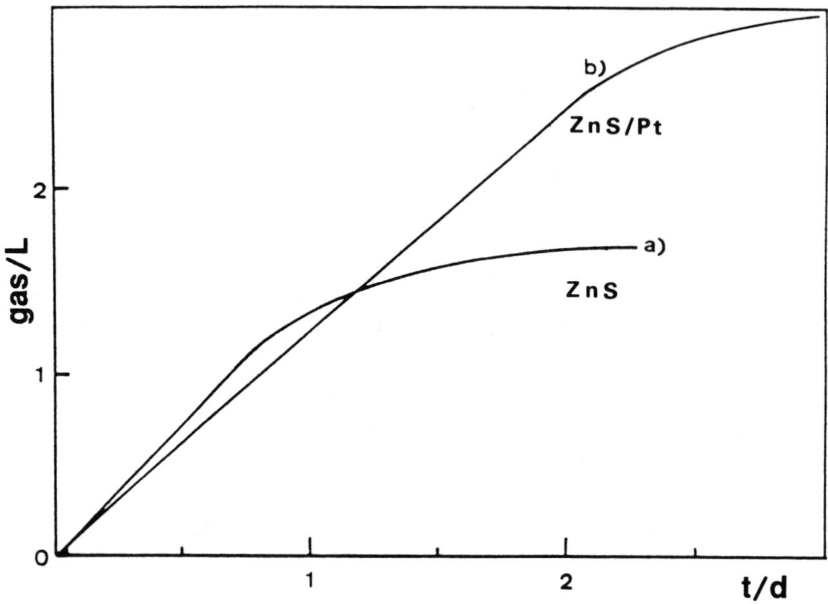

FIGURE 6.   Long-term catalytic activity of (a) nonplatinized and (b) platinized ZnS-B$_2$ in hydrogen production from aqueous sodium formate; $\lambda \geq 254$ nm.

of platinum does not influence the reaction rate, but doubles the time of constant catalytic activity from 1 to 2 d. This reaction is also an example for the suppression of photocorrosion by the adsorbed substrate, since complete darkening of ZnS employed in that study occurs in the absence of formate.

In accordance with this effect of platinum, $\Phi(H_2)$ in the system ZnS/2,5-DHF/H$_2$O is not increased but rather slightly diminished when platinized ZnS is employed.[68]

D. PREPARATION AND CHARACTERIZATION OF THE PHOTOCATALYST

From the results presented in the previous paragraphs of this chapter it follows that catalytic activity and selectivity of a semiconductor photocatalyst should depend on intimate details of the preparation method. In addition to careful description of these synthetic details it is important to characterize the obtained powder by bulk and surface elemental analysis, by determination of crystal structure, specific surface area and surface charge, by ESR and electronic absorption and emission spectroscopy, and by photoelectrochemical experiments to prove the semiconducting properties. The latter are of special importance in order to disfavor a general heterogeneous catalysis mechanism which usually does not involve simultaneous reductive and oxidative interfacial electron transfer steps. Their occurrence may be demonstrated by the effects of electron and hole scavengers added to the reacting system.

However, only a few of these numerous physical methods are available in a preparative oriented research group and therefore cooperation is a typical feature of this multidisciplinary field. Finally, the quantum yield of product formation and the turnover rate give information on the catalytic properties of the powder. Note that the latter and not the former criterion is decisive for the synthetic usefulness of the photocatalyst, since quantum yields are measured at initial reaction stages and do not contain information on the long-term catalyst photostability.

As an example for the characterization of a photocatalyst employed in preparative reactions we summarize some simple measurements performed on zinc sulfide powders.[68,79,96,97]

Zinc sulfide ZnS-A was prepared from zinc sulfate and sodium sulfide, ZnS-B$_1$ from zinc sulfate and thiourea.[94] Both samples contain amorphous and cubic ZnS. While elemental analyses of the bulk materials afforded a ratio Zn/S = 1.00/0.97 for both samples, surface analysis revealed values of Zn/S = 1.00/1.35 and 1.00/1.60 for ZnS-A and ZnS-B$_1$, respectively. This significant excess of surface sulfur may be responsible for the pronounced photostability of these samples as compared to previously employed one, ZnS-B$_2$, which had a surface excess of zinc, Zn/S = 1.00/0.74.[68,97] For ZnS-B$_1$, diffuse reflectance spectroscopy (Figure 7) points to the presence of an impurity with an absorbance maximum in the range of 450 nm. This is supported by emission and excitation spectra. Ultraband-gap excitation induces the appearance of three and two main emission bands (Figure 8A) for ZnS-A and ZnS-B$_1$, respectively. The bands around 400, 390, and 490 nm originate from recombination of the electron-hole pair scavenged at shallow and deep traps within the band gap.[98] The emissions at 525 and 690 nm are due to traces of zinc oxide as indicated by comparison with an authentic sample and by the excitation spectrum of ZnS-A observed at 550 or 720 nm and of ZnS-B$_1$ observed at 600 or 700 nm (Figure 8B). The absence of the 400-nm emission in ZnS-B$_1$ points to the presence of further surface states favoring radiationless charge recombination. Accordingly, the emission intensity of this sample is only one tenth of that of the other. However, the quantum yields of dehydrodimer formation from 2,5-DHF (Section V.A) are identical for both catalysts. This suggests that η_{tr} (Equation 24) has the same value in both cases.

In agreement with the presence of surface states, subband gap excitation at 390 nm (Figure 8A) also gives rise to strong emissions at 500 (ZnS-A) and 600 nm (ZnS-B$_1$). The latter originates from traces of manganese(II), a common impurity in zinc sulfide.[99] The impurity present in ZnS-B$_1$ is much better detectable by the excitation spectrum, λ_{em} = 500 nm, through the maximum at 450 nm (Figure 8B).

It was further reported that grinding of CdS[100] or ZnS[95] decreases the emission intensity by introduction of surface states. The latter are apparently absent in quantized microcrystallites of ZnS prepared at 0°C.[101]

Despite these, although small differences in the surface properties, both catalysts induce the same initial reaction rates in the dehydrodimerization of cyclic unsaturated ethers (see Section V.A). Traces of zinc oxide inhibit this reaction; they can be removed by washing with acetic acid.

The n-type character of these zinc sulfides is demonstrated by the observation of an anodic photocurrent upon illumination of a powder-coated electrode.[96] In the presence of sodium formate the onset of this current occurs at -0.45 V (vs. NHE), which is about 1.4 V more positive than the flat-band potential of zinc sulfide (see Section IV.A). The action spectrum of the photocurrent parallels the wavelength dependence of $\Phi(H_2)$ in the photoreduction of water by formate (Figure 9). Together with the inhibiting effect of sodium nitrate as scavenger for e_{cb}^- these results clearly support the presence of semiconductor photocatalysis for this reaction (see Section IV.A).

As mentioned in Section III.A, adsorption of ions may shift the energy bands. In addition it may alter the adsorption properties of the semiconductor surface toward the substrates and therefore influence the selectivity of the reaction. Since ions can already be adsorbed during catalyst preparation it is recommendable to determine the "point of zero zeta potential" (pzzp).[102] This is conveniently done by measuring the pH change occurring upon titration of the powder suspension adjusted to pH = 10 before. In the case of ZnS-B$_1$ the maximum at pH = 6.1 indicates a weak alkaline surface (pH = 7.9, Figure 10).

While zinc sulfides usually exhibit good catalytic properties without the need of platinization, this is generally necessary for cadmium sulfides. Otherwise photocorrosion predominates and prevents application in preparative reactions. Metallization with noble metals like platinum and palladium partially suppresses this undesired process. The most commonly employed platinization procedure is the photoreduction of hexachloroplatinate at the CdS

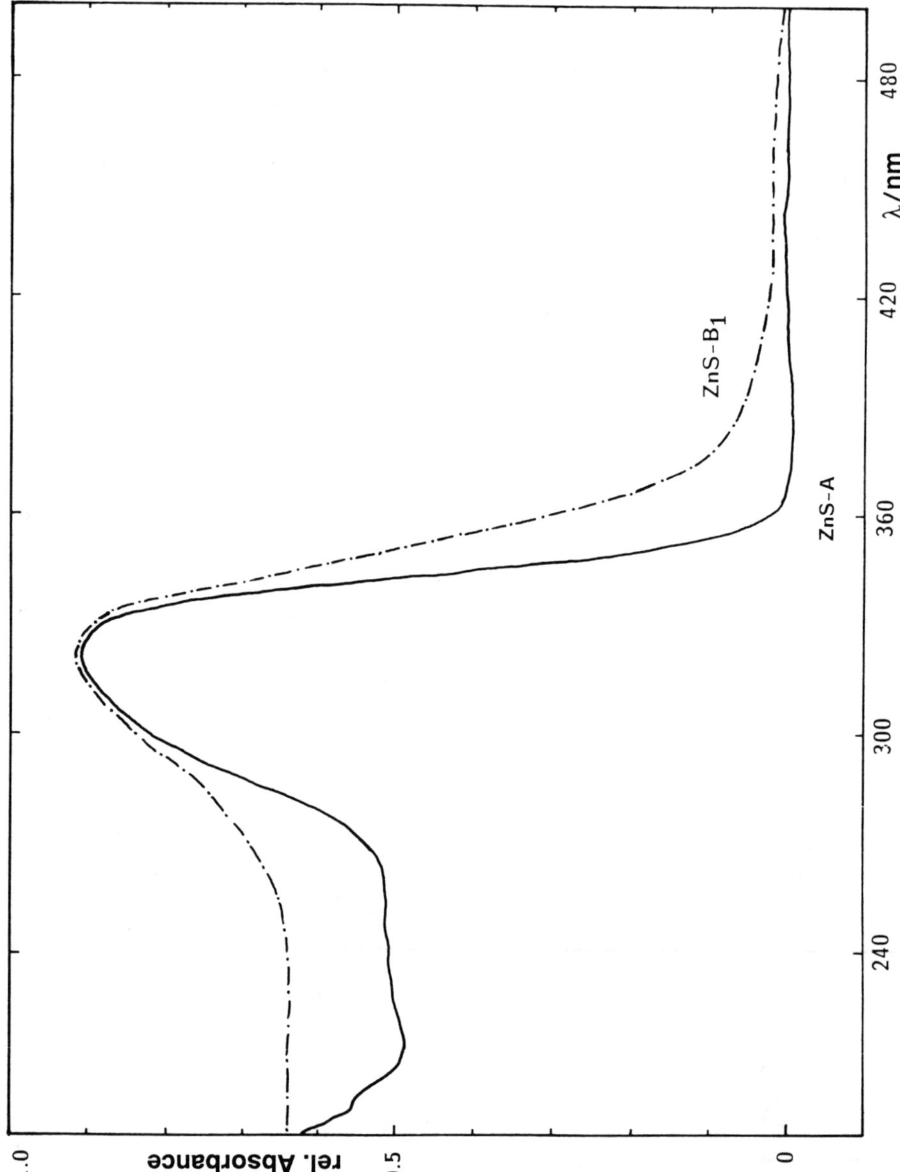

FIGURE 7. Diffuse reflectance spectra of ZnS-A and ZnS-B$_1$.

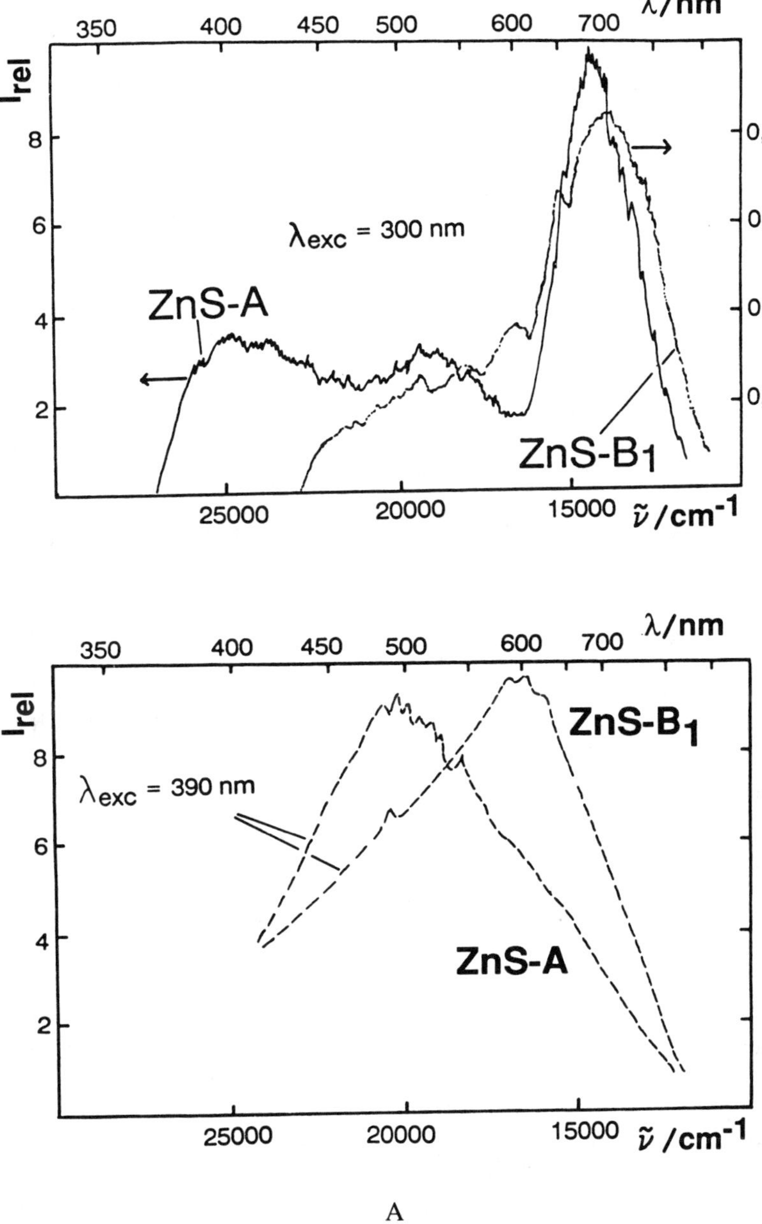

FIGURE 8.   (A) Corrected emission spectra of ZnS-A and ZnS-B$_1$ (note the different intensity scales); (B) corrected excitation spectra of ZnS-A and ZnS-B$_1$.

surface.[103] Size, shape, and density of the produced metal islands determine the catalytic activity of the resulting CdS/Pt powder. These factors are influenced by light intensity,[104] nature of the reducing agent,[105-108] and by the presence of zinc ions.[109] The latter considerably improves the catalytic activity of CdS/Pt powders in the photodehydrodimerization of 2,5-Dihydrofuran (2,5-DHF).[109] Without zinc ions a maximum of reaction rate is observed when the platinum content reaches 4.3 mol% (Figure 11). In the hydrogen-evolving systems CdS/Pt//Na$_2$S/H$_2$O[106] and CdS/Pt/Kaolwool//EDTA/H$_2$O[107] the corresponding maxima occur at similar Pt concentrations but are less pronounced.

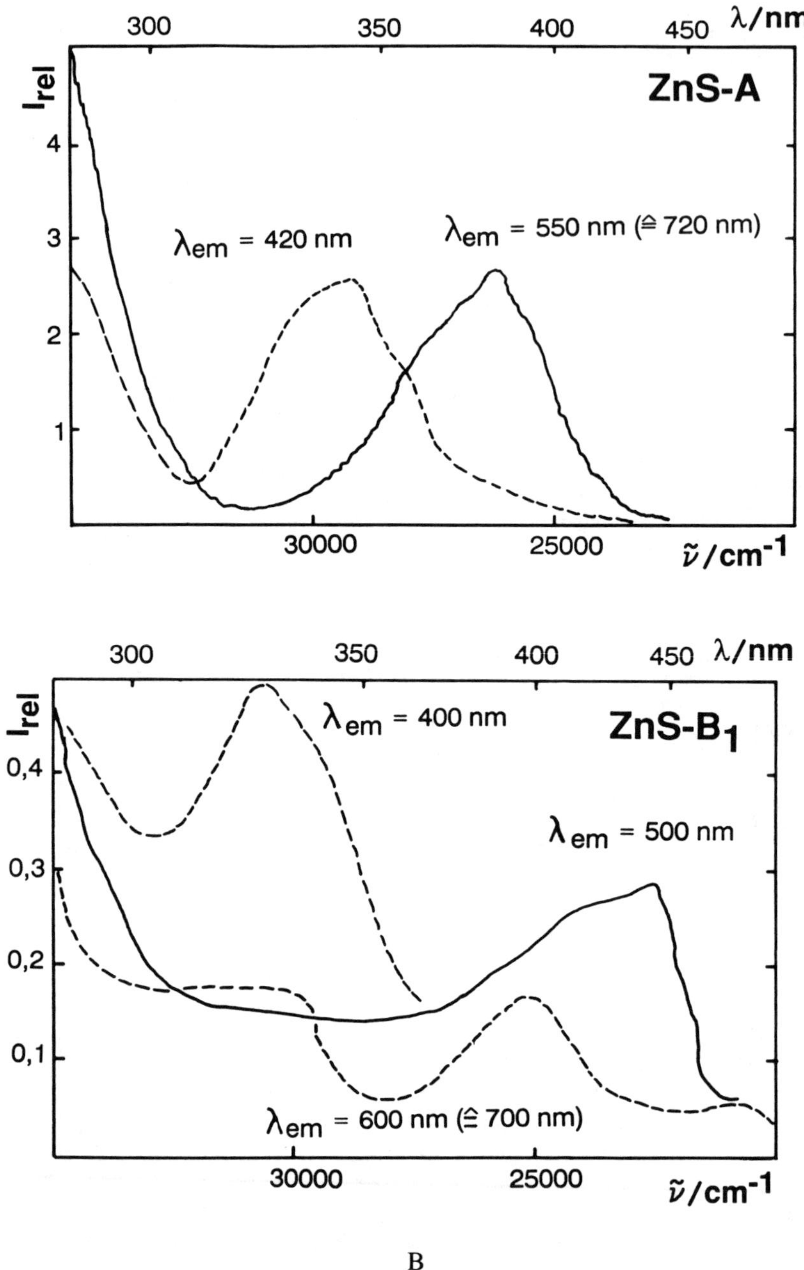

B

IV. PHOTOREDUCTION OF WATER

The catalytic activity of ZnS in the photochemical hydrogen evolution from water was discovered in 1977 during explorative work on the use of metal dithiolenes as homogeneous photocatalysts.[110,111] In the case of zinc dithiolenes it turned out that these soluble complexes were precursors of the heterogeneous catalyst ZnS.[112] Following our initial observation on the catalytic properties of the zinc complexes, the activity of ZnS was independently confirmed by Yanagida et al.[113] Although ZnS in general is quite unstable toward photocorrosion (see Section III.C) it may gain stability for several days when water reduction is performed with mild reducing agents like 2,5-dihydrofuran,[97] sodium formate,[96] and sodium sulfide.[114]

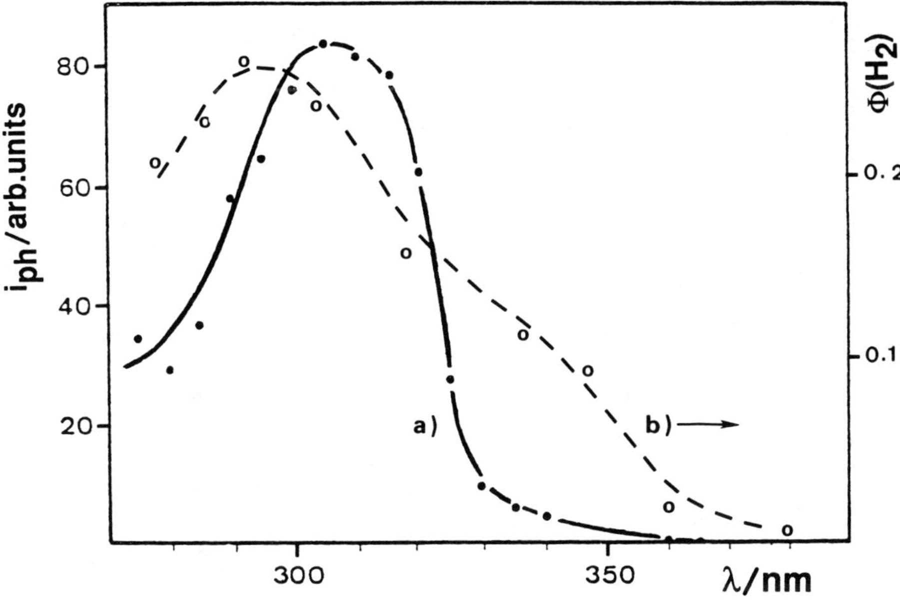

FIGURE 9.   Wavelength dependence of (a) the photocurrent obtained with $ZnS-B_2$ powder deposited on a platinum electrode (0.5 V vs. NHE, 0.015 M HCO_2Na, 0.1 M K_2SO_4) and (b) quantum yield of hydrogen production (mol H_2/Einstein) in the system $ZnS-B_2$/1.84 M HCO_2 Na.

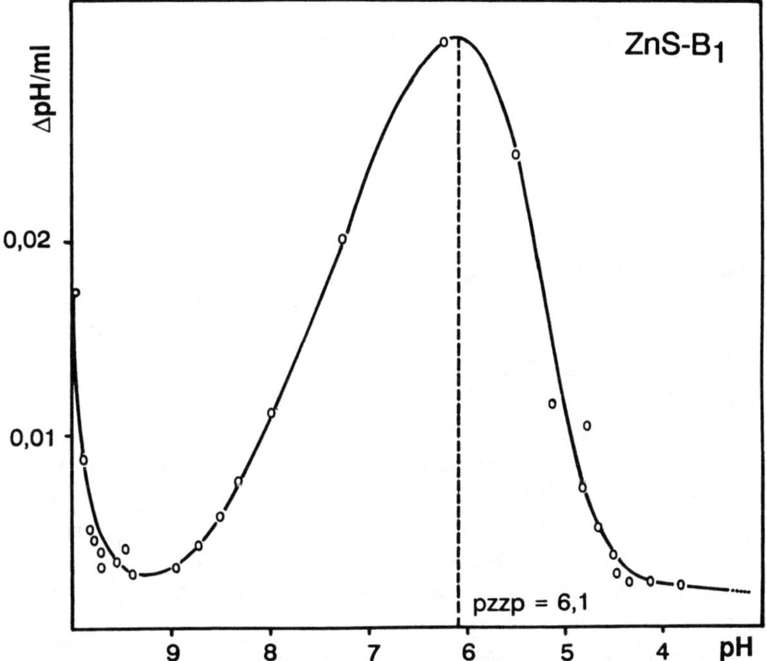

FIGURE 10.   Differential potentiometric titration of a $ZnS-B_1$ suspension of pH 10 with 0.01 M HNO_3.

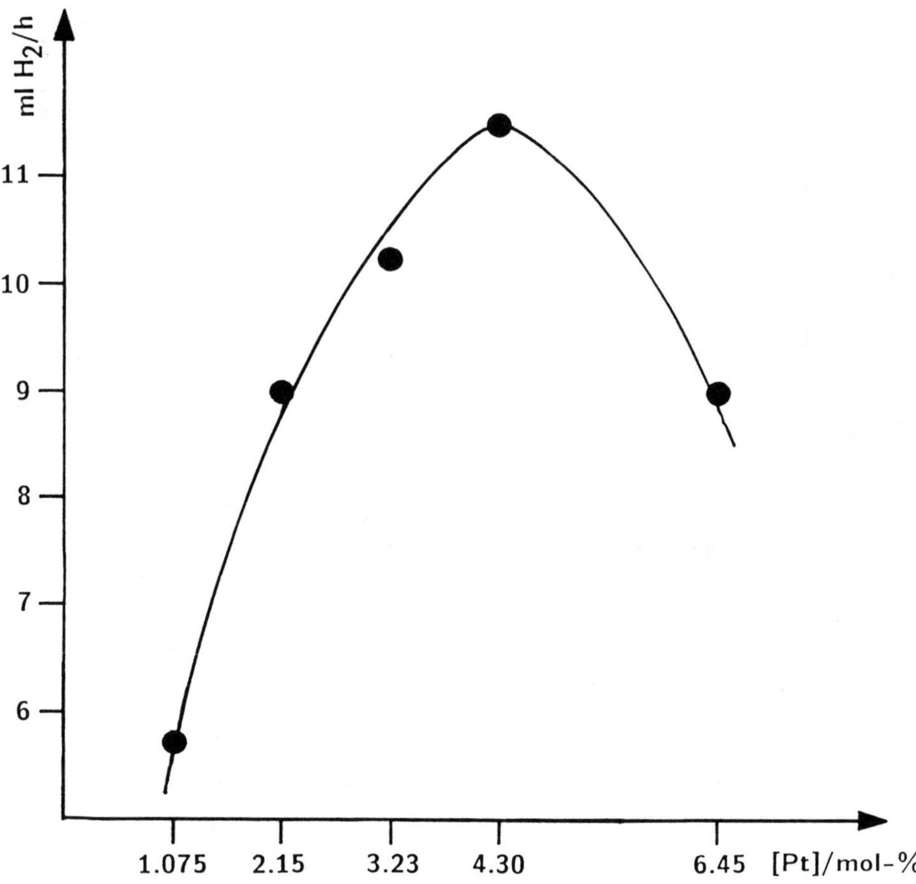

FIGURE 11.   The dependence of the reaction rate of the photodehydrodimerization of 2,5-dihydrofuran (2,5-DHF) on the amount of platinum in CdS/Pt; 2,5-DHF/H_2O = 1/14 (v/v), $\lambda \geq$ 290 nm.

Even carbon monoxide may serve as the source of electrons.[115] In this chapter we summarize the results obtained with formate and carbon monoxide in the presence of ZnS as photo-catalyst. Unless otherwise noted all the results originate from References 96 and 115.

A. SODIUM FORMATE AS REDUCING AGENT

Formic acid and sodium formate have been used as reducing agents for the photofor-mation of hydrogen from water catalyzed by titanium dioxide,[116] platinized[117] or nonplatin-ized cadmium sulfide,[118] and by colloidal zinc sulfide.[119] In general, only small amounts of hydrogen were obtained, no complete material balance[120] was established, and the mechan-istic investigations were rather incomplete. In contrast to that, ZnS-B_2, prepared from zinc sulfate and thiourea,[94] is an efficient photocatalyst which stimulates the formation of liters of hydrogen (Figure 6). The mechanism of this reaction was studied in detail. Analogously prepared CdS does not catalyze the reaction.

ZnS-B_2 consists of particles with an average diameter of 200 nm which form larger aggregates upon suspension in water. Most of the powder is amorphous; the crystalline part has a cubic structure. The specific surface area is 17 m^2/g and the ratio of Zn/S is 1:1 and 1:0.74 in the bulk and at the surface, respectively. The latter drops to 1:0.47 after some hours of irradiation and indicates that formate does not completely inhibit the photocorrosion to Zn(0) and dissolved sulfide. Differential titration (see Section III.D) affords a maximum at pH 8 indicating a slightly acidic surface. Finally, photoelectrochemical experiments with

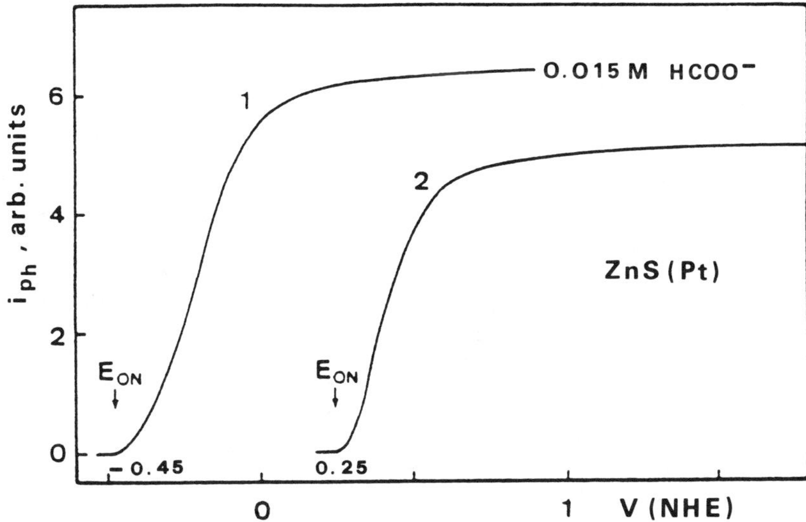

FIGURE 12. Photocurrent-potential curves of ZnS powder deposited on platinum; λ = 310 nm; 0.1 M K$_2$SO$_4$ with (1) and without (2) sodium formate.

a ZnS-B$_2$-coated platinum electrode proves the anticipated *n*-type character. The onset of the photocurrent at -0.45 and $+0.25$ V (vs. NHE) in the presence and absence of formate, respectively, points to the existence of surface states (Figure 12) at these potentials. For the higher saturation current in the presence of formate, electron injection into the conduction band, "current-doubling",[121] is most likely responsible.

The maximal TOR is in the range of four monolayers of formate/s.[122] This high catalytic activity is obtained only with fully hydrated ZnS and in the absence of more than trace amounts of zinc oxide.

A complete material balance established that the product ratio of hydrogen to carbonate is one. Figure 13 shows the concentration changes as function of irradiation time. Note that small amounts of oxalate are produced, also. The composition of the gas phase after 60 min irradiation time is 72% H$_2$, 27% CO, 0% CO$_2$, and 1% unidentified components. Addition of about 20% (v/v) of THF to the aqueous suspension induces a complete disappearance of carbon monoxide. The latter is also absent when the catalyst is prepared photochemically from a zinc dithiolene. In this case, the catalyst contains sulfur and sulfur compounds as trace impurities.[123] Accordingly, addition of a disodium ethylen-1,2-dithiolate to the aqueous suspension of ZnS-B$_2$ also inhibits CO formation. These consequences of minor changes in the composition of catalyst or solution nicely demonstrate that impurities may be a crucial factor for product selectivity.

The stoichiometry of the hydrogen-producing reaction is therefore given by Equation 43.

$$\text{HCO}_2\text{Na} + \text{H}_2\text{O} \xrightarrow{\text{ZnS/h}\nu} \text{H}_2 + \text{HCO}_3\text{Na} \tag{43}$$

The quantum yield $\Phi(\text{H}_2)$ is 0.24 ± 0.02 mol H$_2$ per Einstein at 290 nm and increases to 0.31 ± 0.02 when the light intensity is decreased to one tenth of its original value. Each point of Figure 14 is the average of three measurements and the vertical bar indicates the standard deviation. This, although slight, decrease of $\Phi(\text{H}_2)$ with increasing light intensity most likely originates from increased surface recombination of the electron-hole pair (Section III.B.1).[69]

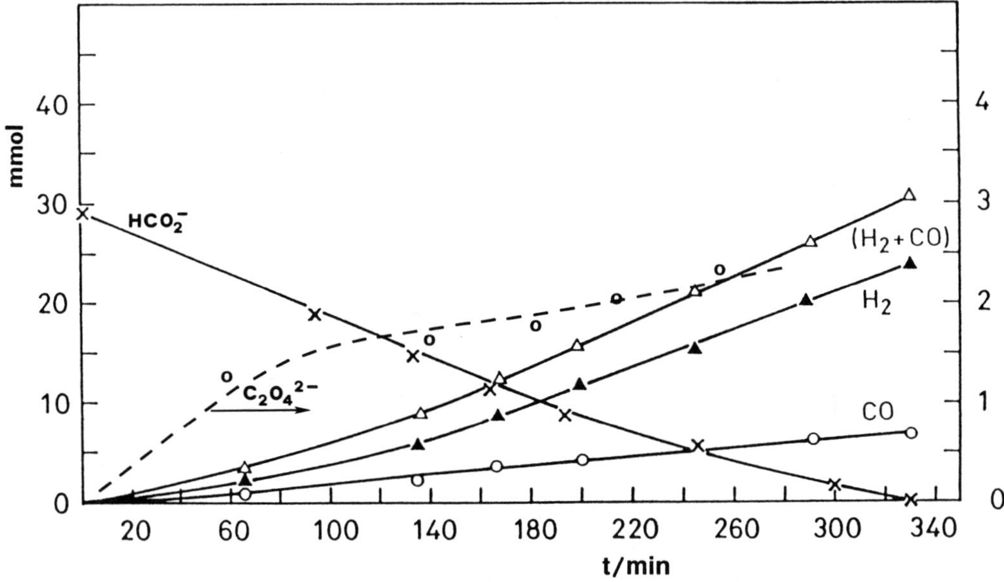

FIGURE 13.   Concentration changes of educts and products observed in the photoreduction of water by sodium formate (0.245 *M*) catalyzed by ZnS-B$_2$; λ ≥ 254 nm; note different concentration range for oxalate.

The reaction rate exhibits a dependence on the formate concentration which resembles a Langmuir adsorption isotherm. Accordingly, a graph of reciprocal rate as function of reciprocal concentration affords a straight line[124] from which an approximate adsorption constant of K = 26 M^{-1} can be calculated (Figure 15).[125]

From these experimental results the following mechanism was proposed. Excitation of ZnS and other photophysical processes (see Section III.A), desorption of products, and hydration of carbon dioxide are omitted for the sake of simplicity. The adsorption equilibria of water and formate (Equations 44 and 45) are supported by the fact that

$$H_2O + \square ZnS \rightarrow H_2O_{ad} \tag{44}$$

$$HCOO^- + \square ZnS \rightarrow HCOO^-_{ad} \tag{45}$$

no reaction occurs with dehydrated zinc sulfide nor with ZnS-B$_2$ in the presence of non-aqueous formic acid, and by the Langmuir-type dependence of rate on the formate concentration (Figure 15), respectively.

The observation that aqueous formic acid reacts three times slower, producing a gas of quite different composition, 15% H$_2$, 76% CO$_2$, 8% CO, and 1% CH$_4$, indicates that adsorbed formate ion is the species which is oxidized by the trapped hole (Equation 46).

$$h^+_{tr} + HCOO^-_{ad} \rightarrow HCOO^{\cdot}_{ad} \tag{46}$$

$$HCOO^{\cdot}_{ad} + H_2O_{ad} \rightarrow CO_2 + H_3O^+_{ad} + e^-_{tr} \tag{47}$$

$$2e^-_{tr} + 2H_2O_{ad} \rightarrow H_2 + 2OH^-_{ad} \tag{48}$$

This reaction is exergonic by about 0.8 eV since E^0 (CO$_2^-$, H$^+$/HCO$_2$) = 1.0 V[126] and the valence band edge is located at about 1.8 V (see Section III.A).

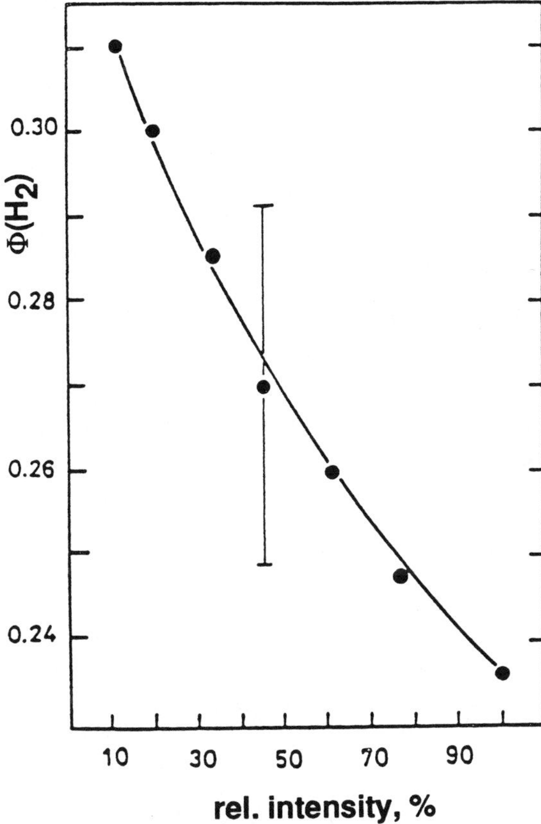

FIGURE 14. Intensity dependence of product quantum yield of the photoreduction of water; λ = 290 nm, experimental conditions as in Figure 9B.

Electron injection from the primary oxidation product $HCO_2\cdot$ into the conduction band (Equation 47) is suggested by the photoelectrochemical "current-doubling" and scavenging experiments with sodium nitrate. The latter is known to scavenge electrons in a diffusion-controlled process according to $NO_3^- + e_{aq}^- \rightarrow NO_3^{2-}$ and $NO_3^{2-} \rightarrow NO_2^- + OH + OH^-$.[127] A Stern-Volmer plot reveals that two species are scavenged (Figure 16). After a linear increase of Φ_0/Φ a plateau is reached at 0.008 M nitrate. At this point CO formation is completely suppressed and therefore CO and half of the hydrogen apparently originate from a common intermediate. This is unlikely to be the $CO_2^{-\cdot}$ radical, whose presence follows from formation of oxalate and spin trapping experiments, since its radiolytic generation in a suspension of ZnS-B_2 does not increase the yield of hydrogen. It is also unlikely that at this low concentration nitrate can scavenge e_{tr}^- in competition with the adsorbed water. More likely is the assumption that it is the formate radical[126] which is completely quenched in the concentration range of the plateau from 0.008 to 0.03 M. This is supported by the fact that under these conditions $\Phi(H_2)$ is decreased to exactly half of its initial value, indicating that formate delivers now only one electron for the reduction of water. At higher concentrations nitrate can successfully compete with water for e_{tr}^-, and $\Phi(H_2)$ starts to decrease again; a value of 0.01 is reached at 1.8 M nitrate.

No hydrogen evolution occurs when formate is replaced by KSCN, usually an efficient hole scavenger. This observation and the appearance of a surface state at -0.45 V in the photoelectrochemical experiment in the presence of formate suggest that hydrogen evolution (Equation 48) occurs via this intraband-gap state. In accord with Equations 46 to 48 the pH

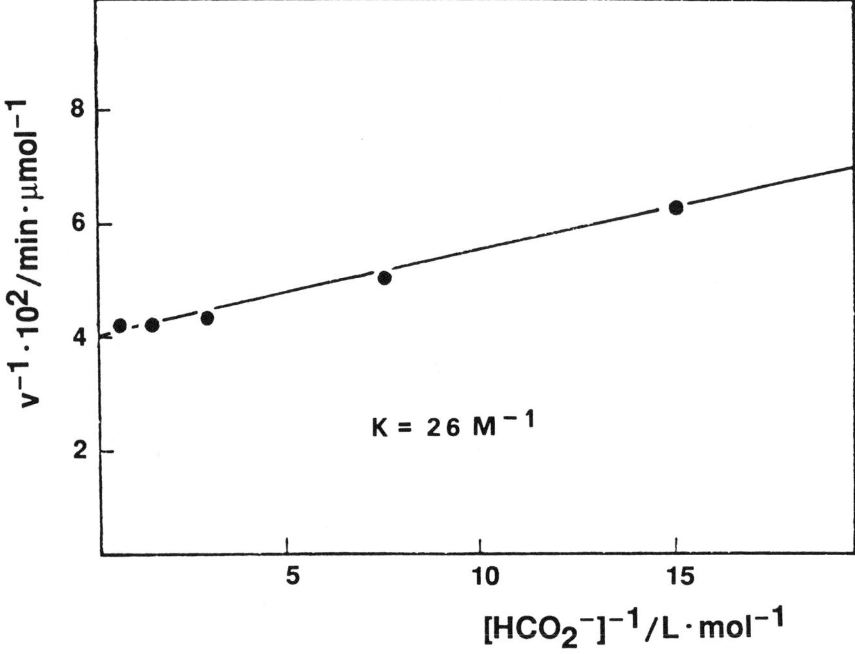

FIGURE 15.   Dependence of reciprocal reaction rate on reciprocal sodium formate concentration in the photoreduction of water catalyzed by ZnS-B$_2$; experimental conditions as in Figure 13.

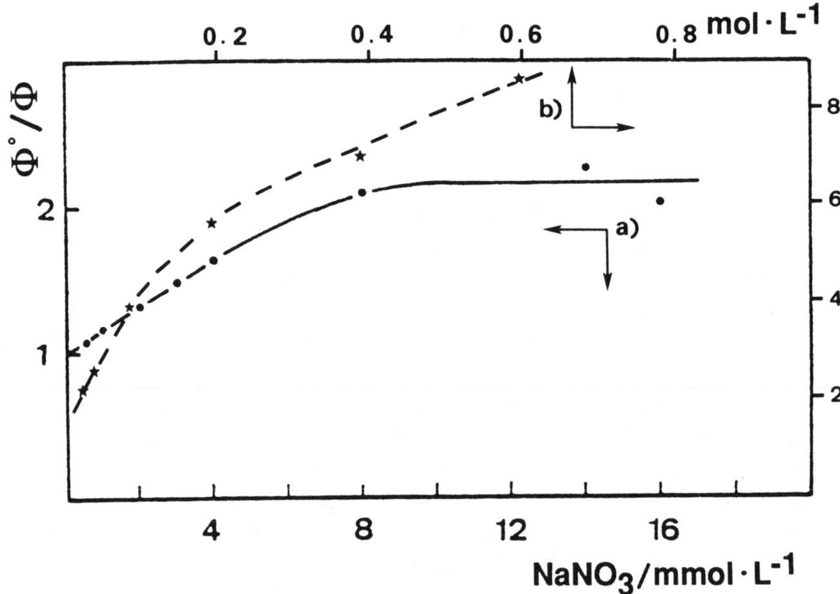

FIGURE 16.   Stern-Volmer plot for the inhibition of hydrogen formation by sodium nitrate; ZnS-B$_2$/1.83 M HCO$_2$Na, λ = 290 nm.

value of the suspension increases from 8 to 9, to 10 to 11, after several hours of irradiation. When KSCN is added to the complete system in a tenfold excess relative to formate, $\Phi(\text{H}_2)$ changes from 0.24 to 0.28. Since the oxidized product $(\text{SCN})_2^-$ could not be detected, it is unclear whether the effect originates from more efficient hole scavenging or by a photophysical effect like an increase of η_{tr}.

Experiments with deuterated substrates support the molecular mechanism for hydrogen evolution (Equation 48) and illustrate the importance of substrate adsorption. Replacing HCO_2^- by DCO_2^- results in a small rate decrease of $r(H)/r(D) = 1.3$. This suggests that the slowest reaction step is the electron injection according to Equation 47. The fact that only H_2 but no HD or D_2 can be detected even in $0.7\,M\ DCO_2^-$ solution is due to the comparative much higher concentration of H_2O (mole fraction $x_H \approx 1$). This observation supports Equation 48 and disfavors a reduction mechanism via adsorbed hydrogen atoms which should afford HD by D-abstraction from DCO_2^-. When in aqueous HCO_2^- water is substituted by D_2O, the rate does not change, but the reaction ceases after a few hours irradiation time due to enhanced photocorrosion, as indicated by complete darkening of ZnS. The isotopic composition of the hydrogen gas changes from an initial value of 85% D_2, 10% HD, and 5% H_2 to 63% D_2, 30% HD, and 7% H_2 when about half a liter of gas has been produced. This relative high content of HD suggests that D_2HO^+ and HDO, formed according to Equations 47 and 48, are so strongly adsorbed that they do not completely exchange with the large excess of solution D_2O.[128] The increasing H/D exchange with increasing consumption of formate is in agreement with Equations 47 and 48 (see also Chapter 5). Accordingly, no change of the isotopic composition is found when D_2O and DCO_2^- are employed.

The CO-producing reaction part is formulated by Equations 49 to 52 and has been discussed in detail elsewhere.[96]

$$HCOO_{ad}^{\cdot} + H_2O_{ad} \rightarrow H_3O_{ad}^+ + COO_{ad}^{-\cdot} \tag{49}$$

$$COO_{ad}^{-\cdot} \rightarrow COO^{-\cdot} \tag{50}$$

$$2CO_2^{-\cdot} \rightarrow C_2O_4^{2-} \tag{51}$$

$$2CO_2^{-\cdot} + 2H^+ \rightarrow CO + CO_2 + H_2O \tag{52}$$

B. CARBON MONOXIDE AS REDUCING AGENT

In contrast to the well-investigated thermal catalysis of the water gas shift reaction only a few studies deal with the photocatalysis of this technically important process. In homogeneous solution a rhodium bipyridyl complex was used as a photocatalyst.[129] In the field of semiconductor photocatalysis the reaction was investigated in detail in the system $TiO_2/$ Pt with gaseous water.[130,131] It was also reported that aqueous suspensions of SiC or CdS are able to produce hydrogen.[132] We have observed that suspensions of ZnS-B_2 at pH 14 catalyze the photoreduction of water by carbon monoxide according to Equation 53.

$$CO + 2OH^- \xrightarrow{\ \ ZnS,h\nu\ \ } CO_3^{2-} + H_2 \tag{53}$$

The quantum yield of hydrogen formation is 0.008 at 320 nm. The maximum TON in the range of 3000 monolayers/20 h indicates the catalytic nature of the process. A typical CO uptake curve is shown in Figure 17. The reaction ceases after 25 h of irradiation. This is due to concomitant photocorrosion to elemental zinc which is not as efficiently inhibited as in the case of sodium formate as reducing agent.

For the TiO_2/Pt-catalyzed reduction it was postulated that the trapped hole oxidizes adsorbed CO and OH^- to CO_2 and a surface hydrogen atom. The latter is supposed to recombine with another one which is formed by the reduction of water through the trapped electron.[130,131] Since the ZnS-catalyzed reaction is very slow at pH 10 or 7, it is assumed that adsorbed hydroxide may also be involved. As further possibility the authors mention that hydroxide may oxidize sulfur impurities to adsorbed sulfur radicals which may function

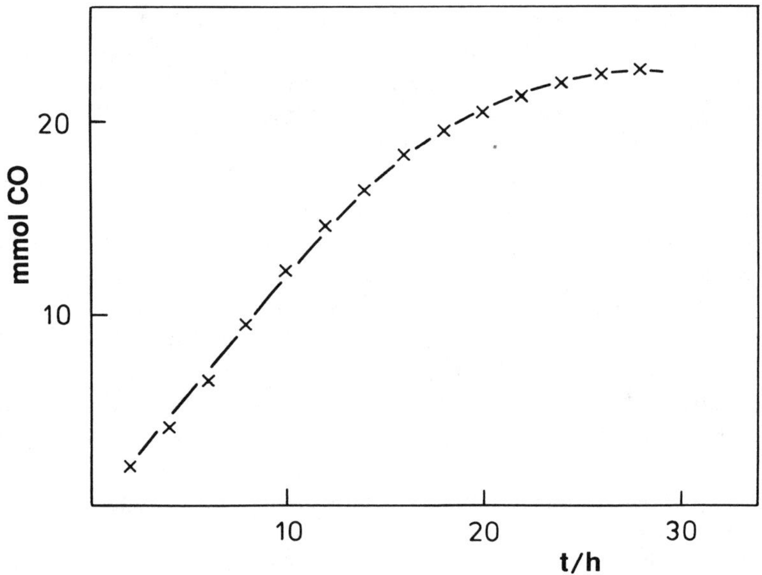

FIGURE 17.   Uptake of carbon monoxide as function of irradiation time in the photoreduction of water catalyzed by ZnS-B$_2$; pH 14, $\lambda \geq 254$ nm.

as active sites for the reaction. Such a type of reaction was recently reported for homogeneous solutions.[133]

V. PHOTOINDUCED C–C AND C–N COUPLING REACTIONS

This chapter summarizes the few reactions which are of preparative significance in the field of semiconductor photocatalysis. It centers on the dehydrodimerization of cyclic enol or allyl ethers and cyclic olefins. In addition, a new type of insertion reaction between 1,2-diazenes and cyclic ethers is discussed.

The few other C–C bond forming reactions all lead to well known compounds and have not been performed on a preparative scale.[92,119,134-136] Unless otherwise noted, the following results originate from Ref. 68, 97, 112, 137, 138.

A. DEHYDRODIMERIZATION OF CYCLIC ETHERS AND OLEFINS
1. Catalyst Characterization

The zinc sulfide powder used in all reactions described in this and the following chapters was prepared as described for ZnS-B$_2$,[94,96] except that the hydrolysis time of thiourea was increased. This affords the catalyst ZnS-B$_1$ which has an inverted Zn/S ratio, namely, 1:1.6 at the surface, while the ratio of the bulk material, 1:0.97, is almost the same as for ZnS-B$_2$ (see Sections II.D and IV.A). The particle diameter varies from 0.5 to 10 μm, and the specific surface area of 100 to 170 m^2/g, as determined from BET measurements, consists of 80 to 90% of inner surface. The most important difference to ZnS-B$_2$ is the complete absence of photocorrosion when ZnS-B$_1$ is irradiated in an aqueous suspension. This unprecedented photostability is not induced by the large specific surface area but rather by surface sulfur compounds, since ZnS-C, obtained from diethyl zinc and H$_2$S, has only a surface area of 34 m^2/g but is also photostable. As already discussed, the presence of surface impurities increases charge recombination (Equations 19a to 19c, 22, and 23) on expense of photocorrosion (Equations 31 to 33; see also Section III.D).

CdS/Pt was prepared from sodium sulfide and cadmium sulfate. The resulting powder

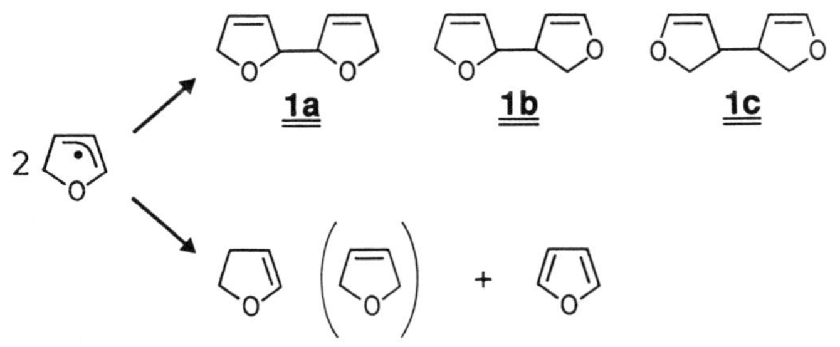

FIGURE 18.   Structure of products obtained by irradiation of zinc sulfide in 2,5-dihydrofuran/water.

had a specific surface area of 120 m²/g and a particle size of 0.5 to 2 μm.[138] Platinization was performed photochemically.[103]

2. Photocatalysis by Zinc Sulfide

When ZnS-B$_1$, suspended in aqueous 2,5-dihydrofuran (2,5-DHF), is illuminated with a high-pressure mercury lamp for 4 d, 4 l of hydrogen and 10 g of the dehydrodimers of the cyclic ether are produced. The latter are easily separated by distillation after filtering off the catalyst powder. Isolated yields are in the range of 30 to 60%. The catalytic nature of the reaction is indicated by the values of TOR which are in the range of 30 monolayers 2,5-DHF/h. Diethyl ether, allyl vinyl ether, methanol, isopropranol, dimethyl amine, and dioxane do not react. In contrast to that, the three latter compounds were photodehydrodimerized using *in situ*-prepared ZnS,[139] and isopropanol by colloidal ZnS.[119]

The structure of the dehydrodimers **1a, 1b, 1c** and their statistical ratio of 1:2:1, respectively, suggest that they are formed by dimerization of an intermediate dihydrofuryl radical (Figure 18). In agreement with this proposal is the production of small amounts of the disproportionation products furan and 2,3-DHF. Accordingly, **1a** to **1c** are also formed when the dihydrofuryl radical is produced via hydrogen abstraction by the OH· radical, which may be generated by γ-radiolysis of water or photolysis of hydrogen peroxide. In these thermal reactions, however, large amounts of oligomers and 3-hydroxytetrahydrofuran are formed. The fact that the latter compound is completely absent in the ZnS-B$_1$-catalyzed reaction suggests that OH is no intermediate in the heterogeneous system.

3. Mechanistic Aspects

Water is a necessary constituent of the system since no reaction occurs when it is absent. In the presence of D$_2$O the initial isotopic composition of hydrogen is 90% D$_2$ and 10% HD. Although the appearance of D$_2$ implicates that water is consumed, this is not the case as indicated by a material balance which included the determination of water. This is an important observation since the formation of D$_2$ in many "sacrificial" systems for water reduction was usually taken as proof for water consumption. Thus the overall reaction obeys the stoichiometry of Equation 54 and some exchange process between water and 2,5-DHF must operate.

$$2RH_2 \xrightarrow[\text{H}_2\text{O}]{\text{ZnS},h\nu} HR\text{--}RH + H_2 \tag{54}$$

In fact, the amount of HD and H$_2$ increases with increasing reaction time on expense of D$_2$ (Figure 19). Since the concentration of D$_2$O is always much larger than that of 2,5-DHF, significant changes of the isotopic composition are observable only when at least 10 to 20 ml of gas was produced.

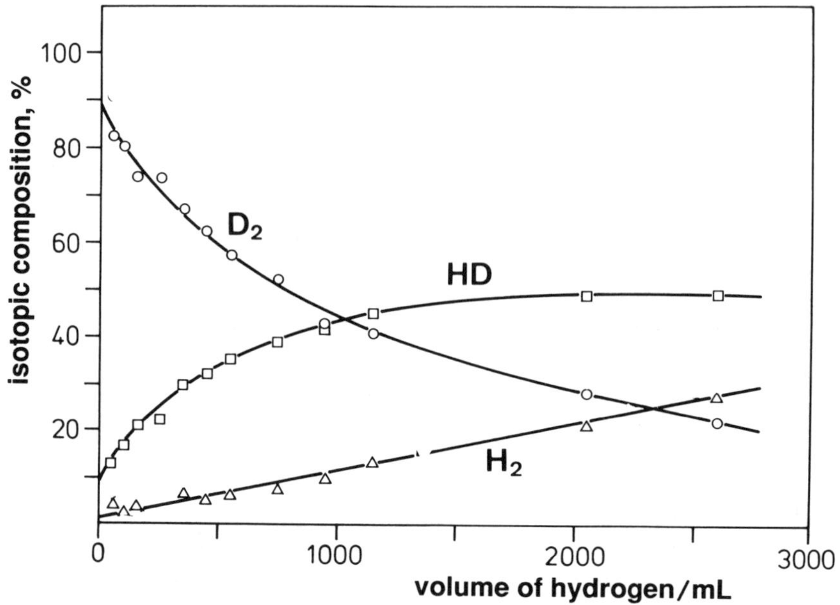

FIGURE 19.   Change of isotopic hydrogen composition in the system ZnS-B₂/2,5-DHF/D₂O; $\lambda \geq 254$ nm.

The reaction rate increases linearly with light intensity, indicating that the product quantum yield is equal to the efficiency of charge generation (Section III.B, Equation 30). No change in the rate is observed by varying the pH value from 4 to 10 and by raising the temperature from 20 to 50°C.

These results suggest that the oxidative primary reaction is formation of the cation radical $RH_2^{+\cdot}$ by the transfer of an electron from adsorbed 2,5-DHF (RH_2) to h_{tr}^+ (Equation 21). Assuming that the band positions of an Al-doped ZnS[53] crystal are also valid for the powder used in these studies, this reaction is more likely than the oxidation of water (Figure 20).

Deprotonation of the radical cation (Equation 55) is driven by the large hydration energy of the proton and may be one reason for the observation that at least about 5% (v/v) of water has to be present to observe a reasonable reaction rate. Since the cyclic

$$2RH_{2ad}^{+\cdot} + 2H_2O_{ad} \rightarrow 2RH_{ad}^{\cdot} + 2H_3O_{ad}^{+} \tag{55}$$

ethers are irreversibly oxidized at a platinum electrode, calculated potentials[140] for the reaction $RH_2 \rightarrow RH\cdot + H^+ + e^-$ are given in Figure 20.

When the second photon is absorbed, a second radical is produced, and statistical dimerization affords the products (Equation 56).

$$2RH_{ad}^{\cdot} \rightarrow HR\text{–}RH_{ad} \tag{56}$$

The two photons adsorbed generate at the same time two conduction band electrons which reduce water according to Equation 57.

$$2H_2O_{ad} + 2e_{tr}^- \rightarrow H_2 + 2OH_{ad}^- \tag{57}$$

Comparison of Equations 55 and 57 reveals that two molecules of water are produced and no net consumption of this substrate of the reductive half-reaction is therefore observed. In the case of D₂O, increasing amounts of HDO are formed upon prolonged irradiation time

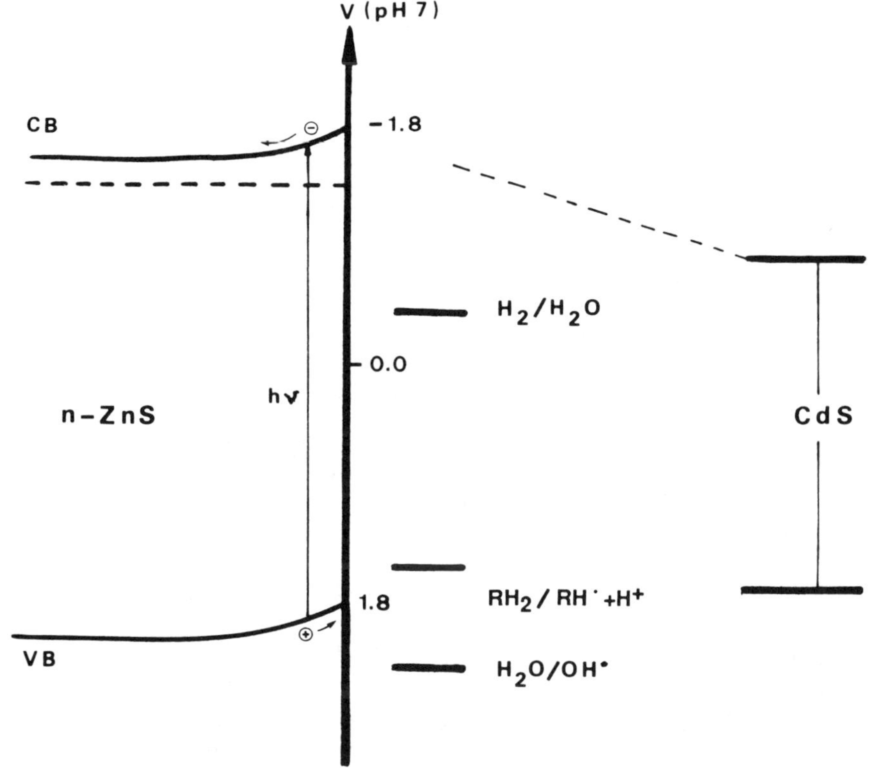

FIGURE 20.   Band positions of ZnS and CdS and redox potentials of the two-electron reduction of water and the one-hole oxidation of water and 2,5-DHF.

and therefore the H/D exchange becomes rationable. In agreement with this mechanism is the fact that no deuterium is incorporated into unreacted 2,5-DHF. The reaction thus differs significantly from the few other synthetic organic photoreactions catalyzed by semiconductor powders, where usually oxygen but not water is reduced by the trapped electrons.[2]

4. Scavenging Experiments

Inhibition studies reveal that both hydrogen and dehydrodimer formation are completely inhibited when electron scavengers like zinc sulfate are present in concentrations equal to that of ZnS-B$_1$. A Stern-Volmer diagram (Figure 21) does not give a straight line but rather resembles a Langmuir adsorption isotherm. Neglecting competitive adsorption by water and 2,5-DHF, an approximate relative equilibrium constant of 360 M^{-1} is obtained. For water adsorption the corresponding value is only 0.5 M^{-1} (*vide infra*) and therefore it becomes rationable why zinc ions can compete so successfully with water for the trapped electron. Zn(0) formed in the scavenging process (Equation 58)

$$Zn_{ad}^{2+} + 2e_{tr}^{-} \rightarrow Zn(0) \tag{58}$$

$$Zn(0) + 2h_{tr}^{+} \rightarrow Zn^{2+} \tag{59}$$

is reoxidized to Zn(II) by trapped holes (Equation 59), since only traces of Zn(0) are produced. Expressed in more physical terms this means that Zn^{2+} adsorption induces the formation of surface states which are active sites for charge recombination. This parallels the photoemission of zinc sulfide which originates from the recombination of electrons trapped at zinc centers with valence band holes.[99]

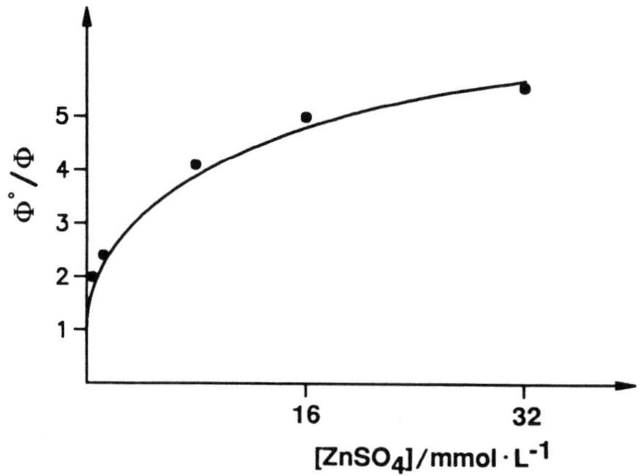

FIGURE 21.   Stern-Volmer plot for the inhibition of hydrogen formation by zinc sulfate in the ZnS-B$_1$-catalyzed photodehydrodimerization of 2,5-DHF; λ = 313 nm, 0.9 M 2,5-DHF.

When cadmium sulfate is employed, complete product inhibition is already observed at concentrations of 0.2 mol% relative to zinc sulfate and is accompanied by strong darkening, i.e., formation of Cd(0) (according to Equation 58). Thus, reoxidation of the metal does not occur in this case. This may originate from the less negative redox potential of cadmium, -0.40 V, as compared to zinc, -0.76 V, which results in a higher and lower driving force for reduction and oxidation, respectively, and therefore the accumulation of the metal becomes rationale. It is further mentioned that surface Cd(0) inhibits hydrogen evolution while trace amounts of Zn(0) rather accelerate it (see Section III.C).

Assuming that the metal ions are completely adsorbed, i.e., they have identical and high adsorption equilibrium constants, one can estimate that complete inhibition occurs when about 10 and 80% of the surface are covered with cadmium and zinc ions, respectively.[141] These differences therefore suggest the presence of different adsorption sites for the two ions. Since usually about 10% of a semiconductor surface is available as active sites[42] for a chemical reaction, it seems probable that Cd^{2+} is adsorbed at these sites but Zn^{2+} at different ones. This assumption would further rationalize the observation that zinc(II) interferes only with the photophysical process of charge recombination, but cadmium(II) with the primary reductive chemical reaction. When all the added Cd^{2+} has been reduced to the metal, an efficient cathodic corrosion of ZnS-B$_1$ occurs.

5. Chemoselectivity

In addition to 2,5-DHF, further unsaturated compounds were dehydrodimerized on a preparative scale. Figure 22 summarizes structures and quantum yields.[142] The dependence of product quantum yield on the substrate structure suggests that a cyclic allyl or enol ether group has to be present in order to observe a reasonable reaction. Further factors like ring size are equally important, since the seven-membered 4,7-dihydro-1,3-dioxepin gives rise to the low value of Φ_p = 0.025. Since Φ_p depends on the rate constants of several photophysical and photochemical processes, it is difficult to decide which one of these reflects the observed changes (see Section III.B).

2,3-Dihydrofuran (2,3-DHF) is also dehydrodimerized to **1a** to **1c**, in agreement with the expectation that the same intermediate dihydrofryl radical should be formed as in the case of 2,5-DHF. The structures of the novel products from 3,4-dihydro-2H-pyran (3,4-DHP) and 4-methyl-5,6-dihyro-2H-pyran (Me-DHP) are summarized in Figure 23.

Since there is no regular relation between Φ_p and the ionization potential of the organic

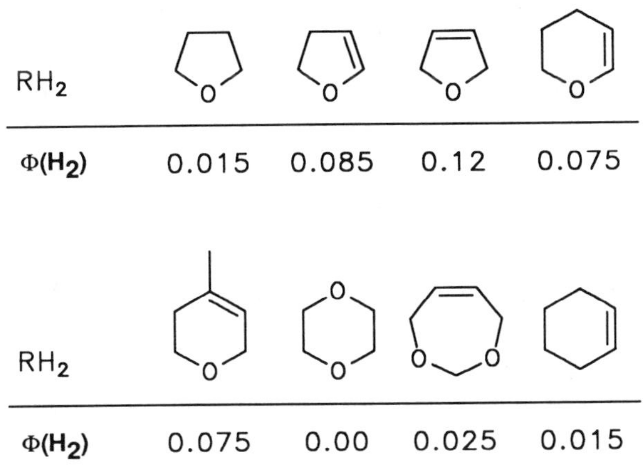

RH_2				
$\Phi(H_2)$	0.015	0.085	0.12	0.075

RH_2				
$\Phi(H_2)$	0.075	0.00	0.025	0.015

FIGURE 22. The variation of product quantum yield with substrate structure.

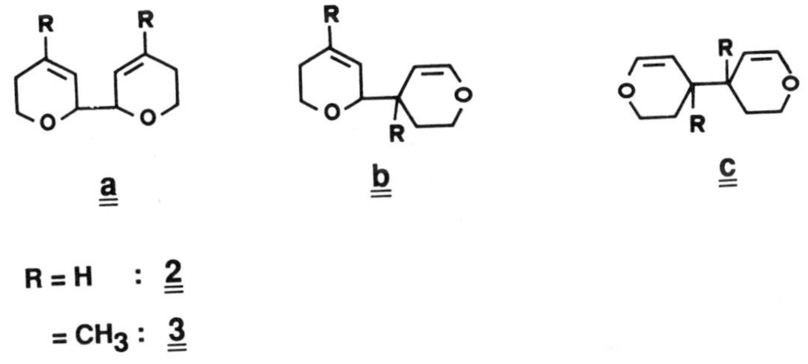

a   **b**   **c**

R = H : **2**

= CH$_3$: **3**

FIGURE 23.   Structure of dehydrodimers obtained from 2-H-dihydropyrans.

substrate, it is probable that the observed chemoselectivity is not governed by the driving force of the oxidative half-reaction (Equation 21). It is more likely that the stability of intermediates and the nature of substrate adsorption onto the zinc sulfide surface are at least equally important. Competition experiments in the system ZnS-B$_1$/THF/2,5-DHF/H$_2$O and the dependence of reaction rate on substrate concentration point to the decisive role of adsorption.

6. Adsorption Effects

Although THF reacts only about ten times slower than 2,5-DHF, the latter is selectively dehydrodimerized even in the presence of an 100-fold excess of the former. At least a 200-fold excess of THF is necessary to observe cross-products and dehydrodimers of THF when the ratio ZnS/2,5-DHF is approximately 0.3. When the latter is 3.8, i.e., ZnS is in molar excess over 2,5-DHF, the major products are the dehydrodimers of THF. This suggests that 2,5-DHF is adsorbed much stronger than THF and that the latter cannot desorb the former even when present in a large excess. Adsorption of THF, the prerequisite for its oxidation, seems to occur only when 2,5-DHF does not occupy all adsorption sites due to an excess of ZnS. These observations suggest that chemoselectivity can be introduced by proper selection of the catalyst to substrate ratio. The important role of adsorption is further corroborated by the dependence of reaction rate on substrate concentration.

Figure 24 shows the dependence of hydrogen evolution rate on the concentrations of H$_2$O, THF, and 2,5-DHF upon polychromatic irradiation. The obtained curves resemble a

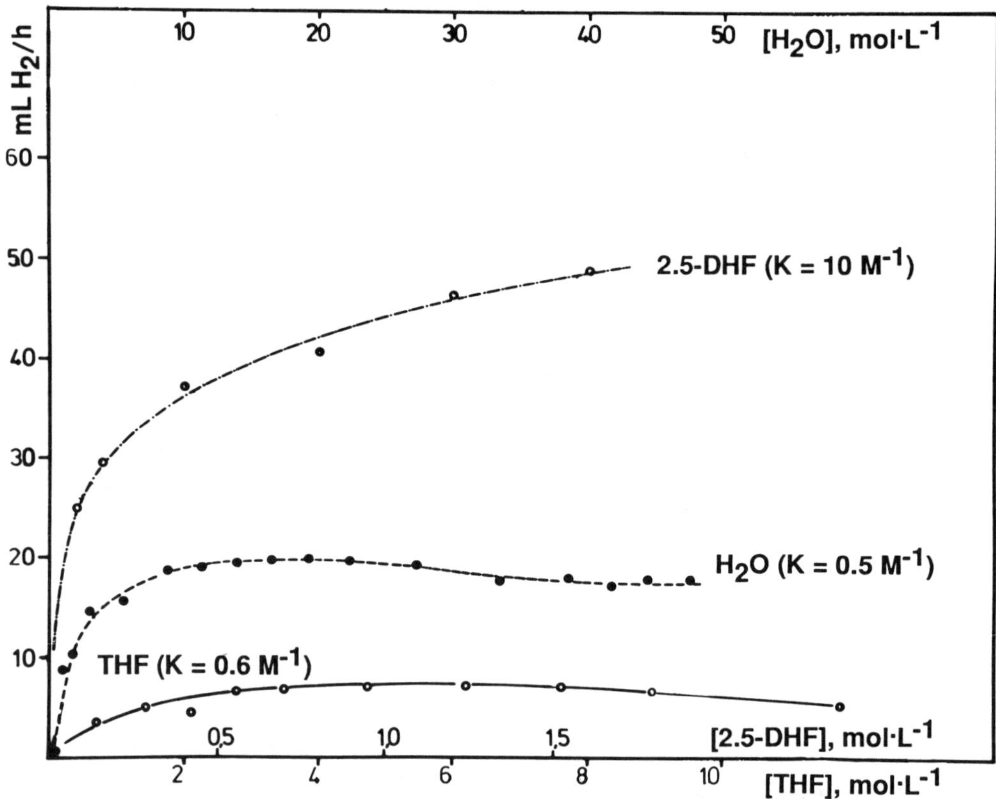

FIGURE 24.   Dependence of reaction rate on the concentrations of substrates in the photodehydrodimerization of THF and 2,5-DHF; ZnS-A, λ ≥ 290 nm.

Langmuir isotherm and indicate that C–C coupling occurs between adsorbed RH radicals (Langmuir-Hinshelwood mechanism). From the linear plot of reciprocal rate as function of reciprocal concentration the approximate relative adsorption constants of 0.5, 0.6, and 10.0 M^{-1} were obtained for H_2O, THF, and 2,5-DHF, respectively. As postulated from the competition experiments, adsorption of 2,5-DHF is strongly favored over that of THF.

7. Olefin Dehydrodimerization

As observed for ether dehydrodimerization, the chemoselectivity of the reaction seems to be governed by steric effects. 1-Octene and cyclooctene are inert, while cyclohexene regioselectively reacts to 3,3′-bicyclohexenyl. This compound has been obtained in low yield together with two further regioisomers by Birch reduction of 1,3-cyclohexadiene.[143] Although the isolated yield of 15% is low, the zinc sulfide-catalyzed reaction is a convenient route to the 3,3′-isomer.

8. Photocatalysis by CdS/Pt

All the reactions discussed above are also photocatalyzed by CdS/Pt but not by the unplatinized sulfide. The rates are much lower than in the case of zinc sulfide and an induction period is observed. The isolated yields are in the range of 62 to 20 and 11% for the cyclic ethers and for cyclohexene, respectively.

Increasing the cadmium amount of homogeneous CdS-ZnS solutions leads to a decrease of reaction rate as shown in Figure 25. The much lower rate induced by CdS/Pt as compared to ZnS-B$_1$ may be due to the more positive potential of the conduction band-edge of the former (Figure 20), which should result in a slower water reduction. In addition to that, the

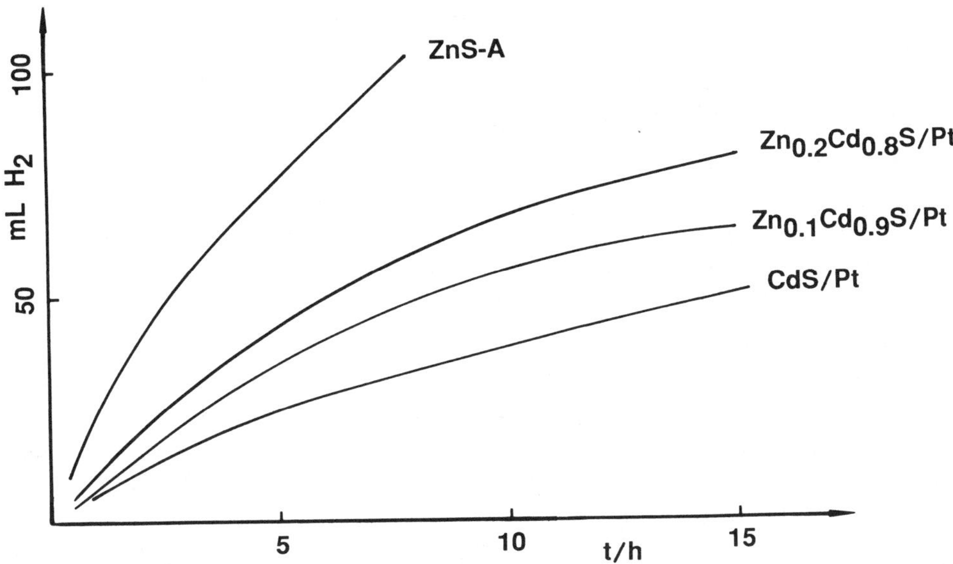

FIGURE 25.   Hydrogen evolution curves for zinc and cadmium sulfide and two homogeneous solutions thereof; 2,5-DHF/H$_2$O = 1/14 (v/v); $\lambda \geq$ 290 nm.

shorter lifetime of the electron-hole pair, 1 ns (CdS) vs. 30 ns (ZnS),[144,145] should induce a smaller stationary concentration of e_{tr}^- and h_{tr}^+ and therefore decrease the rates of both primary reactions.

Scavenging experiments with zinc and cadmium sulfate differ to the results obtained with zinc sulfide and support the assumption that these cations may be reduced at different surface sites. Addition of 14 mol% of $ZnSO_4$ (rel. to CdS) to the reacting suspension results in a short inhibition period; thereafter the reaction continues with the same rate as before. A second and third addition of 28 mol% again induces inhibition periods and consecutive returns to the initial reaction rate. Formation of Zn(0) is evidenced by formation of hydrogen after acidification of the suspension, although no visual darkening of the suspended powder can be observed during the inhibition experiment. Since CdS/Zn is about ten times less reactive, it seems likely that zinc ions are reduced at the CdS/Pt contact. In that case the deposited metal apparently does not alter the catalytic properties of the platinum.

In the case of $CdSO_4$ the rate decreases by about 50% after the first addition, but returns to that low value after second and third additions of the scavenger. A significant darkening of the powder is already observable after the first addition.

These scavenging experiments strongly suggest that reduction of the metal ions occurs at the CdS/Pt contact and at the free CdS surface in the case of zinc and cadmium, respectively. Due to the high porosity of these powders (see Section III.D) attempts failed to locate the metal islands by electron microscopy.

Irrespective of whether the two ions react at identical surface sites or not, their reduction (Equation 58) is much faster than that of water (Equation 57). This may lead to an increase of the stationary hole concentration and favor the transfer of a second hole to an adsorbed substrate. In fact, furan (R) is formed when zinc sulfate is added before irradiating the CdS/Pt/2,5-DHF/H$_2$O suspension (Equations 21, 55, 60, and 61).

$$RH_{ad}^{\cdot} + h_{tr}^+ \rightarrow RH_{ad}^+ \tag{60}$$

$$RH_{ad}^+ + H_2O \rightarrow R + H_3O^+ \tag{61}$$

In agreement with this scavenging mechanism is the decrease of the pH value, since formation of hydroxide according to Equation 45 is suppressed. This differs significantly from the ZnS-B$_1$ system where not only one of the two primary reactions, but both are inhibited through acceleration of charge recombination.

B. INSERTION OF AZOBENZENE INTO C–H BONDS

When azobenzene is added to the reacting system ZnS-B$_1$/H$_2$O/1,4-dioxane/RH$_2$, RH$_2$ = 2,5-DHF, or 3,4-DHP, hydrogen and dehydrodimer formation (Equation 54) are completely suppressed in favor of a new C–N coupling reaction (Equation 62)

$$ArN=NAr + RH_2 \xrightarrow{h\nu,ZnS} ArN(H)\text{–}N(RH)Ar \tag{62}$$

In the case of 3,4-DHP the new product was isolated in a nonoptimized yield of 30%.

The mechanism of this formal insertion of azobenzene into a C–H bond of the cyclic ether is not known at the present. Water may be replaced by an alcohol while aprotic solvents like dioxane prevent the reaction.[146]

VI. PHOTOREDUCTION OF AZOBENZENE, DINITROGEN, AND CARBON DIOXIDE

A. ZnS-CATALYZED PHOTOREDUCTION OF CARBON DIOXIDE

Carbon dioxide has been photoreduced to formate, methanol, and even methane in the presence of semiconductor powders.[147] Usually the yields of products are very low and it is unclear whether these reactions are catalytic or assisted. Henglein et al.[119] reported that the photoreduction of carbon dioxide by methanol or isopropanol is catalyzed by colloidal ZnS. Since the photocatalytic properties of the colloids usually differ from those of the powders — the dehydrodimerization of cyclic ethers (Equation 54) does not occur with the colloid and the powder does not induce H$_2$-evolution from Me$_2$CHOH/H$_2$O — it seemed worthwhile to test if similar differences occur also in CO$_2$-reduction.[148] In fact, no reaction is observed when the system ZnS-B$_1$/H$_2$O/Me$_2$CHOH is irradiated under a constant purge of CO$_2$, while formate and dehydrodimers **1** are produced, when isopropanol is replaced by 2,5-DHF (Figure 26).

Although the reaction is rather slow, it is catalytic as indicated by the production of 1.8 mmol formate per millimole ZnS after 72 h irradiation time. The maximum TON is about 100 monolayers of formate/72 h.[122] Since no reaction occurs at pH 12, it is likely that HCO$_3^-$ is involved in the reduction and the overall process is formulated according to Equation 63.

$$HCO_3^- + 2RH_2 \xrightarrow{h\nu,ZnS} HCO_2^- + HR\text{–}RH + H_2O \tag{63}$$

The reason for the rather low catalytic activity is partly due to the fact that the formate produced is easily back-oxidized (Section IV.A).

Since no oxalate is formed, the reduction does not occur via a one-electron, but rather via a two-electron transfer. This is in accordance with the redox potentials of E^0(CO$_2$/CO$_2^-$) = -2 V[126] and E^0(CO$_2$ + H$^+$ + 2e$^-$ → HCO$_2^-$) = -0.62 V.[119] The same mechanism was proposed for the colloidal system.

The failure of isopropanol to function as a reducing agent in the powder system is further support for the general difference between powder and colloidal systems. Whether this originates from different adsorptive properties or a shift of the band positions cannot be decided at the present.

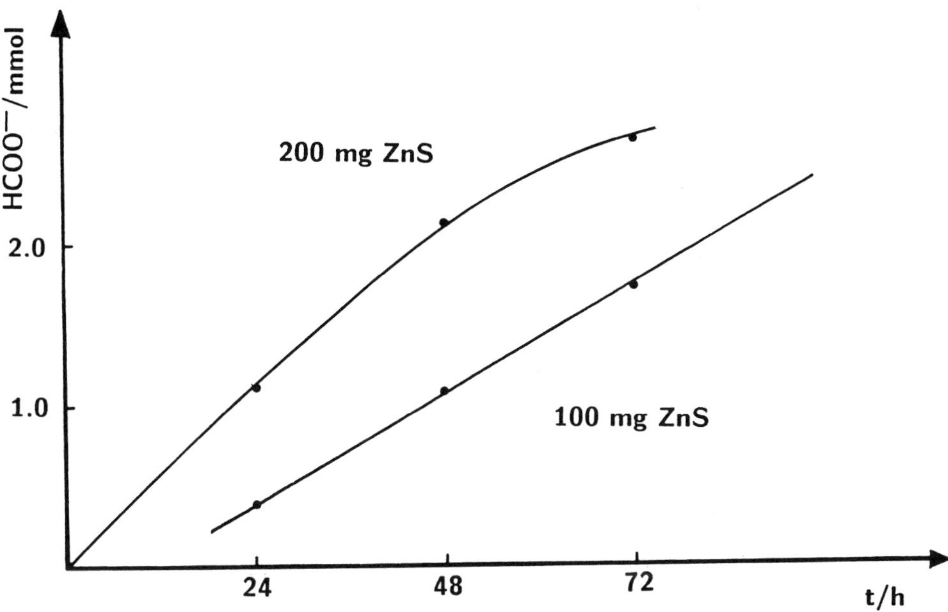

FIGURE 26.   Formate formation curve for the zinc sulfide-catalyzed photoreduction of carbon dioxide in 2,5-DHF/H$_2$O = 7/1 (v/v); $\lambda \geq 290$ nm.

B. CdS/Pt-CATALYZED PHOTOREDUCTION OF AZOBENZENE

Different to zinc sulfide, CdS/Pt induces a photoreduction of azobenzene to hydrazobenzene (76% isolated yield) when irradiated in H$_2$O/2,5-DHF/1,4-dioxane (Equation 64). This is a further example that the reaction of a given substrate can be directed by proper selection of the semiconductor photocatalyst.

$$\text{PhN=NPh} + 2\text{RH}_2 \xrightarrow{\text{h}\nu,\text{CdS/Pt}} \text{PhN(H)–N(H)Ph} + \text{HR–RH} \qquad (64)$$

The reaction is catalytic as indicated by the TON of 190 monolayers/30 h. Since the presence of a protonic solvent is necessary, it seems likely that reduction of azobenzene[150] occurs in a two electron step. Methylorange is also photoreduced by CdS as suggested by UV-spectroscopy.[151]

C. CdS/M-ASSISTED PHOTOREDUCTION OF DINITROGEN

The photoreduction of dinitrogen to ammonia was observed at semiconducting oxides and sulfides like TiO$_2$, SrTiO$_3$, ZnO, Fe$_2$O$_3$, Cr$_2$O$_3$,[4,152-160] and CdS.[160,161] In most cases the amounts of ammonia produced are in the range of a few micromoles and often the nature of the reducing agent is unknown. Although often claimed, most of these reactions are not catalytic. A typical example is the CdS-assisted reaction in the absence of an added reducing agent wherein water oxidation was assumed to be the oxidative primary reaction.[160] This could not be reproduced with cadmium sulfide, prepared as described in Section V.A, which assists dinitrogen reduction only in the presence of a reducing agent like 2,5-DHF.[5] It is therefore proposed that the reaction follows Equation 65.

$$2\text{N}_2 + 6\text{RH}_2 \xrightarrow{\text{h}\nu,\text{CdS/M}} 2\text{NH}_3 + 3\text{HR–RH} \qquad (65)$$

The variation of the amount of NH$_3$ produced was dependent on the presence or absence of

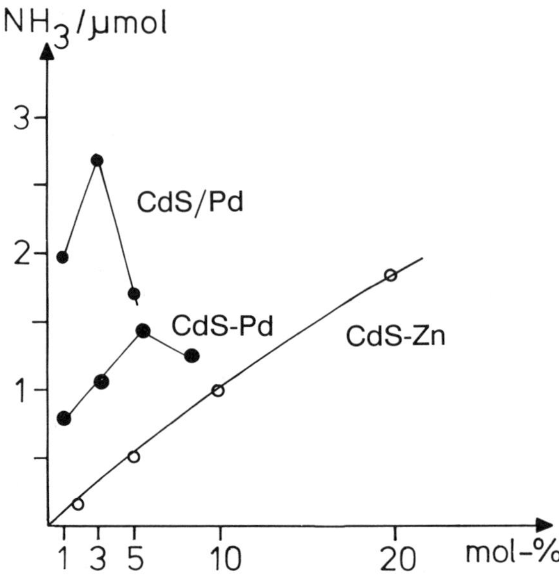

FIGURE 27.   Change of ammonia yield on amount of metal loading in the cadmium sulfide-assisted photoreduction of dinitrogen; CdS/Pt and CdS-M designate conventionally and *in situ* prepared metallized powders.

N_2 during catalyst preparation and suggested that N_2-adsorption from the reacting solution was rather low. In order to promote the latter, different metals were photodeposited onto CdS and CdS/Pt either conventionally (see Section III.D) or by a new *in situ* method (CdS-M samples). Highest yields were obtained with the combinations CdS/M, M=Pd, Pt, CdS-Zn, and CdS/Pt-M, M=Pd, Zn. This differs from the dependence of water reduction on the nature of the metal M involved and suggests that dinitrogen is not reduced via adsorbed hydrogen. More likely it seems a direct two-electron reduction to intermediate 1,2-diazene followed by two further reduction steps to hydrazine and ammonia. The reduction of 1,2-diphenyldiazene to 1,2-diphenylhydrazine supports the second two-electron step (*vide supra*).

Figure 27 summarizes the influence of increasing metal concentration on the yield of ammonia. It demonstrates that conventionally prepared, i.e., isolated, CdS/Pd powder exhibits the same dependence of reactivity on amount of Pd deposited as the *in situ*-prepared sample CdS-Pd; and it further supports the direct reduction of dinitrogen since CdS-Zn powders, which are very poor catalysts for water reduction, gain the same reactivity as the two other samples.

Addition of metal carbonyl complexes to CdS or of $[Ru(NH_3)_5N_2]^{2+}$ and $[Ru(Hedta)Cl]^-$, edta = ethylenediamminetetraacetate to CdS/Pt leads to a slight increase of ammonia yields when 2,5-DHF is present.[162] Recently it was reported that the system CdS/Pt/RuO_2 in the presence of $[Ru(Hedta)N_2]^-$ affords millimole amounts of ammonia and oxygen.[163] This is a surprising result since the photooxidation of water by this doubly loaded CdS was not observed by other workers.[90]

VII. SUMMARY AND OUTLOOK

The light-generated valence band holes and conduction band electrons at the surface of ZnS and CdS/Pt are able to drive photoredox reactions on a preparative scale. Two types of new organic transformations were observed. In one, the radical generated by hole oxidation of a cyclic ether or olefin dimerizes to the product, while water is reduced by the electron.

In the other, the oxidatively produced radical combines with a radical formed by electron reduction of azobenzene. The first process is therefore described as dehydrodimerization, the second as an insertion of a diazene into a C–H bond.

Mechanistic investigations with electron scavengers support the presence of semiconductor photocatalysis. The chemoselectivity peculiar to these catalyst powders seems to be largely governed by the strength of substrate adsorption. Accordingly, chemoselectivity can be controlled by adjusting the catalyst to substrate ratio when two or more compounds react competitively. The activity and selectivity of the catalysts depend on minor details of their synthesis procedure. This is due to the interference of adsorbed ions with the recombination of surface-trapped electrons and holes and with the electron transfer from and to adsorbed substrates. When the latter are electron acceptors the adsorption site seems to depend on their redox potentials.

The results summarized illustrate that suspensions of metal sulfide powders are efficient photocatalysts in synthetic chemistry. Product separation is simple since the heterogeneous catalyst can be removed by filtration.

ACKNOWLEDGMENT

We are grateful to Priv.-Doz. Dr. A. Holzwarth, Max-Planck-Institut für Strahlenchemie, for measuring the luminescence spectra, to Mr. P. Widlok for the BET determinations, to Mr. A. Stiegelschmitt for the surface analysis, to Prof. Dr. H.-J. Freund for ESCA spectra, to the Fonds der Chemischen Industrie, and to the Alfried Krupp von Bohlen und Halbach-Stiftung for financial support.

REFERENCES

1. **Serpone, N. and Pelizzetti, E., Eds.,** *Photocatalysis,* John Wiley & Sons, New York, 1989.
2. **Fox, M. A.,** Selective formation of organic compounds by photoelectrosynthesis at semiconductor particles, *Top. Curr. Chem.,* 142, 71, 1987.
3. **Henglein, A.,** Mechanism of reactions on colloidal microelectrodes and size quantization effects, *Top. Curr. Chem.,* 143, 113, 1988.
4. **Schrauzer, G. N. and Guth, T. D.,** Photolysis of water and photoreduction of nitrogen on titanium dioxide, *J. Am. Chem. Soc.,* 99, 7189, 1977.
5. **Hetterich, W. and Kisch, H.,** Cadmiumsulfid-assistierte Photoreduktion von molekularem Stickstoff, *Chem. Ber.,* 122, 621, 1989.
6. **Hubbard, J. S., Hardy, J. P., and Horowitz, N. H.,** Photocatalytic production of organic compounds from carbon monoxide and water in a simulated martian atmosphere, *Proc. Natl. Acad. Sci. U.S.A.,* 68, 574, 1971.
7. **Reiche, H. and Bard, J. A.,** Heterogeneous photosynthetic production of amino acids from methane-ammonia-water at Pt/TiO$_2$. Implications in chemical evolution, *J. Am. Chem. Soc.,* 101, 3127, 1979.
8. **Heller, A., Degani, Y., Johnson, D. W., and Gallagher, P. K.,** Controlled supression and enhancement of the photoactivity of titanium dioxide (rutile) pigment, *J. Phys. Chem.,* 91, 5987, 1987.
9. **Ollis, D. F., Pelizzetti, E., and Serpone, N.,** Heterogeneous photocatalysis in the environment: application to water purification, in *Photocatalysis,* John Wiley & Sons, New York, 1989, chap. 18.
10. **Basolo, F. and Pearson, R. G.,** *Mechanism of Inorganic Reactions,* 2nd ed., John Wiley & Sons, New York, 1967.
11. **Davies, St. G.,** *Organotransition Metal Chemistry: Applications to Organic Synthesis,* Pergamon Press, Elmsford, NY, 1982.
12. **Gstach, H. and Kisch, H.,** Kobaltinduzierte Synthese von Indolen aus Alkinen und Diaryldiazenen, *Z. Naturforsch. Teil B,* 38, 251, 1983.
13. **Sellmann, D. and Weiss, W.,** Reaktionen an komplexgebundenen Liganden. XXVII. Reaktionen von Distickstoff- zu Diazen-Liganden durch konsekutiven nukleophilen und elektrophilen Angriff, *J. Organometal. Chem.,* 160, 183, 1978.

14. **Plotnikov, J.,** *Allgemeine Photochemie,* 2nd ed., Walter de Gruyter, Berlin, 1936, 362.

15. **Hennig, H., Rehorek, D., and Archer, R. D.,** Photocatalytic systems with light-sensitive coordination compounds and possibilities of their spectroscopic sensitization — an overview, *Coord. Chem. Rev.,* 61, 1, 1985.

16. **Salomon, R. G.,** Homogeneous metal-catalysis in organic photochemistry, *Tetrahedron,* 39, 485, 1983.

17. **Moggi, L., Juris, A., Sandrini, D., and Manfrin, M. F.,** Photocatalysis by transition-metal coordination compounds in homogeneous phase. The role of ligand photodissociation, *Rev. Chem. Intermed.,* 4, 171, 1981.

18. **Wubbels, G. G.,** Catalysis of photochemical reactions, *Acc. Chem. Res.,* 16, 285, 1983.

19. **Wrighton, M. S., Ginley, D. S., Schroeder, M. A., and Morse, D. L.,** Generation of catalysts by photolysis of transition metal complexes, *Pure Appl. Chem.,* 41, 671, 1975.

20. **Carassiti, V.,** What means "photocatalysis", *EPA Newsl.,* 21, 12, 1984.

21. **Kutal C.,** Photochemistry of transition metal-organic systems, *Coord. Chem. Rev.,* 64, 191, 1985.

22. **Teichner, S. J. and Formenti, M.,** Heterogeneous photocatalysis, in *Photoelectrochemistry, Photocatalysis and Photoreactors,* Schiavello, M., Ed., D. Reidel Publishing, Dordrecht, 1985, 457.

23. Glossary of terms used in photochemistry, *EPA Newsl.,* 25, 13, 1985.

24. **Balzani, V., Boletta, F., Ciano, M., and Maestri, M.,** Electron transfer reactions involving light, *J. Chem. Educ.,* 60, 447, 1983.

25. **Bard, A. J.,** Photoelectrochemistry and heterogeneous photocatalysis at semiconductors, *J. Photochem.,* 10, 59, 1979.

26. **Childs, L. P. and Ollis, D. F.,** Is photocatalysis catalytic?, *J. Catal.,* 66, 383, 1980.

27. **Turro, N. J.,** *Modern Molecular Photochemistry,* Benjamin/Cummings, Menlo Park, CA, 1978, 13.

28a. **Julliard, M. and Channon, M.,** Photoelectron-transfer catalysis: its connections with thermal and electrochemical analogs, *Chem. Rev.,* 83, 425, 1983.

28b. **Kavarnos, G. J. and Turro, N.,** Photosensitization by reversible electron transfer: theories, experimental evidence, and examples, *J. Chem. Rev.,* 86, 401, 1986.

29. **Schwendiman, D. P. and Kutal, C.,** Catalytic role of copper(I) in the photoassisted valence isomerization of norbornadiene, *J. Am. Chem. Soc.,* 99, 5677, 1977.

30. **Albini, A.,** Catalysis of photochemical reactions. Opportunities and pitfalls, *J. Chem. Educ.,* 63, 383, 1986.

31. **Mitchener, J. C. and Wrighton, M. S.,** Photogeneration of very active homogeneous catalysts using laser light excitation of iron carbonyl precursors, *J. Am. Chem. Soc.,* 103, 975, 1981.

32. **Kutal, C., Kelley, C. K., and Ferraudi, G.,** Photogenerated catalysis in metal-organic systems. III. Catalyzed valence isomerization of quadricyclane to norbornadiene via photochemical generation of a strong ground-state oxidant, *Inorg. Chem.,* 26, 3258, 1987.

33. **Watanabe, T., Takizawa, T., and Honda, K.,** Photocatalysis through excitation of adsorbates. I. Highly efficient N-deethylation of rhodamine B adsorbed to CdS, *J. Phys. Chem.,* 81, 1845, 1977.

34. **Watanabe, T., Takizawa, T., and Honda, K.,** Photocatalysis through excitation of adsorbates. II. A comparative study of rhodamine B and methylene blue on cadmium sulfide, *J. Phys. Chem.,* 82, 1391, 1978.

35. **Watanabe, T., Takizawa, T., and Honda, K.,** Photocatalysis through excitation of adsorbates. III. Effect of electron acceptors on the efficiency of interfacial charge separation, *J. Phys. Chem.,* 84, 51, 1980.

36. **Tributsch, H.,** Photoelectrocatalysis, in *Photocatalysis,* John Wiley & Sons, New York, 1989, chap. 11.

37. **Fox, M. A.,** Mechanistic photocatalysis in organic synthesis, in *Photocatalysis,* John Wiley & Sons, New York, 1989, chap. 13.

38. **Pichat, P.,** Photocatalytic reactions, in *Photoelectrochemistry, Photocatalysis and Photoreactors,* Schiavello, M., Ed., D. Reidel Publishing, Dordrecht, 1985, 425.

39. **Pichat, P.,** Adsorption and desorption processes in photocatalysis, in *Homogeneous and Heterogeneous Photocatalysis,* Pelizzetti, E. and Serpone, N., Eds., D. Reidel Publishing, Dordrecht, 1985, 533.

40. **Morrison, R. S.,** *The Chemical Physics of Surfaces,* Plenum Press, New York, 1978, 317.

41. **Clark, A.,** *The Theory of Adsorption and Catalysis,* Academic Press, New York, 1970, 199.

42. **Weisz, P. B.,** Effects of electronic charge transfer between adsorbate and solid on chemisorption and catalysis, *J. Chem. Phys.,* 21, 1531, 1953.

43. **Kraeutler, B. and Bard, A. J.,** Heterogeneous photocatalytic decomposition of saturated carboxylic acids on TiO_2 powder. Decarboxylative route to alkanes, *J. Am. Chem. Soc.,* 100, 5985, 1978.

44a. **Bard, A. J.,** Photoelectrochemistry, *Science,* 207, 139, 1980.

44b. **Bard, A. J.,** Design of semiconductor photoelectrochemical systems for solar energy conversion, *J. Phys. Chem.,* 86, 172, 1982.

45. **Aspnes, D. E. and Heller, A.,** Photoelectrochemical hydrogen evolution and water-photolyzing semiconductor suspensions: properties of platinum group metal catalyst-semiconductor contacts in air and in hydrogen, *J. Phys. Chem.,* 87, 4919, 1983.

46. **Hodes, G.,** Photoelectrochemistry of cadmium and other metal chalcogenides in polysulfide electrolytes, in *Energy Resources through Photochemistry and Catalysis,* Grätzel, M., Ed., Academic Press, New York, 1983, 421.
47. **Gerischer, H.,** A mechanism of electron hole pair separation in illuminated semiconductor particles, *J. Phys. Chem.,* 88, 6096, 1984.
48. **Memming, R.,** Photoelectrochemical solar energy conversion, *Top. Curr. Chem.,* 149, 137, 1988.
49. This depends on the doping level and on the potential across the positively charged depletion layer; see, e.g., Reference 48.
50. **Nosaka, Y., Ishizuka, Y., and Miyama, H.,** Separation mechanism of a photoinduced electron-hole pair in metal-loaded semiconductor powders, *Ber. Bunsenges. Phys. Chem.,* 90, 1199, 1986.
51. Different to the case discussed, the energies of the trapped charges, i.e., the potential of the reducing and oxidizing surface sites, may be equal for an exergonic reaction; see Ref. 8.
52. The redox potential may change by almost 1 V when one form of the redox couple is adsorbed much stronger than the other; **Rodriguez, J. F., Harris, J. E., Bothwell, M. E., Mebrahtu, Th., and Soriaga, M. P.,** Surface coordination chemistry of noble-metal electrocatalysts: oxidative addition and reductive elimination of iodide at iridium, platinum and gold in aqueous solutions, *Inorg. Chim. Acta,* 148, 123, 1988.
53. **Fan, F.-R. F., Leempoel, P., and Bard, A. J.,** Semiconductor electrodes. LI. Efficient electroluminescence at ZnS electrode in aqueous electrolytes, *J. Electrochem. Soc.,* 130, 1866, 1983.
54. **White, J. R. and Bard, A. J.,** Electrochemical investigation of photocatalysis at CdS suspensions in the presence of methyl viologen, *J. Phys. Chem.,* 89, 1947, 1985.
55. **Ward, M. D., White, J. R., and Bard, A. J.,** Electrochemical investigation of the energetics of particulate titanium dioxide photocatalysts. The methyl viologen-acetate system, *J. Am. Chem. Soc.,* 105, 27, 1983.
56. **Tributsch, H. and Bennett, J. C.,** Semiconductor-electrochemical aspects of bacterial leaching. I. Oxidation of metal sulphides with large energy gaps, *J. Chem. Tech. Biotechnol.,* 31, 565, 1981.
57. **Finlayson, M. F., Wheeler, B. L., Kakuta, N., Park, K.-H., Bard, A. J., Campion, A., Fox, M. A., Webber, S. E., and White, J. M.,** Determination of flat-band position of CdS crystals, films, and powders by photocurrent and impedance techniques. Photoredox reaction mediated by intragap states, *J. Phys. Chem.,* 89, 5676, 1985.
58. **Meissner, D., Memming, R., and Kastening, B.,** Photoelectrochemistry of cadmium sulfide. I. Reanalysis of the photocorrosion and flat-band potential, *J. Phys. Chem.,* 92, 3476, 1988.
59. **Natan, M. J., Thackeray, J. W., and Wrighton, M. S.,** Interaction of thiols with n-type cadmium sulfide and n-type cadmium selenide in aqueous solutions: adsorption of thiolate anion and efficient photoelectrochemical oxidation to disulfides, *J. Phys. Chem.,* 90, 4089, 1986.
60. **Shiragami, T., Pac, C., and Yanagida, S.,** Nonmetallised CdS-catalysed photoreduction of aromatic ketones to alcohols and/or pinacols, *J. Chem. Soc. Chem. Commun.,* p. 831, 1989.
61. **Al-Ekabi, H. and Serpone, N.,** Mechanistic implications in surface photochemistry, in *Photocatalysis,* John Wiley & Sons, New York 1989, chap. 14.
62. **Marfunin, A. S.,** *Spectroscopy, Luminescence and Radiation Centers in Minerals,* S. 176, Springer-Verlag, Berlin, 1979.
63. The diffusion time of the electron-hole pair to the surface is a few picoseconds; **Serpone, N., Sharma, D. K., Jamieson, M. A., Grätzel, M., and Ramsden, J. J.,** Photophysical and photochemical primary events in semiconductor particulate systems. Colloidal CdS with methyl viologen, *Chem. Phys. Lett.,* 115, 573, 1985.
64. **Egerton, T. A. and King, C. J.,** The influence of light intensity on photoactivity in TiO_2 pigmented systems, *J. Oil Colour Chem. Assoc.,* 62, 386, 1979.
65. These numbers depend on the photoreactor geometry.
66. **Sato, S.,** Photo-Kolbe reaction at gas-solid interfaces, *J. Phys. Chem.,* 87, 3531, 1983.
67. **Al-Ekabi, H. and de Mayo, P.,** Surface photochemistry: CdS photoinduced cis-trans isomerization of olefins, *J. Phys. Chem.,* 89, 5815, 1985.
68. **Kisch, H., Twardzik, G., Hetterich, W., and Künneth, R.,** *Chem. Ber.,* in preparation.
69. **Rosetti, R. and Brus, L.,** Electron-hole recombination emission as a probe of surface chemistry in aqueous CdS colloids, *J. Phys. Chem.,* 86, 4470, 1982.
70. **Myer, G. Y., Leung, L. K., Yu, J. C., Lisensky, G. C., and Ellis, A. B.,** Semiconductor olefin adducts. Photoluminescent properties of cadmium sulfide and cadmium selenide in the presence of butenes, *J. Am. Chem. Soc.,* 111, 5146, 1989.
71. **Uchihara, T., Matsumara, M., and Tsubomura, H.,** Effect of dissolved electron acceptors and platinum loading on the luminescence of CdS powder in aqueous solutions, *J. Phys. Chem.,* 3207, 93, 1989.
72. **Parkinson, B.,** On the efficiency and stability of photochemical devices, *Acc. Chem. Res.,* 17, 431, 1984.
73. **Pichat, P. and Herrmann, J. M.,** Adsorption-desorption, related mobility and reactivity in photocatalysis, in *Photocatalysis,* John Wiley & Sons, New York, 1989, chap. 8.

74. **Pichat, P., Mozzanega, M. N., Disdier, J., and Herrmann, J. M.,** Platinum content and temperature effects on the photocatalytic hydrogen production from aliphatic alcohols over platinum/titanium dioxide, *Nouv. J. Chim.,* 6, 559, 1982.

75. **Lyashenko, L. V., Gorokhovatski, Y. B., Stepanenko, V. I., and Yamposkaya, F. A.,** Photocatalytic oxidation of propane and n-butane on metal oxides, *Teor. Eksp. Khim.,* 93, 35, 1976.

76. **Hussein, F. H. and Rudham, R.,** Photocatalytic dehydrogenation of liquid alcohols by platinized anatase, *J. Chem. Soc. Faraday Trans. 1,* 83, 1631, 1987.

77. **Herrmann, J. M., Courbon, H., and Pichat, P.,** Metal content and temperature effects on the photo-catalytic isotopic exchange cyclopentane-deuterium over Pt or Ni/TiO$_2$ catalysis in the "normal" or "strong metal-support interaction" state, *J. Catal.,* 95, 539, 1985.

78. **Al-Ekabi, H. and de Mayo, P.,** The CdS photoinduced dimerization of N-vinylcarbazole, *Tetrahedron,* 43, 6277, 1986.

79. **Reber, J. F. and Rusek, M.,** Photochemical hydrogen production with platinized suspensions of cadmium sulfide and cadmium zinc sulfide modified by silver sulfide, *J. Phys. Chem.,* 90, 824, 1986.

80. Note that only for nonporous materials the surface area increases with smaller particle size; however, most of the powders employed are very porous.

81. Estimated for a suspension of 10^8 particles/cm^3; see Reference 48.

82. **Lenard, P.,** Über die Lichtwirkung auf Zinksulfid, *Ann. Phys.,* 68(4), 553, 1922.

83. **Weiser, H. B. and Garrison, A. D.,** The role of water in the photochemical decomposition of zinc sulfide, *J. Phys. Chem.,* 31, 1237, 1927.

84. **Platz, H. und Schenk, P. W.,** Coloration of ZnS on exposure to light, *Angew. Chem.,* 49, 822, 1936.

85. **Hopfield, J. J.,** Theory of edge emission phenomena in CdS, ZnS, and ZnO, *J. Phys. Chem. Solids,* 10, 110, 1959.

86. **Birman, J. L. and Shakin, C.,** Electronic energy bands in ZnS: preliminary results, *Phys. Rev.,* 109, 810, 1958.

87. **Birman, J. L.,** Some selection rules for band-band transitions in wurtzite structure, *Phys. Rev.,* 115, 1493, 1959.

88. **Cardona, M.,** Band parameters of semiconductors with zinc blende, wurtzite and germanium structure, *Phys. Chem. Solids,* 24, 1543, 1963.

89. **Cardona, M. and Harbeke, G.,** Optical properties and band structure of wurtzite-type crystals and rutile, *Phys. Rev.,* 137, 1467, 1965.

90. **Lauermann, I., Meissner, D., and Memming, R.,** The problem of light-induced oxygen production at catalyst-loaded CdS-suspensions, *J. Electroanal. Chem.,* 228, 45, 1987.

91. **Yanagida, S., Mizumoto, K., and Pac, C.,** Semiconductor photocatalysis. Cis-trans photoisomerization of simple alkenes induced by trapped holes at surface states, *J. Am. Chem. Soc.,* 108, 647, 1986.

92. **Yanagida, S., Ishimura, Y., Miyake, Y., Shiragami, T., Pac, C., Hashimoto, K., and Sakata, T.,** Semiconductor photocatalysis. ZnS-catalyzed photoreduction of aldehydes and related derivates: two-electron transfer reduction and relationship with spectroscopic properties, *J. Phys. Chem.,* 93, 2576, 1989.

93. **Baral, S., Fojtik, A., Weller, H., and Henglein, A.,** Photochemistry and radiation chemistry of colloidal semiconductors. XII. Intermediates of the oxidation of extremely small particles of CdS, ZnS, and Cd$_3$P$_2$ and size quantization effects (a pulse radiolysis study), *J. Am. Chem. Soc.,* 108, 375, 1986.

94. **Kurian, A. and Suryanarayana,** Studies on a new method of preparation of zinc sulphide useful for luminescent phosphors, *J. Appl. Electrochem.,* 2, 223, 1972.

95. **Anpo, M., Matsumoto, A., and Kodama,** Direct evidence for the participation of extrinsic surface sites in the enhancement of photocatalytic activity of luminescent zinc sulphide catalysts, *J. Chem. Soc. Chem. Commun.,* p. 1038, 1987.

96. **Kisch, H. and Bücheler, J.,** Heterogeneous photocatalysis. VIII. Zinc sulfide catalyzed hydrogen formation from water in the presence of sodium formate, *Bull. Soc. Jpn.,* in press.

97. **Zeug, N., Bücheler, J., and Kisch, H.,** Catalytic formation of hydrogen and C–C bonds on illuminated zinc sulfide generated from zinc dithiolenes, *J. Am. Chem. Soc.,* 107, 1459, 1985.

98. These bands are known as "self activated luminescence"; see Reference 62.

99. **Becker, W. G. and Bard, A. J.,** Photoluminescence and photoinduced oxygen adsorption of colloidal zinc sulfide dispersions, *J. Phys. Chem.,* 87, 4888, 1983.

100. **Matsumura, M., Furukawa, S., Saho, Y., and Tsubomura, H.,** Cadmium sulfide photocatalyzed hydrogen production from aqueous solutions of sulfite: effect of crystal structure and preparation method of the catalyst, *J. Phys. Chem.,* 89, 1327, 1985.

101. **Yanagida, S., Yoshiya, M., Shiragami, T., and Pac, C.,** Semiconductor photocatalysis. Quantitative photoreduction of aliphatic ketones to alcohols using defect-free ZnS quantum crystallites, *J. Phys. Chem.,* 94, 3104, 1990.

102. **Ginley, D. S. and Butler, M. A.,** Flatband potential of cadmium sulfide photoanodes and its dependence on surface ion effects, *J. Electrochem. Soc.,* 125, 1968, 1978.

103. **Kraeutler, B. and Bard, A. J.**, Heterogeneous photocatalytic preparation of supported catalysts. Photodeposition of platinum on TiO_2 powder and other substrates, *J. Am. Chem. Soc.*, 100, 4317, 1978.

104. **Jacobs, J. W. M.**, Photochemical nucleation and growth of Pd on TiO_2 films studied with electron microscopy and quantitative analytical techniques, *J. Phys. Chem.*, 90, 6507, 1986.

105. **Harbour, J. R., Wolkow, R., and Hair, M. L.**, Effect of platinization on the photoproperties of CdS pigments in dispersion. Determination by H_2 evolution, O_2 uptake, and electron spin resonance spectroscopy, *J. Phys. Chem.*, 85, 4026, 1981.

106. **Bühler, N., Meier, K., and Reber, J.-F.**, Photochemical hydrogen production with cadmium sulfide suspensions, *J. Phys. Chem.*, 88, 3261, 1984.

107. **Mau, A. W.-H., Huang, Ch.-B., Kakuta, N., Bard, A. J., Campion, A., Fox, M. A., White, J. R., and Webber, S. E.**, H_2 photoproduction by nafion/CdS/Pt films in H_2O/S^{2-} solutions, *J. Am. Chem. Soc.*, 106, 6537, 1984.

108. **Nakamatsu, H., Kawai, T., Koreeda, A., and Kawai, S.**, Electron-microscopic observation of photodeposited Pt on TiO_2 particles in relation to photocatalytic activity, *J. Chem. Soc. Faraday Trans. 1*, 82, 527, 1985.

109. **Hetterich, W. and Kisch, H.**, Heterogeneous photocatalysis. VII. The influence of zinc ions on the photocatalytic properties of platinized cadmium sulfide, *J. Photochem. Photobiol.*, 51, 1, 1990.

110. **Henning, R., Schlamann, W., and Kisch, H.**, Katalysierte Photolyse von Wasser mittels Übergangsmetallkomplexen, *Angew. Chem.*, 92, 664, 1980.

111. **Kisch, H.**, Catalytic formation of hydrogen and C–C bonds at illuminated zinc sulfide generated by metaldithiolenes, in *3rd Int. Conf. on Photochem. Conversion and Storage of Solar Energy*, Abstracts, Boulder, 1980, 215.

112. **Bücheler, J., Zeug, N., and Kisch, H.**, Zinksulfid als Katalysator der heterogenen Photoreduktion von Wasser, *Angew. Chem.*, 94, 792, 1982.

113. **Yanagida, S., Azuma, T., and Sakurai, H.**, Photocatalytic hydrogen evolution from water using zinc sulfide and sacrificial electron donors, *Chem. Lett.*, p. 1069, 1982.

114. **Reber, J.-F. and Meier, K.**, Photochemical production of hydrogen with zinc sulfide suspensions, *J. Phys. Chem.*, 88, 5903, 1984.

115. **Kisch, H. and Schlamann, W.**, Zinksulfid als Photokatalysator der Konvertierung von Kohlenmonoxid, *Chem. Ber.*, 119, 3483, 1986.

116. **Furlong, D. N., Wells, D., and Sasse, H. F. W.**, Photooxidation of platinum/colloidal titanium dioxide/aqueous solution interfaces. I. Ethylendiaminetetraacetic acid and related compounds, *Aust. J. Chem.*, 39, 759, 1986.

117. **Matsumura, M., Hiramoto, M., Ichara, T., and Tsubomara, H. J.**, Photocatalytic and photoelectrochemical reactions of aqueous solutions of formic acid, formaldehyde, and methanol on platinized CdS-powder and at a CdS electrode, *Phys. Chem.*, 88, 248, 1984.

118. **Willner, I. and Goren, Z.**, Photodecomposition of formic acid by cadmium sulfide semiconductor particles, *J. Chem. Soc. Chem. Commun.*, p. 172, 1986.

119. **Henglein, A., Gutierrez, M., and Fischer, Ch.-H.**, Photochemistry of colloidal metal sulfides. VI. Kinetics of interfacial reactions at ZnS-particles, *Ber. Bunsenges. Phys. Chem.*, 88, 170, 1984.

120. This is a typical feature of many so-called ''sacrificial systems''.

121. **Maeda, Y., Fujishima, A., and Honda, K.**, The investigation of current doubling reactions on semiconductor photoelectrodes by temperature change measurements, *J. Electrochem. Soc.*, 128, 1731, 1981.

122. Calculated for a specific area of 17 m^2/g assuming that one monolayer consist of 10^{14} molecules/cm^2; e.g., Reference 26.

123. **Bücheler, J.**, Dissertation, University of Dortmund, 1987.

124. **Thomas, J. M. and Thomas, W. J.**, *Introduction to the Principles of Heterogeneous Catalysis*, Academic Press, New York, 1967.

125. Omitting competitive adsorption of water, see Chap. 5.

126. The reduction potential of $HCOO_2 \cdot$ is not known but it should not be too different from E_o' (CO_2/CO_2^-) = $-1,9$ V (NHE); see **Koppenol, W. H. and Rush, J. D.**, Reduction potential of the $CO_2/CO_2 \cdot^-$ couple. A comparison with other C_1 radicals, *J. Phys. Chem.*, 91, 4429, 1987.

127. **Gurevich, Y. Y., Pleskov, Y. V., and Rotenberg, Z. A.**, *Photoelectrochemistry*, Consultant Bureau, Plenum Press, New York, 1980.

128. When 0.00166 mol D_2 was produced from a suspension containing 6.6 mol D_2O (13.2 mol D), also 0.0016 mol of HCO_3^- are formed. Assuming that all H of HCO_3^- is completely exchanged with D_2O, one expects that the ratio of H/D in the gas produced should be 1:800. However, experimentally the ratio is found to be 1:4.

129. **Cole-Hamilton, D. J.**, Photocatalysis of the homogeneous water-gas shift reaction, *J. Chem. Soc. Chem. Commun.*, p. 1213, 1980.

130. **Sato, S. and White, J. M.**, Photoassisted water-gas shift reaction over platinized TiO_2 catalysts, *J. Am. Chem. Soc.*, 102, 7206, 1980.

174   *Photochemistry and Photophysics*

131. **Tsai, S. S., Kao, C.-C., and Chung, Y.-W.,** Photoassisted water-gas shift reaction over Pt/TiO$_2$ (100), *J. Catal.,* 79, 451, 1983.

132. **Thewissen, D. H. M. W., Tinnemans, A. H. A., Euwhorst-Reinten, M., Timmer, K., and Mackor, A.,** Photocatalytic reactions over aqueous suspensions of silicon carbide powders, *Nouv. J. Chim.,* 7, 73, 1983.

133. **Sawyer, D. T. and Hojo, M.,** Hydroxide induced reduction of elemental sulfur (S$_8$) to the trisulfur anion radical (S$_3^-\cdot$), *Inorg. Chem.,* 27, 1201, 1989.

134. **Fox, M. A. and Qwen, R. C.,** Mediated photochemical oxidative dimerization, *J. Am. Chem. Soc.,* 102, 6559, 1980.

135. **Yanagida, S., Azuma, T., Kawakami, K., Kizumoto, H., and Sakurai, H.,** Photocatalytic carbon-carbon bond formation with concurrent hydrogen evolution on colloidal zinc sulphide, *J. Chem. Soc. Chem. Commun.,* p. 21, 1984.

136a. **Ohtani, B., Osaki, H., Nishimoto, S., and Kagiya, T.,** A novel photocatalytic process of amine N-alkylation by platinized semiconductor particles suspended in alcohols, *J. Am. Chem. Soc.,* 108, 308, 1986.

136b. **Yanagida, S., Kizumoto, H., Ishimaru, Y., Pac, C., and Sakurai, H.,** Zinc sulfide-catalyzed photochemical conversion of primary amines to secondary amines, *Chem. Lett.,* p. 141, 1985.

137. **Kisch, H.,** Heterogeneous photocatalysis by metal sulfide semiconductor powders, *J. Inf. Rec. Mater.,* 17(5/6), 363, 1989.

138. **Hetterich, W. and Kisch, H.,** Cadmium-Zinksulfide als Katalysatoren der Photodehydrodimerisierung von 2,5-Dihydrofuran, *Chem. Ber.,* 121, 15, 1988.

139. **Yanagida, S., Azuma, T., Midori, Y., Pac, C., and Sakurai, H.,** Semiconductor photocatalysis. IV. Hydrogen evolution and photoredoxreactions of cyclic ethers catalysed by zinc sulphide, *J. Chem. Soc. Perkin Trans. 2,* p. 1487, 1985.

140. Estimated with the help of the following C–H bond dissociation energies; 4.1 eV (THF), 3.7 eV (2,5-DHF); E^0 (H$^+$/H) = −2.1 V.

141. Assuming radii of 0.3 and 0.34 nm for [M(H$_2$O)$_6$]$^{2+}$, M=Zn and Cd, respectively.

142. All quantum yields have the dimensions mol H$_2$/Einstein and not mol H/Einstein as often cited in the literature.

143. **Myers, D. Y., Grabbe, R. R., and Gardner, P. D.,** Radical-anion reactions. The reductive dimerization of 1,3-cyclohexadiene by sodium-ammonia, *Tetrahedron Lett.,* 7, 533, 1973.

144. **Kuczynski, J. P., Milosavljevic, B. H., and Thomas, J. K.,** Photophysical properties of cadmium sulfide in nafion film, *J. Phys. Chem.,* 88, 980, 1984.

145. **Henglein, A. and Gutierrez, M.,** Photochemistry of colloidal metal sulfides. V. Fluorescence and chemical reactions of ZnS and ZnS/CdS co-colloids, *Ber. Bunsenges. Phys. Chem.,* 87, 852, 1983.

146. **Künneth, R. and Kisch, H.,** unpublished.

147. **Eggins, B. R., Irvine, J. T. S., Murphy, E. P., and Grimshaw, J.,** Formation of two-carbon acids from carbon dioxide by photoreduction on cadmium sulphide, *J. Chem. Soc. Chem. Commun.,* p. 1123, 1988.

148. **Kisch, H. and Twardzik, G.,** unpublished.

149. **Willner, I., Maidan, R., Mandler, D., Dürr, H., Dörr, G., and Zengerle, K.,** Photosensitized reduction of CO$_2$ to CH$_4$ and H$_2$ evolution in the presence of ruthenium and osmium colloids: strategies to design selectivity of product distribution, *J. Am. Chem. Soc.,* 109, 6080, 1987.

150. **Klopman, G. and Doddapaneni, N.,** Electrochemical behavior of cis and trans azobenzenes, *J. Phys. Chem.,* 78, 1825, 1974.

151. **Brown, T. G. and Darwent, J. R.,** Photoreduction of methyl orange sensitized by colloidal titanium dioxide, *J. Chem. Soc. Faraday Trans. 1,* 80, 1631, 1984.

152. **Radford, P. R. and Francis, C. G.,** Photoreduction of nitrogen by metal doped titanium dioxide powders: a novel use for metal vapour techniques, *J. Chem. Soc. Chem. Commun.,* p. 1520, 1983.

153. **Endoh, E. and Bard, A. J.,** Heterogeneous photoreduction of nitrogen to ammonia on catalyst-loaded TiO$_2$ powders, *Nouv. J. Chim.,* 11, 217, 1987.

154. **Lichtin, N. N., Vijayakumar, K. M., and Khader, M. M.,** Photoassisted solid-catalyzed reduction of air by aqueous organic materials: a potentially practical method of fixing nitrogen, *Proc. Bienn. Congr. Int. Sol. Energy Soc.,* 9, 1870, 1985.

155. **Lichtin, N. N. and Vijayakumar, K. M.,** Photoassisted solid-catalyzed reduction of N$_2$ by water: evidence for a photostationary state and for catalytic activity of many metal oxides, *J. Indian Chem. Soc.,* 63, 29, 1986.

156. **Tennakone, K., Wickramanayake, S., Fernando, C. A. N., Ilperuma, O. A., and Punchikewa, S.,** Photocatalytic nitrogen reduction using visible light, *J. Chem. Soc. Chem. Commun.,* p. 1078, 1987.

157a. **Augugliaro, V., Lauricella, A., Rizzuti, L., Schiavello, M., and Sclafani, A.,** Conversion of solar energy to chemical energy by photoassisted processes. I. Preliminary results on ammonia production over doped titanium dioxide catalyst in a fluidized bed reactor, *Int. J. Hydrogen Energy,* 7, 845, 1982.

157b. **Augugliaro, V., D'Alba, F., Rizzuti, L., Schiavello, M., and Sclafani, A.,** Conversion of solar energy to chemical energy by photoassisted processes. II. Influence of the iron content on the activity of doped titanium dioxide catalysts for ammonia photoproduction, *Int. J. Hydrogen Energy,* 7, 851, 1982.

158. **Palmisano, L., Augugliaro, V., Sclafani, A., and Schiavello, M.,** Activity of chromium-ion doped titania for the dinitrogen photoreduction to ammonia and for the phenol photodegradation, *J. Phys. Chem.,* 92, 6710, 1988.

159. **Khader, M. M., Lichtin, N. N., Vurens, G. H., Salmeron, M., and Somorjai, G. A.,** Photoassisted catalytic dissociation of H_2O and reduction of N_2 to NH_3 on partially reduced Fe_2O_3, *Langmuir,* 3, 303, 1987.

160. **Miyama, H., Fuji, N., and Nagae, Y.,** Heterogeneous photocatalytic synthesis of ammonia from water and nitrogen, *Chem. Phys. Lett.,* 74, 523, 1980.

161. **Khan, M. M. T., Bhardwaj, R. C., and Bhardwaj, C.,** First catalytic fixation of nitrogen by photocatalytic cadmium sulfide/platinum/rhutenium dioxide particulate system, *Indian J. Chem. Sect. A,* 25, 1, 1986.

162. **Hetterich, W. and Kisch, H.,** unpublished.

163. **Khan, M. M., T., Bardwaj, R. C., and Bardwaj, C.,** Photochemische Reduktion von N_2 zu NH_3 in wäßriger Lösung in Gegenwart von CdS/Pt/RuO$_2$-Partikeln sowie des Anions [Ru(Hedta)N$_2$]$^-$ als Katalysator, *Angew. Chem.,* 100, 1000, 1988.

Chapter 3

HYDROPHOBIC AND LIPOPHOBIC EFFECTS ON PHOTOCHEMICAL AND PHOTOPHYSICAL PROCESSES. EVIDENCE FOR INTERMOLECULAR AGGREGATION AND SELF-COILING

Chen-Ho Tung and Cheng-Ba Xu

TABLE OF CONTENTS

I. INTRODUCTION

Hydrophobic interactions play an important role in biochemical processes such as conformational changes of biopolymers, the binding of a substrate to an enzyme, the association of subunits to form a multisubunit enzyme, the formation of membranes, and higher order organization of biological molecules to form a functional unit in a living system.[1-3] All of these processes possess two features in common. They all involve the association of a number of nonpolar molecules, or groups within larger molecules, and they all occur in aqueous media. To understand life, one must try to understand hydrophobic interactions and molecular assemblies. Micelles are composed of detergent molecules and generally viewed as the simplest models of a mimetic biologic system. However, recent studies demonstrate that in aquiorgano binary mixtures electrically neutral organic molecules tend to associate and the structures of the aggregates are even simpler than micelles.[4,5] The driving force for the aggregate formation is almost solely hydrophobic interactions. These considerations may be taken as the first reason to investigate thoroughly the phenomena of aggregation.

Aggregates which constitute the main theme of this review are composed of nonpolar molecules in aqueous or organic-aqueous mixed solvents. Generally the nonpolar molecules have a long hydrocarbon chain or a large excluded volume. It is commonly observed that there is a relatively small range of concentration below which aggregate are absent and above which virtually all the nonpolar molecules exist as aggregates.[5-9] This relatively small range of concentration is called the critical aggregate concentration (CAC). Although there is no sharp demarcation line between aggregates and micelles, the following criteria for differentiating typical aggregates from micelles are suggested: (1) micelles are formed from amphiphilic surfactants. They have structure, e.g., they have a Stern layer.[10-12] Aggregates are formed from neutral molecules, and perhaps have much less structure; (2) the average aggregation number N (number of molecules per aggregate or micelle) for micelles is usually greater than 40,[10,11] the N value for aggregates is smaller, probably less than 40;[5] (3) generally micelles are formed in aqueous media, while most aggregates studied are formed in the aquiorgano binary system.

The second reason to study and understand aggregation is that the chemical reactivity of organic compounds might be affected by aggregation. For example, under certain circumstances the nature of the substituted effect may be hydrophobic instead of electronic.[13] It has been established[14-21] that the juxtaposition of two reactive long-chain molecules under the influence of hydrophobic interaction results in the rate-enhancing "proximity effect". Furthermore, like micellar solution, the aquiorgano aggregate solution has the potential to be used as media for chemical reactions. Aggregates are able to solubilize hydrophobic molecules. Because of the small size of typical aggregates, this solubilization process allows the organization of solutes on a molecular level, therefore altering the chemical reaction, such that the products or the relative yield of products can be changed relative to homogeneous solution. Aggregates can also be effective in the catalysis of reactions, thereby making their use attractive in synthetic chemistry.

Third, self-coiling of molecules enforced by hydrophobic interactions can be used to synthesize macrocyclic compounds. The formation of a macrocyclic ring from a flexible chain demands a formidable price in terms of entropy.[22-25] In general, chemists use energetically highly favored reactions between the end groups to synthesize large-ring compounds.[26-31] The new approach is to use hydrophobic or lipophobic forces to reduce the entropy expense. Large-ring compounds have been synthesized by the application of this concept.[32,33]

Although there have been great interest and activity in the study and application of hydrophobic effects on chemical reaction in the last decade and promises of even greater activity are evident, lipophobic effects on the behavior of molecules with polar regions in nonpolar solvents are scarcely reported. A recent study demonstrates that lipophobic interactions can force a single polar chain to coil up or push them together and make them form aggregates in nonpolar solvents.[33] These phenomena have also been employed to expedite the formation of macrocyclic entities.

Excellent reviews and collections about hydrophobic effects in chemistry are available.[1,2,4,34] This article will concentrate on hydrophobic effects on photochemical and photophysical processes, with particular emphasis on the studies which are related to the interests in these authors' laboratory. The authors will start with a basic statement about hydrophobic interaction, and then review the current aspects about the investigation of aggregation and self-coiling of molecules with long hydrocarbon chain in aquiorgano binary solution. Finally, the authors will present the preliminary evidences for aggregation and self-coiling of polar molecules in nonpolar solvents driven by lipophobic force.

II. HYDROPHOBIC INTERACTION

A. ORIGIN OF HYDROPHOBIC INTERACTION

Hydrophobic substances are defined as substances which are readily soluble in many nonpolar solvents, but only sparingly soluble in water. These substances are distinct from compounds that have generally low solubility in all solvents and form solids with strong intermolecular cohesion. This distinction is probably especially important from the biological point of view, since it means that molecules expelled from water by hydrophobic interaction will tend to remain in a fluid, deformable state.

The existence of hydrophobic substances and of molecules with a dual nature, containing hydrophobic regions, had been known for a long time, but misconceptions about the origin of the phenomenon of hydrophobic interaction have persisted until recently. In the earlier literature it was believed that the association between hydrocarbon chains in the formation of molecular aggregates such as micelles in aqueous solution arises from "like-to-like" attraction,[35] and it was commonplace to discuss hydrophobic interaction in terms of van der Waals forces between the solute molecules. Clearly, this could not be the whole story. Actually, the attraction of hydrocarbon for itself is essentially the same as its attraction for water. For example, the free energy of attraction between water and octane at 25°C is about -40 erg/cm^2 of contact area. The free energy of attraction of octane for themselves at the same temperature is also about -40 erg/cm^2 of contact area.[2] Thus the peculiarities of the aqueous environment cannot be accounted for by using only van der Waals forces between the solute molecules.

In fact, the attraction of hydrocarbon chains for each other plays only a minor role in the hydrophobic effect. The effect actually arises primarily from the strong attractive forces

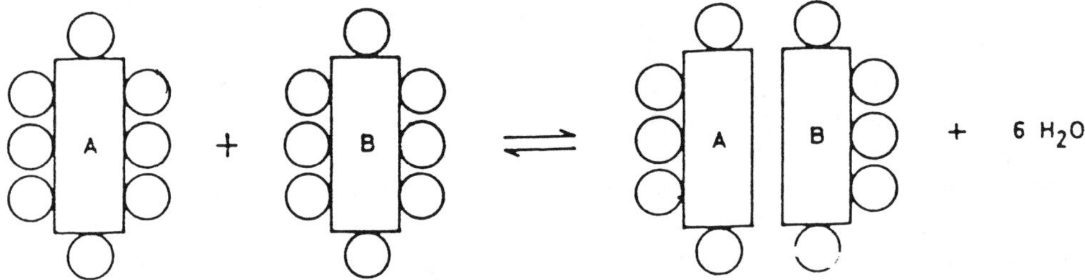

FIGURE 1.   The schematic representation of the formation of hydrophobic interaction between two hydrocarbon molecules A and B. The circles represent water molecules. (Adapted from Reichardt, C., *Solvent effects in organic chemistry*, Verlag Chemie, New York, 1979, 20.)

between water molecules which, being isotropically arranged, must be disrupted or distorted when any solute is dissolved in the water.[36] The large heat of vaporization and surface tension of water imply that it has a high cohesive energy density.[37] Therefore, creating a cavity to accommodate a solute molecule should require a sizable enthalpy input. Although some of this energy should be recovered when the solute is subsequently placed in the cavity, it may be argued that the overall process involves replacing strong attractive water-water interaction by much weaker water-solute ones. This would account for a net positive enthalpy change in the transfer of the solute molecule from the pure liquid alkane to a water solution. On the other hand, when a hydrocarbon chain dissolves in water, it causes some sort of reorganization of nearby water molecules, which leaves them in a more ordered state. A number of conceptual models have been offered which represent attempts to visualize this transformation. Most of those involve formation of short-lived water molecule aggregates near the hydrocarbon chain.[38-40] This implies that transfer of the hydrocarbon chain into water solution involves an entropy decrease. Thus, both enthalpy change and entropy change indicate the thermodynamic disadvantage of a direct contact between hydrophobic and water molecules. However, as two hydrocarbon molecules approach one another they may form an aggregate with simultaneous partial reconstruction of the original undisturbed water structure. This is shown schematically in Figure 1. Due to the contact between A and B, fewer water molecules are now in direct contact with hydrocarbon molecules. Thus, the ordering influence of the hydrophobic molecules will be diminished and the entropy increases. Although thermal energy is required for the destruction of the hydration shells around A and B, the free energy diminishes upon aggregation. Therefore, it is energetically advantageous for nonpolar molecules or nonpolar groups in large molecules in water to aggregate with expulsion of water molecules from the hydration shells.

It was probably Kirkwood who first stressed the importance of the role of the solvent in producing a new "force" between solute molecules other than the direct van der Waals forces.[41] Most chemists or biochemists who are not familiar with the theory of liquids may find the hydrophobic interaction a somewhat mysterious phenomenon. The reason is that we have been accustomed to the factor that a force between two particles is a property of the particles themselves. For example, coulombic force arise from charges situated in the particles; van der Waals forces are due to fluctuations in the locations of the electric charges on the electrons of the two particles; and the strong repulsive forces are a result of the property of the impenetrable "volume" of the particles. When dealing with hydrophobic interaction, we surprisingly claim that the force is mainly dependent on the properties of the solvent, not on the solutes. This causes some difficulties in the appreciation of the origin of these forces.

The concept of hydrophobic interaction described above measures the tendency of two (or more) solute molecules to aggregate in aqueous solutions. Another concept which is often discussed in the literature under the same term, hydrophobic interaction, is represented by the standard free energy of transfer of a solute between water and some other nonaqueous solvent, and it measures the relative tendency of the solutes to prefer one environment over the other.[34] Both quantities have a common feature, i.e., they express some kind of "phobia" for the aqueous environment. Since at present there is no way to estimate the free energy change in the aggregation process, many works attempt to establish the correlation between hydrophobicity and chemical structure of substrate based on the later concept.

B. HANSCH'S π HYDROPHOBIC SUBSTITUENT CONSTANT AND REKKER'S f HYDROPHOBIC FRAGMENTAL CONSTANT

In the quantitative activity-structure relationship of organic compounds, a successful example is the Hammett equation (Equation 1), which has been widely used to calculate a rate or an equilibrium constant of a meta- or para-substituted derivative of C_6H_5-R, with R as the reacting center.[42,43]

$$\log(k_s/k_o) = \delta\sigma \tag{1}$$

where k_0 and k_s represent the rate constant of an unsubstituted structure and a substituted derivative, respectively. The left-hand term can be replaced by log (K_s/K_o), where K represents an equilibrium constant. σ denotes a constant typical of the substituent, reflecting its ability to attract or repel electrons. δ is a constant characteristic of the type of reaction. Hammett standardized Equation 1 to the dissociation of benzoic acid in water at 25°C, δ being taken to be 1,000.

By analogy with the Hammett equation, in the early 1960s Hansch et al. proposed to use the partition coefficients of organic compounds in an organic solvent-water system as an index for the hydrophobic property and correlate the hydrophobicity with the structure as follows:[44-46]

$$\log[P(SX)/P(SH)] = \delta\pi(x) \tag{2}$$

where SX is a derivative of SH; P(SH) and P(SX) represent the partition coefficients of SH and SX, respectively; $\pi(x)$ represents the hydrophobic substituent constant, i.e., the contribution of substituent X to the hydrophobicity of structure SH when X replaces an H atom in SH. The constant δ reflects the characteristics of the solvent pair used in determining the partition coefficient. Standardization here means choosing one of the several solvent systems as reference, so that $\delta = 1.000$ in Equation 2. 1-Octanol-water has been proposed by Hansch and Dunn[47] as the solvent system of choice, and Equation 2 can be transformed into

$$\log[P(SX)/P(SH)] = \pi(x) \tag{3}$$

One of the useful features of P values (when expressed as log P) is their additive-constitutive nature:

$$\log P(S'X_1X_2...X_n) = \log P(SH) + \sum_1^n \pi(x_n) \tag{4}$$

where S' is a constituent part of SH. Thus, the π constants of the substituent can be calculated from partition coefficients of organic compounds. A variety of π values of the important functional groups have been tabulated.[48] However, the π value of a given functional group may vary from one molecule to another. Therefore, in their practical application corrections are needed in some cases. The chief objection to the Hansch equation is the assumption that the CH_3, CH_2, CH, and C contributions to hydrophobicity are identical, i.e., they have the identical π value. For example, according to Equation 4,

$$\log P(C_6H_5CH_2CH_3) = \log P(C_6H_6) + \pi(CH_2) + \pi(CH_3) \tag{5}$$

One can also calculate the hydrophobicity of ethylbenzene by using the following equation:

$$\log P(C_6H_5CH_2CH_3) = \log P(C_6H_5CH_3) + \pi(CH_3)$$

$$= \log P(C_6H_6) + 2\pi(CH_3) \tag{6}$$

Thus, $\pi(CH_3) = \pi(CH_2)$. The same is true for $\pi(CH)$ and $\pi(C)$.

Dealing with the objections to the π system, Rekker proposed to use Equations 7 and 8 to calculate the hydrophobicity of organic compounds:[49]

$$\log P(SX) = f(s) + f(x) \tag{7}$$

$$\log P(S'X_1X_2...X_n) = f(s') + \sum_i^n f(x_n) \tag{8}$$

Equation 8 can also be written as

$$\log P = \sum_1^n a_n f_n \tag{9}$$

In these equations $P(SX)$ represents the partition coefficient of SX in octanol-water, and $f(x_n)$ or f_n represents the hydrophobic fragmental constant of x_n, i.e., the contribution of constituent x_n to the total hydrophobicity. The f values of various functional groups are listed in Reference 49. For any organic compound, summation of the pertinent f values will yield a Σf value that is a good indicator of the hydrophobicity of this compound. A substance with a Σf value greater than zero is hydrophobic, and that with a Σf value smaller than zero is hydrophilic. An example for the application of Equation 9 is as follows:

1. $C_6H_5CH(CH_3)CH_2C_6H_5$ 2. $(CH_3)_3CCH_2OCOCH_3$

$$\log P(1) = 2f(C_6H_5) + f(CH_3) + f(CH_2) + f(CH)$$

$$= 2 \times 1.886 + 0.702 + 0.530 + 0.235 = 5.239 \tag{10}$$

$$\log P(2) = 4f(CH_3) + f(CH_2) + f(C) + f(COO)$$

$$= 4 \times 0.702 + 0.530 + 0.150 + (-1.292) = 2.196 \tag{11}$$

The calculated partition coefficients of the above compounds are in agreement with the experimentally determined values.

Collander proposed that with any given solute the partition coefficient measured in solvent system a should be capable of being related to the value measured in system b using the following equation:[50,51]

$$\log P_a = \delta \log P_b + q \tag{12}$$

where P_a and P_b are the partition coefficients measured in solvent system a and b, respectively. δ and q are constants which are characteristic of the solvent systems employed. An example of such a relationship is given in Equation 13, which connects the partition values in the octanol-water system with those in the iso-butanol-water system:

$$\log P_{octanol} = 1.24 \log P_{isobutanol} - 0.42 \tag{13}$$

The combination of Equation 12 with Equation 9 results in Equation 14.

$$\Sigma f_a = \delta \Sigma f_b + q \tag{14}$$

where Σf_a represents the summation of a series of fragmental constants valid for solvent system a, and Σf_b the summation of the fragmental constants of the same fragments in solvent system b. Equation 14 can be used to transform information on the hydrophobic fragmental constants of a solute for any given solvent system into the data pertinent to another system.

Both Hansch's π substituent constants and Rekker's f constants directly reflect hydrophobic interactions and have already established themselves in an impressively wide range of applications, e.g., in structure-property correlations between such constants and pharmaceutical properties or physiological activities of a variety of chemical species,[52,53] as well as in studies on oil-water distribution coefficients and in liquid chromatography.[54,55]

Recently Jiang investigated the hydrophobicities of a series of aqueous-organic solvent mixtures by using the kinetic method and found that the hydrophobic interaction for a given solute decreases in the following order:[56] ethylene glycol (EG)-H_2O > dimethyl sulfoxide (DMSO)-H_2O > ethylene glycol methyl ether (MEG)-H_2O > dimethyl formamide (DMF)-H_2O ≈ 1,2-dimethoxyethane (DME)-H_2O > 1,4-dioxane(DX)-H_2O > ethanol-H_2O ≈ acetone-H_2O ≈ acetonitrile-H_2O > t-butanol-H_2O. This order agrees well with the hydrophilicities of the organic components in the mixtures. The calculated Σf values of the organic solvents from their Rekker's f hydrophobic fragmental constants are EG, -1.92; DMSO, -1.35; MEG, -1.30; DMF, -1.01; DME, -0.70; DX, -0.49; ethanol, -0.26; acetone, -0.30; acetonitrile, -0.36; t-BuOH, $+0.77$.

Additives may influences hydrophobic interaction obviously. Salts such as NaCl and LiCl, increase the hydrophobic interaction, i.e., "salt out" nonpolar materials dissolved in water. On the other hand, urea and N-methylacetamide may change the water structure so that they normally decrease hydrophobic interaction, i.e., "salt in" nonpolar material.[57]

III. EVIDENCE FOR AGGREGATE FORMATION AND SELF-COILING FROM KINETICS STUDIES

A variety of approaches have been employed to explore and examine the structural and dynamic features of molecular assemblies, including electron micrography, NMR, neutron scattering, laser Raman and infrared spectroscopies, and luminescence spectroscopy. However, until recently almost only the kinetics method has been used to provide the evidence for aggregate formation and self-coiling of "neutral" long-chain molecules.[4]

In general, to the authors' knowledge there might be four common types of effects that aggregates produce on reactions: cage, local concentration, viscosity, and polarity effects.

Cage effect — It involves the ability of aggregates to hold two reactive intermediates together long enough for a reaction to occur. For example, homocleavage of a single bond produces two geminate radicals which can either recombine or diffuse apart. In aggregates, a large cage effect may be observed relative to homogenous solvents whose magnitude cannot be explained by the microviscosity inside aggregates. The main reason for this observation might be that the hydrophobicity of the solutes inhibits diffusion into the aquiorgano or aqueous phase, thereby increasing the time spent by the radical intermediates in the restricted space of aggregates.

Local concentration effect — It results from the high concentration of hydrophobic solutes inside the aggregates. This effect is commonly observed for bimolecular reactions between hydrophobic solutes. In some cases, bimolecular reaction can occur in aggregate solutions where the same total macroscopic concentration in organic solvents would yield no reactions. On the other hand, hydrophilic solutes would be located outside the aggregates. The bimolecular reactions between hydrophobic solute and hydrophilic solute could be retarded.

Microviscosity effect — The microviscosity inside aggregates is unusually higher than that of the bulk solution. This phenomenon leads to modification of chemical and physical processes which are sensitive to viscosity. The microviscosity inside aggregates at a temperature above their phase transition temperature is analogous to that in micelles, but below this temperature is as large as in solid states.[5] In some cases, the viscosity effect on a chemical or physical process can be used to calibrate the microviscosity of the aggregates.

Polarity effect — The polarity inside aggregates is analogous to that of the hydrocarbon solvent, much less polar than the bulk aquiorgano phase. Therefore, the reaction affected by polarity will have different reactivity in aggregates compared to that in bulk solution. The average location of a solute can be quite different for solutes with different hydrophobicities. Thus, the actual polarity and viscosity observed by solutes with different hydrophobicities will be different.

Self-coiling of a long-chain molecule would favor the interaction between its terminal groups. Furthermore, the polarity and/or microviscosity inside the hairpin looping would be different from the bulk solution. Thus, self-coiling of a long-chain molecule with function groups may also lead to modification of chemical and physical processes of the groups.

A performance of the effects mentioned above would provide direct and convincing evidence for aggregate formation and self-coiling.

A. RETARDATION OF HYDROLYSIS OF ESTER WITH LONG CHAIN

Menger and Portony[6] first reported the formation of aggregate of hydrocarbon long-chain molecules in aqueous solution in their study on the hydrolysis of esters. The rate constant for basic hydrolysis of p-nitrophenyl laureate decreases with increasing the initial concentration of ester inthe range of 1 to 5×10^{-6} M, and stays constant above 5×10^{-6} M. The simplest explanation for this observation is that the substrate molecules hydrophobically bind to one another, thereby enclosing the ester groups within the associated hydrocarbon chains. Thus the hydrolysis is retarded. The rate inhibition is appreciable. The second-order rate constant for basic hydrolysis of p-nitrophenyl acetate is 800 times larger than that for p-nitrophenyl laureate at 1.0×10^{-5} M. Later, Blyth and Knowles[14] suggested that such rate retardation could also be attributed to the coiling-up of the long alkyl chains. Murakami et al.[7] also rationalized their results in terms of aggregation and self-coiling, and they specifically suggested that only monomeric species are sufficiently reactive. Figure 2 shows the concentration dependence of the rate constant for alkaline hydrolysis of p-nitrophenyl n-alkanoates.[7] The rate constants for hydrolysis of p-nitrophenyl acetate and p-nitrophenyl hexanoate are independent of their concentrations. However, the rate constants for p-nitrophenyl decanoate, dodecanoate, and hexadecanoate are evidently small relative to those with short chain length, and show concentration dependence. The curves for esters with long chains in Figure 2 display a break. The concentration corresponding to the break point is defined as the CAC. At concentration below CAC, the ester exists only in its monomeric form, and the retardation of hydrolysis is due to self-coiling of the long chain, which masks the ester group from hydroxide attack. Thus the rate constant for hydrolysis is independent of the concentration of the substrate. Above CAC aggregates are formed, and the increase in concentration leads to a reduction in the fraction of monomeric ester and hence in the overall hydrolysis rate. The driving force for self-coiling and aggregation is hydrophobic interactions as manifested by the following facts: (1) a substrate with a longer chain length

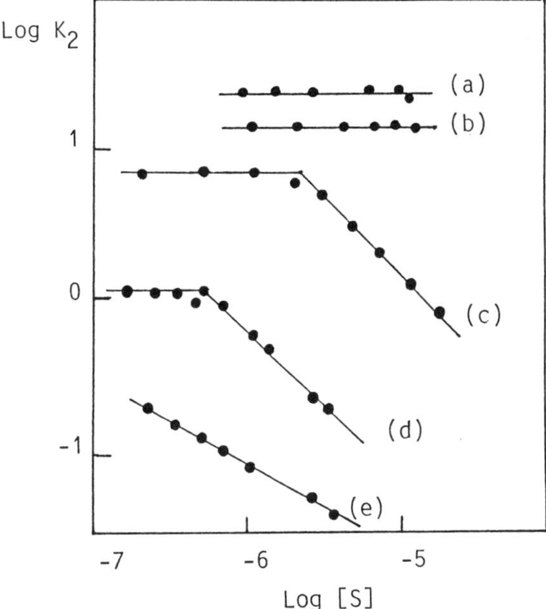

FIGURE 2.   Correlations of second-order rate constant k_2 with initial ester concentration [S] for hydrolysis of *p*-nitrophenyl carboxylates: acetate (a), hexanoate (b), decanoate (c), dodecanoate (d), and hexadecanoate (e) at 40.0°C in 1.0% (v/v) (the former four esters) and 10.9% (hexadecanoate) aqueous dioxane. (Adapted from Murakami et al.[7])

has a smaller CAC value, indicating that its hydrophobicity is greater;[7,15] (2) the addition of urea to the solution results in the increase in the CAC value;[6,7] (3) the addition of inorganic salts to the solution gives the reduction in hydrolysis rate, because the salting-out effects decrease the fraction of monomeric substrate.[7]

In order to confirm further the intermolecular aggregation behavior of long-chain molecules, Murakami et al.[7] carried out the measurements of surface tension for aqueous solutions of *p*-nitrophenyl decanoate as a function of substrate concentration. The plot of surface tension vs. concentration shows clearly a break point characteristic of aggregate formation, and the CAC value thus obtained is in good agreement with the one obtained by the kinetics method.

Jiang investigated the kinetics of hydrolysis of *p*-nitrophenyl ester of *n*-alkanoic acids with various chain lengths in ten aquiorgano binary solvent systems mentioned in the above section.[56] The fraction of water in the mixture is denoted as Φ. The ratios of hydrolysis rate constants of acetate-to-octanoate fall in the range of 2 to 4 in all solvent systems. If the alkyl chain is long enough (say, greater than three carbon atoms), the steric hindrance would be constant. Unless aggregation and self-coiling occur, the hydrolysis rate constant of hexadecanoate would be the same as that of octanoate. The experimental results show that this is indeed the case in all the binary solvent systems with $\Phi = 0.5$, except those for DMSO-H$_2$O and EG-H$_2$O. The ratios of the rate constants of octanoate-to-hexadecanoate are 24 and 68 in DMSO-H$_2$O and EG-H$_2$O, respectively, but in the other eight solvents these ratios all lie in the range from 1.0 to 1.2. This observation suggests that for the other eight solvents with Φ of 0.5 the aggregating power is too small for aggregate formation. At $\Phi = 0.7$, however, these ratios range all the way from 680 for EG-H$_2$O to roughly unity for isobutanol-H$_2$O. Attempts to correlate these ratios with various solvent polarity parameters all failed, but a quantitative correlation between this ratio and the hydrophilicity (Σf) of the organic component in the solvent mixture was established.

$$\log(k_8/k_{16}) = (-1.17 \pm 0.13)\Sigma f + (0.50 \pm 0.12)$$

$$n = 10, r = 0.953, \text{ confidence level} > 99.9\% \tag{15}$$

B. LOCAL CONCENTRATION EFFECT

Aggregate formation is expected to give a local concentration effect or a proximity effect in a bimolecular reaction in which the two reactants are both located in aggregate. Blyth and Knowles[17] measured the second-order rate constants of aminolysis in aqueous solution of the p-nitrophenyl esters of long- and short-chain n-alkanoic acids by long- and short-chain alkylamines. Using an amine with short chain, ethylamine, the aminolysis rate constant of p-nitrophenyl decanoate is 13 times smaller than that for p-nitrophenyl acetate at ester concentration well below the CAC. This difference in the aminolysis rates is attributed to greater steric hindrance for amine attack on the decanoate ester than that on the acetate. However, for aminolysis by decylamine, the rate constant of the former ester is much greater than that of the later ester. The ratio of $k_{\text{decylamine}}/k_{\text{ethylamine}}$ for p-nitrophenyl decanoate is 317, despite the larger steric hindrance expected in the reaction of decylamine compared with ethylamine. Obviously the remarkable large rate enhancement for the reaction of long-chain ester with long-chain amine is due to formation of aggregate, which results in enhancement of local concentration. Supporting this proposal, the ratio of the rate constant for aminolysis of decanoate by decylamine to that of acetate by ethylamine in aqueous solution is much reduced when the reactions are performed in 50% (v/v) aqueous 1,4-dioxane, a mixture solvent in which hydrophobic interactions are expected to be markedly decreased.

Jiang also demonstrated the local concentration effect of aggregate in his study on the transesterification reaction between para-substituted phenyl esters of n-alkanoic acids and thiols with various chain lengths in DMSO-H_2O, acetone-H_2O, and ethanol-H_2O binary solvents.[21] In the DMSO-H_2O solvent mixture with $\Phi > 0.4$, a large rate enhancement can be realized for the reaction between long-chain esters and long-chain thiols, due to formation of coaggregate between the two reactants. This effect decreases with increasing volume content of DMSO, and there is no local concentration effect in acetone-H_2O and ethanol-H_2O even with large Φ values.

C. LARGE RING "NEIGHBORING-GROUP" PARTICIPATION

Jiang proposed that long-chain molecules can be forced to fold and interact intramolecularly by hydrophobic forces.[58-60] They measured the hydrolysis rate constants of ω-substituted carboxylic esters with various chain lengths in DMSO-H_2O with $\Phi = 0.50$. The hydrolysis rate constant of ω-sulfhydryl-substituted p-nitrophenyl hexadecanoate is 124 times greater than that for p-nitrophenyl hexadecanoate. The initial substrate concentration they used were far below the CAC values. Thus, the rate enhancement was attributed to a 17-membered ring neighboring-group participation involving the ω-sulfhydryl and the carbonyl group. The ω-sulfhydryl-substituted heptadecanoate and tridecanoate show the similar phenomenon. However, the ω-sulfhydryl undecanoate does not give the neighboring-group participation due to the larger strain for a 12-membered ring. Hopefully, this large-ring neighboring-group participation may be used to synthesize macrocyclic entities from flexible chain.

D. EFFECTS OF ADDING AMYLOSE

Particularly interesting evidence for aggregate formation and self-coiling of long-chain molecules in aquiorgano mixtures was provided by Jiang et al. in their study on the hydrolyis of para-substituted phenyl esters of *n*-alkanoic acids in the presence of amylose.[61-65] It has been well established[66-69] that amylose can form inclusion complex with molecules bearing linear hydrocarbon chains. The inclusion process is accompanied by conformational changes of the amylose polymer chain from random coils to interrupted helices. An interesting feature of the helical cavities is that the size can be adjusted according to the substrate. In the absence of amylose the hydrolysis rate constant of *p*-chlorophenyl hexadecanoate in DMSO-H_2O with $\Phi = 0.5$ is much smaller relative to that of ester with a short chain, due to the aggregate formation of the long-chain ester. The addition of amylose to the solution results in the decrease in the ratio of rate constant for long-chain ester to that for short-chain ester, suggesting the destruction of aggregates. On the other hand, formation of an inclusion complex between amylose and substrate should completely inhibit neighboring-group participation. Indeed, in the presence of amylose, the hydrolysis rate constant of ω-sulfhydryl-substituted *p*-nitrophenyl hexadecanoate in DMSO-H_2O ($\Phi = 0.5$) is close to that for *p*-nitrophenyl hexadecanoate.[60]

IV. PHOTOPHYSICAL AND PHOTOCHEMICAL EVIDENCE FOR AGGREGATION AND SELF-COILING

The evidence for aggregation and self-coiling based on kinetics studies is further strengthened by luminescence spectroscopy. Luminescence probes which incorporate into molecular assembly may undergo characteristic changes in their luminescence intensity, spectral distribution, and decay time. Therefore, they are suited for the investigation of the structural and dynamic feature of assembly.[70] Fluorescence probe studies have not only provided new evidence for aggregation and self-coiling, but have also characterized the aggregates.

A. EXCIMER FORMATION OF ALKYL 2-NAPHTHOATES WITH LONG CHAINS

The fluorescence spectra of alkyl 2-naphthoates with various chain lengths in DMSO-H_2O and EG-H_2O mixtures have been reported.[9]

$$NpCO_2C_nH_{2n+1},$$

$$A_n, n = 4, 6, 8, 12, 16; Np = 2\text{-naphthyl}$$

The 2-naphthoates with short chain, A_4 and A_6, only show monomer fluorescence in both solvent mixtures, while the fluorescence spectra of long-chain alkyl 2-naphthoates are dominated by excimer emission. Figure 3 shows the fluorescence spectra of A_{12} at room temperature in EG-H_2O as a function of Φ. Formation of excimer requires either a high-enough concentration of the ground state or a long-enough lifetime of the excited-state substrate, so that the excited molecule can encounter a ground-state molecule during its lifetime. A_4, as the model of the 2-naphthoate, has essentially the same lifetime in EG-H_2O with various Φ values. Evidently, formation of excimer for 2-naphthoates with long chains should be attributed to the local concentration enhancement of the ground-state molecule, induced by intermolecular aggregation.

Formation of aggregate depends both on the concentration of substrate and the solvent aggregating power which usually parallels the solvent hydrophilicity.[4] For a particular substrate the increase in Φ value of the aquiorgano binary mixture results in the increase in solvent aggregating power.[71] Figure 4 gives the ratio of the intensity of excimer fluorescence (I_D, $\lambda_{em} = 420$ nm) to the sum of the intensities of the excimer and the monomer fluorescence (I_M, $\lambda_{em} = 340$ nm) as a function of Φ at a fixed A_{12} concentration. A clearly discernible break is shown in the plot. The Φ value corresponding to the break is defined as the critical solvent composition (CΦ) for aggregate formation. Figure 5 shows the plot of $I_D/(I_D + I_M)$

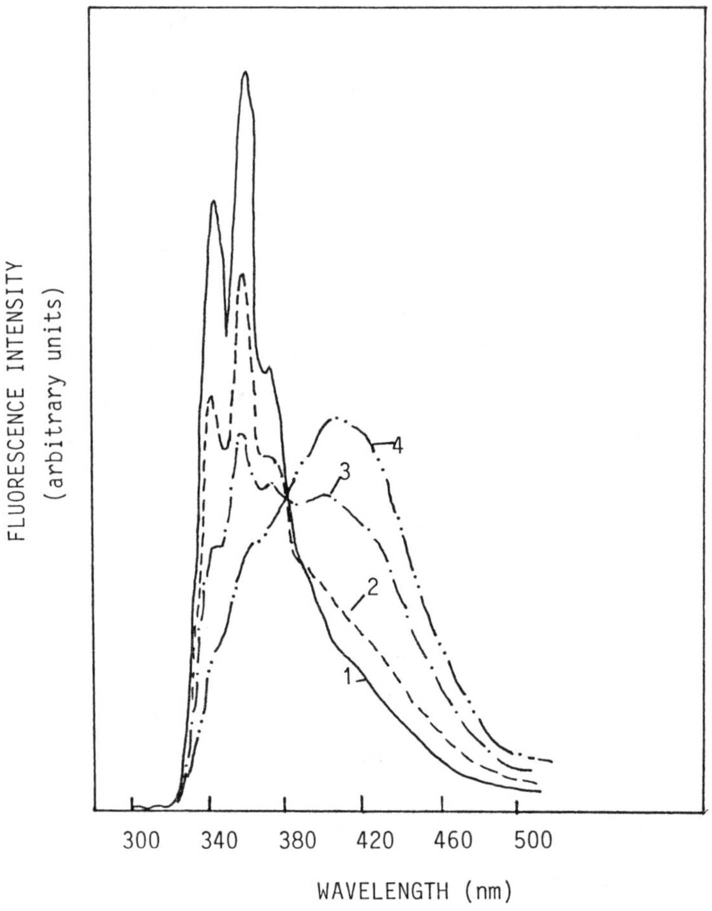

FIGURE 3. Fluorescence spectra of dodecyl 2-naphthoate in ethylene glycol-water. The excitation wavelength λ_{ex} is 280 nm. The concentration of dodecyl 2-naphthoate is $5 \times 10^{-5}\ M$. The fractions of water in the solvent mixtures are the following: 1, 0; 2, 0.2: 3, 0.3; 4, 0.4.

against the substrate concentration at a fixed Φ value. This plot also exhibits a break, indicating the occurrence of a CAC. The $C\Phi$ and CAC values of various alkyl 2-naphthoates are given in Table 1. Both $C\Phi$ and CAC decrease as the chain length of the 2-naphthoate increase.

The behavior of A_n in DMSO-H_2O is essentially identical with that in EG-H_2O, but the $C\Phi$ and CAC values in EG-H_2O are slightly smaller than that in DMSO-H_2O. However, in ethanol-H_2O A_{12} only emits monomer fluorescence until Φ reaches 0.7. These results are in good agreement with the observation in the kinetics study on *p*-nitrophenyl ester of alkanoic acid. This again indicates that for solvent mixtures with different organic components at the same Φ value, the solvent aggregating power decreases with the inceasing lipophilicities of the organic components, which can be estimated from the hydrophobic fragmental constants.[56]

The addition of amylose to the solution of A_n in DMSO-H_2O is expected to destroy the aggregates, since amylose may form an inclusion complex with A_n.[66-69] Indeed, the monomer fluorescence of A_{16} in DMSO-H_2O with $\Phi > C\Phi$ and at the concentration greater than CAC increases at the sacrifice of excimer fluorescence when amylose is added. This assertion may serve as an additional evidence for aggregate formation of A_{16} in the absence of amylose. It is interesting to note that the smallest amount of amylose, [Amylose]$_s$, required for the

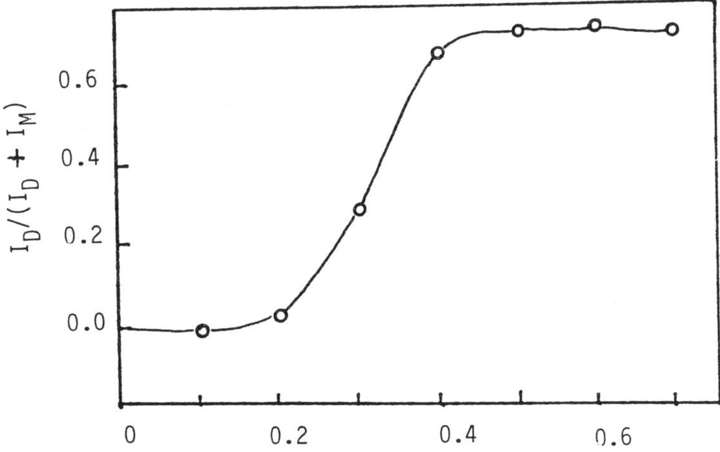

FIGURE 4. The ratio of the intensity of excimer fluorescence to the sum of the intensities of the excimer and monomer fluorescence $I_D/(I_D + I_M)$ for dodecyl 2-naphthoate in ethylene glycol-water as a function of the fraction of water in the solvent mixture. The concentration of dodecyl 2-naphthoate is 5×10^{-5} M.

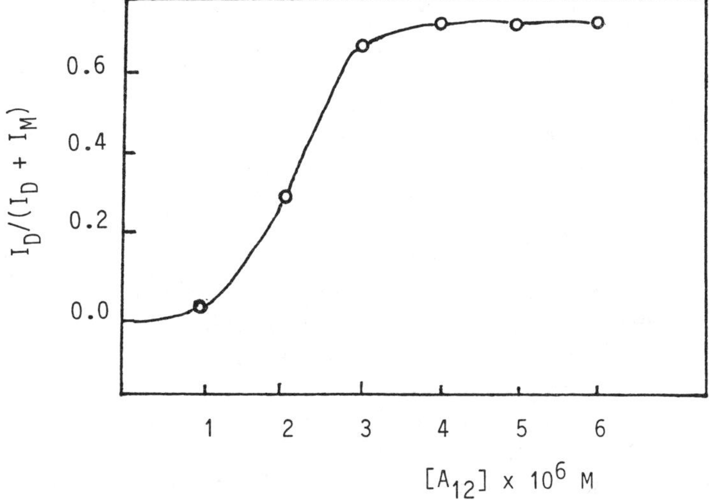

FIGURE 5. The ratio of the intensity of excimer fluorescence to the sum of the intensities of the excimer and monomer fluorescence $I_D/(I_D + I_M)$ as a function of concentration of dodecyl 2-naphthoate A_{12} in ethylene glycol-water. The fraction of water in the solvent mixture is 0.6.

disappearance of A_{16} excimer (i.e., A_{16} only emits monomer fluorescence) depends on the concentration of A_{16}. The ratio of $[A_{16}]$ to $[\text{amylose}]_s$ is a constant at fixed Φ value. This ratio yields about 20 glucose residences of amylose required to wrap up a single piece of A_{16} chain.

In order to evaluate the orientation of the substrates inside aggregate, the interaction between the ground-state 2-naphthoates and the kinetics of excimer formation was investigated. Excimers are either formed by diffusion together of excited and unexcited chromophores or originate from pairs of chromophores which exist prior to excitation. The absorption

TABLE 1

Critical Solvent Composition for Aggregate Formation and Critical Aggregate Concentration of Alkyl 2-Naphthoates A_n (n Represents the Chain Length of the Alkyl Group) in Ethylene Glycol-Water (EG-H_2O) and Dimethyl Sulfoxide-Water (DMSO-H_2O)

Compound	Critical composition of solution mixture[a]		Critical aggregate concentration[b]	
	EG-H_2O	DMSO-H_2O	EG-H_2O	DMSO-H_2O
A_4	0.90	0.90	—	—
A_8	0.40	0.55	$1 \times 10^{-5}\,M$	$3 \times 10^{-5}\,M$
A_{12}	0.30	0.40	$2 \times 10^{-6}\,M$	$5 \times 10^{-6}\,M$
A_{16}	0.20	0.25	$1 \times 10^{-7}\,M$	$1 \times 10^{-7}\,M$

[a]   The concentrations of the substrates $[A_n]$ are $5 \times 10^{-5}\,M$.
[b]   The fraction of water in the solvent mixture is 0.6.

spectra of A_{16} and A_{12} in DMSO-H_2O ($\Phi = 0.5$) are essentially identical with that of A_4. In ^{1}HNMR spectra of A_{16} and A_{12} in DMSO-d_6-D_2O the aromatic protons show the same chemical shifts and splitting pattern as those of A_4. These observations suggest that inside aggregates there exists no strong interaction between the naphthoate chromophores in the ground state. Furthermore, the monomer emission at 340 nm decays biexponentially. The lifetime of the longer living component is 20 ns and the shorter living one 1.5 ns. The analysis of the decay of the excimer at 420 nm yields growing-in time of 1.5 ns and decay time of 20 ns, both of which correspond with the components of the monomer emission. Thus the excimers are formed via a dynamic process.[72] The rate constant of excimer formation and that of excimer dissociation were calculated from the decay parameters of excimer and monomer emission to be 3.8 and 1.2×10^8 s^{-1}, respectively, at room temperature. The measurements of these two rate constants as a function of temperature in the range of 20 to 70°C yield the activation energy of excimer formation and of excimer dissociation to be 2.7 and 6.6 kcal/mol, respectively. The enthalpy change and entropy change for excimer formation are -3.9 kcal/mol and -13 cal/mol, respectively. On the basis of the absorption spectra, the ^{1}HNMR, and the kinetics of excimer formation one can conclude that inside aggregates the chromophores in ground state do not orientate themselves in excimer conformation.

The ratios of $I_D/(I_D + I_M)$ were also measured as a function of temperature. Figure 6 shows the plots of the logarithm of the ratio of the fluorescence quantum yields, ln (Φ_{fD}/Φ_{fM}), in excimer and monomer regions for A_{12} and A_{16} vs. 1/T. These plots have a "high temperature region" where the ratio of Φ_{fD}/Φ_{fM} increases with decreasing temperature, and a "low temperature region" where the ratio decreases dramatically with decreasing temperature. From the tangent at the curve in the high temperature region, the values of enthalpy change for excimer formation, ΔH, of A_{12} and A_{16} are obtained[73] to be -3.8 and -4.0 kcal/mol, respectively, which are in good agreement with the values derived from the time-resolved measurements mentioned above.

The O–O band of monomer fluorescence is at $\lambda = 345$ nm, and the excimer band peak at 414 nm. The balance of the red shift of 4800 cm^{-1} must be attributed to ΔH and the repulsion energy, F_R, between the ground-state molecules at the equilibrium excimer configuration.[75]

$$h(v_M - v_D) = -\Delta H + E_R \tag{16}$$

The E_R was calculated to be 10 kcal/mol.

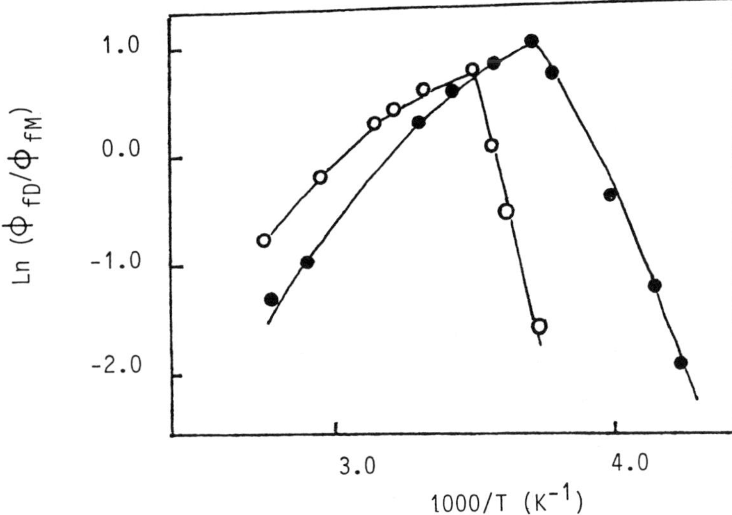

FIGURE 6.   Plot of the logarithm of the ratio of fluorescence quantum yields in excimer and monomer region vs. 1/T for dodecyl 2-naphthoate (●) and hexadecyl 2-naphthoate (○).

The tangent at the curve in Figure 6 in the low temperature region might yield the activation energy for excimer formation, E_{DM}. However, as shown in Figure 6, below certain temperature the excimer fluorescence disappears abruptly. This, in turn, yields a very large E_{DM} value. Obviously, there exists a phase transition in aggregate. Above the phase transition temperature the excited chromophore can diffuse and encounter a ground-state chromophore. However, below this temperature, though the bulk solution has good fluidity, the microviscosity inside aggregates is assumed to be very high, and the excimer cannot be formed via diffusion. The phase transition temperature for A_{16} and A_{12} aggregate is 5 and $-20°C$ respectively.

B. CHARACTER OF AGGREGATES FORMED BY LONG-CHAIN ALKANE

The alkyl 2-naphthoates have been used as the fluorescence probe for aggregate formation of long-chain n-alkane (C_n, n represents the number of carbon atoms in the alkane molecule).[9] The solution of A_{16} in DMSO-H_2O ($\Phi = 0.5$) at concentration of $5 \times 10^{-5} M$ mainly emits the excimer fluorescence due to the formation of aggregate (Figure 7). The addition of n-octadecane (C_{18}) to the solution results in the enhancement of monomer emission and reduction in excimer fluorescence. In the presence of a sufficient amount of C_{18}, A_{16} only emits monomer fluorescence. The addition of n-tetradecane (C_{14}) causes a similar result. However, the addition of hydrocarbons with chain length shorter than decane brings no effect on the fluorescence spectrum of A_{16}. The fluorescence behaviors of A_{12} in DMSO-H_2O upon addition of alkane are similar to that of A_{16}. These observations provide evidence for coaggregation of A_n with C_n. Since the chromophores in coaggregates are diluted by C_n, the excimer formation is retarded. A quenching experiment also demonstrates the existence of the coaggregate. The fluorescence of A_4 in DMSO-H_2O ($\Phi = 0.5$) is quenched by sodium nitrite with the quenching constant of $4 \times 10^9 \ M^{-1} \ s^{-1}$. However, in the same solvent mixture and in the presence of C_{18} ($1 \times 10^{-3} \ M$) the quenching process of A_{12} ($5 \times 10^{-5} M$) fluorescence by sodium nitrite is inhibited. The quenching constant is reduced by one order of magnitude ($5 \times 10^8 \ M^{-1} \ s^{-1}$). Undoubtedly this is caused by incorporating A_{12} into C_{18} aggregates which protect the excited state of A_{12} from quenching by sodium nitrite dissolved in water.

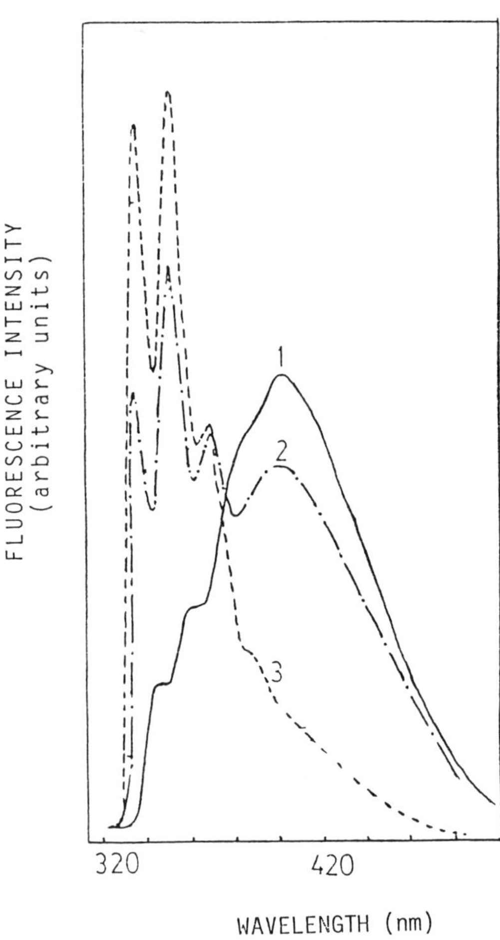

FIGURE 7. Fluorescence spectra of dodecyl 2-naphthoate (5×10^{-5} M) in dimethyl sulfoxide-water in the presence of n-octadecane. The fraction of water in the solvent mixture is 0.5. The concentrations of n-octadecane are the following: 1, 0 M; 2, 6×10^{-4} M; 3, 1×10^{-3} M.

The aggregates of C_n have been characterized by fluorescence probe techniques. First, the microscopic polarity within aggregates and the CAC values were measured. The fluorescence spectra of 2-naphthoates are sensitive to solvent polarity. In nonpolar solvents, such as cyclohexane, A_{12} or A_{16} emits fine structural fluorescence with four peaks at 336, 352, 370, and 388 nm (shoulder), respectively. The maximum emission is at 352 nm. In polar solvents, such as methanol, a structureless emission with $\lambda_{max} = 364$ nm was observed instead of the fine-structural fluorescence. A_{12} or A_{16} is unable to form aggregates and only shows monomer emission in DMSO-H_2O ($\Phi = 0.5$) at a concentration below its CAC. Because of the large polarity of the mixed solvent, the fluorescence spectrum of A_{12} is structureless ($\lambda_{max} = 372$ nm). The addition of alkane, such as C_{14}, to the solution results in the change of A_{12} fluorescence spectrum from a structureless one to a fine-structural one with four peaks at 336, 352, 370, and 390 nm (shoulder). This spectrum is very similar to that in cyclohexane, indicating the similarity of the polarities in C_{14} aggregates and in cyclohexane. 3-Methyl-1-hexadecylindole (1-In-16) was also used as a probe to detect the polarity inside the aggregates.[75-78] The fluorescence λ_{max} of 1-In-16 in DMSO-H_2O ($\Phi =$

TABLE 2

Critical Aggregate Concentration (CAC) and Average Aggregate Number N of *n*-Alkane Aggregates in Dimethyl Sulfoxide-Water Solvent (the Fraction of Water in the Solvent Mixture is 0.5)

	Hexadecane	Tetradecane	Dodecane	Decane	SDS[a]	HDTBr[b]
CAC ($\times$ 10^4 M)	1.0	2.0	3.0	4.0	—	9.0[c]
N	24	21	18	14	32[a]	88[c]

[a] SDS represents sodium dodecylsulfate. The literature N value of SDS in dimethyl sulfoxide (DMSO-H_2O) is 35 (see Reference 82).

[b] HDTBr represents hexadecyltrimethylammonium bromide. The CMC and N values for HDTBr were measured in aqueous solution by using A_{12} as probe.

[c] The literature CMC of HDTBr is 8.0 $\times$ 10^{-4} M and the literature N value of HDTBr is 82 (see Reference 81).

0.5) solution at concentration below its CAC is at 366 nm. In the same solvent mixture but in the presence of C_{14} the fluorescence λ_{max} blue-shifts to 327 nm, which is almost identical with that in cyclohexane (322 nm). This suggests that the microscopic polarity inside C_{14} aggregates is indeed as small as that of cyclohexane.

The fluorescence spectra of A_{16} in DMSO-H_2O (Φ = 0.5) were recorded as a function of C_n concentration.[5] The change from a structureless fluorescence to a fine structural one occurs at CAC of C_n. The CAC values of C_n thus determined are listed in Table 2. Under a specific condition, the value of CAC of a substrate depends on its structural feature. For example, the CACs of C_n decrease with increasing chain length.

Aggregate number (N) refers the number of long-chain molecules which are involved in one aggregate, and is one of the most important parameters of aggregate. Atik et al.,[79] Lianos and Zana,[80] and Bolt and Turro[81] successfully measured the N values of a number of cationic and anionic surfactant micelles by using the fluorescence probe method which involves pyrene excimer formation and time-resolved fluorescence analysis of micelle-solubilized pyrene. The N values of C_n aggregates in DMSO-H_2O (Φ = 0.5) were determined by a similar method but using A_{12} as the probe.[5] Based on poison distribution of A_{12} in an aggregate population the time evolution of A_{12} monomer fluorescence is described by the following equation:[79]

$$I(t) = I(0)\exp\{-k_1 t - n[1 - \exp(-k_e t)]\} \tag{17}$$

where $I(0)$ and $I(t)$ are the intensities of A_{12} monomer fluorescence at t = 0 and at t = t, respectively; n is the average number of A_{12} molecules in each C_n aggregate (mean occupancy number); k_1 is the decay rate constant of A_{12} monomer in singly occupied aggregate; and k_e is the rate constant of A_{12} intraaggregate excimer formation. According to Equation 17, the A_{12} monomer fluorescence decay curve shows two decay components: a fast decay due to A_{12} in multiple occupied aggregates and a slow decay due to A_{12} in singly occupied aggregates. In the slow decay, because of large t after pulsed excitation, the time dependence of A_{12} monomer fluorescence is

$$I(t) = I(0)\exp[-(n + k_1 t)] \tag{18}$$

Thus, from the intercepts of the fast and slow decay profiles one can obtain n. Then the N value can be calculated from Equation 19.

$$N = n\{[C_n] - CAC\}/[A_{12}] \tag{19}$$

The N values of C_n aggregates calculated from Equation 19 are listed in Table 2. In order to check the feasibility of A_{12} as a probe in this method, the N values for micelles of hexadecyltrimethylammonium bromide (HDTBr) in water and sodium dodecyl sulfate (SDS) in DMSO-H_2O ($\Phi = 0.5$) were measured. The data obtained are in good agreement with literatures.[81,82]

Table 2 shows that the aggregate formed from hydrocarbon with the longer chain length has the larger aggregation number. All the N values of the aggregates are much smaller than those of micelles formed by surfactant in aqueous solution. For example, the N value of HDTBr in aqueous solution is 82,[81] while that of the aggregate of C_{16}, which has the same alkyl chain length as HDTBr, in DMSO-H_2O ($\Phi = 0.5$) is only 24. Similarly, the N value of SDS in aqueous solution is 65,[82] while that of the aggregate of C_{12}, which has the same alkyl chain length as SDS, in DMSO-H_2O ($\Phi = 0.5$) is only 18. Molecular structure of substrate and the shape of aggregate could be the factors to influence aggregation number. However, solvent nature may be one of the most important factors. For example, a change of solvent from water to DMSO-H_2O ($\Phi = 0.5$) resulted in the decrease in aggregation number from 65 to 35 for SDS.[82]

A number of methods for measuring the microviscosity of molecular assemble have been reported.[78,83] The fluorescence depolarization technique has been used to evaluate the microviscosity within the C_n aggregates. The degree of polarization, P, is defined as

$$P = (I_\parallel - I_\perp)/(I_\parallel + I_\perp) \tag{20}$$

where $I_\parallel$ and $I_\perp$ stand for the intensities of the emitted light polarized parallel and perpendicular to the polarization of the incident light, respectively. The relation between the fluorescence polarization degree of a probe and the viscosity (η) of the environment around the probe molecule is given in the following equation:[78,84]

$$(1/P - 1/3) = (1/P_0 - 1/3)(1 + kT\tau/\eta V_0) \tag{21}$$

where P_0 is the limiting fluorescence polarization degree when $T/\eta \to 0$, k is the Boltzmann constant, T is the temperature (K), τ is the fluorescence lifetime (ns), and V_0 is the volume of the probe molecule. A_{16} was employed as a probe. The fluorescence lifetimes and polarization degrees of A_{16} in glycerol at various temperatures were determined ($\lambda_{ex} = 300$ nm, $\lambda_{em} = 350$ nm). Since the viscosities of this solvent under the used temperatures are known,[85] P_0 and V_0 of A_{16} were calculated from Equation 21 to be 0.25 and 1.7×10^{-23} cm^3/molecule, respectively.

In Figure 8 is shown the fluorescence polarization degrees of A_{16} in C_{14} and C_{18} aggregates as a function of temperature. There exists a characteristic temperature for each curve at which the P value is abruptly changed. The appearance of the break of the plot is the typical behavior of phase transition. The phase transition temperatures for C_{18}, C_{16}, C_{14}, C_{12}, and C_{10} are 0, 0, -10, -40, and $-60°C$, respectively. The microviscosities in C_n aggregates were evaluated from Equation 20 and listed in Table 2. Above phase transition temperature the microviscosities in C_n aggregates fall in the range from 220 to 260 cp, which are analogous to those in micelles,[10] while below phase transition temperature the microviscosities in aggregates are as large as that in solid state.[86]

C. DRIVING FORCE FOR AGGREGATION

The proposal that hydrophobic interaction is the driving force for aggregation was further confirmed by fluorescence probe methods.[9] As Φ increases in aquiorgano binary mixtures, both polarity and hydrophilicity of the solvent increase. Attempts to correlate aggregate formation with solvent polarity failed. For example, the fluorescence of pyrene-3-carboxaldehyde (PyCHO) shows red shifts with increasing solvent polarity and a linearity between

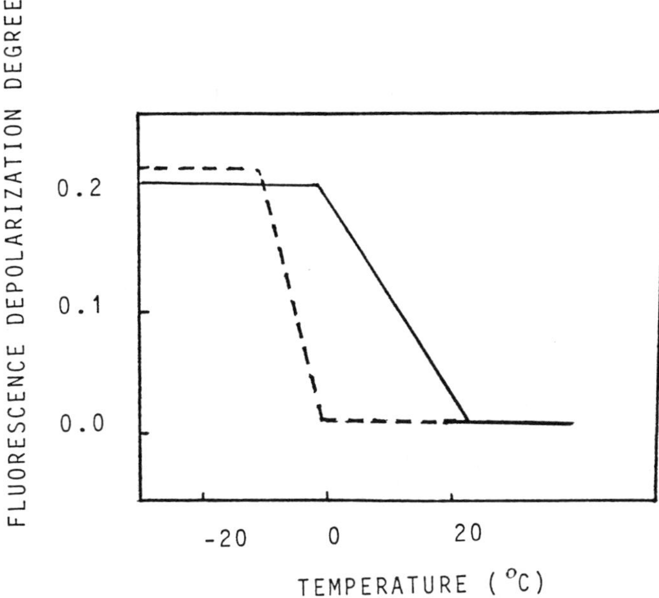

FIGURE 8. Fluorescence polarization degree of hexadecyl 2-naphthoate in *n*-octadecane (————) and tetradecane (— — — —) aggregates formed in dimethyl sulfoxide-water as a function of temperature. The fraction of water in the solvent mixture is 0.5.

the wavelength of the fluorescence maximum (λ_{max}) and the dielectric constant of solvent exists.[87] A linear relationship between λ_{max} of PyCHO and Φ value in DMSO-H$_2$O mixtures was found, suggesting that the polarity of the mixed solvents increases gradually with the increase in Φ value. This is in contrast with Figure 4 where an abrupt change of $I_D/(I_M + I_D)$ with increase in Φ is shown. Thus there is no correlation between the aggregation of the substrate and the solvent polarity. On the other hand, for solvent with a particular aggregating power, quantitative correlation between the hydrophobicity (Σf) of substrate and its aggregating ability was observed. Figures 9 and 10 show the plots of CAC and N of C_n in DMSO-H$_2$O ($\Phi = 0.5$) against their Rekker's hydrophobic constants (Σf), respectively. Both plots are linear, suggesting that aggregation of C_n indeed relies on their hydrophobicity.[88] Furthermore, the addition of LiCl, a salt which increases the solvent hydrophobic effect, to the solution leads to the reduction of CΦ and CAC.[9] For example, in DMSO-H$_2$O ($\Phi = 0.5$) the CAC of A$_{12}$ is 5×10^{-6} M, but in the presence of 5 M LiCl it reduces to 1×10^{-7} M. The CΦ of A$_{12}$ (at concentration of 5×10^{-5} M) in DMSO-H$_2$O is 0.40, but in the presence of 5 M LiCl it begins to aggregate at $\Phi = 0.20$.

D. ALKYL SALICYLATES AS FLUORESCENCE PROBE FOR AGGREGATION

Alkyl salicylate exhibits triplex fluorescence in a variety of organic solvents.[89,90] The unusual fluorescence originates from three different ground-state species in equilibrium (Scheme 1). In the *cis* conformer I the phenolic hydrogen is H-bonded to the carbonyl

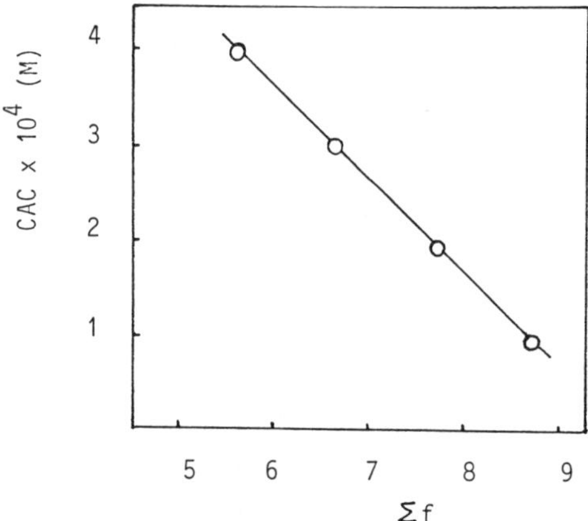

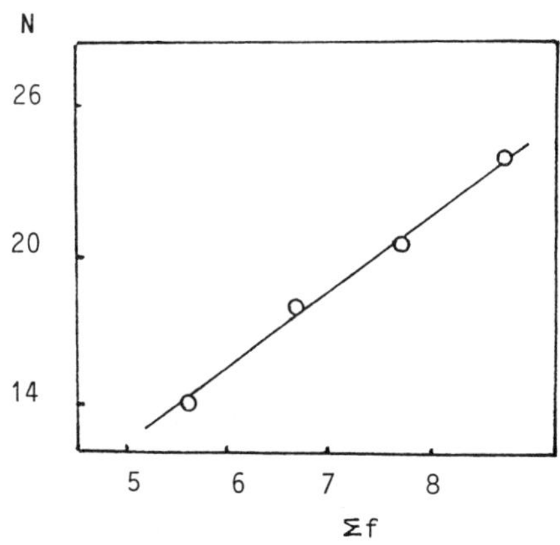

FIGURE 9.   Linear relationship between critical aggregate concentration CAC of *n*-alkanes in dimethyl sulfoxide-water and their Rekker's hydrophobic constants Σf. The fraction of water in the solvent mixture is 0.5.

FIGURE 10.   Linear relationship between average aggregate number N of *n*-alkane aggregates in dimethyl sulfoxide-water and their Rekker's hydrophobic constants Σf. The fraction of water in the solvent mixture is 0.5.

oxygen. In conformer II intramolecular H-bond between the phenolic hydrogen and the ether oxygen is formed. The *trans* conformer III may form intermolecular H-bond with solvent. Following excitation of the *cis* form, proton transfer is rapid and virtually complete, and the resulted species emits at $\lambda_{max} = 450$ nm. In conformer II and III no proton transfer occurs, and they emit at $\lambda_{max} = 330$ and 360 nm, respectively. The excited states of the three conformers do not equilibrate. The ground-state population of the three conformers are affected by solvent. In H-bonding solvents the *trans* conformer III is stabilized by H-bonding with the solvent, thus the fluorescence has greater intensity at $\lambda_{max} = 360$ nm. In non-H-bonding solvents conformer I and II are the preferred species due to the strong intramolecular H-bond.

The fluorescence spectra of alkyl salicylates with various chain lengths in DMSO-H_2 and EG-H_2O were studied.[91] As Φ value increases, the ratio of the fluorescence intensity, I_{450}/I_{360}, for methyl salicylate and butyl salicylate decreases due to the increase in H-bond formation ability of the solvent. On the contrary, the fluorescence bond at $\lambda_{max} = 450$ nm dominates the emission spectra of alkyl salicylates with long alkyl chains at high Φ value. Figure 11 gives the plots of $I_{450}/(I_{450} + I_{360})$ for hexadecyl salicylate and dodecyl salicylate in DMSO-H_2O against Φ value. A clear break in the plot is shown, which is typical of aggregate formation. Although there is a great deal of water in the bulk solution, no H-bonding solvent exists within the aggregates. Therefore, a dramatic change in the fluorescence spectra occurs at Φ value corresponding to $C\Phi$. In addition to the fluorescence distribution change, the fluorescence quantum yield also can serve as an evidence for aggregate formation. It is established[92] that the fluorescence quantum yield of alkyl salicylate decreases with increasing solvent polarity. However, in DMSO-H_2O or EG-H_2O, the fluorescence quantum yield of hexadecyl and dodecyl salicylate is evidently enhanced at $\Phi > C\Phi$. For example, in DMSO-H_2O ($\Phi = 0.5$) the fluorescence quantum yield of hexadecyl salicylate is eight times greater than that in acetonitrile.[91] This suggests that the polarity inside aggregates is very small.

E. 1,3-DIALKYLINDOLES AS FLUORESCENCE PROBE FOR AGGREGATION

The fluorescence of 1,3-dialkylindoles has been shown to be sensitive to the polarity of environment.[75-78] The fluorescence band maximum (λ_{max}) for 1,3-dimethylindole (1-In-1) appears at 322 nm in cyclohexane, and shifts to 375 nm in water. In DMSO-H_2O mixed

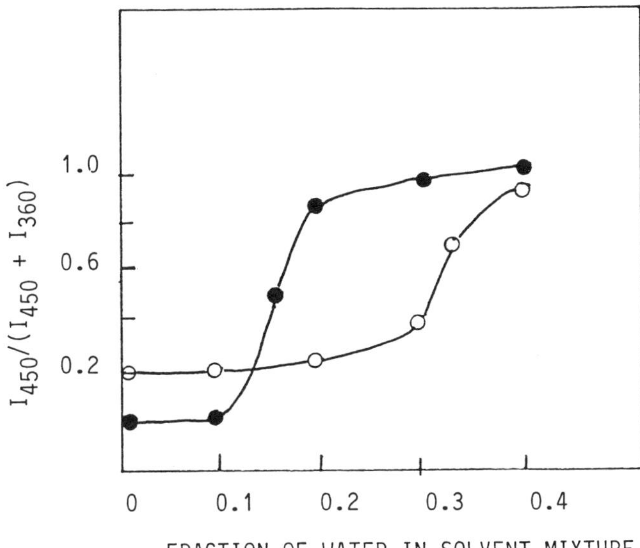

FIGURE 11. Plot of the ratio of the fluorescence intensity from Conformer I to the sum of the fluorescence intensities from Conform I and III, $I_{450}/(I_{450} + I_{360})$, for hexadecyl (●) and dodecyl (○) salicylates in dimethyl sulfoxide-water as a function of the fraction of water in the solvent mixture. The concentration of both the salicylates is 2×10^{-5} *M*.

solvent as Φ value increases from 0 to 0.8, the fluorescence λ_{max} of 1-In-1 shifts from 355 to 372 nm. For the indoles with long alkyl chain such as 1-hexadecyl-3-methylindole (1-In-16), the fluorescence λ_{max} is identical with that of 1-In-1 in the Φ ranging between 0 and 0.4. However, in the vicinity of $\Phi = 0.4$, the fluorescence λ_{max} of 1-In-16 shifts to 327 nm, which is close to that in cyclohexane (Figure 12). This reflects that 1-In-16 forms aggregate at $\Phi > 0.4$, and the polarity inside the aggregates is quite small. The fluorescence lifetime of 1,3-dialkylindoles is also sensitive to solvents: 15 ns in water and 4 ns in cyclohexane. The time evolution of 1-In-16 emission at 340 nm in DMSO-H$_2$O with several Φ values was measured. At Φ smaller than 0.4, the fluorescence of 1-In-16 decays monoexponentially and the fluorescence lifetime is 9 ns, closely resembling the corresponding data for 1-In-1 in the same solvent. However, at Φ greater than 0.4 the fluorescence of 1-In-16 decays biexponentially. The lifetime of the longer living component is 7 to 9 ns depending on the Φ value, and the shorter living one about 1 ns. The dual exponential character of the fluorescence decay was attributed to the different location of the indole. The emission with longer lifetime is due to the indole in bulk solution and the emission with shorter lifetime is due to the indole inside aggregates. This was confirmed by the fluorescence quenching experiment. In the solution of 1-In-16 in DMSO-H$_2$O ($\Phi = 0.5$) and in the presence of NaNO$_2$, a water-soluble quencher for indoles, only the shorter living fluorescence was detected. The ratio of the fraction of the longer living component to the shorter living one and the fluorescence quantum yield of 1-In-16 in bulk solution and inside aggregates enables the calculation of the distribution of 1-In-16 between the two phases. Figure 13 gives the fraction of 1-In-16 in bulk solution as a function of Φ. Evidently, in the vicinity of $\Phi = 0.3$, the plot shows a break, and the inflection point of the curve represents the CΦ for 1-In-16 in DMSO-H$_2$O.

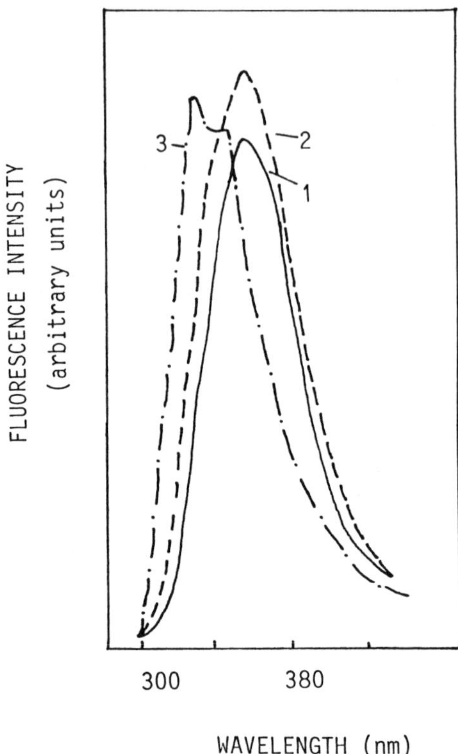

FIGURE 12.   Fluorescence spectra of 1-methyl-3-hexadecylindole (1×10^{-5} M) in dimethyl sulfoxide-water solvent. The fractions of water in the solvent mixtures are the following: 1, 0; 2, 0.2; 3, 0.5.

F. SINGLET ELECTRONIC ENERGY TRANSFER BETWEEN ALKYL NAPHTHOATES AND 9-ANTHRYLMETHYL *n*-ALKANOATES

Intermolecular energy transfer and fluorescence quenching are expected to be sensitive to aggregate formation, since the mutual chromophore separation, orientation, and mobility inside the aggregates are apparently different from those in bulk solution. The efficiency of singlet energy transfer between A_n and 9-anthrylmethyl alkanoate (E_n) in DMSO-H_2O was measured.[93]

$$A–CH_2–O–CO–R \qquad\qquad (A = 9\text{-anthryl})$$

$$E_1, R = CH_3; \quad E_{10}, R = C_{10}H_{21}; \quad E_{17}, R = C_{17}H_{35}$$

Like the A_n with long chain,[9] excitation of E_{10} or E_{17} in DMSO-H_2O gives excimer fluorescence if Φ is greater than their $C\Phi$, due to the formation of aggregates. For the solution of A_4 + E_1, and the solution of A_4 + E_{10} (or E_{17}) in DMSO-H_2O, selective excitation of the naphthalene chromophore at 280 nm only gives A_4 monomer fluorescence irrespective of Φ value. Under the same condition the solution of A_{12} + E_1 exhibits A_{12} excimer fluorescence and no E_1 fluorescence can be detected, suggesting that A_{12} forms aggregate, but does not coaggregate with E_1. On the contrary, the solution of A_{12} + E_{10} (or E_{17}) in DMSO-H_2O shows the structured fluorescence of anthracene chromophore which is superimposed on the excimer fluorescence of A_{12}. This observation suggests that A_{12} coaggregates with E_{10} and energy transfer from A_{12} excited state to E_{10} occurs.

The excitation of the solution of A_{12} + C_{14} in DMSO-H_2O at 280 nm gives only A_{12}

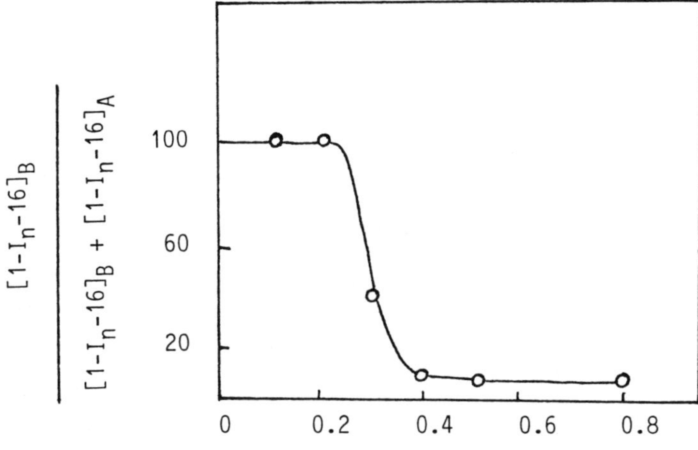

FIGURE 13.   Plot of $[1\text{-}In\text{-}16]_B/\{[1\text{-}In\text{-}16]_B + [1\text{-}In\text{-}16]_A\}$ in dimethyl sulfoxide-water as a function of the fraction of water in the solvent mixture. $[1\text{-}In\text{-}16]_B$ and $[1\text{-}In\text{-}16]_A$ represent the concentration of 1-methyl-3-hexadecylindole located in the bulk solution and in the aggregates, respectively. The total concentration of 1-methyl-3-hexadecylindole was $1 \times 10^{-4}\,M$.

nomoner fluorescence if the concentrations of A_{12} and C_{14} are controlled so that each C_{14} aggregate only contains one A_{12} molecule.[9] The addition of E_{10} results in the reduction of the A_{12} fluorescence and the appearance of the monomer fluorescence of E_{10} (Figure 14). Obviously, A_{12} and E_{10} form triplex coaggregate with C_{14} and efficient energy transfer from A_{12} excited state to E_{10} takes place. The quantum efficiency of energy transfer (x) is defined as the fraction of the number of einsteins transferred to E_{10} in the number of einsteins absorbed by A_{12}.

$$x/(1 - x) = \Phi(A_{12})S(E_{10})/\Phi(E_{10})S(A_{12}) \tag{22}$$

where $\Phi(A_{12})$ and $\Phi(E_{10})$ present the fluorescence quantum yields of A_{12} in the absence of E_{10} and E_{10}, respectively. $S(A_{12})$ and $S(E_{10})$ are the areas under the fluorescence emission curves of A_{12} and E_{10}, respectively, measured for the system of A_{12} and E_{10} in C_{14} aggregate. When the (0, 0) band of A_{12} fluorescence spectrum in A_{12} and E_{10} coaggregates in Figure 14 is normalized to that of A_{12} fluorescence spectrum in A_{12} aggregates and the latter spectrum is subtracted, the structured fluorescence spectrum of E_{10} is obtained. Thus, $S(A_{12})$ and $S(E_{10})$ can be measured. Experimental results reveal that the value of x is as high as 50%.

The energy transfer rate for A_{12} + E_{10} in C_{14} aggregate was determined from the A_{12} fluorescence decay curves.[93] The concentrations of A_{12} and E_{10} were controlled so that the ratio of $[A_{12}]$ to [aggregate] is much smaller than 1 and that of $[E_{10}]$ to [aggregate] is around 1. In the absence of E_{10}, the fluorescence of A_{12} decays monoexponentially with the decay rate k_0. The decay curve in the presence of E_{10} is nonexponential and is characterized by a transient fast decay in the earlier time region and a single exponential decay in the longer time region. The decay curve was computer fitted to Equation 23.

$$I(t) = I(0)\exp\{-k_1 t - n[1 - \exp(-k_e t)]\} \tag{23}$$

where k_1 is the decay rate constant of A_{12} including all the decay processes but the intraaggregate energy transfer process, k_e is the intraaggregate energy transfer rate, and n is a constant relative to the concentration and distribution of both A_{12} and E_{10} in aggregates.

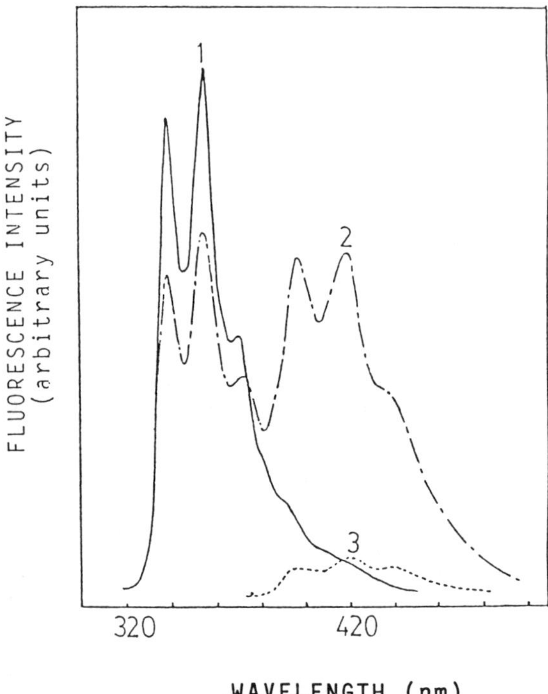

FIGURE 14.   Fluorescence spectra of dodecyl 2-naphthoate (4×10^{-5} M) in tetradecane (7.5×10^{-4} M) aggregates in dimethyl sulfoxide-water in the presence and absence of 9-anthrylmethyl undecanote E_{10}. The fraction of water in the solvent mixture is 0.5. (1) In the absence of E_{10}, $\lambda_{ex} = 280$ nm; (2) in the presence of 4×10^{-5} M E_{10}, $\lambda_{ex} = 280$ nm; (3) in the presence of 4×10^{-5} M E_{10}, $\lambda_{ex} = 350$ nm.

The fitting yielded k_1 and k_e. The difference between k_0 and k_1 is the interaggregate energy transfer rate. The value of k_e is $(4.0 \pm 0.2) \times 10^8$ s^{-1}. At aggregate concentration of 7×10^{-4} M the interaggregate energy transfer rate varies from 3.5×10^6 to 3.6×10^7 s^{-1} at the E_{10} concentration ranging 1×10^{-5} to 1×10^{-4} M.

G. ENHANCEMENT OF EXCIMER FORMATION OF PYRENE SUBSTITUENTS BY SURFACTANTS BELOW THEIR CRITICAL MICELLE CONCENTRATIONS

Herkstroeter et al.[94] studied the interactions of pyrene-substituted ionic molecules, Py-3-$(CH_2)_3COO(CH_2)_2N^+(CH_3)_3$ $CH_3SO_4^-$ (Py-N$^+$, Py = pyrenyl), with surfactant sodium dodecyl sulfate (SDS) in aqueous solution at SDS concentration well below its CMC. Py-N$^+$ shows the pyrene characteristic well-resolved absorption and monomer fluorescence spectra. The addition of sodium methyl sulfate, a salt with the same counterion as Py-N$^+$, to the solution has no effect on the absorption and fluorescence spectra of Py-N$^+$. However, the addition of SDS results in profound changes in these spectra even if the concentration of SDS is more than two orders of magnitude below its CMC. The absorption of Py-N$^+$ in the presence of SDS shows band broadening as well as a small shift to the red, and the emission spectrum shows both excimer and monomer fluorescence. The excitation spectrum monitored at the monomer fluorescence maximum at 377 nm is well resolved and differs from the more poorly resolved and red-shifted spectrum monitored in the range of the excimer fluorescence band at 500 nm. All these observations suggest that the hydrophobic interaction and ionic attraction between Py-N$^+$ and SDS bring the pyrene moieties into contact with one another to form aggregate in ground state, and the excimer fluorescence originates from the excited-state aggregate of pyrene moieties. When the concentration of SDS is high enough for micelle formation, the pyrene substituents prefer solvation within the micelles, where, as in free solution, ground-state interactions do not occur. However, excimer still can be formed via a dynamic process wherever there are two or more pyrene molecules within the individual micelles. In this case, the absorption spectrum of Py-N$^+$ becomes sharp again.

H. INTRAMOLECULAR EXCIMER FORMATION OF 1,16-BIS(2-NAPHTHYL)-3,14-DIOXAHEXADECANE AND POLYMETHYLENE BIS-2-NAPHTHOATES — EVIDENCE FOR SELF-COILING

In bichromophoric systems the formation of an intramolecular excimer requires the conformational changes in the linking chain to allow the chromophores to approach one another, and the conformation distribution prior to excitation plays an important role in excimer formation.[95,96] Hydrophobic interaction may bring the two chromophores into proximity in ground state, thus promoting intramolecular excimer formation.

Jiang and co-workers studied the fluorescence behaviors of 1,16-bis-(2-naphthyl)-3,14-dioxahexadecane in DMSO-H$_2$O and in dioxane-H$_2$O mixtures with different Φ values.[97]

$$N_PCH_2CH_2O(CH_2)_{10}OCH_2CH_2N_P,$$

NDH, N_P = 2-naphthyl

In DMSO-H$_2$O with $\Phi < 0.3$ NDH (10^{-5} to 10^{-4} M) only shows the monomer emission, but when $\Phi > 0.3$ the excimer emission appears. Figure 15 shows the ratio of excimer to monomer fluorescence intensities as a functionof NDH concentration at several Φ values. Each plot displays a break, which is typical of aggregation. The concentration corresponding to the break point is the CAC. At concentration below CAC, NDH exists only in its monomeric form, and the excimer formation is due to self-coiling. Thus, the ratio of I_D/I_M is independent on [NDH] and is defined as R_m. At concentration above CAC the degree of aggregation increases with [NDH]. At high-enough concentration the excimer mainly originates from the aggregate, and its emission dominates the fluorescence spectra. For a particular aquiorgano binary solvent system, the hydrophobic force increases with Φ value, and the substrate ability for aggregation and self-coiling would become greater at higher Φ value. This is manifested by the following facts: (1) larger R_m for the intramolecular coiling process; (2) steeper slopes of the lines at a concentration above CAC for the intermolecular aggregation process; (3) smaller CAC values. These conclusions are strengthened by the curves shown in dioxane-H$_2$O in Figure 15.

The fluorescence behavior of polymethylene bis-2-naphthoates in DMSO-H$_2$O is very similar to that of NDH.[98]

$$N_PCO_2(CH_2)_nCO_2N_P \qquad B_n, \qquad n = 2, 3, 4, 5, 10; \qquad N_P = 2\text{-naphthyl}$$

Figure 16 shows the ratio of I_D/I_M for B_5 (1×10^{-4} M) as a function of Φ. At $\Phi > 0.4$ the excimer emission dominates the fluorescence spectra, and the ratio of I_D/I_M is independent on [B_5] in the concentration range of 1×10^{-7} to 1×10^{-4} M. This observation suggests that B_5 molecules self-coil and the excimer is intramolecular.

I. DIMER FORMATION OF 1,3-DINAPHTHYLPROPANES

It is well established that 1,3-di(1-naphthyl)propane (1-DNP) and 1,3-di(2-naphthyl)propane (2-DNP) in organic solution form intramolecular excimers.[99] However, in

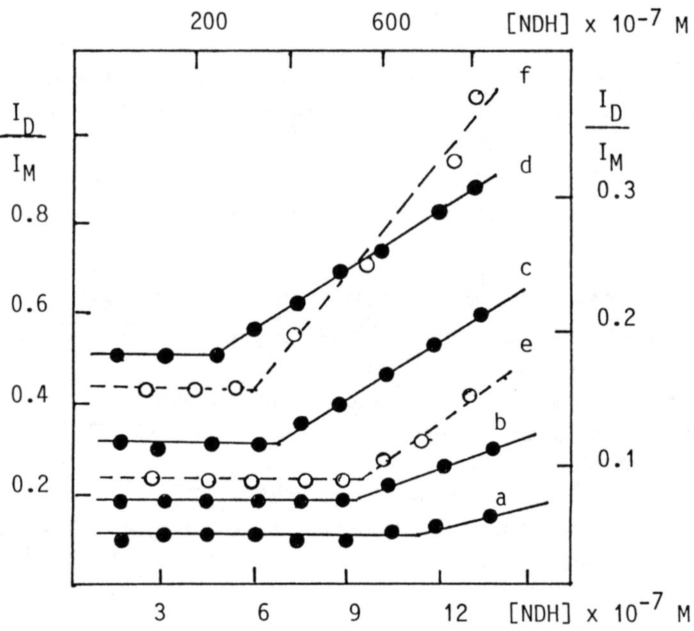

FIGURE 15.   The ratio of fluorescence intensities of excimer to monomer I_D/I_M as a function of concentration of 1,16-bis-(2-naphthyl)-3,14-dioxahexadecane NDH at different fractionof water in the solvent mixtures. Values on the left are for the dimethyl sulfoxide-water system and values on the right the dioxane-water system. Solid lines a to d are for the dimethyl sulfoxide-water system, and dashed lines e and f for dioxane-water. Concentration units in the bottom are for the dimethyl sulfoxide-water system, and those on the top are for the dioxane-water system. The fractions of water in the solvent mixtures are the following: a, 0.35; b, 0.40; c, 0.45; d, 0.50; e, 0.60; f, 0.70. (Adapted from Jiang et al.[97])

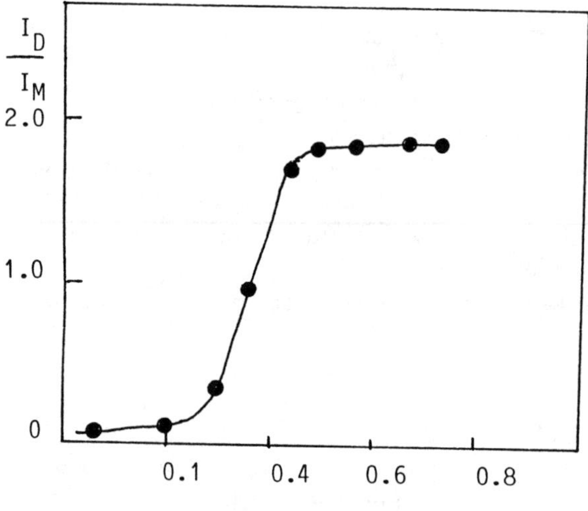

FIGURE 16.   Plot of the ratio of fluorescence intensities of excimer to monomer I_D/I_M for pentamethylene bis-2-naphthoate in dimethyl sulfoxide-water as a function of the fraction of water in the solvent mixture. The concentration of pentamethylene bis-2-naphthoate is 1×10^{-4} M.

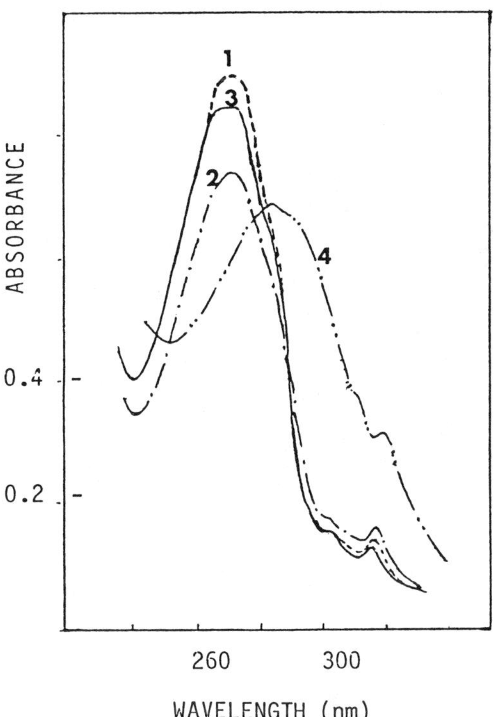

FIGURE 17. Absorption spectra of 1,3-di(2-naphthyl)propane (2,2'-DNP, 1 × 10^{-5} *M*) and 2-methylnaphthalene (2-MN, 2 × 10^{-5} *M*) in ethylene glycol-water. Φ represents the fraction of water in the solvent mixture. (1) 2-MN, Φ = 0; (2) 2-MN, Φ = 0.5; (3) 2,2'-DNP, Φ = 0; (4) 2,2'-DNP, Φ = 0.5.

DMSO-H_2O or EG-H_2O their behavior is quite different from that in other organic solvents.[100] Figure 17 shows the absorption spectra of 2-DNP and the model compound 2-methylnaphthalene (2-MN) in EG-H_2O with Φ = 0 and 0.5. The spectrum of 2-DNP at Φ >0.2 is evidently red-shifted, suggesting the existence of the strong interaction between pairs of the naphthyl groups in ground state. 1-DNP in DMSO-H_2O and EG-H_2O also shows the spectrum shifts as 2-DNP does. The fluorescence spectra of 2-DNP in EG-H_2O with Φ = 0 and 0.5 are given in Figure 18 together with those of 2-MN. 2-MN only emits monomer fluorescence and the spectrum is constant irrespective of Φ. The fluorescence spectrum of 2-DNP at Φ = 0 is consisted of monomer and excimer emission as it is in other organic solvents. However, at Φ >0.2 a new structural emission with maxima at 345, 362, 363, and 400 nm (shoulder) appears, and the excitation spectrum for the new emission corresponds well to the absorption spectrum shown in Figure 17, curve 4. This emission spectrum is very analogous to the dimer fluorescence spectrum of poly(2-vinylnaphthalene) in poor solvents, reported by Irie et al.[101] Thus, the absorption and fluorescence spectra of 2-DNP in EG-H_2O with Φ >0.2 are attributed to the dimer of the naphthyl groups. A similar fluorescence spectrum for 1-DNP was detected. These observations suggest that the hydrophobic interactions force the intramolecular two naphthyl groups in DNP together to form a sandwich arrangement in ground state.

In order to estimate the magnitude of the hydrophobic force acting on the naphthyl ring, the fluorescence spectra of *cis*-1,4-di(2-naphthyl)cyclohexane (*cis*DNCH) in DMSO-H_2O were studied.[102] The twisted-boat conformation of the *cis*DNCH molecule is favorable to intramolecular dimer formation. The potential energy of the twisted-boat conformation for

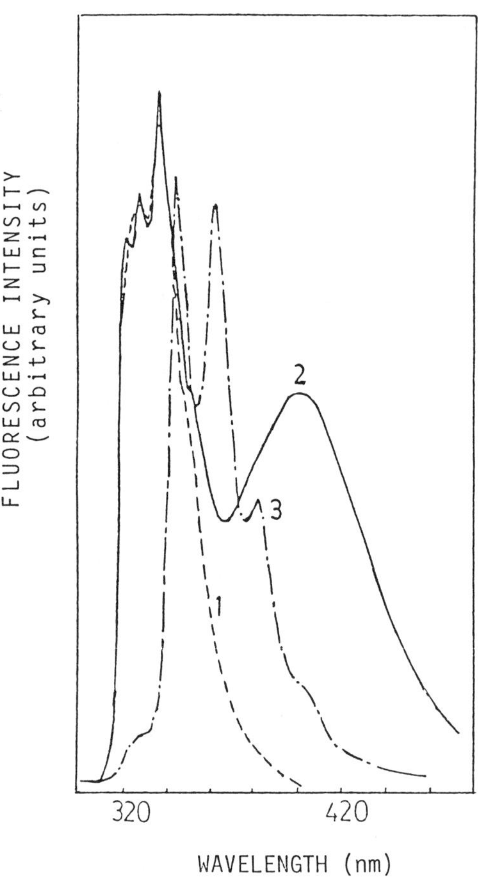

FIGURE 18.   Fluorescence spectra of 1,3-di(2-naph-thyl)propane (2,2′-DNP, 1×10^{-5} M) and 2-meth-ylnaphthalene (2-MN, 2×10^{-5} M) in ethylene glycol-water. Φ represents the fraction of water in the solvent mixture. (1) 2-MN, $\Phi = 0.5$; (2) 2,2′-DNP, $\Phi = 0$; (3) 2,2′-DNP, $\Phi = 0.5$.

cycloexane is *circa* 5.5 kcal/mol relative to the chair conformation. The repulsion energy between the two naphthyl rings in a sandwich arrangement in ground state is more than 10 kcal/mol.[9] To get a twisted-boat conformation for *cis*DNCH in ground state, the hydrophobic force acting on the naphthyl rings should be at least 15 kcal/mol. This conformation would give rise to a intramolecular dimer fluorescence when the naphthyl is excited. The dimer fluorescence is indeed detected when the concentration is large enough. However, the plot of the ratio of dimer to monomer fluorescence intensities, I_D/I_M, vs. *cis*DNCH concentration turns out that the dimer fluorescence is intermolecular due to formation of aggregate. The absence of intramolecular dimer emission at a concentration below CAC is inconsistent with a twisted-boat conformation of the cyclohexane ring in *cis*DNCH. Furthermore, the intra-molecular excimer emission for *cis*DNCH in any solvent used could not be observed. This is due to the slow process for the conformation conversion of the cyclohexane ring as compared to the deactivation process of the locally excited naphthyl chromophore. The slow conformation conversion of *cis*DNCH at room temperature is also confirmed by its ^{1}HNMR, which clearly shows the axial and equatorial protons of the cyclohexane ring at different chemical shifts.

J. INTRAMOLECULAR [2 + 2] CYCLOADDITION OF α,ω-DICINNAMOXY-ALKANES

Jiang studied the photochemical intramolecular cycloaddition of the 1,2-dicinnamoxy-ethane (1, n = 2) and the 1,10-dicinnamoxydecane (2, n = 10) in various solvents.[32] The irradiation of 1 and 2 resulted in the β-truxinates (3, n = 2; 4, n = 10) and the δ-truxinate (5, n = 10). In acetonitrile the yields were good for 3 (90%) but very poor for the macrocyclic products 4 + 5 (6 to 7%). However, in DMSO-H_2O with Φ = 0.7 the yield of 4 + 5 was dramatically increased to 90%. Obviously, hydrophobic interactions force the molecule of 2 to self-coil, thus increasing the intramolecular ring-closure probability.

Ph

X

Ph

hv

Ph

X

Ph

+

Ph

X

Ph

1 x = -OCO$(CH_2)_2$OCO-

2 x = -OCO$(CH_2)_{10}$OCO-

3 x = -OCO$(CH_2)_2$OCO-

4 x = -OCO$(CH_2)_{10}$OCO-

5 x = -OCO$(CH_2)_{10}$OCO-

K. CONFORMATION OF LINEAR POLYMER CHAIN

For linear polymers the conformation of the polymer chain can be affected by hydro-phobic interaction. Winnik et al.[103] examined the fluorescence spectroscopy of pyrene-labeled (hydroxypropyl)cellulose (HPC-Py) in solution in various alcohols and in water. The molecular weight of HPC is *circa* 1×10^5 and each molecule contains on average four pyrene or each 26 glucose units contain one pyrene (HPC-Py/26). In methanol HPC-Py/26 shows the pyrene characteristic emission of both monomer and excimer. The ratio of I_D/I_M is a constant over a range of concentrations of HPC-Py/26 from 10 ppm ([pyrene] = 1.2×10^{-6} *M*) to 180 ppm, suggesting that the excimer is intramolecular. The excitation spectra for the emission of monomer and excimer are identical, and the maxima correspond to those in the UV absorption spectra. The decay for the monomer emission can be fitted to a sum of two exponential terms with decay times of 19 and 114 ns. The excimer profile shows both a growing-in and a decaying component (11 and 67 ns, respectively). Its decay parameters are substantially different from those of the monomer. All these observations are consistent with dynamic formation of excimer, with little preassociation of ground-state pyrene groups. The complexity of the fluorescence decay reflects the distribution of pyrene separations on the polymer and the fact that various polymer chains in sample contain different numbers of pyrene. In contrast, HPC-Py/26 in water presents a relative weak monomer emission and strong excimer fluorescence. The excitation spectra for the monomer and for the excimer are clearly different. The general features of the spectra are similar, but the fomer is blue-shifted about 4 nm. It is the latter that corresponds to the UV absorption spectrum of the polymer in water solution. Furthermore, in the fluorescence decay measurements no rising component can be detected in the excimer decay profile. These observations all indicate that the excimer originate from pairs or aggregates of pyrene groups that exist prior to excitation. They examined the effect of unlabeled HPC on the fluorescence of HPC-Py/26 and concluded: (1) in aqueous solution HPC tends to associate even at very low polymer concentration, and the association is a property of the polymer itself and is not brought about by the presence of the pyrene groups; and (2) at the lowest concentration of polymer, the value of I_D/I_M is still much larger than the value found in methanol, suggesting

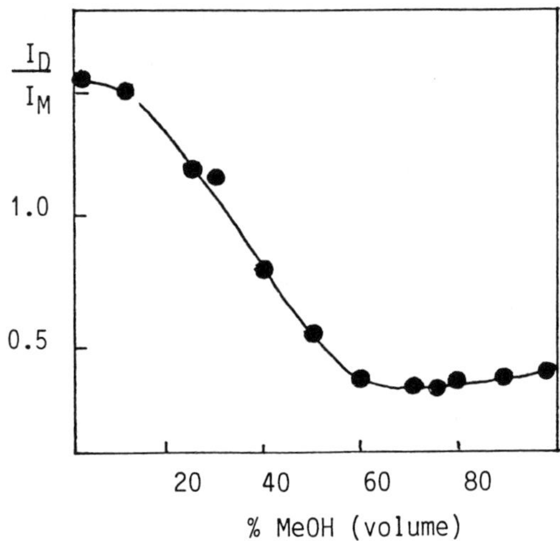

FIGURE 19.   Plot of the ratio of fluorescence intensities of excimer to monomer I_D/I_M for pyrene-labeled (hydroxypropyl)cellulose HPC-Py/26 as a function of the water/methanol composition (v/v). (Adapted from Winnik et al.[103])

a significant portion of the intense excimer emission is of intramolecular origin. This, in turn, implies that HPC assumes a different conformation in water than in alcohol solvents. HPC is a very hydrophobic polymer. The driving force for association and conformation difference in water may be some combination of hydrophobic interactions and highly co-operative intermolecular hydrogen bonding. They also measured the ratios of I_D/I_M for HPC-Py/26 in water-methanol mixtures. Figure 19 shows the plot of I_D/I_M as a function of the water/methanol composition (v/v). The plot exhibits a break at water/methanol of 40/60.

Herkstroeter et al.[94] reported the fluorescence study of pyrene substituents covalently bonded to polyelectrolytes in aqueous solution. These polyelectrolytes are polymethacrylate with anionic or cationic side chains. As compared with pyrene-substituted ionic molecule mentioned in Section IV.G, the well-resolved vibrational bands in the absorption spectra of the pyrene-labeled polyelectrolytes become somewhat diffuse and red-shifted. This requires ground-state interactions of pyrene moieties. In addition to monomer fluorescence, strong excimer emission was observed. The ratio of I_D/I_M in the fluorescence spectrum and the absorption spectrum did not change upon further dilution, suggesting that the interactions between pyrene moieties is intramolecular. The relatively low pyrene content in polymer (each 24 structure units of the polyelectrolyte contain one pyrene) means that few pendant pyrene substituents will have another pyrene substituent as a nearest neighbor. The polymers have to fold to bring nonnearest-neighbor pendant fluorophores into contact. These ground-state interactions were further substantiated with excitation spectra. The excitation spectra monitored at excimer fluorescence wavelengths were red-shifted relative to that for monomer fluorescence. The driving force for pyrene moiety aggregation is hydrophobic interactions.

Char et al.[104] studied the fluorescence behavior of pyrene-end-labeled poly(ethylene glycol) (Py-PEG-Py) with three different molecular weights. These polymers show both monomer and excimer fluorescence in aqueous solution. Figure 20 gives the plot of the logarithm of the ratio of the excimer-to-monomer fluorescence intensities (I_D/I_M) against the logarithm of the polymer concentration. The I_D/I_M for samples of different molecular weights shows similar trends with concentration. The concentration region can be clearly divided into two parts, a plateau region and a power-law region with a positive slope. The plateau region is due to intramolecular excimer formations and the power-law region is due to the

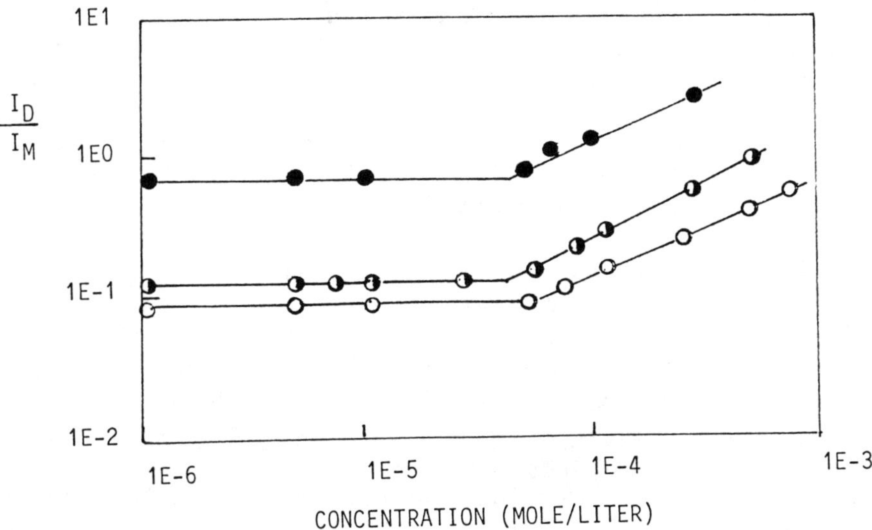

FIGURE 20.   The log-log plot of the ratio of fluorescence intensities of excimer to monomer I_D/I_M as a function of polymer concentration in water for pyrene-end-labeled poly(ethylene glycol) Py-PEG-Py samples of different molecular weights at 298 K: (●) Py-PEG-Py(4800); (◑) Py-PEG-Py(9200); (○) Py-PEG-Py(11200). (Adapted from Char.[104])

TABLE 3

Relative Comparison of the Measured Ratio of the Excimer-to-Monomer Fluorescence Intensities I_D/I_M and the Predicted Cyclization Rate Constant k_{cy} for Samples with Different Molecular Weights

Mol wt of Py-PEG-Py to be compared	$I_D/I_M(1)^a/I_D/I_M(2)$	$k_{cy}(1)^b/k_{cy}(2)$
9200(1)-11200(2)	1.39	1.34
4800(1)-9200(2)	5.41	2.65
4800(1)-11200(2)	7.48	3.56

[a] I_D/I_M values are taken from the plateau region in Figure 20.

[b] $k_{cy} \propto N^{-3/2}$ where N is the length of the polymer chain.

Adapted from Char et al.[104]

association of pyrene groups from different polymer chains as well as the cyclization of pyrene groups of a single chain. According to the Wilemski-Fixman theory,[105] for a non-draining Rouse chain the cyclization rate constant k_{cy} is proportional to $N^{-3/2}$, where N is the number of repeat units in the chain. The I_D/I_M ratio in the plateau region in Figure 20 should be proportional to k_{cy}. The ratios of the observed I_D/I_M values between the polymers with different molecular weights are listed in Table 3. In this table are also given the ratios of the k_{cy} for different samples predicted from the Wilemski-Fixman theory. For Py-PEG-Py (9200) to Py-PEG-PY(11200) the I_D/I_M ratio is close to the k_{cy} ratio, while for Py-PEG-Py(4800) to the other two samples the observed I_D/I_M ratios are much higher than the theory predicting k_{cy} ratios. This implies that the end-to-end cyclization for Py-PEG-Py(4800) is not simply a diffusion-limited process as assumed in the Wilemski-Fixman theory. The unusually high I_D/I_M ratio for Py-PEG-Py(4800) is attributed to the hydrophobic attractions between pyrene groups. Since the hydrophobic interactions caused by the pyrene groups can be compensated by the hydrophilic interactions between the PEG chain and water, and the later interactions are proportional to PEG chain length, the difference between the observed I_D/I_M ratio and the calculated value is most pronounced for the low molecular weight polymer.

V. LIPOPHOBIC EFFECTS ON PHOTOCHEMICAL AND PHOTOPHYSICAL BEHAVIOR OF MOLECULES WITH POLAR CHAIN IN NONPOLAR SOLVENTS

It has recently been established that in nonpolar solvents the "inverse hydrophobic" or the lipophobic interactions can force molecules with polar regions to aggregate or to self-coil.[33] For example, substances with a poly(ethylene glycol) (PEG) chain have been widely used as surfactants. Since PEG chains play a hydrophilic role in surfactant molecules, one might expect that in nonpolar solvents a single PEG chain, driven by lipophobic interactions, could coil up or form aggregates. A fluorescence probe study indicates that this is indeed the case for PEG oligomers in appropriate temperature ranges.

In fluorescence probe techniques, among the many possible fluorescence chromophores pyrene has been extensively used due to its unusually long fluorescence lifetime, its ability to form excimer, and the sensitivity of its structured fluorescence spectrum to the polarity of the environment.[106] However, in the study of fluorophore-end-labeled PEG oligomer-nonpolar solvent systems, pyrene appears not to be a proper candidate as a fluorescence probe, since its large bulk might undergo lipophilic interaction which would compensate the lipophobic interaction between the PEG chain and the solvent. Furthermore, for the pyrene-end-labeled PEG oligomers, the fluorescence spectra are expected to be dominated by intramolecular excimer emission both in nonpolar and polar solvent with low viscosity because of the long fluorescence lifetime of pyrene and the short chain length of PEG oligomer. Therefore, in the study of PEG oligomer-nonpolar solvent systems, naphthalene was used as the probe, becuase it has a relative small bulk, reasonable fluorescence quantum yield, and photochemical reaction potential. The molecules used in the reported study have the following structures and are abbreviated as N-P_n-N:

$$N_pCOO\text{--}(CH_2CH_2O)_n\text{--}CON_p \qquad N_p = 2\text{-naphthyl}$$

$$N\text{--}P_n\text{--}N \ (n = 2, 3, 4, 5, 6, 7, 12)$$

For comparison the behavior of polymethylene bis-2-naphthoate (B_n) was also studied.

Through the examination of excimer formation, delayed excimer fluorescence, excimer phosphorescence, fluorescence depolarization, and photochemical dimerization, the aggregation and self-coiling of N-P_n-N in nonpolar solvents are demonstrated.

A. FLUORESCENCE SPECTRA

The fluorescence spectrum of N-P_3-N in isopentane at ambient temperature, which is typical of the other naphthalene-end-labeled PEGs in nonpolar solvents, is shown in Figure 21. The fluorescence spectra of the model compounds, butyl 2-naphthoate (A_4) and B_{10}, are also given in Figure 21. When the (O,O) band of the N-P_3-N emission spectrum is normalized to the (O,O) band of the A_4 fluorescence spectrum and the latter spectrum is subtracted, a structureless excimer emission band is resolved with a maximum at *circa* 400 nm. At a concentration below $1 \times 10^{-3} M$, the ratio of fluorescence intensities of excimer to monomer I_D/I_M is independent of concentration, suggesting the excimer is intramolecular. Thus the excimer formation is attributed to the self-coiling of PEG chain. The excitation spectra for N-P_3-N excimer and monomer emission are identical, and the maxima correspond to that in the UV absorption spectrum, suggesting the absence of strong interaction between the naphthoate chromophores in the ground state. The emission at the monomer region (335

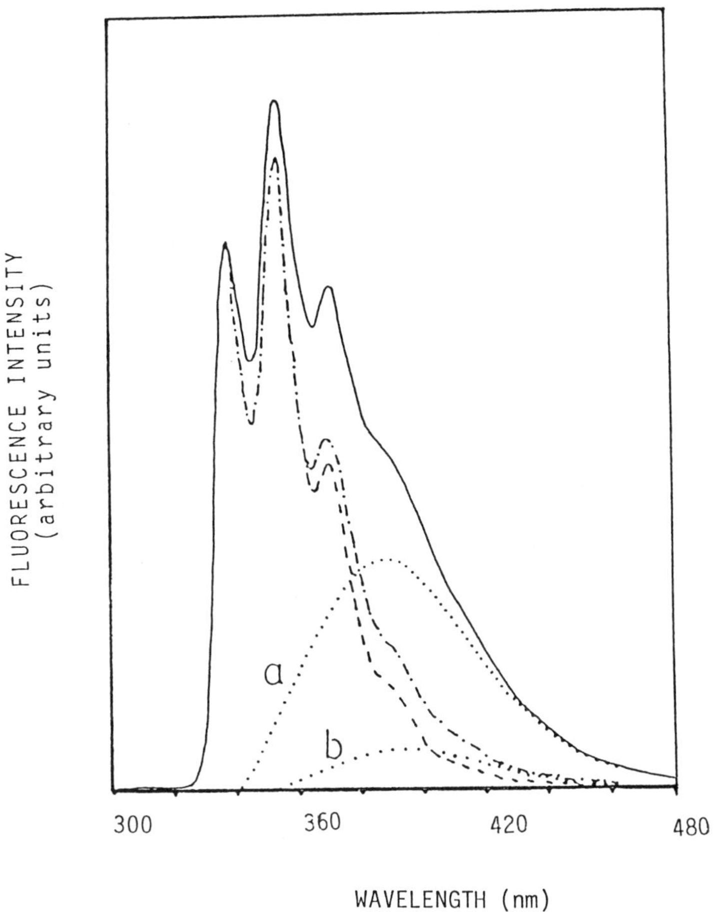

FIGURE 21. Fluorescence spectra of triethylene-glycol bis-2-naphthoate (———),
butyl 2-naphthoate (— — —), and decamethylene bis-2-naphthoate (· — · —)
in isopentane at room temperature. The concentration of the chromophore for all
samples is 2×10^{-4} M. Curves a and b represent the excimer emission of
triethyleneglycol bis-2-naphthoate and decamethylene bis-2-naphthoate, respectively.

nm) decays biexponentially with lifetimes of 2 and 10 ns. The analysis of the decay at 400
nm yields a growing in time of 2 ns and a decay time of 10 ns. All these results suggest
that the excimer is formed through a dynamic process.[72] Obviously N-P$_n$-N in isopentane
at room temperature tend to assume the self-coiling conformation, and the two end groups
are in proximity but do not associate with each other to form a sandwich pair in the ground
state.

The ratios of I_D/I_M of N-P$_n$-N in isopentane have been measured at temperatures ranging
from -90 to $45°C$, and are shown in Figure 22. All of the measurements were made at
constant N-P$_n$-N concentration (1×10^{-4} M). At higher temperatures the fluorescence
spectra consist of predominantly monomer emission for all of the compounds. However, at
lower temperatures significant differences in the behavior of N-P$_n$-N with long chains and
those with short chains are apparent. The plots for N-P$_2$-N and N-P$_3$-N have a "high
temperature region" where I_D/I_M increases with decreasing temperature, and a "low temperature region" where I_D/I_M decreases with decreasing temperature. This temperature dependence of I_D/I_M can be well understood by consideration of excimer formation through a
dynamic process.[72] The plots for molecules with n >4 bear an analogy to that of N-P$_3$-N

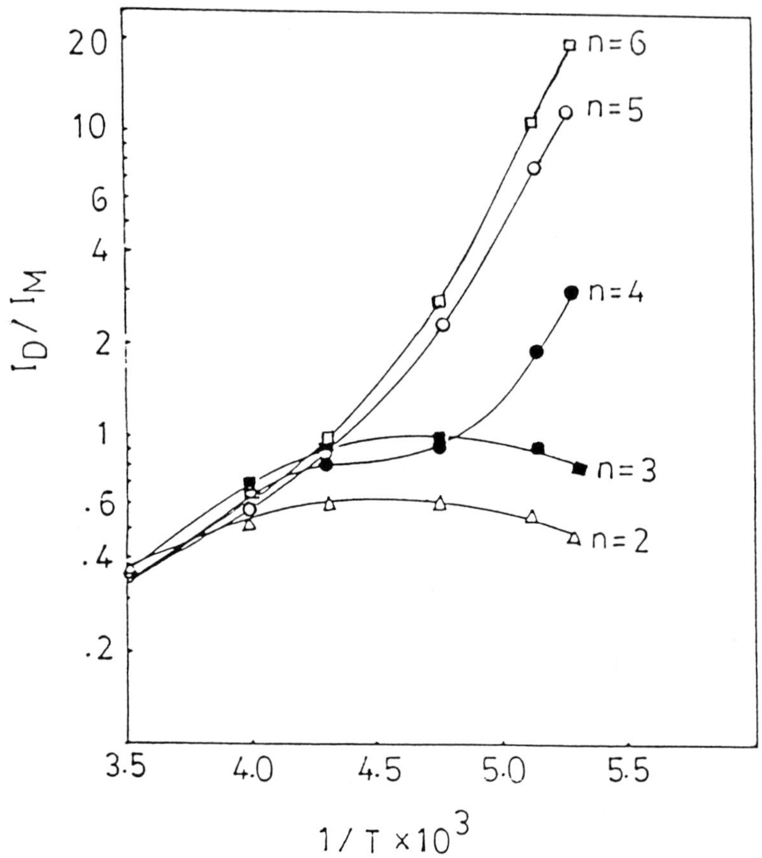

FIGURE 22.   Plot of the ratio of fluorescence intensities of excimer to monomer I_D/I_M as a function of temperature for naphthalene-end-labeled poly(ethylene glycol) $N\text{-}P_n\text{-}N$ (n represents the number of structure unit in poly[ethylene glycol] chain) in isopentane. $[N\text{-}P_n\text{-}N] = 1 \times 10^{-4}\ M$.

in high temperature region, but below a particular temperature the ratio of I_D/I_M increases drastically with decreasing temperature. In order to clarify the reason for the intense excimer emission in low temperature region the ratio of I_D/I_M was examined as a function of $N\text{-}P_n\text{-}N$ concentration. As shown in Figure 23, the plot of I_D/I_M vs. $[N\text{-}P_n\text{-}N]$ displays a break, which is a typical behavior of aggregation. Thus the increase in the excimer emission at low temperature was attributed to intermolecular aggregation, which induces the enhancement of local chromophore concentration. The concentration corresponding to the break point is the CAC. At a concentration below CAC $N\text{-}P_n\text{-}N$ exists only in its monomeric form and the excimer formation is due to self-coiling. Thus, I_D/I_M is independent of $[N\text{-}P_n\text{-}N]$ and is defined as R_m. It is noticed that R_m decreases with increasing PEG chain length. This result is probably due to the fact that the naphthoate groups inside the coil are further diluted by PEG chain for $N\text{-}P_n\text{-}N$ with a longer chain. Above CAC the degree of aggregation increases with $[N\text{-}P_n\text{-}N]$. At high-enough concentration the excimer mainly originates from the aggregate and its emission dominates the fluorescence spectra. At this stage the excitation spectra for the monomer and for the excimer are clearly different. The spectra for the excimer are slightly but evidently red-shifted. In addition, in the fluorescence decay measurements no rising component can be detected in the excimer decay profile, suggesting that the growing in time is less than 0.2 ns (the time resolution of the used instrument). These observations indicate that the excimer originates from pairs of naphthoate groups which exist prior to

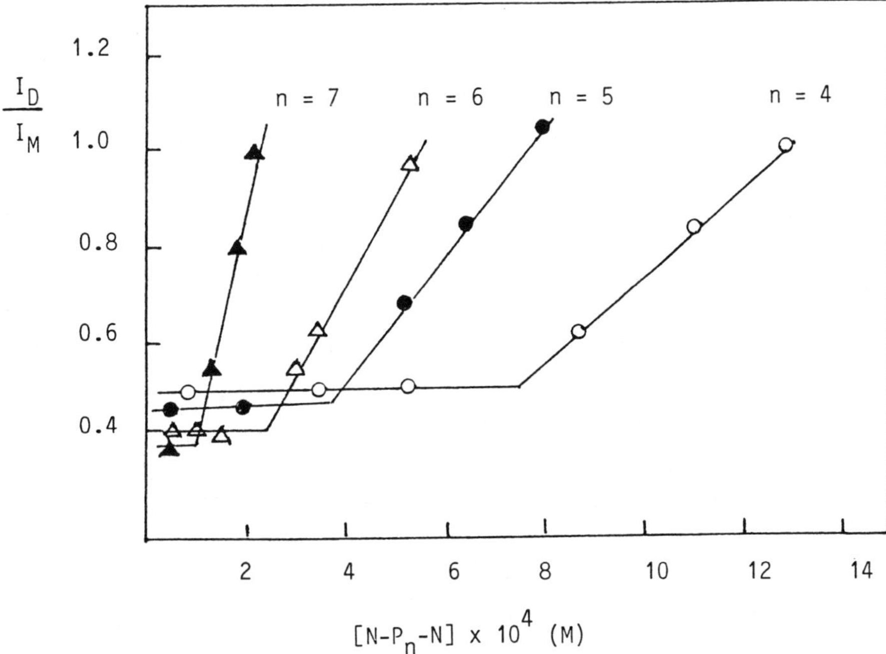

FIGURE 23.   Plot of the ratio of fluorescence intensities of excimer to monomer I_D/I_M as a function of concentration for naphthalene-end-labeled poly(ethylene glycol) N-P$_n$-N (n represents the number of structure unit in poly[ethylene glycol] chain) in isopentane at $-23°C$.

excitation. In other words, inside the aggregate the chromophores are in the excimer conformation in the ground state. This is confirmed by the fact that in isopentane glass (77 K) where the rotation of the chromophores is frozen, N-P$_n$-N mainly emits excimer fluorescence.

The driving force for self-coiling and aggregation is probably the lipophobic interaction, since in tetrahydrofuran N-P$_n$-N emits only monomer fluorescence both at high temperature and low temperature. Furthermore, B$_5$ and B$_{10}$ which have the nonpolar chains show exclusively monomer fluorescence in isopentane solution at low temperature, and very weak excimer fluorescence at room temperature. The lipophobic interaction occurs only when polar (lipophobic) molecules are dissolved in nonpolar medium. The N-P$_n$-N with longer PEG chains is expected to undergo greater lipophobic force and possesses higher aggregate ability. This is supported by the following observations. First, at a given temperature (Figure 23) the N-P$_n$-N with longer chains has smaller CAC values and greater slopes of the lines at concentrations above CAC. Second, at a given concentration the N-P$_n$-N with longer chains forms an aggregate at a higher temperature (Figure 22).

B. EFFECTS OF UNLABELED PEG ON EXCIMER FORMATION OF N-P$_n$-N

To provide more evidence for aggregation the effect of unlabeled PEG on the excimer formation of N-P$_n$-N in isopentane was studied. In the low temperature and high concentration region the excimer emission dominates the fluorescence spectrum of N-P$_5$-N (Figure 23). The addition of unlabeled poly(ethylene glycol) with a molecular weight of 600 results in the enhancement of monomer emission and reduction in excimer fluorescence. In the presence of sufficient unlabeled PEG, the fluorescence spectrum is dominated by monomer emission. Obviously this is due to the fact that the unlabeled PEG coaggregates with N-P$_5$-N, and the chromophores are diluted within the coaggregate.

C. FLUORESCENCE DEPOLARIZATION AND SINGLET ENERGY MIGRATION

The relative orientations of the absorption dipole and emission dipole of a chromophore in a rigid matrix are fixed, so that if the chromophore absorbs polarized radiation, its emission will retain memory of the excitation polarization. The singlet energy migration from one chromophore to another of different orientation results in fluorescence depolarization. For randomly oriented chromophores the fluorescence polarization degree, P, varies from $-1/3$ to $1/2$. The mutual chromophore separation in aggregate favors excitation migration, and the fluorescence is expected to be depolarized. Representative values of the fluorescence polarization at $\lambda_{em} = 355$ nm, $\lambda_{ex} = 337$ nm for N-P_n-N, B_n, and A_4 in isopentane, 2-methyl-tetrahydrofuran (MTHF), and ethanol glasses (77 K) are listed in Table 4. In ethanol and MTHF glasses all the compounds have the P values ranging from 0.20 to 0.39, suggesting no significant energy migration. However, in isopentane glass, while A_4 and B_n show almost the same P values as in ethanol and MTHF glasses, the fluorescences of N-P_n-N are significantly depolarized, suggesting the existence of singlet excitation migration. This observation provides additional evidence for aggregate formation.

D. INTRA- AND INTERMOLECULAR PHOTODIMERIZATION REACTIONS OF THE END NAPHTHALENE GROUPS

Photoirradiation of alkyl 2-naphthoate resulted in a "cubane-like" photodimer as the unique product (Scheme 3) in spite of the fact that six isomeric dimers are formally pos-

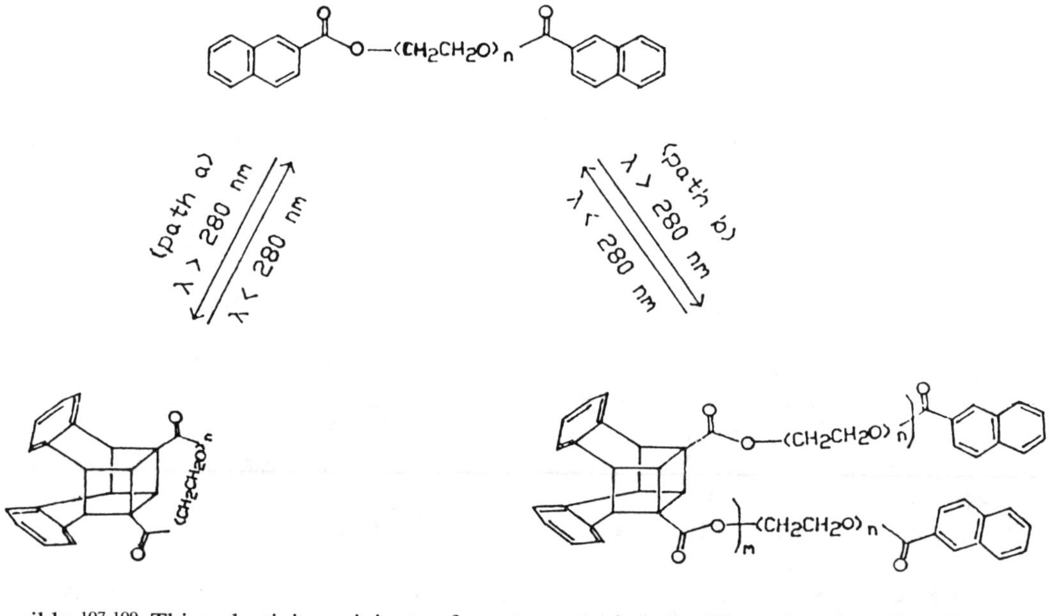

sible.[107-109] This selectivity originates from two restrictions. First, the photodimerization occurs only between the substituted rings. Second, in the dimer the substituents are in the head-to-tail orientation. The photoirradiation of a bichromophoric compound, like N-P_n-N, can lead either to intra- or intermolecular reactions. The intramolecular reaction gives macrocyclic ring-closure products, while the intermolecular reaction results in polymers. Self-coiling of the chain linking the two terminal chromophores will increase the intramolecular ring-closure probability, while aggregation will enhance the intermolecular polymerization.

Irradiation with $\lambda > 280$ nm of 1×10^{-4} M solution of N-P_n-N in isopentane at room temperature only leads to formation of intramolecular ring-closure photodimers as shown in Scheme 3 (path a). No polymerization products (path b) were detected. The yields of the

TABLE 4

Fluorescence Polarization Degree of Naphthalene-End-Labeled Poly(ethylene glycol) N-P$_n$-N (n Represents the Number of Structure Unit of Poly[ethylene glycol] Chain), Polymethylene Bis-2-Naphthoates B$_n$ (n Represents the Number of Carbon Atom in Polymethylene), and Butyl 2-Naphthoate A$_4$ in Isopentane, 2-Methyltetrahydronfuran (MTHF), and Ethanol Glasses at 77 K

Glass	N-P$_2$-N	N-P$_3$-N	N-P$_4$-N	N-P$_5$-N	N-P$_6$-N	N-P$_7$-N	A$_4$	B$_5$	B$_{10}$
Isopentane	0.07	0.02	0.02	0.05	0.06	0.08	0.34	0.28	0.36
Ethanol	0.20	0.24	0.22	0.23	0.26	0.21	0.39	0.27	0.30
MTHF	0.23	0.26	0.25	0.27	0.28	0.26	0.37	0.22	0.20

Note: The emission (λem) and excitation (λex) wavelength are 355 and 337 nm, respectively. [N-P$_n$-N] = [B$_n$] = 0.5 × [A$_4$] = 1 × 10^{-4} M.

TABLE 5

The Conversion of Naphthalene-End-Labeled Poly(ethylene glycol) N-P$_n$-N (n Represents the Number of Structure Unit of Poly[ethylene glycol]) and Decamethylene Bis-2-Naphthoate B$_{10}$ in Isopentane and Methanol at Room Temperature after Irradiation for 0.5 h with a 450-W Hanovia Lamp (the Concentration of All the Sample was ca. 1 × 10^{-4} M)

Solvent	N-P$_2$-N (%)	N-P$_3$-N (%)	N-P$_5$-N (%)	N-P$_{12}$-N (%)	B$_{10}$ (%)
Isopentane	58	51	32	29	5
Methanol	30	24	18	5	4

intramolecular products are 100% based on the consumption of the starting materials. Irradiation of N-P$_n$-N solution in methanol at the same condition gives the same products, but the quantum yields are much lower than those in isopentane. Table 5 gives the conversions of N-P$_n$-N and B$_{10}$ in these two solvents after irradiation for 0.5 h with a 450-W Hanovia high pressure mercury lamp in a merry-go-round apparatus. The conversions of all N-P$_n$-N in methanol are smaller than those in isopentane, particularly for N-P$_n$-N with longer chain. Furthermore, B$_{10}$, which has the similar length but nonpolar chain as N-P$_3$-N, shows solvent-independent conversion, and the conversions are one order of magnitude smaller than that of N-P$_3$-N in isopentane. Obviously the high yield of N-P$_n$-N in isopentane is attributed to self-coiling of the PEG chains. The above results demonstrate the potential application of self-coiling to the synthesis of macrocyclic entities. For example, large-ring compounds including thc 42-mcmbcrcd ring (thc intramolccular dimcr of N-P$_{12}$-N) havc bccn synthcsizcd by the application of lipophobic force.

At −25°C irradiation of N-P$_n$-N solution in isopentane at a concentration above CAC results in polymers (path b in Scheme 3) in addition to the intramolecular photodimers (path a). The polymers were obtained as precipitate, fractionated by HPL, and identified by ^{1}H-NMR and mass spectroscopies. The intermolecular dimer and trimer for N-P$_5$-N were detected. The absence of higher polymer in the products is probably due to the insolubility of the dimer and trimer in the reaction solution at the reaction temperature. They precipitated out in the course of irradiation, which prevents them from further growing. Irradiation of N-P$_n$-N solution in methanol at the same condition gives the intramolecular photodimers only, and no polymers were formed. These experimental results confirm that N-P$_n$-N in isopentane forms aggregates at low temperature.

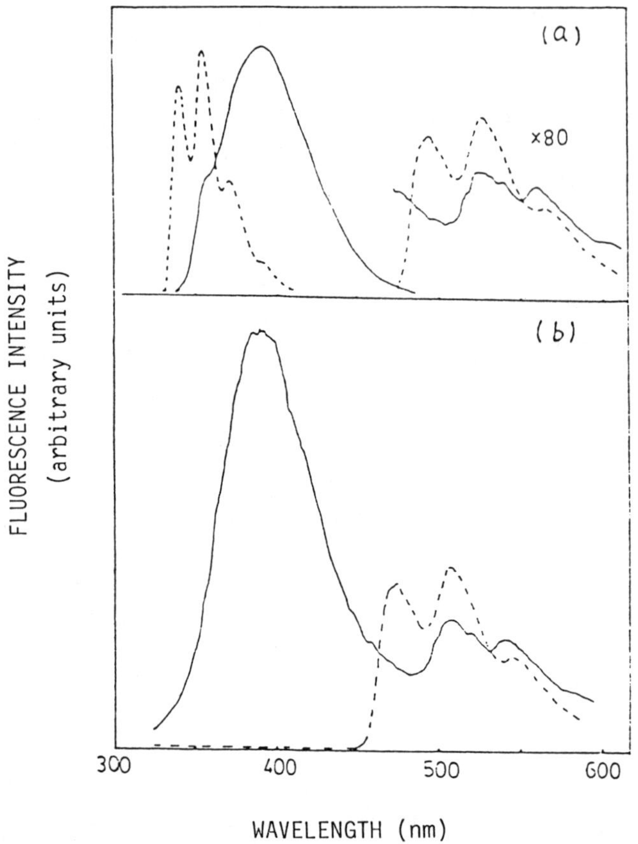

FIGURE 24. Emission spectra of pentaethylene-glycol bis-2-na-
phthoate (——) and butyl 2-naphthoate (— — —) in isopentane glass
(77 K). (a) Prompt emission; (b) delayed fluorescence and phosphorescence. The concentration of pentaethylene-glycol bis-2-naphthoate
and butyl 2-naphthoate is 1 and 2×10^{-4} M, respectively.

E. DELAYED EXCIMER FLUORESCENCE AND TRIPLET EXCIMER PHOSPHORESCENCE

Although singlet excimers of aromatic hydrocarbons have been extensively studied, until
recently relatively little has been reported on triplet excimers. The main reason for this lies
with the experimental difficulties. Formation of an excimer via a dynamic process requires
a fluid medium. However, phosphorescence detection of a triplet excimer in fluid solutions
is difficult. Lim et al.,[110-113] and Takemura et al.[114,115] have observed the weak excimer
phosphorescence from fluid solutions of naphthalene and related compounds, using a home-
made spectrophosphorimeter which combines the high intensity of a repetitively pulsed flash
lamp (excitation source) and the high sensitivity of photon counting (detection system). To
form a triplet excimer in a rigid matrix the two chromophores have to be arranged in the
excimer conformation prior to excitation. Recently Burkhart et al.[116,117] and Kim and Weber[118]
reported the excimer phosphorescence from poly(vinylnaphthalene) in solid film and in
MTHF glass, respectively.

Figure 24 shows the emission spectra of N-P$_5$-N, which is typical of the N-P$_n$-N, and
the model compound A$_4$ in isopentane glass (77 K). The significant differences in both
phosphorescence and fluorescence between N-P$_5$-N and A$_4$ are evident. First, the phosphorescence spectrum of A$_4$ is typical of monomer phosphorescence of naphthoate derivatives.
The phosphorescence of N-P$_5$-N shifts to lower energy by about 45 nm, and its general

features are analogous to the triplet excimer phosphorescence of naphthalene and related compounds reported by Lim et al.[110-113] Second, A_4 emits monomer fluorescence only and no delayed fluorescence was detected at concentrations below 1×10^{-3} M. On the other hand, for N-P_5-N the excimer fluorescence dominates the fluorescence spectrum, although the monomer fluorescence is still observed as a shoulder. Furthermore, N-P_5-N emits intense delayed excimer fluorescence. These observations are consistent with aggregation of N-P_n-N. In the aggregate the mutual chromophore separation is expected to favor triplet and/or singlet excitation migration. Some chromophores in the aggregate might be arranged in a triplet excimer conformation (triplet excimer site), some in a singlet excimer conformation (singlet excimer site), and others in monomeric form. The singlet and triplet excimer sites may function as an energy trap in singlet and triplet energy migration, respectively. The singlet excited state of the chromophore in a triplet excimer site and monomeric conformation may undergo intersystem crossing, energy migration, and give rise to monomer fluorescence. Excitation of the chromophores in singlet excimer sites exclusively leads to excimer fluorescence. Thus the singlet excimer originates from two pathways: (1) the excitation of the chromophores in singlet excimer sites, and (2) the excitation of the chromorphores in triplet excimer sites as well as the monomeric conformation, followed by excitation migration to a singlet excimer site. As a result, the excimer fluorescence is much stronger compared with monomer fluorescence in the prompt fluorescence emission spectra. Similarly, the triplet excimer may originate from the excitation of the chromophores in triplet excimer sites followed by intersystem crossing, or from the migration of the monomer triplet state which was generated at the monomeric sites. Since triplet excimer sites act as energy traps in triplet migration, the concentration of these need not be high in order to observe excimer phosphorescence. Their low concentration makes it possible that two triplet states, which are formed by intersystem crossing of singlet excited chromophores in the monomeric conformation, migrate together and annihilate each other with the creation of one singlet excited and one ground state. The resultant singlet excited state may migrate to the singlet excimer site to give rise to delayed excimer fluorescence.

It is well established[110,119] that the conformation of the naphthalene triplet excimer is "L-shaped" or "butterfly-shaped" rather than a sandwich-pair geometry favored by the singlet excimer. In the L-shaped triplet excimer the long axes of the two naphthalene rings are very nearly parallel, but the short axes are highly nonparallel. To demonstrate this conformational difference between triplet and singlet excimers the following two experiments were performed. First, the isopentane solution of photodimer of N-P_5-N obtained in the previous section was prepared. The solution was cooled to 77 K and was subsequently reirradiated with $\lambda = 250$ nm to regenerate the naphthoates which are in sandwich-pair arrangement. The glass exhibits delayed excimer fluorescence only (no phosphorescence was detected). This observation suggests that the naphthoate groups in a sandwich-pair arrangement do not form triplet excimer. In the second experiment, the glassy solution of N-P_5-N in isopentane (77 K) was thoroughly irradiated with $\lambda > 280$ nm so that all naphthoate groups in the sandwich-pair arrangement are transformed to the dimerization product. The resultant glass showed only triplet excimer phosphorescence and weak monomer fluorescence (no delayed excimer fluorescence was observed). These experimental results confirm that some naphthoate groups in the aggregate are in sandwich-pair arrangement, some in an L-shaped conformation, and others in a monomeric conformation, which give delayed excimer fluorescence, triplet excimer phosphorescence, and delayed monomer fluorescence, respectively.

VI. CONCLUSION

We have reviewed the hydrophobic and lipophobic effects on the behavior of molecules with flexible chains in solutions. It has been demonstrated by photophysical and photochemical probe techniques as well as kinetics methods that in aqueous or aquiorgano solutions hydrophobic interactions force single nonpolar chains to coil up or push them together and make them form aggregates. The average aggregate number for *n*-alkane aggregate in DMSO-H_2O is in the range of 15 to 30, depending on alkane chain length. The microscopic polarity within the aggregates is similar to that in cyclohexane. At a certain temperature the aggregate undergoes phase transition. Above phase transition temperature the microviscosity within the aggregate is slightly greater than in typical micelles, but below this temperature the microviscosity within the aggregates is as large as that in solid phase. Molecules with polar chains in nonpolar solvents also tend to self-coil or to associate, driven by lipophobic interaction. The aggregation can affect the chemical reactivity of organic compounds, and the self-coiling of molecules can be used to synthesize macrocyclic compounds. However, the facts to support aggregation and self-coiling are still limited in scope and depth, and many problems are waiting to be addressed and explored, particularly for the case of li-pophobic effects. It is hoped that more basic research in this area will continue.

ACKNOWLEDGMENT

We acknowledge the support of the National Science Foundation of China. Tung acknowledges the efforts of all his colleagues and students over the years in making all of this possible. Thanks also go to Ms. Shuang Liu for her valuable help in preparing the manuscript.

REFERENCES

1. **Ben-Nain, A.,** *Hydrophobic Interactions,* Plenum Press, New York, 1980.
2. **Tanford, C.,** *The Hydrophobic Effect: Formation of Micelles and Biological Membranes,* 2nd ed., John Wiley & Sons, New York, 1980.
3. **Tanford, C.,** Organizational consequences of the hydrophobic interaction, *Proc. Indian Acad. Sci. (Chem. Sci.),* 98, 343, 1987.
4. **Jiang, X.-K.,** Hydrophobic-lipophilic interaction. Aggregation and self-coiling of organic molecules, *Acc. Chem. Res.,* 21, 362, 1988.
5. **Xu, C.-B. and Tung, C.-H.,** Hydrophobic effects on photophysical and photochemical processes. VI. Character of aggregates formed by hydrocarbon with long chain in poor solvents, *Acta Chim. Sin. Engl. Ed.,* No. 1, 72, 1989.
6. **Menger, F. M. and Portony, C. E.,** The effect of urea and other reagents on the reactivity of associated p-nitrophenyl laurate, *J. Am. Chem. Soc.,* 90, 1875, 1968.
7. **Murakami, Y., Aoyama, Y., and Kida, M.,** Kinetic consequences of molecular aggregation in the hydrolysis of p-nitrophenyl carboxylates, *J. Chem. Soc. Perkin Trans. 2,* p. 1947, 1977.
8. **Jiang, X.-K., Hui, Y.-Z., and Fan, W.-Q.,** The effect of hydrophobic-lipophilic interactios on chemical reactivity. I. New evidence for intermolecular aggregation and self-coiling, *J. Am. Chem. Soc.,* 106, 3839, 1984.
9. **Tung, C.-H. and Liu, Y.-Y., and Xu, C.-B.,** Hydrophobic effects on photophysical and photochemical processes. I. Aggregation and excimer formation of alkyl 2-naphthoates with long chain in poor solvents, *Acta Chim. Sin. Engl. Ed.,* No. 1, 52, 1988.
10. **Fendler, J. H. and Fendler, E. J.,** *Catalysis in Micellar and Macromolecular Systems,* Academic Press, New York, 1975, 1.
11. **Fendler, J. H.,** *Membrane Mimetic Chemistry,* John Wiley & Sons, New York, 1982, 6.
12. **Bunton, C. A. and Savelli, G.,** *Advances in Physical Organic Chemistry,* Vol. 22, Gold, V. and Bethell, D., Eds., Academic Press, London, 1986, 213.
13. **Fan, W.-Q. and Jiang, X.-K.,** Effect of hydrophobic-lipopholic interactions on chemical reactivity. VI. The first example of correlation of hydrophobic-lipophilic substituent constants with chemical reactivity for a simple substrate-solvent system, *J. Am. Chem. Soc.,* 107, 7680, 1985.
14. **Blyth, C. A. and Knowles, J. R.,** The kinetic consequences of intermolecular attraction. II. The hydrolysis of a series of fatty acid p-nitrophenyl esters catalyzed by a series of N-n-alkylimidazoles. A very simple esterase model, *J. Am. Chem. Soc.,* 93, 3021, 1971.

15. **Guthrie, J. P.,** Aggregation of p-nitrophenyl alkanoates in aqueous solution: a caution concerning their use in enzyme model studies, *J. Chem. Soc. Chem. Commun.,* p. 897, 1972.
16. **Knowles, J. R. and Parsons, C. A.,** Proximity effects in bimolecular reactions in solutions, *Chem. Commun.,* p. 755, 1967.
17. **Blyth, C. A. and Knowles, J. R.,** The kinetic consequences of intermolecular attraction. I. The aminolysis of p-nitrophenyl decanoate and acetate by n-decylamine and ethylamine, *J. Am. Chem. Soc.,* 93, 3017, 1971.
18. **Maugh, T. H. and Bruice, T. C.,** Hydrolysis and aminolysis of o-acylhydroxyquinolines. Intracomplex general base-catalyzed aminolysis, *J. Am. Chem. Soc.,* 93, 6584, 1971.
19. **Oakenfull, D. G.,** Effects of hydrophobic interaction on the kinetics of the reactions of long chain alkylamines with long chain carboxylic esters of 4-nitrophenol, *J. Chem. Soc. Perkin Trans. 2,* p. 1006, 1973.
20. **Oakenfull, D. G. and Fenwick, D. E.,** Kinetic of the reaction of the hydrophobic ester 4-nitrophenyl decanoate with long chain N-alkylimidazoles above and below their critical micelle concentration, *Aust. J. Chem.,* 27, 2149, 1974.
21. **Fan, W.-Q., Jiang, X.-K., Hui, Y.-Z., and Li, M.-Z.,** The effect of hydrophobic-lipophilic interactions on chemical reactivity. V. The proximity effect in the reaction of long chain esters with long chain thiols., *Acta Chim. Sin.,* 43, 839, 1985.
22. **Illuminati, G. and Mandolini, L.,** Ring closure reactions of bifunctional chain molecules, *Acc. Chem. Res.,* 14, 95, 1981.
23. **Mandolini, L.,** *Advanced Physical Organic Chemistry,* Vol. 22, Gold, V. and Bethell, D., Eds., Academic Press, London, 1986, and references therein.
24. **Oneal, H. E. and Benso, S. W.,** Entropies and heat capacities of cyclic and polycyclic compounds, *J. Chem. Eng. Data,* 15, 266, 1970.
25. **Page, M. I. and Jencks, W. P.,** Entropic contributions to rate accelerations in enzymic and intramolecular reactions and the chelate effect, *Proc. Natl. Acad. Sci. U.S.A.,* 68, 1678, 1971.
26. **Corey, E. G., Nicolaon, K. C., and Melvin, L. S.,** Synthesis of novel macrocyclic lactones in the prostaglandin and polyether antibiotic series, *J. Am. Chem. Soc.,* 97, 653, 1975 and references therein.
27. **Porter, N. A., Magnin, D. R., and Wright, B. T.,** Free radical macrocyclization, *J. Am. Chem. Soc.,* 108, 2787, 1986.
28. **Gerlach, H. and Thalmann, A.,** Formation of esters and lactones by silver ion catalysis, *Helv. Chim. Acta,* 57(8), 2661, 1974.
29. **Narasaka, K., Maruyama, K., and Mukaiyama, T.,** A useful method for the synthesis of macrocyclic lactone, *Chem. Lett.,* p. 885, 1978.
30. **Masamune, S., Kamata, S., and Schilling, W.,** Syntheses of macrolide antibiotics, III. Direct ester and lactone synthesis from S-tert-butyl thioate (thiol ester), *J. Am. Chem. Soc.,* 97, 3515, 1975.
31. **Lloyd, K. and Yong, G. T.,** Use of acylamino acid esters of 2-mercaptopyridine in peptide synthesis, *Chem. Commun.,* p. 1400, 1968.
32. **Jiang, X.-K., Hui, Y.-Z., and Fei, Z.-X.,** A successful application of the concept of making use of hydrophobic forces to prepare large-ring compounds, *J. Chem. Soc. Chem. Commun.,* p. 689, 1988.
33. **Tung, C.-H. and Wang, Y.-M.,** Lipophobic effects on photochemical and photophysical behavior of molecules with polar chain in nonpolar solvents — evidence for intermolecular aggregation and self-coiling, *J. Am. Chem. Soc.,* 112, 6322, 1990.
34. **Rekker, F.,** *The Hydrophobic Fragmental Constant,* Part I, Elsevier, Amsterdam, 1977.
35. **McBain, J. W. and Salmon, C. S.,** Colloidal electrolytes. Soap solutions and their constitution, *J. Am. Chem. Soc.,* 42, 426, 1920.
36. **Muller, N.,** Search for a realistic view of hydrophobic effects, *Acc. Chem. Res.,* 23, 23, 1990.
37. **Hildebrand, J. H., Prausnitz, J. M., and Scott, R. L.,** *Regular and Regulated Solutions,* Van Nostrand Reinhold, New York, 1970, 85.
38. **Frank, H. S. and Evans, M. W.,** Free volume and entropy in condensed systems. III. Entropy in binary liquid mixtures, partial molal entropy in dilute solutions; structure and thermodynamic in aqueous electrolytes, *J. Chem. Phys.,* 13, 507, 1945.
39. **Frank, H. S. and Wen, W. Y., III.** Ion-solvent interaction. Structural aspects of ion-solvent interaction in aqueous solutions: a suggested picture of water structure, *Discuss. Faraday Soc.,* 24, 133, 1957.
40. **Nemethyl, G. and Scheraga, H. A.,** Structure of water and hydrophobic bonding in proteins. I. A model for the thermodynamic properties of liquid water, *J. Chem. Phys.,* 36, 3382 and 3401, 1962.
41. **Kirkwood, J. G.,** *A Symposium on the Mechanism of Enzyme Action,* McElray, W. D. and Glass, R., Eds., John Hopkins University Press, Baltimore, MD, 1954.
42. **Hammett, L. P.,** Some relations between reaction rates and equilibrium constants, *Chem. Rev.,* 17, 125, 1935.
43. **Jaffe, H. H.,** A reexamination of the Hammett equation. *Chem. Rev.,* 53, 191, 1953.
44. **Hansch, C., Maloney, P. P., Fujita, T., and Muir, R. M.,** Correlation of biological activity of phenoxyacetic acids with Hammett substituent constants and partition coefficients, *Nature,* 194, 178, 1962.

45. **Hansch, C., Muir, R. M., Fujita T., Malonry, P. P., Geiger, F., and Streich, M.,** The correlation of biological activity of plant growth regulators and chloromycetin derivatives with Hammett constants and partition coefficients, *J. Am. Chem. Soc.,* 85, 2817, 1963.

46. **Hansch, C. and Fujita, T.,** Analysis. A method for the correlation of biological activity and chemical structure, *J. Am. Chem. Soc.,* 86, 1616, 1964.

47. **Hansch, C. and Dunn, W. J., III,** Linear relations between lipophilic character and biological activity of drugs, *J. Pharm. Sci.,* 61, 1, 1972.

48. **Rekker, F.,** *The Hydrophobic Fragmental Constant,* Part I, Elsevier, Amsterdam, 1977, chap. 2.

49. **Rekker, F.,** *The Hydrophobic Fragmental Constant,* Part I, Elsevier, Amsterdam, 1977, chap. 3.

50. **Collander, R.,** The distribution of organic compounds between isobutanol and water, *Acta Chem. Scand.,* 4, 1085, 1950.

51. **Collander, R.,** Partition of organic compounds between higher alcohols and water, *Acta Chem. Scand.,* 5, 774, 1951.

52. **Hansch, C. and Anderson, S. M.,** The effect of intramolecular hydrophobic bonding on partition coefficients, *J. Org. Chem.,* 32, 2583, 1967.

53. **Wooldridge, K. R. H.,** *Progress in Pharmaceutical Research,* Vol. 4, Blackwell Scientific, Oxford, 1982.

54. **Fujita, T.,** *Progress in Physical Organic Chemistry,* Vol. 14, Taft, R. W., Ed., John Wiley & Sons, New York, 1983, chap. 2.

55. **Koopmans, R. E. and Rekker, R. F.,** High-performance liquid chromatography of alkylbenzenes. Relationship between lipopholicities as determined from octanol-water partition coefficients or calculated from hydrophobic fragmental data and connectivity indexes; lipophilicity predictions for polyaromatics, *J. Chromatogr.,* 285, 267, 1984.

56. **Jiang, X.-K., Hui, Y.-Z., and Fang, W.-Q.,** The effect of hydrophobic-lipophilic interactions on chemical reactivity. II. Solvent effects on the aggregation and self-coiling of long-chain molecules, *Acta Chim. Sin.,* 42, 1276, 1984; *Engl.Ed.,* p. 111, 1985.

57. **Menger, F. M. and Venkataram, U. V.,** A microscopic hydrophobicity parameter, *J. Am. Chem. Soc.,* 108, 2980, 1986.

58. **Jiang, X.-K., Fan, W.-Q., and Hui, Y.-Z.,** The effect of hydrophobic-lipophilic interactions on chemical reactivity. IV. A case of 17-membered-ring "neighboring-group" participation: compelling evidence for self-coiling, *J. Am. Chem. Soc.,* 106, 7202, 1984.

59. **Jiang, X.-K., Fan, W.-Q., Hui, Y.-Z., Gu, J.-H., and Cheng, M.-F.,** The effect of hydrophobic-lipophilic interactions on chemical reactivity. VII. 14- and 18-membered-ring neighboring-group participation, *Acta Chim. Sin.,* 45, 900, 1987. *Engl. Ed.,* p. 98, 1987.

60. **Cheng, X.-E., Hui, Y.-Z., Gu, J.-H., and Jiang, X.-K.,** The neighboring group participation by the carbonyl group in the hydrolysis of flexible keto-esters and its inhibition when substrates are placed inside the helical cavities of amylose, *J. Chem. Soc. Chem. Commun.,* p. 71, 1985.

61. **Hui, Y.-Z., Wang, S.-J., and Chen, Y.-D.,** The microenvironmental effects of the helical conformations of amylose and its derivatives. VII. The influences of solvent composition of the Me_2SO-H_2O mixture on the catalytic or inhibitory effects of amylose on the hydrolysis of p-nitrophenyl dodecanoate, *Acta Chim. Sin.,* 42, 1161, 1984.

62. **Hui, Y.-Z., Wang, S.-J., and Jiang, X.-K.,** Catalytic hydrolysis of esters of p-nitrophenol bound by amylose in the Me_2SO-H_2O mixed-solvent system, *J. Am. Chem. Soc.,* 104, 347, 1982.

63. **Hui, Y.-Z., Wang, S.-J., and Jiang, X.-K.,** The microenvironmental effects of the helical conformations of amylose and its derivatives. VI. Catalytic hydrolysis of esters of p-nitrophenol bound by amylose in the $DMSO$-H_2O mixed-solvent system, *Acta Chim. Sin.,* 40, 1148, 1982.

64. **Jiang, X.-K., Hui, Y.-Z., and Fan, W.-Q.,** Microenvironmental effects of helical conformations of amylose — catalytic effects of amylose on the hydrolysis of p-substituted phenol esters and structure effects of substrates, *Acta Chim. Sin.,* 43, 57, 1985; *Engl. Ed.,* p. 217, 1986.

65. **Jiang, X.-K., Li, X.-Y., and Huang, B.-Z.,** The effect of hydrophobic-lipophilic interactions on chemical reactivity. X. The competition between the aggregation of long-chain esters and the formation of inclusion complexes of these esters with amylose, *Proc. Indian Acad. Sci. (Chem Sci.),* 98, 423, 1987.

66. **Hui, Y.-Z., Cheng, X.-N., Gu, J.-H., and Jiang, X.-K.,** Microenvironmental effects of the helical conformations of amylose and its derivatives — the formation of inclusion complexes with a substrate by amylose in the $DMSO$-H_2O mixed-solvent system, *Sci. Sin. Engl. Ed. Ser. B,* 25, 698, 1982.

67. **Yamamoto, M. and Sano, T.,** Interaction of amylose with iodine. I. Characterization of cooperative binding isotherms for amylose, *Bull. Chem. Soc. Jpn.,* 55, 1886, 1982.

68. **Hui, Y.-Z., Russell, J. C., and Whitten, D. G.,** Host-guest interactions in amylose inclusion complexes: photochemistry of surfactant stilbenes in helical cavities of amylose, *J. Am. Chem. Soc.,* 105, 1374, 1983.

69. **Bulpin, P. V., Cutler, A. N., and Lips, A.,** Cooperative binding of sodium myristate to amylose, *Macromolecules,* 20, 44, 1987.

70a. **Turro, N. J., Gratzel, M., and Braun, A. M.,** Photophysical and photochemical processes in micellar systems, *Angew. Chem. Int. Ed. (Engl.),* 19, 675, 1980.

70b. **Thomas, J. K.**, Radiation-induced reactions in organized assemblies, *Chem. Rev.*, 80, 283, 1980.

71. **Jiang, X.-K., Li, X.-Y., and Huang, B.-Z.**, The effect of hydrophobic-lipophilic interactions on chemical reactivity. IX. Putting the spotlight on lipophilic forces in solvent-effect studies, *Proc. Indian Acad. Sci. (Chem. Sci.)*, 98, 409, 1987.

72. **Birks, J. B.**, *Photophysics of Aromatic Molecules*, John Wiley & Sons, London, 1970, 312.

73. **Todeco, R., Gelan, J., Martens, H., Put, J., and De Schryver, F. C.**, A kinetic scheme for intramolecular excimer formation in bis(2-naphthylmethyl)ether, involving different starting conformations, *J. Am. Chem. Soc.*, 103, 7304, 1981.

74. **Turro, N. J.**, *Modern Molecular Photochemistry*, Benjamin/Cummings, London, 1978, 137.

75. **Tung, C.-H., Zhen, Z., and Liu, J.-Y.**, Hydrophobic effects on photophysical and photochemical processes. VIII. Aggregation and fluorescence spectra of long chain N-alkylindoles in poor solvents, *Acta Phys. Chem. Sin.*, 5, 415, 1989.

76. **Schore, N. E. and Turro, N. J.**, Novel fluorescent probe for micellar systems. 1,3-Dialkylindoles, *J. Am. Chem. Soc.*, 97, 2488, 1975.

77. **Turro, N. J., Tanimoto, Y., and Gabor, G.**, Functional detergent probes of micelle structure. Absorption, fluorescence and quenching measurements, *Photochem. Photobiol.*, 31, 527, 1980.

78. **Turro, N. J. and Tanimoto, Y.**, The effects of aliphatic alcohols on fluorescence quenching and fluorescence polarization of luminescence probes in hexadecyltrimethylammonium bromide micelles, *Photochem. Photobiol.*, 34, 157, 1981.

79. **Atik, S. S., Nam, M., and Singer, L. A.**, Transient studies on intramicellar excimer formation. A useful probe of the host micelle, *Chem. Phys. Lett.*, 67, 75, 1979.

80. **Lianos, P. and Zana, R.**, Use of pyrene excimer formation to study the effect of NaCl on the structure of sodium dodecyl sulfate micelles, *J. Phys. Chem.*, 84, 3339, 1980.

81. **Bolt, J. D. and Turro, N. J.**, New probes for surfactant solutions: phosphorescent labelled detergents, *Photochem. Photobiol.*, 35, 305, 1982.

82. **Almgren, M., Swarnp, S., and Lofroth, J. E.**, Effect of formamide and other organic polar solvents on the micelle formation of sodium dodecyl sulfate, *J. Phys. Chem.*, 89, 4621, 1985.

83. **Turro, N. J., Aikawa, M., and Yekta, A.**, A comparison of intermolecular and intramolecular excimer formation in detergent solutions. Temperature effects and microviscosity measurements, *J. Am. Chem. Soc.*, 101, 772, 1979.

84. **Shinitzky, M., Dianoux, A. C., Gitler, C., and Weber, G.**, Microviscosity and order in the hydrocarbon region of micelles and membranes determined with fluorescence probes. I. Synthetic micelles, *Biochemistry*, 10, 2106, 1971.

85. **Riddick, J. A. and Bunger, W. B.**, *Techniques of Chemistry, Organic Solvents*, Vol. 11, John Wiley & Sons, New York, 1970, 202.

86. **Tung, C.-H., Ma, G.-Z., Guo, S.-Y., Wu, S.-K., and Xu, H.-J.**, Study on intramolecular energy migration of bichromophoric compounds by fluorescence depolarization technique, *Acta Chim. Sin.*, 43, 1092, 1985.

87. **Kalyanasundaram, K. and Thomas, J. K.**, Solvent-dependent fluorescence of pyrene-3-carboxaldehyde and its applications in the estimation of polarity at micelle-water interface, *J. Phys. Chem.*, 81, 2176, 1977.

88. **Xu, C.-B.**, Effect of Hydrophobic Interaction on Photochemical and Photophysical Processes, Ph.D. thesis, Institute of Photographic Chemistry, Academia Sinica, 1989, 86.

89. **Ford, D., Thistlethwaite, P. J., and Woolfe, G. J.**, The fluorescence behavior of methyl and phenyl salicylate, *Chem. Phys. Lett.*, 69, 246, 1980.

90. **Acuna, A. U., Amat-Guerri, F., Catalan, J., and Gonzalez-Tablas, F.**, Dual fluorescence and ground state equilibria in methyl salicylate, methyl 3-chlorosalicylate and methyl 3-tert-butylsalicylate, *J. Phys. Chem.*, 84, 629, 1980.

91. **Ouyang, X.-X., and Tung, C.-H.**, Hydrophobic effects on photophysical and photochemical processes. II. Aggregation and fluorescence spectra of long chain alkyl salicylates in poor solvents, *Acta Chim. Sin.*, 46, 191, 1988.

92. **Smith, K. K. and Kaufmann, K. J.**, Solvent dependence of the nonradiative decay rate of methyl salicylate, *J. Phys. Chem.*, 85, 2895, 1981.

93. **Xu, C.-B. and Tung, C.-H.**, Hydrophobic effects on photochemical process. IX. Enhancement of singlet-singlet energy transfer from alkyl 2-naphthoates to 9-anthrylmethyl n-alkanoates via aggregation, *Acta Chim. Sin.*, 48, 174, 1990.

94. **Herkstroeter, W. G., Martic, P. A., Hartman, S. E., Williams, J. L. R., and Farid, S.**, Unique hydrophobic interactions of pyrene in aqueous solution as effected by polyelectrolytes and surfactants, *J. Polym. Sci. Polym. Chem. Ed.*, 21, 2473, 1983.

95. **Morawetz, H.**, Fluorescence studies of conformational mobility and of the mutual interpenetration of flexible chain molecules, *Pure Appl. Chem.*, 52, 277, 1980.

96. **De Schryver, F. C., Moens, L., Van der Auweraer, M., Boens, N., Monnerie, L., and Bokobza, L.,** Kinetic and thermodynamic aspects of excimer formation in 2,4-diphenylpentanes as polystyrene models, *Macromolecules,* 15, 64, 1982.

97. **Jiang, X.-K., Hui, Y.,-Z., and Fei, X.-Z.,** The effect of hydrophobic-lipophilic interactions on chemical reactivity. VIII. Large-ring intramolecular excimer formation brought about by hydrophobic forces, *J. Am. Chem. Soc.,* 109, 5862, 1987.

98. **Xu, C.-B., Liu, Y.-Y., Shou, H.-S., and Tung, C.-H.,** Hydrophobic effects on photophysical and photochemical processes. IV. The conformation of polymethylene bis-2-naphthoates in poor solvents and their intramolecular excimer, *Acta Chim. Sin. Engl. Ed.,* p. 141, 1988.

99. **Chandross, E. A. and Dempster, C. J.,** Intramolecular excimer formation and fluorescence quenching in dinaphthylalkanes, *J. Am. Chem. Soc.,* 92, 3586, 1970.

100. **Xu, C.-B. and Tung, C.-H.,** Hydrophobic effects on photophysical and photochemical processes. V. The dimer formation of 1,3-dinaphthylpropanes in poor solvents, *Acta Chim. Sin.,* 46, 686, 1988.

101. **Irie, M., Kamijo, T., Aikawa, M., Takemura, T., and Hayashi, K.,** Absorption and fluorescence spectra of the intramolecular dimer in polyvinylnaphthalene, *J. Phys. Chem.,* 81, 1571, 1977.

102. **Xu, C.-B. and Tung, C.-H.,** Hydrophobic effects on photophysical and photochemical processes. New fluorescence from molecular aggregates of 1,4-di(2-naphthyl)cyclohexane, *J. Phys. Lett.,* 166, 128, 1990.

103. **Winnik, F. M., Winnik, M. A., Tazuke, S., and Ober, C. K.,** Synthesis and characterization of pyrene-labeled(hydroxypropyl)cellulose and its fluorescence in solution, *Macromolecules,* 20, 38, 1987.

104. **Char, K., Frank, C. W., Gast, A. P., and Tang, W. T.,** Hydrophobic attraction of pyrene-end-labeled poly(ethylene glycol) in water and water-methanol mixtures, *Macromolecules,* 20, 1833, 1987.

105. **Wilemski, G. and Fixman, M.,** General theory of diffusion-controlled reactions, *J. Chem. Phys.,* 58, 4009, 1973.

106. **Kalyanasundaram, K. and Thomas, J. K.,** Environmental effects on vibronic band intensities in pyrene monomer fluorescence and their application in studies of micellar systems, *J. Am. Chem. Soc.,* 99, 2039, 1977.

107. **Collin, P. J., Roberts, D. B., Sugowdz, G., Wells, D., and Sasse, W. H. F.,** The formation of photodimers from dimethyl naphthalene-1,8-dicarboxylate and methyl naphthalene-2-carboxylate, *Tetrahedron Lett.,* p. 321, 1972.

108. **Kowala, C., Sugowdz, G., Sasse, W. H. F., and Wunderlich, J. A.,** The X-ray analysis and molecular structure of the photodimer of methyl naphthalene-2-carboxylate, *Tetrahedron Lett.,* p. 4721, 1972.

109. **Teitei, T., Wells, D., and Sasse, W. H. F.,** Photochemical syntheses. X. Photodimers of derivatives of naphthalenecarboxylic acids, *Aust. J. Chem.,* 29, 1783, 1976.

110. **Lim, E. C.,** Molecular triplet excimers, *Acc. Chem. Res.,* 20, 8, 1987.

111. **Subudhi, P. C. and Lim, E. C.,** Phosphorescence from fluid solutions of dinaphthylpropane: evidence for a conformational difference between singlet and triplet excimers, *J. Chem. Phys.,* 63, 5491, 1975.

112. **Subudhi, P. C. and Lim, E. C.,** Formation and decay of triplet excimers as revealed by their absorption spectra: intramolecular excimers of 1,1'-dinaphthylalkanes, *Chem. Phys. Lett.,* 54, 59, 1978.

113. **Okajima, S., Subudhi, P. C., and Lim, E. C.,** Triplet excimer formation and phosphorescence in fluid solutions of diarylalkanes: excimer phosphorescence of dinaphthylalkanes and monomer phosphorescence of diphenylpropane, *J. Chem. Phys.,* 67, 4611, 1977.

114. **Takemura, T., Baba, H., and Shindo, Y.,** Excimer phosphorescence of naphthalene in fluid solution, *Chem. Lett.,* p. 1091, 1974.

115. **Takemura, T., Aikawa, M., Baba, H., and Shindo, Y.,** Kinetics study of triplet excimer formation in fluid solution by means of phosphorimetry, *J. Am. Chem. Soc.,* 98, 2205, 1976.

116. **Burkhart, R. D., Avilés, R. G., and Magrini, K.,** Triplet luminescence properties of poly(1-vinylnaphthalene) solid films, *Macromolecules,* 14, 91, 1981.

117. **Starzyk, F. C. and Burkhart, R. D.,** Triplet emission from poly(3,6-dibromo-N-vinylcarbazole): spectra and kinetics, *Macromolecules,* 22, 782, 1989.

118. **Kim, N. and Weber, S. E.,** Effect of molecular weight on triplet exciton processes. IV. Delayed emission of solid poly(2-vinylnaphthalene), *Macromolecules,* 13, 1233, 1980.

119. **Agosta, W. C.,** Transannular and interannular effects in 1,8-(1',8')-naphthalynaphthalene and related compounds, *J. Am. Chem. Soc.,* 89, 3505, 1967.

Chapter 4

PHOTOINITIATORS FOR FREE RADICAL POLYMERIZATION

E. A. Lissi and M. V. Encinas

TABLE OF CONTENTS

APPENDIX 1: SUMMARY OF ABBREVIATIONS

Ri	initiation rate
k_p	propagation-specific rate constant
k_t	termination-specific rate constant
M	monomer
Rp	propagation rate
Mn	number average polymer size
Φ	free radical M production quantum yield
I_a	absorbed light intensity
I_0	incoming light intensity
[A]	photoinitiator concentration
OD	optical density
τ	excited-state lifetime
S	solvent
Q	additive
P	product
Φ_j	quantum yield of the j process
α_j	average number of M radical produced per j process
MMA	methyl methacrylate
ABIN	α,α'-azo-bis-isobutyronitrile
DTBP	di-*tert*-butylperoxide
β	fraction of radicals that react in the primary solvent cage
TX	thioxanthone
AQ	anthraquinone
TEA	triethylamine

There is extensive literature on the mechanisms and rates of UV-visible radiation-mediated polymerization. The scope of this chapter is limited to free radical vinyl polymerization in homogeneous, fluid solutions. There are several reviews[1-4] and books[5,6] covering different aspects of this broad subject. The present chapter will make emphasis on the peculiar aspects of the radiation-induced processes that differentiate them from the thermally-initiated polymerization.

I. DEPENDENCE OF THE POLYMERIZATION RATE AND MEAN POLYMER MOLECULAR WEIGHT UPON THE RADIATION INTENSITY AND SAMPLE OPTICAL DENSITY

The rate and type of initiation process will determine all the characteristics of the polymerization: the polymerization rate, the average polymer molecular weight, the polymer weight distribution, the polymer end groups, and the produced polymer tacticity. We will be mostly concerned with the relationship between absorbed light intensity and polymerization rate. Free radical-mediated vinyl polymerization is a chain process whose minimal mechanism involves the following steps:

$$\text{Initiation systems} \rightarrow \text{M·} \qquad \text{Initiation } R_i \qquad (1)$$

$$\text{M·} + \text{M} \rightarrow \text{M·} \qquad \text{Propagation } k_p \qquad (2)$$

$$\text{M·} + \text{M·} \rightarrow \text{polymer} \qquad \text{Termination } k_t \qquad (3)$$

The initiation (Equation 1) comprises a series of elementary steps leading to the formation of a monomer-derived free radical, and will be characterized by its rate R_i. $M\cdot$ represents any monomeric or polymeric free radical with the odd free radical in the monomer. Propagation and termination steps are elementary processes characterized by their specific rate constants k_p and k_t, respectively.

This simplified mechanism, under steady-state conditions and assuming that propagation and termination rate constants are independent of the macroradical size, leads to a polymerization rate (Rp) given by

$$Rp = -d[M]/dt = (R_i/k_t)^{1/2}k_p[M] \tag{4}$$

and to a number average polymer size (Mn) given by

$$Mn = \sigma k_p[M]/(R_i k_t)^{1/2} \tag{5}$$

where σ is a factor between 1 and 2 that is determined by the combination to disproportionation ratio in process 3. The term photoinitiated polymerization applies to processes in which Reaction 1 is driven by light absorption. Under these conditions

$$R_i = \Phi I_a \tag{6}$$

where I_a is the absorbed light intensity (in Einsteins per unit volume and unit time), and Φ is the free radical $M\cdot$ production quantum yield. Equation 4 can be then expressed in terms of these parameters as

$$Rp = (\Phi I_a/k_t)^{1/2}k_p[M] \tag{7}$$

Equation 4 is valid for each differential volume dV of the reaction cell. If any of the parameters varies from one dV to another, the average polymerization rate will be given, in absence of mixing, by

$$Rp = (1/V)\int (Ri/k_t)^{1/2}k_p[M]dV \tag{8}$$

In processes initiated by dark reactions taking place in batch reactors, the parameters in Equation 4 can be considered as constant if significant temperature gradients do not occur. On the other hand, in any photochemical system the light intensity changes with the depth of penetration of the incoming radiation. Under these conditions, if mixing is not relevant (i.e., $M\cdot$ are not transported from one dV to another), the expression that gives the average polymerization rate is

$$Rp = (1/V)(\Phi/k_t)^{1/2}k_p[M]\int I_a^{1/2}\,dV \tag{9}$$

if the only inhomogeneity is due to light intensity gradients. Similarly, the expression for the average polymer size is, under the same conditions,

$$Mn = \sigma k_p(\Phi k_t)^{-1/2}[M]\int (I_a)^{1/2}\,dV \Big/ \int I_a\,dV \tag{10}$$

Inhomogeneity in I_a over the polymerization volume in homogeneous solution can be due to inhomogeneity in the incoming radiation beam and/or to high optical densities of the samples. For homogeneous light beams filling all the volume of a polymerization system of low optical density, I_a can be considered as constant and given by

$$I_a = \epsilon I_o[A] \tag{11}$$

where ϵ is the photoinitiator extinction coefficient, I_0 is the incoming light intensity (in Einstein per unit surface and unit time), and A stands for the photoinitiator concentration. Under these conditions, the average polymerization rate will be given by

$$Rp = (\Phi\epsilon I_o[A]/k_t)^{1/2}k_p[M] \tag{12}$$

and, in terms of the sample optical density (OD), if a cylindrical reaction vessel of large l parallel to the light beam is considered

$$Rp = (\Phi ODI_o)^{1/2}(lk_t)^{-1/2}k_p[M] \tag{13}$$

These equations predict that the polymerization rate must change with $\epsilon^{1/2}$, $I_o^{1/2}$, and $[A]^{1/2}$. Also, Equation 13 shows that, for a given reactor, the polymerization rate increases with $(OD)^{1/2}$, while, for different reactors at fixed OD, the polymerization rate decreases when the size of the reactor increases.

Photopolymerizations carried out in samples of finite optical densities must include the light inhomogeneity in the quantitative analysis of the obtained data. Kinetic expressions, both for the unstirred and well-stirred reactors, have been derived by Shultz and Joshi[7] and by Lissi et al.[8-10] The treatment of Shultz and Joshi did consider the influence of light absorption by components other than the photoinitiator (e.g., solvent, monomer) upon the initiator order as determined by conventional kinetics. The treatment of Lissi et al. only considers light absorption by the photoinitiator but includes its autoquenching in the general mechanism. If the initiation mechanism is reduced to

$$A + h\nu \rightarrow A^* \tag{14}$$

$$A^* \rightarrow A \tag{15}$$

$$A^* \rightarrow 2R\cdot \tag{16}$$

$$A^* + A \rightarrow \text{nonradical products} \tag{17}$$

$$R\cdot + M \rightarrow M\cdot \tag{18}$$

integration of Equations 9 and 10 leads, for the unstirred reactor, to

$$Rp = 2(\Phi I_o)^{1/2}(ODlk_t)^{-1/2}(1 + k_{17}\tau[A])^{-1/2}k_p[M]\{1 - \exp(-OD/2)\} \tag{19}$$

$$Mn = \sigma l^{1/2}(1 + k_{17}\tau[A])^{1/2}(\Phi ODI_o k_t)^{-1/2}k_p[M]\{1 - \exp(-\tfrac{1}{2}OD)\}/[1 - \exp(-OD)] \tag{20}$$

Equation 19 shows that, at a given photoinitiator concentration (and hence at a given optical density), the order in incoming light intensity is $\frac{1}{2}$, independently of the autoquenching extent and sample optical density. Deviations from the expected $\frac{1}{2}$ value for the order in light intensity cannot then be explained in terms of screening effects. Similarly, Equation 20 shows that the average molecular weight of the produced polymer changes as

$I_0^{-1/2}$ under any given set of conditions. Equations 19 and 20 also show that the autoquenching effect {taken into account by the factor $(1 + k_{17} \tau [A])$} is separable from that due to the optical density effect. Equation 19 shows that, due to the autoquenching effect, the poly merization rate order in photoinitiator concentration changes from zero (at low [A]) to $-1/2$ {when $(k_{17} \tau [A])$ is larger than one}. Similarly, from the autoquenching effect, Mn appears as independent of [A] at low photoinitiator concentrations, but increases with $[A]^{1/2}$ at large [A]. It is interesting to note then that the influence of the photoinitiator concentration upon Mn at large [A] is opposite in the autoquenching effect to that in the optical density effect. Furthermore, while the autoquenching effect is independent of the excitation wavelength, the absorbance effect is extremely sensitive to it. This difference can then be employed to ascertain the origin of anomalous dependencies of Rp with [A]. Similarly, it can be mentioned that, if the optical density effect is disregarded, inhibition due to degradative chain transfer to the photoinitiator leads, at high [A], to Rp independent of [A] and Mn values proportional to $[A]^{-1}$. These differences, as well as the independence of the degradative chain transfer effect on the employed wavelength, makes it possible to differentiate it from the absorbance and autoquenching effects. Lissi and Zanocco[8] have shown that the general optical density effect can explain anomalous data attributed to autoquenching, degradative chain transfer, or left unexplained.

Equations 19 and 20 show that if autoquenching and chain transfer to the initiator are neglected, the only factors that determine the polymerization rate and polymer molecular weight are Φ and the sample optical density. For a given photoinitiator, the expected order changes from 0.5 at low [A] to $-1/2$ at high [A]. The maximum polymerization rate is obtained at an optical density equal to 2.51, irrespective of I_0, 1, Φ, and monomer concentration. The polymer molecular weight changes as $[A]^{-1/2}$ both at low and high [A] values, but a smaller dependence with the initiator concentration is obtained at intermediate optical densities of the sample.

Two further conclusions can be derived from the above considerations. In the first place, and due to the spatial inhomogeneity in initiation rates and hence in steady-state macroradical concentrations, the polydispersity of the produced polymer increases with the sample optical density. Second, relative values of Φ, and hence information regarding the relative efficiency of a photoinitiator, can only be obtained at very low optical densities or working under matched conditions. Only under these conditions polymerization rates will be directly related to Φ values.

The above derived equations assume monochromatic light and disregard macroradical diffusion. If polychromatic light is employed, the only noticeable effect will be a less well-defined maximum at an optical density of 2.51, but the general features of the dependence of Rp and Mn on [A] will be maintained. With regard to macroradical molecular diffusion in unstirred reactors, a typical macroradical lifetime is about 10^{-2} s, and its diffusion coefficient can be taken as *circa* 10^{-6} cm²/s. The average displacement is then of the order of 10^{-4} cm and can be completely disregarded.

Different expressions can be employed to relate the average initiation rate to the photoinitiator concentration. Equation 21

$$Ri = \Phi I_o \epsilon [A] \tag{21}$$

can be employed only at very low optical densities of the polymerization sample. Equation 22

$$Ri = \Phi (I_o/l)\{1 - \exp(-OD)\} \tag{22}$$

can be employed, at any optical density, in well-stirred reactors. In unstirred reactors, the general expression is

TABLE 1
**Percentage of Error in the Evaluation of
Initiation Rates Due to Improper Data
Treatment**

Optical density	Equation 21	Equation 22
0.01	0.6	<0.001
0.1	5.3	0.02
0.2	10.6	0.08
0.5	28	2.0
1.0	61	8.0
2		19
4		32

From Lissi and Zanocco.[8]

$$Ri = 4\Phi I_o\{1 - \exp(-\tfrac{1}{2}OD)\}^2/(lOD) \tag{23}$$

The error introduced by application of Equation 21 or 22 to data obtained in unstirred reactors at different optical densities is given in Table 1. The correspondence of the experimental data to the equations derived for the unstirred[9] and well-stirred reactors[10] has been confirmed in the photopolymerization of methyl methacrylate by benzoin methyl ether irradiated at 366 nm.

II. GENERAL PHOTOINITIATION SCHEME

In the following sections the authors will analyze the dependence of Φ values with the structure of the photoinitiator and the effect of the monomer, solvent, and additives. Regarding these points, it is relevant to note that photoinitiation rates are considerably more sensitive to these factors than thermal processes. This higher sensitivity is mainly due to the significant changes that in the primary unimolecular photoprocesses can produce small modifications in the photoinitiator structure, and to the large number of bimolecular photoprocesses that can take place between the excited molecule and the solvent, the monomer, and/or the additives.

A general photoinitiation scheme must comprise the following processes:

$$A + h\nu \rightarrow {}^1A \tag{24}$$

$${}^1A \rightarrow P_{25} \tag{25}$$

$${}^1A \rightarrow A \tag{26}$$

$${}^1A \rightarrow {}^3A \tag{27}$$

$${}^1A + A \rightarrow P_{28} \tag{28}$$

$${}^1A + S \rightarrow P_{29} \tag{29}$$

$${}^1A + M \rightarrow P_{30} \tag{30}$$

$${}^1A + Q \rightarrow P_{31} \tag{31}$$

$$^3A \rightarrow A \tag{32}$$

$$^3A \rightarrow P_{33} \tag{33}$$

$$^3A + A \rightarrow P_{34} \tag{34}$$

$$^3A + S \rightarrow P_{35} \tag{35}$$

$$^3A + M \rightarrow P_{36} \tag{36}$$

$$^3A + Q \rightarrow P_{37} \tag{37}$$

where S = solvent, M = monomer, Q = additive, and P represents one or several reaction products generated in one (or several) reaction paths that can involve either one or several reversible and/or irreversible steps, and Equations 26 and 32 include all radiative and non-radiative unimolecular deactivation processes. The value of Φ will be given by

$$\Phi = \sum_j \Phi_j \alpha_j \tag{38}$$

where the summatory is over all the processes leading to P, Φ_j measures the j process quantum yield, and α_j measures the average number of $M\cdot$ radicals produced per j process. It has to be considered that, for a given A, all the Φ and α values could be dependent on the temperature, wavelength, solvent, and monomer considered.

Φ values can be determined from Rp measurement if ϵ, I_0, k_p, and k_t [or $k_p/(k_t)^{1/2}$] are known:

$$\Phi = \frac{Rp^2}{(k_p/k_t^{1/2})^2[M]^2 I_a} \tag{39}$$

Alternatively, Φ values can be obtained from Rp and Mn measurements:

$$\Phi = \frac{\Phi_m}{\sigma Mn} \tag{40}$$

where Φ_m is the number of monomer units polymerized per quantum absorbed (Rp/I_a).

Equation 39 can be applied in the absence of termination by primary radicals and degradative chain transfer. Application of Equation 40 requires the knowledge of σ and the absence of degradative and nondegradative chain transfer. Furthermore, as written, Equations 39 and 40 can only be applied at very low optical densities. At finite optical densities, in unstirrer reactors, Equations 19 and 20 must be applied to obtain Φ values from Rp measurements.

The advantage of these procedures is that they lead to "absolute" Φ values. Relative Φ values can be simply evaluated if a reference photoinitiator of Φ_{ref} is employed:[11]

$$\Phi = \Phi_{ref} \frac{Rp^2 - Rp_o^2}{(Rp)_{ref}^2 - Rp_o^2} \tag{41}$$

where Rp and $(Rp)_{ref}$ are the measured polymerization rates under matched (i.e., equal optical densities) conditions for the photoinitiator considered and the reference compound,

TABLE 2
Values of $(k_p/k_t^{1/2})$ and Bulk Monomer
Concentrations

Monomer	$k_p/k_t^{1/2}$ $(M\,s)^{-1/2}$	[M] (M)
Methyl methacrylate	0.055	9.46
Styrene	0.0102	8.72
Vinyl acetate	0.0317	10.86
Methylacrylate	0.34	11.1
Acrylonitrile	0.059	15.2

and Rp_0 is the dark polymerization rate. The advantage of Equation 41 is that it applies independent of the optical density of the samples and the homogeneity of the light beam. Furthermore, its application does not require determination of the light intensity at the absorbed wavelength. However, it can only be applied for monochromatic light or when the compound considered and the reference compound present absorption bands of similar shape. Furthermore, it cannot be applied if either the initiator employed or the reference initiator enter in degradative chain transfer processes or in the presence of primary termination.

For different monomers polymerized under matched conditions, employing a given photoinitiator, relative initiation efficiencies (Φ_I and Φ_{II}) are related to the observed polymerization rates through

$$\frac{\Phi_I}{\Phi_{II}} = \frac{(Rp^2 - Rp_o^2)_I}{(Rp^2 - Rp_o^2)_{II}} \times \frac{\{[M](k_p/k_t^{1/2})\}_{II}^2}{\{[M](k_p/k_t^{1/2})\}_I^2} \tag{42}$$

in absence of primary termination and degradative chain transfer processes. Values of $(k_p/k_t^{1/2})$ and [M] in bulk monomer, for the most frequently employed monomers, are collected in Table 2.

III. UNIMOLECULAR PHOTOINITIATIONS

Photoinitiation processes can be divided in *unimoleculars,* where the production of free radicals only involves the photoinitiator (Processes 25 and 33) and *bimoleculars,* where the free radical production results from the interaction between an excited molecule and one (or several) ground-state molecule (processes 28 to 31 and 34 to 37). These bimolecular processes can represent a simple elementary step and/or take place by complex mechanisms.

Molecules whose unimolecular photoreactions lead to the production of free radicals can act as photoinitiators through the occurrence of Reactions 25 and/or 33. For these photoinitiators, and in the absence of other free radical forming processes,

$$\Phi = \alpha_{25}\Phi_{25} + \alpha_{33}\Phi_{33} \tag{43}$$

If a new reaction channel is introduced, i.e., by addition of Q, it would increase or decrease the value of Φ depending upon the relative α values of the processes whose Φ values are reduced, and those of the new reaction paths. For example, it has been reported that, in the polymerization of methyl methacrylate, addition of the triethylamine increases the polymerization photoinitiated by biacetyl,[12] but reduces it when it is initiated by α,α-dimethoxy phenylacetophenone.[13]

Unimolecular bond breaking processes following the photon absorption can produce two free radicals (i.e., in simple photocleavages) or a biradical (intramolecular atom transfer or cleavage of cyclic compounds). In the next section we shall discuss photoinitiations following bond cleavages that lead to two free radicals. Unimolecular photoprocesses leading to biradicals will be described in a later section.

A. UNIMOLECULAR PHOTOCLEAVAGES

Simple photocleavages leading to a free radical pair can be represented by

$$R - R' + h\nu \rightarrow R\cdot + R'\cdot \tag{44}$$

or

$$R - X - R' + h\nu \rightarrow R\cdot + X + R'\cdot \tag{45}$$

In the first process, the bond cleavage leads directly to the formation of two free radicals. In the second process, expulsion of a molecule takes place simultaneously (or in a time scale faster than that of the reactions of the primary free radicals) to the primary bond cleavage.

Among the compounds that photoinitiated by the first type of cleavage stand the halogenated organic compounds, simple peroxides, and carbonyl compounds. Among those in which one (or more) neutral molecule is expulsed "simultaneously" to the photochemical cleavage stand the azo-compounds and some peroxidated compounds (i.e., peresters).

1. Halogenated Organic Compounds

Halogens can photocleave when photolyzed in solution, and the photopolymerization of methyl methacrylate (MMA) has been studied using both bromine[14] and iodine[15] as photoinitiators. However, the multiple role that these compounds can play both as chain transfer and degradative chain transfer agents[16] leads to polymer of low molecular weight and to very complex polymerization kinetics.

Carbon-halogen bonds are generally broken after light absorption both in simple halogenated alkanes

$$-CH_2-CH_2-X + h\nu \rightarrow -CH_2-CH_2\cdot + X\cdot \tag{46}$$

and in compounds where the halogen atom is located in position α to a chromophore

$$A-CH_2-X + h\nu \rightarrow A-CH_2\cdot + X\cdot \tag{47}$$

In the first case (Equation 46) the excitation is $p\sigma^*$ type, localized almost totally in the carbon-halogen bond. Efficient cleavage does occur but the wavelengths of the transition (λ_{max} = 203 and 258 nm for X = Br and I, respectively) render these processes of very limited applicability as free radical sources in vinyl photopolymerization. On the other hand, processes such as those represented by Equation 47 can take place at wavelengths determined mainly by the chromophore A, and have been further investigated.[17-20] In particular, both for aliphatic and aromatic ketones, processes such as

$$CH_3-CO-CH_2X + h\nu \rightarrow CH_3-CO-CH_2\cdot + X\cdot \tag{48}$$

and

$$Ph-CO-CH_2X + h\nu \rightarrow Ph-CO-CH_2\cdot + X\cdot \tag{49}$$

TABLE 3
Photoinitiation Efficiencies of α-Halogenated
Carbonyl Compounds

Compound	Monomer	Φ	Reference photoinitiator[a]
Chloroacetone	MMA	0.68	DTBP
	Styrene	0.44	DTBP
Bromoacetone	MMA	0.48	DTBP
	Styrene	0.16	DTBP
α-Bromoacetophenone	MMA	0.29	AIBN
	Styrene	0.15	AIBN

[a]   DTBP = di-*tert*-butyl peroxide; AIBN = azobisisobutyronitrilo.

From Encinas et al.[22]

occur with high quantum yields from short lifetime excited states,[21-24] leading to relatively high Φ values (see Table 3). The lower values of Φ measured for styrene can then be interpreted in terms of triplet quenching by the monomer (Process 36) taking place with low α_{36} values. For these ketones, it has been suggested that the photocleavage must, at least partially, take place from the excited triplet. For example, McGimpsey and Scaiano[24] have reported that photocleavage of α-bromoacetophenone occurs with a quantum yield of 0.41, from a short-lived (≤ 0.1 ns) triplet.

Processes such as Reactions 48 and 49 can take place with high yields if the energy of the excited state is higher than the bond cleavage endothermicity. Low Φ values obtained for 1,4-dibromo-2,3-butanedione have been explained in terms of the unfavorable energetics of the process (triplet energy ≈ 56 kcal; bond dissociation energy ≈ 67 kcal).[22]

2. Hydroperoxides, Peroxides, and Disulfides

Early data (up to 1966) on photoinitiation by peroxides, hydroperoxides, and disulfides have been reviewed by Oster and Yang.[1] Simple peroxides and hydroperoxides of the general formula $R\text{–}O\text{–}O\text{–}R'$ (R and R' = H or alkyl group) show absorptions of the $n\sigma^*$ type leading to repulsive states.[25] The absorption takes place in the 200 to 300 nm region with increasing absorption coefficients at lower wavelengths. Being the excited state repulsive, excited molecules cleave the O–O bond

$$R\text{–}O\text{–}O\text{–}R' + h\nu \rightarrow RO\cdot + \cdot OR' \tag{50}$$

with nearly unitary efficiency[26] almost independently of the media. Similar high photocleavage yields can be expected also by energy transfer processes from sensitizers that absorb at longer wavelengths.[27-30] This efficient apparently endothermic energy transfer has been explained in terms of the repulsive nature of the peroxide first excited state. Also, it has been proposed that in the benzophenone/hydrogen peroxide-induced photopolymerization, the initiation process involves the photoreduction of the triplet carbonyl by hydrogen peroxide,[31] according to

$$Ph_2CO + H_2O_2 \rightarrow Ph_2\dot{C}OH + HO_2\cdot \tag{51}$$

The radicals produced in the peroxide bond cleavage, either hydroxyl and/or alkoxy radicals, are reactive species that can interact with most monomers, either by hydrogen abstraction and/or double bond addition.[32-35] High photocleavage yields even in the presence of monomer ($\Phi_{25} + \Phi_{33} \approx 1$), together with the high reactivity of the produced radicals (that in nonviscous solvents, where cage recombinations are minimized, imply that $\alpha_{25} \approx \alpha_{33} \approx 1$), must lead to very high initiation efficiencies for these types of compounds. However, there are very few quantitative data regarding the initiation efficiencies of peroxides and hydroperoxides. For the later compounds, the data treatment can be complicated by degradative and nondegradative chain transfer processes,[36] that for hydrogen peroxide have been reported to be pH dependent.[37]

Ghosh et al.[38] have studied the photopolymerization of MMA at 30°C employing hydrogen peroxide as photoinitiator. A normal kinetic behavior was observed in bulk solution, while in diluted solution the kinetic law was extremely dependent upon the employed solvent, the observed orders in monomer ranging from 0.9 (chloroform as solvent) to 2.7 (benzene as solvent). Similarly, the order in initiator was found to range from 0.44 (chloroform) to nearly zero (benzene). This anomalous behavior was interpreted in terms of a strongly solvent-dependent degradative transfer to the initiator.

Ho et al. have employed hydrogen peroxide as photoinitiator of isoprene polymerization both under steady[39] and intermittent[40] illumination. Under steady-state conditions, the polymerization produces hydroxyl-terminated polymer with a rate law given by

$$Rp = k[M]^{0.94}[H_2O_2]^{0.43} \qquad (52)$$

in acetone and tetrahydrofurane solutions. This rate law was explained in terms of a simple mechanism including quenching by the monomer to explain the 0.94 order in monomer. Efficient polymerization was observed even under oxygen, as a result that, at least for the initiation step, can be explained in terms of the short lifetime of the excited hydrogen peroxide and the high reactivity toward the monomer of the initially produced hydroxyl radicals.

A styrene-oxygen copolymer has been employed as photoinitiator of styrene polymerization.[41] The value of Φ was 0.66 times that obtained by *tert*-butyl peroxide. The lower value indicates an increased cage recombination for the macroradicals.

Photopolymerizations initiated by peroxy bond cleavage sensitized by a chromophore absorbing at longer wavelengths have been mostly carried out by introducing both the peroxide and the chromophore in the same compound. These compounds comprise a *tert*-butyl perester and a polycyclic aromatic[42] or ketonic[30,43-45] chromophore as sensitizer. The light-absorbing group is coupled to the *tert*-butyl perester either directly or through a spacer of variable length, and the energy initially localized in the chromophore is tranferred to the perester leading to the peroxide bond cleavage. The process can be represented by

$$\text{Chrom-spac-CO–O–O-}\textit{tert}\text{-Bu} + h\nu \rightarrow \text{Chrom-spac-CO–O}\cdot + \textit{tert}\text{-BuO}\cdot \qquad (53)$$

Peresters attached to benzophenone chromophores of general formula

with R = H, CH$_3$, or CH$_3$O decompose with quantum yield near unity in both polar and nonpolar solvents when irradiated at the wavelength of the carbonyl group absorption, giving benzoyloxy and *tert*-butoxy radicals.[30] The benzoyloxy radical undergoes photodecarboxylation or abstracts a hydrogen atom from the solvent. Photodecomposition quantum yields of various peresters are collected in Table 4. The decomposition rates of the peresters are

TABLE 4
Photopolymerization Rates of MMA and Styrene Photoinitiated by Benzophenone Peresters[a]

Photoinitiator	[Initiator] ($M \times 10^3$)	Monomer or solvent	[Monomer] (M)	Φ_{33}	$Rp \times 10^5$ (M/s)	Ref.
I	20	Bz		0.94		30
	20	Styrene	8.73	Very low		44
	20	Styrene	4.37	0.12		44
	20	Styrene	3.28	0.14		44
	20	Styrene	2.19	0.20		44
	20	Styrene	1.09	0.25		44
	20	MMA	4.68	0.80		44
	20	MMA	2.34	0.88		44
	20	MMA	Bulk		70	44
	0.3	Styrene	3.5	0.14	0.65	44
	0.6	Styrene	3.5	0.14	0.92	44
	1.1	Styrene	3.5	0.14	1.29	44
	2.5	Styrene	3.5	0.14	1.47	44
	8	Styrene	3.5	0.14	1.83	44
	10	Styrene	2.0	0.25	0.95	44
	10	Styrene	3.5	0.14	1.50	44
	10	Styrene	5.0	0.12	2.05	44
II	2.4	Bz		1.20		43
	2.4	MMA	Bulk		144	43
	2.4	MMA	6.0		57.5	43
	2.4	MMA	3.0		44.5	43
	2.0	Styrene	6.0		2.7	43
	2.0	Styrene	3.0		4.2	43
III	2.4	MMA	Bulk		79	43

	Structure					
IV	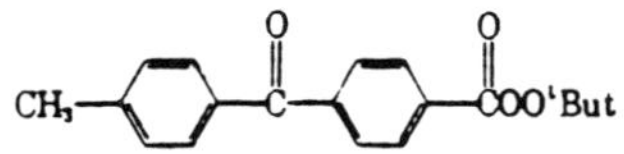 CH$_3$—C$_6$H$_4$—CO—C$_6$H$_4$—COO^{t}But	2.4	Bz		0.80		43
		2.4	MMA	Bulk		63	43
V	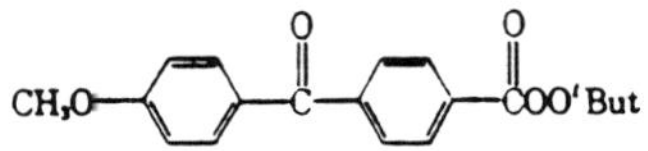 CH$_3$O—C$_6$H$_4$—CO—C$_6$H$_4$—COO^{t}But	2.4	Bz		0.75		43
		2.4	MMA	Bulk		64	43
VI	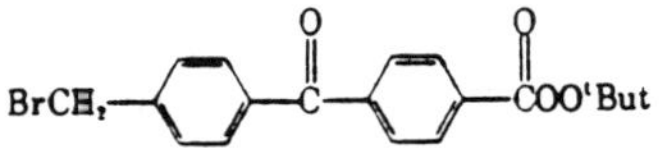 BrCH$_2$—C$_6$H$_4$—CO—C$_6$H$_4$—COO^{t}But	2.4	MMA	Bulk		9	43

a $\gamma_{irradiation}$ = 366 nm.

in the order of 1.5×10^4 s^{-1}. Product quenching studies suggest that the triplet lifetimes of the peresters are in the 1- to 3-ns range.[30] The observed decomposition rates will then be determined mainly by the intramolecular energy transfer from the initially excited carbonyl group to the O–O linkage.

A series of benzophenone perester derivatives have been employed as photoinitiators.[43,44] In agreement with the proposed cleavage, all the obtained polymers present benzophenone groups linked to the polymer ends. The data given in Table 4 show that the perester decomposition is reduced by the monomer, this effect being much more important in the presence of styrene than when MMA is employed. This result has been interpreted in terms of benzophenone-localized excitation quenching by the monomers. Indeed, benzophenone triplets are quenched by styrene with a rate constant of 3.3×10^9 M^{-1} s^{-1},[46] the process taking place through energy transfer from the $n\pi^*$ benzophenone triplet ($E_T = 67.5$ kcal) to styrene ($E_T = 62$ kcal). Data of Table 4 also show that the polymerization of styrene initiated by each of the benzophenone peresters is much slower than the polymerization of MMA. Order in monomers is considerably smaller than one, and in styrene polymerization photoinitiated by the diperester, the polymerization rate decreases with the monomer concentration. These results have been explained as due to triplet quenching by the monomer.[43] However, simple quenching by the monomer cannot explain orders in monomer below 0.5, if a propagation rate proportional to the monomer concentration holds. Quenching by MMA is less efficient, leading to higher polymerization rates for this monomer than for styrene.[43,44] Furthermore, the MMA polymerization rate always increases with the monomer concentration, although orders as low as 0.43 were reported.[43]

For all the peresters considered, the order in photoinitiator is 0.5 at low perester concentrations. At higher photoinitiator concentrations, the polymerization rate goes through a maximum.[43,44] This behavior, together with the decrease observed in the average polymerization degree, can be attributed to the optical density effect discussed previously.

The polymer p-(p'-vinylbenzoyl)peroxybenzoic acid t-butylester (VII)

$$\text{+CH-CH}_2\text{+}_n$$

C=O

C=O

O

O

H$_3$C–C–CH$_3$

CH$_3$

VII

is an efficient radical source when irradiated at the absorption band of the benzophenone carbonyl group.[47] The decomposition quantum yield of the polymeric perester is only 0.66 times that of the model compound p-(p'-methylbenzoyl)tert-butyl benzoate (IV), a result

TABLE 5
Photopolymerization of Vinyl Monomers
Initiated by Peresters[a]

Initiator		Rp × 10⁴ (M/s)	
		MMA	Styrene
VIII	1-pyrene-CO-⟨⟩-CO-O-O-t-Bu	1.54	0.244
IX	1-pyrene-CO-O-CH₂-⟨⟩-CO-O-O-t-Bu	1.1	0.195
X	9-fluorenone-4-CO-O-CH₂-⟨⟩-CO-O-O-t-Bu	1.92	0.098
XI	9-anthracene-CO-O-CH₂-⟨⟩-CO-O-O-t-Bu	0.56	0.122

[a] In bulk monomer.

From Abu-Abdoun et al.[42]

that can be ascribed, at least partially, to an increase in cage recombination of the primary radicals.[47] However, the polymerization of styrene induced by (VII) is higher than that induced by the model compound (IV) and shows autoacceleration at lower conversions with formation of insoluble, cross-linked polymer.[47] The higher rate and early autoacceleration is due to hindered termination by the macroradicals with reduced mobility in the polymer gel. This implies that chains initiated by the macroradical will grow by longer times than those initiated by the small *tert*-butoxy radical.

Copolymers of styrene with low percentages of (VII) decompose with high efficiency, generating radicals that behave as those produced from typical perester photoinitiators.[48] In absence of monomers the photodecomposition gives cross-linked polymers. In the presence of monomers, such as MMA, graft copolymers are produced.

Abu-Abdoun et al.[42] have reported the photoinitiation by *tert*-butyl peresters whose activity derives from the presence of polycyclic aromatic hydrocarbons or fluorenone. The structures of the compounds employed and the photopolymerization rates obtained are given in Table 5. For some of these compounds, particularly compounds VIII and X, the dependence of the polymerization rate with the monomer concentration was indicative of significant quenching of their excited states by styrene.

Allen et al.[45] have reported the photopolymerization activity of a series of *t*-butylperester derivatives bearing a fluorenone group as chromophore (Table 6). The lowest excited singlet state of fluorenone is $\pi\pi^*$ in nature.[49] Substitution of 9-fluorenone notably increases the fluorescence quantum yields. This result, together with the blue shift and increased extinction coefficients observed in the absorption spectra, has been explained in terms of increasng rigidity of the molecule. While 9-fluorenone itself is a very poor photoinitiator of MMA polymerization, the three derivatives considered are efficient photoinitiators, the polymers produced bearing intact 9-fluorenone groups. High photocleavages of peresters leading to high photoinitiation efficiencies are in agreement with the high photopolymerization rates obtained even in the presence of isopropanol or triethylamine, compounds that are efficient photoreductors of the fluorenone chromophore.

Aromatic peroxides can behave widely different than the aliphatic peroxides previously considered. For example, ϕ values lower than 10^{-3} have been reported for the polymerization of acrylonitrile, MMA, styrene, and vinyl acetate sensitized by di-benzoyl peroxide.[50] Higher efficiencies have been reported for benzoyl peroxide-initiated polymerization in presence of different additives such as triethylamine,[51] cetyl pyridinium bromide,[52] cetyl trimethyl am-

TABLE 6
Fluorescence Quantum Yields (Φ_F), Photoreduction Quantum Yield (Φ_{red}), and Polymerization Rates of Fluorenone Derivatives[a]

Compound	$\Phi_F \times 10^4$	Φ_{red}	$Rp^b \times 10^4$ (*M*/s)
XII 9-Fluorenone	1.6	0	0
XIII 4-*t*-Butylperoxycarbonyl-7-nitro-9-fluorenone	0.64	—	5.33
XIV 2-*t*-Butylperoxycarbonyl-9-fluorenone	2.6	0.245	5.23
XV 2,7-Di-*t*-butylperoxycarbonyl-9-fluorenone	6.6	0.479	5.16

[a] In 2-propanol as solvent.
[b] Bulk MMA, 0.25% w/w of photoinitiators.

From Allen et al.[45]

monium bromide,[53] and Michler's ketone.[54] However, the polymerization kinetics presents strong nonidealities and is extremely solvent dependent. Active radicals are considered to be produced by photolysis of benzoyl peroxide/additive complexes.

Simple (dibenzyl disulfide [XVI]) and strained (4-[1,2-dithiolan-3-yl]valeramide [XVII]) disulfides, dibenzoyl disulfides (XVIII), and more complex disulfides (bis[isopropyl-xantogen]disulfide [XIX]) initiated polymerization by S–S bond cleavage.[55-57]

$$\text{Ph–CH}_2\text{–S–S–CH}_2\text{–Ph} + h\nu \rightarrow 2\text{Ph–CH}_2\text{–S}\cdot \qquad (54)$$

(XVI)

(55)

XVII

$$\text{R–Ph–CO–S–S–CO–Ph–R} + h\nu \rightarrow 2\text{R–Ph–CO–S}\cdot \qquad (56)$$

(XVIII) R = Cl, Br, CH_3, OCH_3, NO_2, CN

$$\begin{array}{ccccccccccc} CH_3 & & & & & & CH_3 & & & CH_3 & \\ | & & & & & & | & & & | & \\ CH & –O– & C & –S–S– & C & –O– & CH & + h\nu \rightarrow & 2 & CH–O– & C–S\cdot \\ | & & \| & & \| & & | & & & | & \| \\ CH_3 & & S & & S & & CH_3 & & & CH_3 & S \end{array} \qquad (57)$$

XIX

Compounds XVI and XVIII were shown both to initiate and terminate MMA and styrene polymerization.[55] In the photopolymerization initiated by compound XVII, the diradical formed by its ring opening through homolytic fission was found to attack styrene producing a copolymer with high content of monomeric units of XVII.[56] Compound XIX was shown to possess interesting properties for the design of block copolymers, i.e., it acts as a terminator and a chain transfer agent as well as a photoinitiator. The obtained polymers contain two

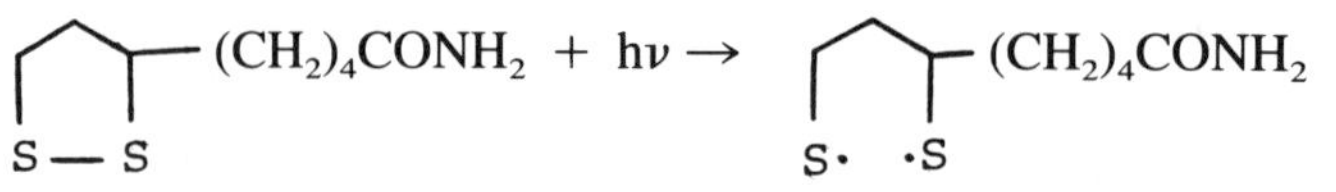

reactive isopropyl xanthate groups bonded at their chain ends, which could act as macro-photoinitiators.[57]

Photolysis of disulfides is considerably less well understood than that of peroxides. Also, the reactivity of sulfur-centered radicals is considerably smaller than that of alkoxy radicals, rendering primary termination processes much more likely. Values of $(\Phi_{25} + \Phi_{33})$, as well as α_{25} and α_{33}, are going to be more dependent upon the sensitizer structure and monomer employed.

For compound (XIX) it has been reported that the polymerization efficiency increases, in homogeneous solutions, in the following order:[57] methyl acrylate > MMA $\gg$ vinyl acetate > styrene. These differences are due to degradative chain transfer to the initiator, termination by primary radicals, and to monomer-sensitive α values. In agreement with these considerations it was reported that Φ values notably increase when the monomer concentration increases.

3. Azo Compounds

The azo group ($-N=N-$) presents a $n\pi^*$ band of relatively low molar absorptivity in the 350 nm region. Whereas arylazo compounds (i.e., azobenzene) are stable, alkyl azo compounds are dissociated by light to give alkyl-free radicals. Photocleavage for small alkyl azo compounds take place with high quantum yields in gas phase, where azo compounds have been extensively employed as free radical sources.[58] In solution, smaller free radical quantum yields are obtained and the photochemistry becomes more complex.[58,59] Simple alkyl azo compounds have been only employed as photoinitiators in few cases. Among larger azocompounds, α,α'-azo-bis-isobutyronitrile (ABIN) is the most commonly employed. For this compound, a photodecomposition quantum yield of 0.4 has been reported[60] at 25°C, according to

$$
\begin{array}{ccc}
\text{CN} & \text{CN} & \text{CN} \\
| & | & | \\
\text{CH}_3\text{--C--N=N--C--CH}_3 + h\nu \rightarrow & & 2\text{CH}_3\text{--C}\cdot + \text{N}_2 \\
| & | & | \\
\text{CH}_3 & \text{CH}_3 & \text{CH}_3
\end{array}
\qquad (58)
$$

where the second N–C bond is ruptured simultaneously in a very short time scale. The rather high initiation efficiency of the initiator can be related to the formation of the resonance-stabilized cyanoalkyl radicals. The same considerations apply to the use of the water-soluble compound 2,2′-azo-bis(2-amidino)propane hydrochloride

$$
\begin{array}{ccc}
\text{CH}_3 & \text{CH}_3 & \text{CH}_3 \\
| & | & | \\
\text{CH}_3\text{--C--N}=\!=\text{N--C--CH}_3 + h\nu \rightarrow & & 2\text{CH}_3\text{--C}\cdot + \text{N}_2 \\
| & | & | \\
\text{C} & \text{C} & \text{C} \\
/\!/+\backslash & /\!/+\backslash & /\!/+\backslash \\
\text{NH}_2\ \ \text{NH}_2 & \text{NH}_2\ \ \text{NH}_2 & \text{NH}_2\ \ \text{NH}_2 \\
\text{Cl}^- & \text{Cl}^- & \text{Cl}^-
\end{array}
\qquad (59)
$$

Simple aromatic azo compounds such as azobenzene do not photocleave and hence are very poor photoinitiators. However, Nuyken et al.[61,62] have reported the use of several monomeric and polymeric azophenyl derivatives as photoinitiators. 1-Methyl-1-(3-vinyl-phenylazo)ethyl acetate, diphenyl(3-vinylphenylazo)methyl acetate, and α-(phenylazo)-α-(4-vinylphenyl)benzyl acetate were prepared and copolymerized with styrene. The resulting copolymers were used as photoinitiators for the polymerization of methacrylonitrile. Graft copolymer and polymethacrylonitrile were obtained, showing that both the macro- and the small radical produced by cleavage of the azo-containing polymer were able to initiate the methacrylonitrile-free radical polymerization.

Azo compounds have the advantage that they absorb at relatively long wavelengths, have very little or no chain transfer capacities, low induded decompositions, and a decomposition rate very little sensitive to the solvent or the presence of additives. For AIBN it has been reported that Φ values in the polymerization of MMA and styrene are almost independent of carbon tetrachloride, triethylamine, and diethylamine addition, compounds that notably modify the initiation efficiency of different carbonyl compounds.[12]

B. CARBONYL COMPOUNDS

Simple organic carbonyl compounds can photocleave by the process known as Norrish type I that leads to the formation of an alkyl and an acyl free radical

$$R\text{–}CO\text{–}R' + h\nu \rightarrow R\text{–}CO\cdot + R'\cdot \qquad (60)$$

where R is an hydrogen atom (aliphatic aldehydes), an alkyl group (aliphatic ketones), or an aryl group (acetophenone derivatives) and R' is an aliphatic carbon-centered radical. Both for aliphatic[63] and aromatic[64] ketones, as well as for the aldehydes,[65] the rate of the type I photocleavage step is determined by the substitution at the α carbon.

1. Aliphatic Carbonyl Compounds

Aliphatic ketones and aldehydes present a $n\pi^*$ lower excited state. The photochemistry of these compounds, particularly the type I photocleavage, can occur both from the singlet and triplet states.[63,65] If the polymerization is initiated by the radicals produced by this type of process, the expected initiation efficiencies will be determined by the photocleavage quantum yield in absence of quenching, and by the capacity of the monomer (or other added quenchers) to deactivate the excited states. The first factor is determined by the degree of α substitution, and ranges from nearly zero in acetone, to nearly one in di-*tert*-butylketone.[63] The second factor, the extent of excited state quenching by the monomer, will be determined by the monomer and the excited-state lifetimes. All these factors have been extensively discussed.[11,12,66] The data obtained for the aliphatic monocarbonyl compounds that initiate by a type I cleavage are collected in Table 7. In absence of additives and significant self-quenching,

$$\Phi = \alpha_{25}\Phi_{25} + \alpha_{30}\Phi_{30} + \alpha_{33}\Phi_{33} + \alpha_{36}\Phi_{36} \qquad (61)$$

The monomer can quench both the singlet and the triplet excited states. The quenching process will be more relevant for long-lived carbonyl compounds and when styrene is employed as monomer. However, the fact that for long-lived excited states, where Φ_{30} and Φ_{36} must have the higher values, Φ values are very small, indicates that $\alpha_{30} \approx \alpha_{36} \approx 0$, both for MMA and styrene. The results of Table 7 show that increasing α substitution increases the initiation efficiency. For these ketones, α_{25} and α_{33} can be considered to be equal to $(1-\beta)$, where β is the fraction of radicals reacting in the primary solvent cage, and Φ_{25} and Φ_{33} are determined by the rate of the type I photocleavage and the rate of quenching by the monomer. For all the ketones considered, Φ values for MMA are larger than those found for styrene, the difference being larger when the initiator efficiency decreases. For example, the ratio between the initiation efficiencies for MMA and styrene is nearly 15 for

TABLE 7
Initiation Efficiencies in the Polymerization Sensitized by Alkyl Carbonyl Compounds

Compound	τ_T (ns)[a]		Φ_{MMA}[b]	$\Phi_{styrene}$[b]
Acetone	1.4×10^4	(67)	$\approx 0.0016^c$	
3-Pentanone	136	(67)	0.006^c	
2,4-Dimethyl-3-pentanone	8	(68)	0.22^c	
3,3-Dimethyl-2-butanone	2	(68)	0.70	0.24
2,2,4,4-Tetramethyl-3-pentanone	<1	(69)	1.10^c	0.36
Propanal	>250	(65)	0.016	≈ 0.0002
2-Methylpropanal	2	(65)	0.014	≈ 0.0014
3-Hydroxy-3-methyl-2-butanone	≤ 7	(70)	1.30	0.80
3-Hydroxy-2-butanone	≤ 25	(70)	0.66	—
Methoxy acetone			0.042	0.0028
Phenylacetone			0.30	0.10
1,1-Diphenyl-acetone			0.04	0.030
1,3-Diphenyl acetone	≈ 0.1	(71)	0.22	0.04

[a] Number given in parenthesis indicates the reference.
[b] In bulk, from Encinas et al.,[22] DTBP employed as reference.
[c] From Lissi et al.[11]

TABLE 8
Quenching of Singlet and Triplet Excited States of 2-Heptanone by Monomers[a]

Monomer	$k_{30} \times 10^{-7}$ $(M^{-1}\,s^{-1})$	$k_{36} \times 10^{-8}$ $(M^{-1}\,s^{-1})$
Styrene	3.9	40
Vinyl toluene		38
α-Methylstyrene	3.9	14
Butyl methacrylate		8
Methyl methacrylate	2.9	7
Acrylonitrile	8.6	5
Vinyl acetate	0.9	0.18^b
Allyl chloride		0.1^b

[a] In benzene as solvent.
[b] Values obtained with 4-methyl-2-pentanone.

Data from Encinas and Lissi.[66]

methoxy-acetone and only 1.6 for 3-hydroxy-3-methyl-2-butanone. In the latter compound, quenching by the monomer is minimized by the short lifetime associated to the α-hydroxyl substitution.[70]

Quenching of singlet and triplet excited states of aliphatic carbonyl by several monomers has been reported by Encinas and Lissi,[66] and the data are collected in Table 8. With regard to singlet quenching, the data show that the faster quencher of the monomers considered is acrylonitrile, supporting a mechanism dominated by charge transfer interactions. The fraction of singlets quenched in bulk monomer solutions ranges from 0.11 for vinyl acetate to 0.54 when the photolysis is carried out in acrylonitrile. Triplet quenching can take place by energy transfer and/or charge transfer complex formation. The data indicate that the main factor

TABLE 9
Photoinitiation Efficiencies of Various
Monomers Polymerization Sensitized by 2,4-
Dimethyl-3-Pentanone

Monomer	Φ^a	k_{36} [M], 1/s^b
Styrene	0.22	35
Acrylonitrile	0.24	7.6
Methyl methacrylate	0.22	6.6
Butyl methacrylate	0.50	5
Vinyl acetate	0.96	0.3
Vinyl acetate + *n*-hexane	1.84	0.17
	2.0	0.097

[a]   Initiation efficiencies relative to DTBP.
[b]   For the triplet quenching of 2-heptanone.

Data from Encinas and Lissi.[66]

that determines the value of the rate constant for the interaction between the carbonyl triplet and an olefin is the triplet energy of the latter compound. The nearly diffusional quenching observed for the styrene derivatives can be explained by the exothermicity of the triplet energy transfer. For olefins with triplet energies higher than those of the excited donor the main factors that control the reaction rate are the triplet energy difference, the charge density at the olefinic double bond, the electronegativity of the excited triplet, the polarity-polarizability of the solvent, and the steric hindrance to the charge transfer complex formation. The relevance of the charge transfer interactions is stressed by the different reactivity of the alkyl ketones and benzophenone toward different types of olefins. Kochevar and Wagner have explained the larger reactivity of aromatic ketones toward alkyl-monosubstituted monoolefins in terms of their higher electronegativity.[72] On the other hand, comparison of the data of Table 8 with that of Kuhlman and Schnabel for the quenching of benzophenone triplets[46] indicates that quenching of 2-heptanone is 10 and 15 times faster for methyl methacrylate and acrylonitrile, respectively. The faster quenching rates for the alkyl ketone can be due to its higher triplet energy and to the withdrawing character of the substituents. For olefins with low electron density at the π system, attack by the excited triplet involves the nucleophilic carbon atom of the carbonyl group,[73] and the electron-withdrawing character of the phenyl group would reduce the reactivity of benzophenone. The possibility of carbonyl quenching both by nucleophilic and electrophylic interactions is stressed by data showing that fast processes occur both for olefins of very low ionization potential or very high electron affinities.[65,72-75] The direct relationship between triplet quenching rate and Φ values for the photopolymerization initiated by 2,4-dimethyl-3-pentanone is evidenced from the data presented in Table 9. As expected, in all these systems where significant quenching by the monomer takes place in a process not leading to active free radicals, the order in monomer is considerably smaller than one.[66]

The results obtained for the phenyl-acetones (Table 7) do not conform to the simple pattern previously discussed. This is not a consequence of inhibition since these compounds have no effect upon the polymerization initiated by AIBN.[22] In particular, the low value measured for 1,1-diphenylketone can be due to termination by the unreactive diphenylmethyl radicals, since the photocleavage yield measured for this ketone is 0.3, a value considerably larger than the reported Φ.

TABLE 10
Photoinitiation Efficiencies of Various Monomers
Photoinitiated by Aliphatic Carbonyl Polymers[a]

Photoinitiator	XX	XXI	3,3-Dimethyl-2-butanone
Acrylonitrile	0.12	2.6	0.003
Methyl methacrylate	0.17	0.29	0.51
Styrene	0.01	0.24	0.03
Vinyl acetate	0.016	0.03	0.05

[a] $[M] = 4\ M$, in benzene. Values of initiation efficiencies were calculated from the Rp reported data by Equation 39.

From Naito et al.[50]

The photodegradation of poly(3-methyl-3-buten-2-one) (XX)

$$
\begin{array}{ccc}
& & CH_3 \\
& & | \\
CH_3 & & (-CH_2-C-CH_2-)_n \\
| & & \bullet \\
(-CH_2-C-)_n + h\nu \rightarrow & & + \\
| & & \bullet \\
C{=}O & & C{=}O \\
| & & | \\
CH_3 & & CH_3 \\
XX & &
\end{array}
\tag{62}
$$

and poly(4,4-dimethyl-1-penten-3-one) (XXI)

$$
\begin{array}{ccc}
& & (-CH_2-CH-CH_2-)_n \\
(-CH_2-CH-)_n & & | \\
| & & C{=}O \\
C{=}O & & \bullet \\
| \quad + h\nu \rightarrow & & + \\
CH_3-C-CH_3 & & \bullet \\
| & & CH_3-C-CH_3 \\
CH_3 & & | \\
& & CH_3 \\
XXI & &
\end{array}
\tag{63}
$$

proceeds predominantly through a Norrish type I mechanism via the triplet state with relatively high (*circa* 0.3 for XX and 0.5 for XXI) photocleavage yields.[50] The use of these polymeric initiators produces graft copolymer (from the macroradical) and homopolymer. Relative initiation efficiencies between the macroinitiators and a small model compound (3,3-dimethyl-2-butanone) are extremely dependent upon the monomer employed (Table 10). These differences have been explained in terms of different degrees of quenching by the monomer between the small and polymeric-excited carbonyl compounds. This conclusion is supported by results obtained by Scaiano et al.[76] on the quenching of triplet macromolecules by small molecules. In this work it was shown that intramolecular triplet energy migration in a macromolecule containing carbonyl chromophores decreases its reactivity toward a variety of small molecules, including conjugated dienes, 1-methylnaphthalene, stable nitroxides, and oxygen. In particular, the relatively large acrylonitrile polymerization rate observed for XXI would indicate that this compound, due to steric hindrance and/or reduced penetration of the quencher inside the macromolecular coil, could be particularly protected from acrylonitrile quenching.

2. α-Dicarbonyl Compounds

Photoinitiation by α-dicarbonyl compounds, both aliphatic and aromatic, has been explained in terms of a simple bond cleavage process.[77-79] From flash and steady-state photolysis measurements it has been concluded that benzil operates primarily by photofragmentation.[79] In the absence of hydrogen donors, the low efficiency of benzil was explained in terms of the high rate of self-termination by the produced benzyl radicals.[79] However, efficient cleavage of benzil contrasts with the low quantum yields reported for this process when the photolysis is carried out at room temperature employing long wavelength radiation.[80] The apparent discrepancy between both sets of data is due to the short (254 nm) wavelength employed in the photopolymerization studies, since it has been shown that benzil, although unreactive from its lowest triplet, efficiently decomposes from higher excited states that can be reached by short wavelength irradiation and/or two photon processes.[81]

2,3-Butanedione has been reported to photoinitiate by carbon-carbon photocleavage.[77,82] However, it has been lately shown that initiation by aliphatic α-dicarbonyl compounds is mediated, at least in nonprotic solvents, by its interaction with the monomer and/or additives.[12] In protic solvents (methanol or water), biacetyl is solvated giving an α-substituted alkanone.[83,84] Irradiation of the solvated species at 300 nm in the presence of vinyl acetate or vinyl pyrrolidone produces a significant amount of polymer, while its irradiation at longer wavelengths, absorbed only by the unsolvated molecules, is completely ineffective.[83] These results show that only the solvated molecules are able to produce active free radicals, as expected for the photolysis of α-substituted alkanones.

3. Aromatic Carbonyl Compounds

Type I photofragmentation of aromatic carbonyl compounds of the type $Ph-CO-CR_1R_2R_3$ is extremely sensitive to the groups R_1, R_2, and R_3. In acetophenone ($R_1 = R_2 = R_3 =$ H), photocleavage is a negligible process and significant photopolymerization requires bimolecular processes leading to free radical formation.[85] Photocleavage yields of substituted acetophenones are determined both by the substitution at the benzene ring and the α-carbon atom.[64,86] The effect of ring substitution is mainly related to changes in the character (from $n\pi^*$ to $\pi\pi^*$) of the first excited triplet.[64] Ketones unsubstituted in the benzene ring decompose almost exclusively from the lower ($n\pi^*$) triplet with rates that are mainly determined by the stability of the resulting radicals. This stability is achieved by multiple substitution and/or substitution by aromatic groups.

4. Benzoin Derivatives

Benzoin, benzoin ethers, and benzoin esters photocleave according to

$$\begin{array}{ccc} \overset{O}{\underset{\|}{}} \; \overset{R_1}{\underset{|}{}} & & \overset{O}{\underset{\|}{}} \; \overset{R_1}{\underset{|}{}} \\ Ph-C-C-Ph + h\nu \rightarrow & Ph-C\cdot \; + \; \cdot C-Ph \\ | & | \\ R_2 & R_2 \end{array}$$

XXII

(64)

R_1 = H; R_2 = H, alkyl, aryl (deoxy benzoin)
R_1 = H; R_2 = OH (benzoin)
R_1 = H; R_2 = O–R (benzoin ethers)
R_1 = H; R_2 = O–CO–R (benzoin esters)
R_1 = R_2 = O–R_3 (α,α-dialkoxy phenylacetophenones)

in a process that is mainly driven by the stability of the resulting benzyl radical. These compounds stand among the most widely investigated and extensively used photoinitiators due to their high photocleavage efficiency and relatively long wavelength absorption (≈ 300 to 380 nm) at which they can be excited. However, there are noticeable differences between the initiation efficiencies of different benzoin derivatives. Benzoin esters are generally poor photoinitiators, while benzoin ethers and α,α-dialkoxyphenylacetophenones present large Φ values and stand among the most extensively commercially used photoinitiators.[87-91]

Rates of α-cleavage and the quantum yields of the process in an inert solvent (benzene) are collected in Table 11 for a series of benzoin derivatives. This table shows that, while all photocleavage yields are in the 0.3 to 0.45 range, photocleavage rate constants (and hence the lifetimes of the excited triplet states) vary over four orders of magnitude. In particular, the type I photocleavage is relatively slow for benzoin esters, whereas is extremely fast ($k > 10^9$ s^{-1}) for benzoin ether.[96] It has been proposed that the transition state of the α-cleavage possesses a significant ionic character and that it is stabilized, with the concomitant increase in cleavage rate, by nonbonding electrons stabilizing substituents.[92,94]

In spite of their similar photocleavage yields in benzene, the wide range of triplet lifetimes of the benzoin derivatives leads to considerable differences in their ability to initiate the polymerization of vinyl monomers. In particular, for those long-lived compounds, such as benzoin and benzoin esters, the α-cleavage competes with triplet quenching by monomers. The values of Φ then become determined by the excited triplet/monomer interaction extent. Data reported in the literature for the triplet quenching rate constants by monomers and polymerization rates are given in Table 11. For a series of compounds whose efficiencies are reported by a given research group relative values conform to a single pattern that can be explained in terms of photocleavage yields and quenching by the monomer. However, absolute Φ values reported by different groups for a given compound show noticeable differences. For example, for the initiation of MMA polymerization sensitized by benzoin methyl ether values of 0.48,[96] 0.075,[13] and 0.65,[97] have been reported. These large differences can be partly due to different experimental conditions (sample absorbance, monomer concentration, light intensity) and differences in the parameters or procedures used to evaluate the initiation efficiency. However, they can also reflect the uncertainties associated with the calculation of absolute Φ values. The errors involved in these types of calculations could arise from the occurrence of partial primary radical termination, the presence of impurities that can modify the polymerization rate, and/or from the difficulties associated with the determination of the absorbed light intensity in conditions of poor incident light monochromacity.

The data given in Table 11 show that excited benzoin derivatives react with almost diffusion-controlled rates with styrene and also possess a high reactivity toward MMA and acrylonitrile. These relative quenching efficiencies are reflected in the reported polymerization rates. Benzoin and benzoin esters are poor initiators of styrene polymerization due to extensive quenching by the monomer. The fraction of triplet that avoids quenching by the monomer and that photocleaves is given by

$$\Phi_{33} = k_{33}/(k_{33} + k_{36}[M]) \tag{65}$$

For benzoin, Φ_{33} values calculated employing Equation 65 are 0.03 for styrene and 0.33

TABLE 11
Photocleavage and Initiation Efficiency Data for Aromatic Carbonyl Compounds

	Media	[M] (M)	ϕ_{33}[a]	$k_{33} \times 10^{-8}$ ($1/M$ s)	$k_{36} \times 10^{-8}$ ($1/M$ s)	$Rp \times 10^{5}$ (M/s)	Φ_m	Φ	Ref.
Deoxybenzoin	Benzene		0.44	0.016					92
	Benzene			0.013					93
	Styrene				65				93
	MMA				10				93
	Acrylonitrile				7				93
	Vinylacetate				0.17				93
α-Methyl deoxybenzoin	Benzene		0.44	0.21					92
	Benzene			0.11					93
	Styrene				15				93
	MMA				2				93
	Acrylonitrile				3, 8				93
	Vinylacetate				<0, 1				93
α-Phenyl deoxybenzoin	Benzene		—	0.1					92
	Benzene		—	0.037					93
	Styrene				1.9				93
	MMA				0.13				93
	Acrylonitrile				1.8				93
	Vinylacetate				0.06				93
Benzoin	Benzene		0.41	12					94
	Benzene			1.1					95
	Styrene	5			81.6[b]	1.2	10.9	0.05	96
	MMA	5			4.8[b]	14	127	0.235	96
	Acrylonitrile	5			<0.1[b]	24	218		96
	Vinylacetate	5			<0.1[b]	6.3	57.3		96
Benzoin acetate	Benzene		0.33	0.53					94
	Benzene	—		0.53					95
	Styrene	5			48	≤0.1	<0.9		96
	MMA	5			8	1.5	13.6	0.0027	96
	Acrylonitrile	5			13	3.9	35.5		96
	Vinylacetate	5			0.67	3.9	35.5		96

Benzoin methyl ether	Benzene		0.44	>100					94
	Styrene	5			17	2.7	24.5	0.25	96
	MMA	5			2	20	182	0.48	96
	Vinylacetate	5			<1	14	127		96
	Acrylonitrile	5			<1	25	227		96
Benzoin isopropyl ether	Benzene		0.44	>100					94
	Styrene	5				2.3	21	0.18	96
	MMA	5				19	173	0.4	96
	Vinylacetate	5				9.1	82.7		96
	Acrylonitrile	5				26	236		96
Benzoin methyl ether	MMA	Bulk					530	0.65	97
	Methylacrylate	Benzene					2560	0.42	97
Deoxy benzoin	MMA	Bulk				5	5.1	0.001	13
α,α-Dimethoxy-deoxybenzoin	MMA	Bulk				3, 5	48.2	0.062	13
Benzoin methyl ether	MMA	Bulk				45.5	55.7	0.075	13
Benzoin ethyl ether	MMA	Bulk				46	56.2	0.095	13
Benzoin	MMA	Bulk				47	57.4	0.1	13
Benzoin isopropyl ether	MMA	Bulk				46	56.2	0.095	13
α,α-Dimethoxydeoxy benzoin	MMA	Bulk				11	44		98
1-Hydroxy-1,1-dimethyl phenyl ketone	MMA	Bulk				10	40		98
1-Hydroxy-cyclohexyl phenyl ketone	MMA	Bulk				13	52		98
α,α-Dimethoxy-deoxybenzoin	MMA	Bulk				42.6	50	0.07	99
	Decyl methacrylate	Bulk				120	141	0.07	99
	Butyl acrylonitrile	Bulk				57			99
1-Hydroxy-1,1-dimethyl phenyl ketone	MMA	Bulk			0.07				100
	Styrene	Bulk			71				100
	Acrylonitrile	Bulk			0.1				100
	Benzene	Bulk	0.2—0.3	0.2					100
α,α-Dimethoxy deoxy benzoin	MMA	7		>100					101
α,α-Diethoxy deoxy benzoin	MMA	7			~1	31			101
	Bz		~10						101
α,α-Di-*sec*-butyl deoxy benzoin	MMA	7				45			101
α,α-Diisopropyl deoxy benzoin	MMA	7				45			100
	Bz			0.8					101
α,α-Methyl-propyldeoxy benzoin	MMA	7				41			101
α,α-Diclohexyl deoxy benzoin	MMA	7				53			101
	Bz			>10					101

TABLE 11 (continued)
Photocleavage and Initiation Efficiency Data for Aromatic Carbonyl Compounds

	Media	[M] (M)	ϕ_{33}[a]	$k_{33} \times 10^{-8}$ ($1/M$ s)	$k_{36} \times 10^{-8}$ ($1/M$ s)	$Rp \times 10^{5}$ (M/s)	Φ_m	Φ	Ref.
α,α-Ethyl-*sec*-butyl deoxy benzoin	MMA	7				40			101
α-Phenyl-α-*p*-methylphenyl acetophenone	MMA	7		>1		22			101
α,α-Diphenyl-*p*-methylacetophenone	MMA	7		0.12	7	5			101
α,α-Diphenyl-*p*-methoxyacetophenone	MMA	7		0.01	12	0			101
α-OH-Me-benzoin sulfonic esters	Bz		0.43	1.0					102

[a]   Using R-SH as radical scavenger.
[b]   Taking $k_{33} = 1, 2 \times 10^9$ from Kuhlmann and Schnabel.[96]

for MMA.[96] These considerations show that benzoins with long-lived triplets will be efficient photoinitiators only of those monomers with low quenching capacities (such as acrylonitrile and vinyl acetate), while benzoin derivatives of short lifetimes (such as benzoin ether) will efficiently decompose even in the presence of high concentrations of monomers that act as diffusionally controlled quenchers. Furthermore, the fact that for all these compounds the polymerization decreases with the extent of quenching by the monomer points to rather low α_{36} values.

Polymerization efficiencies are smaller than photocleavage yields calculated by Equation 65 due to free radical primary recombination and inefficient addition of the primary radicals to the monomer. Cage recombination of the primary radicals has been invoked to explain photocleavage yields in benzene smaller than one.[92,94] The efficiency of the primary radicals that scape cage recombination to add to the monomer will be extremely dependent upon the produced radical stability and the monomer reactivity and concentration. The nature of the radicals produced in the α-cleavage of benzoin derivatives has been inferred by product studies.[94,103-106] The benzoyl radical leads to benzaldehyde and benzil through hydrogen abstraction and dimerization, respectively. Substituted benzyl radicals lead mainly to pinacols by dimerization.

Benzoin ester can also react by intramolecular cyclization to a benzofuran derivative with concurrent heterocyclic acid elimination.[107]

$$\underset{\underset{H}{|}}{\overset{\overset{O\;\;OR_2}{\|\;\;|}}{Ph-C-C-Ph}} + h\nu \rightarrow R_2\text{-OH} + Ph- \!\!\left\langle\!\!\bigcirc\!\!\right\rangle \tag{66}$$

The α,α-dimethoxy-2-phenylmethyl radical, produced with high yields in α,α-dimethoxy-2-phenylacetophenone photolysis, cleaves to methylbenzoate and methyl radicals. This step is the major reaction pathway of the initially formed α,α-dimethoxybenzyl radicals.[87] Thus the high reactive methyl radical produced can contribute to enhance the photoinitiation efficiency of the parent compound.[3,108]

The structure and/or multiplicity of the initially produced radicals have also been elucidated by ^{1}HNMR-CIDNP,[109-111] ^{13}CNMR-CIDNP,[112] and ESR spin trapping experiments.[113-115] Furthermore, the relatively low reactivity of both the benzoyl and α-substituted benzyl radicals allows their detection by UV transient spectroscopy.[96,116] In the photolysis of α,α-dimethoxy-2-phenylacetophenone, both radicals were detected by their absorption in the near-UV and visible region.[116] The short-lived transient (lifetime 750 ns) was identified as the benzoyl radical, while the long-lived transient (lifetime >10 μs) was identified as the α,α-dimethoxybenzyl radical.

The initiation efficiency can be expressed as

$$\Phi = \Phi_{33}(1 - \beta)f \tag{67}$$

where f is the fraction of radicals that once avoided primary recombination added to the monomer. For a series of benzoin derivatives it has been reported that $\Phi \approx 0.3\ \Phi_{33}$ in 5 M methyl methacrylate.[96] The difference between both values might be either due to cage recombination of radical pairs, as evidenced by Lewis and Magyar for phenylpropiophenone,[117] or it might be due to the fact that one sort of the radicals generated by α-cleavage is much more reactive toward monomers than the other.

There are several reports aimed to identify the radicals that initiate the polymerization after the type I photocleavage of benzoin derivatives. Efficient trapping of both radicals by

TABLE 12
Bimolecular Rate Constants for the Radical Addition to the Monomer[a]

Initiator	Monomer	$k_{(R\cdot+M)} \times 10^{-5}$ (1/M s)	Ref.
Benzoin methyl ether	Styrene	1.6	96
	MMA	0.9	96
	Vinyl acetate	1.5	96
	Acrylonitrile	0.2	96
Benzoin acetate	Vinyl acetate	1.2	96
Benzoin	Vinyl acetate	1.2	96
α-Phenyldeoxybenzoin	MMA	0.5	95
	Acrylonitrile	0.015	95
1-Hydroxy-1,1-dimethylphenyl ketone	Styrene	4.7	100
	MMA	6.3	100
	Acrylonitrile	16	100

[a]   In benzene as solvent.

diamagnetic radical scavengers suggests that both types of radicals could efficiently add to the olefinic double bond of vinyl monomers in the same way as they add to N-oxides and nitroso compounds.[114,118] This conclusion is consistent with the results reported by Hutchison and Ledwith[119] and Pappas et al.[97,120] These authors found that the benzoin methyl ether (^{14}C-labeled in various positions) initiated the MMA polymerization and the benzoyl and α-methoxy benzyl radicals appear as equally effective in initiating the polymerization. However, these results have been questioned on the basis of studies utilizing benzoin methyl ether as photoinitiator in styrene polymerization.[121] The obtained results indicate that only the benzoyl radical was efficient in initiating the polymerization, whereas benzyl ether radicals dimerized or terminated the polymerization. Similar results were obtained for α-methoxy and α,α-dimethoxy benzoin in the presence of 1,1-diphenylethylene[122] and styrene, where it was proposed that the monomer reacts with benzoyl radicals but not with benzyl ether radicals.[103,123] However, studies of Pappas and Asmus[124] give evidence for the initiation of MMA polymerization by the α-methoxybenzyl radical photogenerated by hydrogen abstraction from ^{14}C-labeled benzyl methyl ether by benzophenone triplet. The apparent discrepancy between these results can be explained in terms of the effect of temperature, monomer reactivity and concentration, and absorbed light intensity on the efficiency of radical initiation when monomer addition is competing with primary radical recombination and/or termination. Phan[125] analyzed the products formed in MMA polymerization photoinitiated by α,α-dimethoxy-2-phenylacetophenone and concluded that, under his experimental conditions, 30% of the benzoyl radicals and 15% of the α,α-dimethoxybenzyl radicals were involved in the initiation step. A comprehensive study of the factors determining the efficiency of primary radical addition to the monomer has been carried out by Lipscomb and Tarshiani,[126] who have examined the effects of light intensity, temperature, monomer concentration, and solvent on the kinetics of styrene photopolymerization in bulk and in diluted solutions using benzoin isobutyl ether as photoinitiator. The results suggest that at low temperature, high intensity, and/or low monomer concentration, the benzoyl radical is the dominant initiating species, and the benzyl ether radical was consumed mainly in the termination step. Meanwhile, at low intensity, high temperature, and/or high monomer concentration, both radicals would initiate the polymerization.

Kinetic parameters for the addition of primary radicals to monomer have been obtained in a limited number of systems.[96,100] The data obtained are collected in Table 12. These data were obtained from the pseudo-first-order decay of one of the sensitizer-derived radicals and/or recording the formation of styrene-derived radicals, and show that the reactivity of the radicals generated in the different systems are rather similar. However, due to the difficulty to identify the different primary radicals' absorption, the results are not conclusive about the type of radical involved in the process. Regarding the values of the different addition rates to the monomer, the data show that acrylonitrile is the less reactive monomer, the difference being larger for the more stabilized radical derived from α-phenyl deoxybenzoin.

Phan[125,127] has investigated the effect of molecular oxygen on the photopolymerization of MMA initiated by α,α-dimethoxy-2-phenylacetophenone and 1-hydroxycyclohexyl phenyl ketone. The observed decrease in polymerization rate could be due to quenching of the excited photoinitiator, trapping of the primary radicals, or reaction of the macroradicals to produce a less reactive peroxyl radical. However, due to the short lifetimes of the sensitizers considered, the first possibility must be disregarded and the effect related to free radical-oxygen reactions. Product analysis shows that benzoyl radicals generated from both initiators are efficiently trapped by oxygen, yielding benzoic acid among the main products. In benzene, the quantum yields of disappearance of α,α-dimethoxy-2-phenylacetophenone are higher in the presence of oxygen, but those of 1-hydroxycyclohexyl phenyl ketone are not affected. The former result was attributed to oxygen trapping of the primary radicals, reducing so the back recombination extent. When methyl methacrylate is employed as solvent, yields of nonpolymeric products derived from the primary radicals increase, resulting in fewer radicals available for polymerization. This effect leads to a fivefold decrease in polymerization quantum efficiency.

Polymers containing photodissociable benzoin methyl ether moieties directly connected to the polymer backbone through ester linkages have been reported to be efficient photoinitiators.[128] Irradiation of this type of polymer gives benzoyl-free radicals and polymer-bound α-methoxybenzyl-free radicals, according to

$$\tag{68}$$

In the presence of styrene, both types of radicals initiate the monomer polymerization, leading to a mixture of polystyrene and graft-type copolymers. Also, extensive cross-linking was observed as a consequence of processes mediated by the polymeric-bound α-methoxybenzyl radicals. A copolymer of α-methylolbenzoin methyl ether acrylate and styrene was also employed as photoinitiator of MMA polymerization in benzene. In this later system, MMA homopolymer and graft copolymer were produced and extensive cross-linking was observed. The conversion and degree of cross-linking of the produced polymers can be controlled by adjusting the benzoin methyl ether side chains in the polymeric photoinitiators.

The macrophotoinitiators present a greater efficiency than nonpolymeric benzoin ether derivatives such as benzoin, benzoin methyl ether, and α,α-dimethoxy-2-phenylacetophenone. Under conditions where these small photoinitiators give less than 8% conversion, the macroinitiators give more than 20% conversion. This larger polymerization rate was attributed to a decrease in the termination step due to shielding of the chain carrying free radical ends as a consequence of partial gelation of the polymerization systems.

Sulfonic esters of α-hydroxymethylbenzoin have been used as sulfonic acid-releasing photoinitiators for acid-hardening systems.[3] Gaur et al.[102] have studied the mechanism of photofragmentation for several sulfonic esters, and found photocleavage quantum yields in the 0.38 to 0.49 range (Table 11). The primary cleavage takes place from a short-lived triplet, giving a benzoyl radical and a benzoyl-β-sulfonyloxyketyl radical that undergoes an extremely fast elimination of sulfonic acid togive the benzoyl methyl radical:

$$
\begin{array}{c}
\underset{\substack{\displaystyle | \\ CH_2-O-SO_2-R}}{\overset{\substack{O\ \ OH \\ || \ \ | }}{Ph-C-C-Ph}} \;+\; h\nu \;\rightarrow\; Ph-C\cdot \;+\; \underset{\substack{\displaystyle | \\ CH_2-O-SO_2-R}}{\overset{\substack{O\quad OH \\ || \quad\ | }}{\cdot C-Ph}}
\end{array}
$$

$$
XXIII \qquad\qquad\qquad\qquad \overset{\substack{O \\ ||}}{Ph-C} \;+\; R-SO_3H \qquad (69)
$$
$$
\overset{|}{CH_2\cdot}
$$

Studies of the photodecomposition in the presence of 1,1-diphenylethylene as a model substrate for vinyl monomers indicate that both benzoyl and benzoyl methyl radicals are highly reactive toward the olefinic double bond. The fraction of the generated free radicals which add to the olefin was calculated from the analysis of the products. Values of 0.61 for benzoyl radicals and 0.59 for benzoyl methyl radicals were obtained when R = methyl. With R = 4-methylbenzyl, the values obtained were 0.57 and 0.53 for benzoyl and benzoyl methyl radicals, respectively.[102] The value of Φ for these compounds can be estimated as *circa* 0.5. Rates of MMA polymerization measured by the gel time method show that the efficiency of these photoinitiators is very little dependent on the group R, and that they are similar to those of benzoin methyl ether or α,α-dimethoxy-2-phenylacetophenone.[129]

5. α-Hydroxyacetophenone Derivatives

α-Hydroxy-substituted acetophenones can efficiently photocleave from the triplet state even when lacking the α-phenyl group. Representative compounds of this class are 1-hydroxy-1,1-dimethylethyl phenyl ketone (XXIV) and 1-1-hydroxycyclohexyl phenyl ketone (XXV). For both compounds, α-cleavage is the main process from the excited triplet:

$$
\begin{array}{c}
\overset{\substack{CH_3 \\ |}}{\underset{\substack{|| \ | \\ O\ OH}}{Ph-C-C-CH_3}} \;+\; h\nu \;\rightarrow\; Ph-C\cdot \;+\; \overset{\substack{CH_3 \\ |}}{\underset{\substack{| \\ OH}}{\cdot C-CH_3}} \qquad (70)
\end{array}
$$
$$
XXIV
$$

$$
\begin{array}{c}
\underset{\substack{|| \ | \\ O\ OH}}{Ph-C-C}\!\!\bigcirc \;+\; h\nu \;\rightarrow\; Ph-C\cdot \;+\; \cdot C\!\!\bigcirc \qquad (71)
\end{array}
$$
$$
XXV
$$

The radicals produced in the photocleavage have been evidenced by ^{1}HNMR-CIDNP,[3] chemical trapping,[3,130] and laser flash photolysis experiments.[131] Values of α-cleavage rate constants and free radical quantum yields are given in Table 11. As with the benzoin derivatives, the excited triplets of this type of compound are efficiently quenched by the monomers, being the quenching rate by styrene diffusionally controlled. Smaller quenching rate constants have been determined for MMA and acrylonitrile.[100]

Lougnot and Fouassier have employed compounds XXIV and XXV as photoinitiators of MMA polymerization.[98] The data obtained (Table 11) show that these compounds have almost the same efficiency than α,α-dimethoxy-2-phenylacetophenone, compound employed as reference.

6. Dialkoxy Acetophenones

Dialkoxy acetophenones are efficient photoinitiators in spite of the fact that their triplet states undergo, with comparable efficiencies, type I photocleavage and intramolecular hydrogen abstraction:

$$(72)$$

Although it has been suggested that the biradical produced by intramolecular hydrogen abstraction was involved in the initiation process,[108,132] it has been lately shown, by ^{1}HNMR-CIDNP experiments conducted in the presence of MMA, that the initiation is mainly due to the free radicals generated by α-cleavage.[133] Similar conclusions have been derived from spin trapping experiments employing TEMPO in the presence of 1,1-diphenylethylene.[122]

Fouassier and Lougnot[101] have measured polymerization rates photoinitiated by several dialkoxyacetophenones. Polymerization rates were similar to those obtained employing α,α-dimethoxy-2-phenylacetophenone, a very efficient photoinitiator of the benzoin family (Table

11). However, significant differences are observed between the different compounds considered. These differences have been explained in terms of steric and/or electronic effects of the bulky alkoxy substituents upon the free radical yields. These effects were correlated to the stabilization of a transition state having a partial charge transfer character.[101] The role of charge transfer complex stabilization by the methoxy groups is stressed by the data shown in Table 11, indicating that diphenyl acetophenone photocleavage is slower than that of α,α-dimethoxyacetophenone, in spite of the high stability of the radical produced by photocleavage of the former compound. The low initiation yields of these compounds are then related to increased triplet quenching and the low reactivity of the produced free radicals.

Introduction of a *p*-methoxy group further reduces the photocleavage rate constant of α,α-diphenylacetophenone, with the concomitant decrease in the photoinitiation efficiency. This effect can be attributed to shifting from a $n\pi^*$ triplet to a less reactive $\pi\pi^*$ triplet.

7. Phosphines

Acylphosphine oxides and acylphosphonates are effective photoinitiators,[134,135] due to their efficient α-photocleavage:[136-140]

$$
\begin{array}{ccc}
\text{O} & & \text{O} \\
\| & & \| \\
\text{R}_1\text{--C--P--R}_2 + h\nu \rightarrow \text{R}_1\text{--C·} & + & \text{·P--R}_2 \\
\| \quad | & \| & | \\
\text{O} \quad \text{R}_3 & \text{O} & \text{R}_3
\end{array}
\qquad (73)
$$

XXVII

$$
\begin{array}{ccc}
\text{O} & & \text{O} \\
\| & & \| \\
\text{R}_1\text{--C--P--OR}_2 + h\nu \rightarrow \text{R}_1\text{--C·} & + & \text{·P--OR}_2 \\
\| \quad | & \| & | \\
\text{O} \quad \text{OR}_3 & \text{O} & \text{OR}_3
\end{array}
\qquad (74)
$$

XXVIII

α-Cleavage quantum yields, triplet yields, and excited-state lifetimes for several acylphosphine oxides and acylphosphonates are compiled in Table 13.

Acyl and phosphinyl radicals have been identified by CIDNP,[142] or chemical trapping with 2,2,6,6-tetramethylpiperidin-1-oxyl.[140] The results obtained indicate that the α-photocleavage takes place from the triplet state, and that the process is nearly solvent independent. Photolysis of compounds where R_2 and/or R_3 are aromatic groups generates a strong transient absorption with maxima around 330 nm that has been ascribed to the phenyl phosphinyl radicals.[137-139]

Studies on the photodecomposition of diphenyl-2,4,6-trimethylbenzoylphosphide oxide in the presence of 1,1-di-*p*-tolylethylene, a model sustrate of vinyl monomers, showed that both radicals add to the olefinic double bond.[143] The addition of the diphenylphosphinyl radicals was determined to be twice more effective than that of the 2,4,6-trimethylbenzoyl radicals.

The reactivity of the phosphinyl radicals toward vinyl monomers was assessed by measuring the rate constant of the process

$$
\begin{array}{ccc}
\text{O} & & \text{O} \\
\| & & \| \\
\text{R}_2\text{--P·} + \text{M} \rightarrow & & \text{R}_2\text{--P--M·} \\
| & & | \\
\text{R}_3 & & \text{R}_3
\end{array}
\qquad (75)
$$

TABLE 13
Photocleavage and Polymerization Data for Acylphosphine Oxides[a]

$$R_1-\overset{\overset{\displaystyle O}{\|}}{C}-\overset{\overset{\displaystyle O}{\|}}{\underset{R_3}{P}}-R_2$$

	Φ_{33}	τ_S (ns)	τ_T (ns)	Rp × 10[4b] (1/M s)	Ref.
R_1 = 2,4,6-trimethylphenyl R_2 = R_3 = phenyl	0.4	<0.3			139
				1.22	141
				0.21[c]	141
R_1 = 2,4,6-trimethylphenyl R_2 = R_3 = methoxy	0.3	0.7	<20		139
R_1 = t-butyl R_2 = R_3 = phenyl	1.0	30			139
R_1 = t-butyl R_2 = R_3 = methoxy	0.3	11	30		139
R_1 = phenyl R_2 = R_3 = ethoxy		<1	24	0.01[c]	139
R_1 = 2,4,6-trimethylphenyl R_2 = methyl R_3 = methoxy	0.5				137
R_1 = t-butyl R_2 = R_3 = ethoxy	0.3				137
R_1 = 2,6-dimethoxyphenyl R_2 = R_3 = phenyl				1.18	141
R_1 = phenyl R_2 = R_3 = methoxy				0.02	141
R_1 = 2,4,6-trimethylphenyl R_2 = R_3 = ethoxy				0.53	141
Benzoil methyl ether				0.67	141

[a]   Benzene as solvent.
[b]   Bulk MMA, initiator = 2.5 10^{-4} M.
[c]   Styrene polymerization.

TABLE 14
Bimolecular Rate Constants for the Addition of
Phosphinyl Radicals to Monomers

	$k_{(R\cdot+M)}$ × 10^{-7} (1/M s)	
Monomer	O = $\dot{P}(Ph)_2$[a]	O = $\dot{P}(OCH_3)_2$[b]
Methacrylonitrile	5	9.2
Styrene	6	22
Methyl methacrylate	8	5.8
Acrylonitrile	2	0.58
Methyl acrylate	3.5	1.7
Butyl vinyl ether	0.4	2.1
Vinyl acetate	0.16	0.29

[a]   Measured by the transient absorption at 330 nm.
[b]   Measured from the growth of styrene-derived radical and its
      change in presence of added monomers.

From Sumiyoshi and Schnabel.[137]

These measurements were carried out either by following the decrease of the radical absorbance at 330 nm (aromatic substituted radicals) or by measuring the extent of styrene-derived radicals.[137,138] The data, given in Table 14, show that phosphinyl radicals are highly reactive toward monomers, dialkoxy phosphinyl radicals being the most reactive members of the family. All rate constants obtained are larger than those obtained for the reaction of carbon-centered radicals generated in the photolysis of hydroxy alkylphenones or benzoin derivatives. The high reactivity of the phosphinyl radicals results from their tetrahedral structure.[137,139]

Baxter et al.[141] have reported the polymerization rate of MMA and styrene using acyl phosphine oxides and acyl phosphonates as photoinitiators (Table 13). The data show that, while acylphosphine oxides are more efficient than benzoyl methyl ether, acylphosphonates are very poor photoinitiators.

C. UNIMOLECULAR PHOTOPROCESSES LEADING TO BIRADICAL FORMATION

Photocleavages of cyclic compounds or intramolecular atom transfer processes lead to biradical formation. If both radical centers are able to initiate the polymerization, a different kinetic law and polymer molecular weight can be expected. In particular, for monomers such as styrene whose termination is dominated by combination, very high polymer molecular weights could be obtained even at very high absorbed intensities. However, the experimental results do not fulfill this expectative.

Cyclic peroxides, for example, ergosterol peroxide, have been used as photoinitiators for the polymerization of styrene.[144] The relationship between the number-average molecular weight and the rate of polymerization was normal, indicating that initiation occurs via a monoradical produced from the cyclic peroxide.

Carbonyl compounds that predominantly photodecompose by an intramolecular hydrogen abstraction (Norrish type II mechanism) are other potential sources of biradicals with capacity to initiate polymerization. These studies have been performed employing both aliphatic[11] and aromatic[145] carbonyl compounds. The results obtained employing methyl methacrylate as monomer, together with the triplet and biradical lifetimes in benzene and the triplet quenching rate constant by the monomer, are given in Table 15. For aliphatic ketones decomposing predominantly by a type II mechanism, both singlet[63,148] and triplet[63,149] lifetimes are conditioned by the degree of substitution at the γ-carbon atom. The triplet lifetime is determined by the rate of the process

$$^{3}A \rightarrow {}^{3}\text{Biradical} \tag{76}$$

that generates a triplet correlated biradical whose lifetime is determined by the spin flip rate.[147,150,151] Similarly, intramolecular γ-hydrogen abstraction also takes place from the excited singlet of aliphatic ketones, producing a short-lived singlet biradical that can react by back hydrogen transfer, cyclization, and/or fragmentation:

$$\tag{77}$$

The short lifetime[150] of the singlet biradical precludes its participation in the initiation process.

TABLE 15

Photoinitiation Data for the Polymerization of Methyl Methacrylate Photoinitiated by the Biradicals from the Norrish Type II[a]

Ketone	% conversion	Φ	τ_T (ns)	$k_{36} \times 10^{-9}$ (1/M s)	τ_{Bir} (ns)	α_{78}	Ref.
2,4-Dimethyl-3-pentanone[b]	12.6	0.22	8				11
4-Methyl-2-pentanone	1.2	0.002	130			≤0.21	11
2-Heptanone	1.5	0.003	5	0.7	330[c]	≤0.06	11
5-Methyl-2-hexanone	2.5	0.0086	1	1.4	540[c]	≤0.12	11
Acetophenone[b]	2.2						145
Butyrophenone	5.5		117				145
Valerophenone	6.7		7.6		34[d]		145
γ-Methylvalerophenone	10.2		1.9		35[d]		145

[a] In benzene as solvent.
[b] Included as comparison.
[c] From Naito[146] in hexane as solvent.
[d] From Scaiano.[147]

The data given in Table 15 show that, for these ketones, the polymerization efficiency increases when the triplet lifetimes decrease, i.e., they increase when the degree of substitution at the γ-carbon atom increases. These results imply that the initiating species is the triplet biradical, the relationship between efficiency and triplet lifetime being due to the competition between biradical formation and triplet quenching by the monomer. Furthermore, these results imply that α_{36} must be considerably smaller than α_{76}. Initiation efficiencies for the aliphatic ketones are given in Table 15. From these data, the triplet quantum yield, and the fraction of triplets that avoids quenching by the monomer, it is possible to estimate the values of α_{76}. These values are included in Table 15, and they are considerably smaller than one. For ketones of similar triplet lifetimes, those decomposing by a type I mechanism are then much more efficient photoinitiators than those in which a type II process is competitive to the α-cleavage.

Aliphatic triplet biradicals derived from intramolecular hydrogen abstraction have lifetimes of the order of 300 to 500 ns. Lifetimes of biradicals derived from aromatic ketones are nearly ten times shorter. The efficiency of the triplet biradical to initiate the polymerization would result in the competition between the spin flip and addition to the olefinic double bond. The relatively low efficiencies observed are then due to the short lifetimes and their relatively low reactivities toward the addition of carbon-carbon double bonds.[147] The data reported in Table 15 would indicate that process

$$^3\text{Biradical} + \text{M} \rightarrow \text{M·} \qquad (78)$$

takes place, for aliphatic biradicals, with a rate constant of *circa* $2 \times 10^5\ M^{-1}\ \text{s}^{-1}$.

The relation between the molecular weight of the polymer produced and the polymerization rate obtained for the ketones decomposing by a type II mechanism has been compared to that obtained for ketones reacting by α-cleavage and di-*tert*-butylperoxide.[11] No appreciable differences were observed between the three series of data, suggesting that the polymerization law of biradical initiated kinetics is indistinguishable from that of the classical free radical vinyl polymerization. To date, not a suitable explanation has been provided for this rather unexpected result.

IV. BIMOLECULAR PHOTOPROCESSES

Photoinitiation efficiencies can be modified both by photosensitizer/monomer, photosensitizer/solvent and photosensitizer/additive interactions. In the previous sections, interaction with the monomer, due to low values of α_{36}, decreased the values of Φ. However, in several systems it has been observed that the initiation is mediated by the monomer/sensitzer interaction.

A. PHOTOINITIATION DUE TO SENSITIZER/MONOMER INTERACTION

Monomer-mediated initiation has been reported employing α-dicarbonyl compounds as photoinitiators.[12,22,152] 2,3-Butanedione irradiated at low temperature in an inert solvent does not produce significant amounts of free radicals,[153] and its triplet lifetime is longer than 0.1 ms.[154,155] 2,3-Pentanedione partly decomposes by a Norrish type II intramolecular hydrogen abstraction, but presents a lifetime in benzene of the same order of magnitude.[156] Recent reports[155,157,158] have described the photoaddition of biacetyl triplets to unsaturated substrates, an exciplex formation being proposed as primary photochemical intermediate. The values of triplet quenching constants reported in Table 16 imply that, in bulk monomer, most dicarbonyl triplets must be quenched by the monomer. The rather large initiation efficiencies reported for methacrylic monomers (Table 16) must arise from the interaction monomer-excited dicarbonyl. In agreement with this, the initiation efficiency has been reported to

TABLE 16
Data for the Polymerization of Various Monomers Photoinitiated by α-Dicarbonyl Compounds[a]

Compound	Monomer[b]	Φ_{30}	$k_{36} \times 10^{-3}$ $(1/M\ s)$	Φ_{36}	α_{36}	Φ^{c}	Ref.
2,3-Butanedione	MMA	0.17	2.3	0.91		0.12	12
	Butyl MA	0.11	3.3	0.91		0.16	12
	Styrene	0.24	100	0.99		0.01	12
2,3-Pentanedione	MMA	≤0.08				0.12	12
	Butyl MA	≤0.04				0.16	12
	Styrene	≤0.04				—	12
Benzil	MMA			0.89	0.0007	0.001	152
	Butyl MA			0.85	0.02	0.017	152
	Isobutyl MA			0.95	0.01	0.01	152
	Cyclohexyl MA			0.90	0.011	0.01	152
	Vinyl acetate			0.43	0.047	0.02	152
	Phenyl acrylate					0.001	152
	Styrene			0.97	0.259	0.25	152
	p-Methylstyrene			0.98	0.29	0.28	152
p,p′-Dimethoxybenzil	Styrene			0.97	0.15	0.15	152
	MMA			0.91	0.0002	0.0002	152

[a] In benzene as solvent; $\lambda_{irradiation}$ = 366 nm.
[b] MA = methacrylate.
[c] In bulk, taken ABIN as reference photoinitiator.

increase with the monomer concentration.[12] Furthermore, the fact that the initiation efficiency is larger than the fraction of quenched singlet (Table 16) implies that at least partially the initiation must involve quenching of the dicarbonyl triplet, according to

$$^{3}(CH_{3}\text{–}\underset{\overset{\|}{O}}{C}\text{–}\underset{\overset{\|}{O}}{C}\text{–}CH_{3})^{*} + M \rightarrow CH_{3}\text{–}\underset{\overset{|}{OH}}{C}\text{—}\underset{\overset{\|}{O}}{C}\text{–}CH_{3} + \dot{M}(\text{–H}) \qquad (79)$$

However, the fact that Φ values even in bulk monomer are considerably smaller than two indicates that Reaction 79 must be sided by nonradical generating interactions with the monomer double bond. The larger Φ obtained for butyl methacrylate than for MMA could reflect the faster photoreduction expected for the former compound given its longer alkyl chain.

The data in Table 16 show that styrene, in spite of being a better quencher of excited 2,3-butanedione, is polymerized less efficiently. This difference implies that α_{36} (and probably also α_{30}) is considerably smaller for styrene than for the methacrylate esters, a difference that is due to the presence of labile allylic hydrogens in the methacrylate derivatives and to the higher energy of the excited triplets for these compounds, which makes energy transfer to the styrene a much more likely process.

Photoinitiation by aromatic α-dicarbonyl compounds is also monomer dependent,[152] but the initiation mechanism appears as completely different to that of the aliphatic α-dicarbonyl compounds. The values of Φ obtained for several monomers employing benzil and 4,4′-dimethoxybenzil are included, together with the quenching rate constants, fraction of quenched triplets, and fraction of triplet quenched that initiated the polymerization, in Table 16.

The interaction-excited benzil-methacrylic monomers appear as highly inefficient with regard to the initiation process, suggesting an interaction through a charge transfer complex with only a relatively small fraction of the complex decomposing through hydrogen transfer.

Reactions of excited carbonyls and olefinic double bonds go through a charge transfer mechanism taking place with a rate constant strongly dependent on the free energy of formation of the donor-acceptor ion pair. In the interaction of benzil with electron-rich olefins (styrene), the carbonyl excited triplet acts as electron acceptor

$$^3A + M \rightarrow (^3A^{\delta-}...M^{\delta+}) \rightarrow \begin{array}{l} \nearrow \quad A^- + M^+ \\ \quad \text{Biradical} \\ \searrow \quad \text{Products} \end{array} \qquad (80)$$

The biradical will result from the electrophilic attack to the *n*-oxygen orbital to the olefin. The extent of photoproducts will be determined by the polarity of the solvent. The high efficiency of styrene and *p*-methyl styrene must arise from initiation by one of the intermediates produced. An initiation mediated by the formation of charge transfer complexes is compatible with the high efficiency of styrene, the monomer of lower ionization potential. Also, the smaller efficiency of *p*-dimethoxy benzil is compatible with the decrease in electronaffinity associated with the methoxy group. The initiation efficiency of the different monomers correlates with the reactivity of their double bonds toward *tert*-butoxyl radicals, oxygen-centered free radicals which are frequently invoked as simple models of carbonyl triplets. Their reactivity and reaction rates are strongly influenced by the extent of electron charge contributions to the stability of the critical configuration.

Monomer-benzil interaction through the proposed reaction scheme could lead to polymerization by either a biradical and/or an ion pair mechanism. Nevertheless, the average molecular weight obtained and the molecular weight distribution are similar to those expected from a normal free radical mechanism. This result, that in the system considered could be due to the ineffectiveness of one of the free radical centers to add to the monomer (most likely the one located in the carbon atom of the benzil moiety[119]), appears as a general feature of the photoinitiator systems involving the formation of biradicals.

B. BIMOLECULAR PROCESSES INVOLVING THE SOLVENT AND/OR ADDITIVES

Quenching by the solvent and/or additives can promote or inhibit the polymerization depending on the fraction of active free radicals produced in the interaction step, relative to the other processes with which the interaction is competing. Decrease of the initiation efficiency of a photoinitiator by addition of a quencher that does not produce active free radicals is simply expressed in a Stern Volmer-like equation

$$\Phi^0/\Phi = 1 + K_{SV}[Q] \qquad (81)$$

where Φ^0 and Φ are the efficiencies measured in absence and presence of the additive, respectively.

Compounds that do not produce active free radicals in the absence of additives can become efficient photoinitiators by addition of quenchers. Among the quenchers more effective in promoting free radical formation stand those reacting by simple hydrogen transfer processes or by charge transfer complex formation. For this type of photoinitiated polymerization, if only triplet processes are considered and primary termination is disregarded, the value of Φ will be given by

$$\Phi = \Phi_{27}\alpha_{37}\{k_{37}[Q]/(k_{37}[Q] + \tau^{-1})\} \qquad (82)$$

TABLE 17
Quenching of Benzophenone Triplets
by Vinyl Monomers in Benzene

Monomer	$k_{36} \times 10^{-9}$ (1/M s)	Ref.
Styrene	3.2	46
α-Methylstyrene	2.7	46
Vinylpyrrolidone	0.36	46
Methyl methacrylate	0.069	46
Acrylonitrile	0.034	46
Vinyl acetate	0.0054	46
Methyl acrylate	0.01	162
Ethyl acrylate	0.05	162
Butyl acrylate	0.013	162
Decyl methacrylate	0.07	162

where τ is the triplet lifetime in the absence of quencher. The expected order in quencher of the polymerization process ranges from 0.5 at low [Q], to zero at high Q concentrations.

1. Direct, Free, Radical-Like Hydrogen Abstraction

Excited carbonyl compounds can abstract a hydrogen atom from suitable donors such as alcohols or tetrahydrofurane (THF) by a bimolecular step that can be simply represented by

$$\text{>C=O* + R–H} \rightarrow \text{>Ċ–OH + R·} \tag{83}$$

This process can be visualized as an elementary, radical-like process. The reaction rates correlate reasonably with the rate constants measured for typical free radicals such as *tert*-butoxyl radicals,[159,160] and presents only a moderate solvent dependence. The reactivity of different substrates is mainly determined by the strength of the hydrogen-substrate bond being broken, and the preexponential A factors are compatible with those expected for a single, elementary hydrogen abstraction.[161]

For most carbonyl compounds, processes such as Process 83 involving alkanols or substituted aromatic hydrocarbons (i.e., cumene) are slower than quenching by the monomers, leading to very low initiation efficiencies. Among the most studied carbonyl compounds stand benzophenone, a sensitizer that requires an additive to lead to significant Φ values. Values of quenching rate constants of benzophenone triplets by monomers are given in Table 17. In Table 18 are given quenching rate constants for substrates that can be considered to react by a simple hydrogen abstraction process, together with polymerization rates obtained employing the benzophenone-coinitiator systems. In these systems, initiation results from the occurrence of processes like Processes 84 to 86

$$\text{(Ph–C–Ph)* + Ph–CH–Ph} \rightarrow 2\text{Ph–Ċ–Ph} \tag{84}$$

XIX

$$\text{(Ph–C–Ph)* + Ph–CH}_2\text{–O–CH}_3 \rightarrow \text{Ph–Ċ–Ph + Ph–ĊH} \tag{85}$$

TABLE 18

Quenching of Benzophenone Triplet by Hydrogen Donors and Polymerization Rates by the Benzophenone-Coinitiator System

Monomer or solvent	Hydrogen donor	$Rp \times 10^5$ $(1/M\ s)$	$k_{37} \times 10^{-6}$ $(1/M\ s)$	Ref.
MMA 4.7 M	Benzyl methyl ether 1 M	7.18		124
	Benzyl methyl ether 0.1 M	2.53		124
Benzene	Benzyl methyl ether		1	163
MMA 4.7 M	Benzhydrol 0.2 M	0.13		124
	Benzhydrol 0.1 M	0.23		124
Benzene	Benzhydrol		9	124
Vinyl acetate	THF 5 M	22		95
Acrylonitrile	THF 5 M	16		95
MMA	THF 5 M	9.5		95
Benzene	THF		3	46
	Isopropanol		1.8	164

$$(\text{Ph–C–Ph})^* + \underset{O}{\bigcirc} \rightarrow \text{Ph–}\overset{\cdot}{\text{C}}\text{–Ph} + \underset{O}{\bigcirc}. \qquad (86)$$

that produce a ketyl (XIX) radical and a coinitiator-derived free radical.

In the presence of THF, polymerization rates decrease when the quenching rates by the monomer increase,[95,165] as expected from the competition between hydrogen abstraction and monomer quenching. Similarly, when benzyl methyl ether is employed as hydrogen donor, polymerization rates increase when the additive concentration increases, and the polymerization follows a conventional free radical mechanism in which the initiating species is the ether-derived free radical.[124] Benzhydrol, in spite of its higher quenching rate constant and ketyl radical quantum yield of 2,[166] is considerably less efficient than the other additives. This result, as well as the decrease in polymerization rate associated with the increase in benzhydrol concentration, is due to the low reactivity of the ketyl radicals that are engaged mainly in bimolecuar free radical processes, either between them or with monomer-derived radicals.[124,167] The presence of benzhydrol then reduces the polymerization rate even below that obtained in the absence of additives.[164] The slow MMA polymerization rate observed in the absence of additives can be explained in terms of monomer photoreduction by benzophenone triplets.

The rate constants for the ketyl/monomer addition have been determined for several monomers. The values obtained are 5.5, 3.8, and 9 $\times$ 10^3 M^{-1} s^{-1} for vinyl acetate, acrylonitrile, and MMA, respectively.[95] These values are almost two orders of magnitude smaller than those reported for the addition of radicals produced in the photocleavage of benzoin derivatives. This low reactivity makes these radicals very inefficient in promoting the polymerization under normal irradiation conditions.[46,124,167]

2. Processes Mediated by Charge Transfer Complexes

Charge transfer complexes can take place between ground-state molecules and/or involve an excited molecule (exciplex). Most of the photoinitiations explained in terms of charge transfer complexes involve free radical production following exciplex formation. However, ground-state complex photolysis has also been invoked as a source of active free radicals. MMA photopolymerization using photoinitiator systems containing chloro-derivatives of acetic acid in combination with dimethylaniline has been explained in terms of the photolysis of an acid-dimethylaniline complex whose initiation efficiency increases with the number of chlorine atoms.[168]

Free radical formation following the interaction of an excited carbonyl and an additive frequently occurs after formation of a charge transfer complex intermediate. In these systems, the minimum mechanism needed to explain the results must consider the steps leading to the complex formation and its decomposition:

$$\text{>C=O* + R–H} \rightarrow \left[\text{>Ċ–O}^- ... \text{H–R}^{.+} \right] \rightarrow \text{>Ċ–OH + R·} \tag{87}$$

decay processes

From the charge transfer complex and/or the caged or free ions, free radicals can be produced simply by bond cleavage, hydrogen transfer, or proton transfer. The rate of the process, as well as the free radical yield, is extremely dependent upon the electron affinities and ionization potentials of the compounds involved. Also, the behavior of the system strongly depends on the solvent characteristics.

Carbon tetrachloride addition increases Φ values of acetone and ethyl pyruvate.[12] Carbon tetrachloride quenches both the excited singlet and triplet states of carbonyl compounds by an exciplex mechanism that leads to the formation of chloride atoms and trichloromethyl radicals.[169,170]

$$A* + CCl_4 \rightarrow A + Cl_3C· + Cl· \tag{88}$$

A similar mechanism has been proposed for the polymerization of MMA sensitized by 1-benzyl-1,4-dihydronicotinamide in the presence of carbon tetrachloride.[171] The carbon tetrachloride-induced polymerization can be considered then to arise simply from a homolytic bond dissociation through the charge transfer complex.

Most photopolymerizations mediated by charge transfer complexes involve radicals produced by a formal hydrogen transfer. This process can take place in the charge transfer complex, or be due to electron transfer followed by proton transfer. The most widely studied systems comprise amines as hydrogen donors.[2,79,172]

3. Benzophenone in the Presence of Aliphatic Amines

The photoreduction of benzophenone triplet by aliphatic amines has been the subject of several studies. Based on photoreduction quantum yields and phosphorescence quenching studies, it was initially proposed that the original complex resulting from the interaction of the triplet and amine (i.e., triethyl amine) could decay by proton transfer, generating the ketyl XXIX and the amine-derived radical XXX, or by spin inversion followed by back electron transfer to generate the ground-state reactants:[173-175]

$$>C=O* + >CH-N(R)_2 \longrightarrow [>Ċ-O^- ... >CH-Ṅ^+(R)_2] \tag{89}$$

$$>C=O + >CH-N(R)_2 \qquad >Ċ-OH + >Ċ-N(R)_2 \qquad >Ċ-O^- + >CH-Ṅ^+(R)_2$$

XXIX   XXX

Competition between proton transfer and back electron transfer was invoked to explain the rather low photoreduction yields obtained under steady-state irradiation. Subsequently, nanosecond laser flash photolysis studies have shown that the primary quantum yield of the

TABLE 19
Benzophenone Triplet Quenching by Amines[a]

Amine	$k_{37} \times 10^{-9}$ (1/M s)	Ref.
Triethylamine	3	166
Di-*n*-propylamine	3.4	166
sec-Butylamine	0.23	166
	0.3[b]	166
tert-Butylamine	0.064	166
	0.11[b]	166
Cyclohexylamine	0.33	166
Methyldiethanolamine	1.3	180
1,4-Diazabicyclo[2.2.2]octane	0.56	180
Methyl *p*-amino benzoate	2	180

[a]   Benzene as solvent unless it is noted.
[b]   In acetonitrile.

ketyl radical is nearly one, indicating that proton transfer must be considerably faster than back electron transfer.[166] A mechanism involving fast electron transfer followed by an efficient proton transfer has received substantial support by direct measurement of the electron transfer process using picosecond absorption techniques.[176-179] These studies also allow the measurement of the intraion-pair proton transfer in the subnanosecond region. Further studies have evidenced that the photoreduction proceeds by rapid electron transfer to form the solvent-separated ion pair. This process is followed by diffusion to form a contact ion pair from which proceeds the proton transfer:[178,179]

$$>\!C\!=\!O^* \; + \; >\!CH\!-\!N\!< \; \rightarrow \; \rightarrow \; >\!\dot{C}\!-\!O^- \; + \; >\!CH\!-\!\dot{N}^+\!<$$

contact ion pair
↓

$$>\!\dot{C}\!-\!OH \; + \; >\!\dot{C}\!-\!N\!< \tag{90}$$

The dynamics of the interconversion between these ion pairs, as well as the yield of free and noncaged radicals, is highly dependent on the medium polarity. The reaction rate of the overall quenching process correlates with the ionization potential of the amine, tertiary amines being more effective than the corresponding secondary and primary derivatives (Table 19). However, the ketyl quantum yields are nearly one, irrespective of the amine substitution degree.[166] Quenching of other aromatic carbonyl compounds, such as fluorenone,[181,182] anthraquinone,[183] and thioxanthone,[184] also takes place through a charge transfer mechanism.

A peculiar feature in the quenching of benzophenone by aliphatic amines is the reduction of a second molecule of the carbonyl in the ground state by the α-aminoalkyl radical:[177,185]

$$>\!C\!=\!O \; + \; >\!CH\!-\!\dot{C}H\!-\!N(R)_2 \; \rightarrow \; >\!\dot{C}\!-\!OH \; + \; >\!C\!=\!CH\!-\!N(R)_2 \tag{91}$$

Similar processes have been reported to occur for different benzophenone derivatives,[166,177,185] benzil,[186,187] thioxanthone,[187] and biacetyl.[188] In the presence of monomers, this process competes with the addition of the α-aminoalkyl radical to the monomer double bond.[187]

TABLE 20
Data on the Acrylamide Polymerization Sensitized by Ionic Benzophenone Derivatives in the Presence of Amines[a]

p-Substituents	Amine	$k_{37} \times 10^{-8}$ (1/M s)	$k_{36} \times 10^{-8}$ (1/M s)	Rp $\times 10^{-3}$ (M/ s)	Φ_m	Φ	Ref.
$-CH_2SO_3^- Na^-$	Triethanolamine	12					199
	Triethylamine	11			200		200
	—		8.8		65		200
$-CH_2-N^+(CH_3)_3Cl^-$	Triethanolamine	16					199
	Triethanolamine			2.9	870	0.23	200
	Methyldiethanolamine			2.3	700	0.15	200
	Triethylamine	5		1.65	400	0.06	200
	—		4.9				200

[a] Solvent water. Polymerization with 0.05 M of amine and monomer concentration 0.7 M.

initiation has been interpreted in terms of complex formation between the anionic triplet and acrylamide from which it is generated an active biradical.[200] Another significant difference between both types of photoinitiators results from the dependence of the polymerization rate on the monomer concentration. The orders in monomer determined in the presence of amines were 1.0 for the cationic compound and 1.6 for the sulfonic derivative, suggesting a more complex initiation mechanism for the later type of compounds.[200]

Recent works[202-205] have related the photoinitiator activity of water-soluble benzophenones to the photoreduction properties of the amines. Polymerization rates of a monoacrylic resin and 2-hydroxyethyl methacrylate are given in Table 21. These data show that efficient polymerization of 2-hydroxyethyl methacrylate is obtained only when tertiary amines are employed in the absence of oxygen, showing a strong correlation with the amine ionization potential.[204] For a given amine, anionic benzophenone derivatives appear as more efficient than cationic compounds, a result opposite to that reported in acrylamide polymerization.[200]

The results obtained have been explained in terms of an initial electron transfer from the amine to the lowest triplet of the benzophenone derivatives, followed by proton transfer to produce the ketyl and amino-derived free radicals. This proposal is supported by the observed relationship between the photoreduction quantum yield of the ionic derivative by 2-N,N-diethanolamine and the polymerization quantum yield of the monoacrylic resin.[203] Polymerization rates measured with different amines also correlate with the intensity of the end of the transient absorption of $\approx$540 nm, suggesting a close relationship between ketyl radical yield and photopolymerization efficiency.[205]

6. Fluorenone in the Presence of Amines

The photophysics of fluorenone in solution exhibits peculiar features as a consequence of the sensitivity of the relative position of its excited states to the solvent polarity. The lowest singlet excited state is $\pi\pi^*$ in character in polar solvents, while it is predominantly $n\pi^*$ in nonpolar media. The lowest triplet remains in predominantly $n\pi^*$ character in both types of media.[49,206,207] The strong displacement of the relative energy levels generates significant changes both in intersystem crossing rate and the triplet quantum yield with the solvent polarity. Triplet quantum yield decreases when the solvent polarity increases, from values near 1 (i.e., in cyclohexane) to 0.46 in acetonitrile.[49,207]

Fluorenone is photoreduced in the presence of amines; the rate of the process depends on the amine ionization potential.[173,181,208-211] Values obtained by measuring the decrease in photoreduction yields elicited by stilbene addition range from $\approx 10^4$ M^{-1} s^{-1} for primary amines to nearly diffusion control for alkylanilines.[210] The excited singlet is also efficiently quenched by secondary and tertiary amines, as revealed by fluorescence quenching measurements. Representative data collected in Table 22 also give overall photoreduction yields. These last values are dependent both on the amine type and concentration and the solvent. Photoreduction yields by tertiary alkyl amines, at low amine and fluorenone concentrations in nonpolar solvents, approach values of $\approx$1. The 9-hydroxy-9-fluorenyl radical (XXXIX), produced by hydrogen abstraction from the amine

$$\text{(93)}$$

XXXIX

has been detected in polar solvents.[213] Hydroxysubstituted tertiary amines present higher photoreduction rates, with overall yields of nearly two. On the other hand, alkylanilines, in spite of their high reactivity, do not produce significant fluorenone photoreduction.

TABLE 21

Photopolymerization Data of Vinyl Monomers Initiated by Water-Soluble Benzophenones in the Presence of Amines

p-Substituent	Monomer or solvent	Amine	Rp (%)	Φ_m [a]	IP (eV)	Absorption[b]	Φ_{red} [c]	Ref.
XXXV $-OCH_2CH(OH)CH_2N(CH_3)_3Cl^-$	HEMA[d]	N-Methyldiethanolanine	2.31		7.2			205
	HEMA	Tri-n-butylamine	2.06		7.4			205
	HEMA	Triethylamine	0.1		7.85			205
	HEMA	Diethylamine	0.03		8.40			205
	HEMA	Dicyclohexylamine	0.01		9.20			205
XXXVI $-OCH_2CH_2CH_2N(CH_3)_3SO_3^-CH_3$	2-Propanol	Tri-n-butylamine				0.039		205
	2-Propanol	Triethylamine				0.036		205
	2-Propanol	Diethylamine				0.034		205
	2-Propanol	Dicyclohexylamine				0.031		205
XXXVI $-OCH_2CH_2CH_2N(CH_3)_3SO_3^-CH_3$	Acrylic resin[e]	N,N'-Dimethylaniline		19,700			0.60	203
XXXV $-OCH_2CH(OH)CH_2N(CH_3)_3Cl^-$	Acrylic resin	N,N'-Dimethylaniline		19,500			0.40	203
XXXIV $-CH_2SO_3^-Na^+$	Acrylic resin	N,N'-Dimethylaniline		24,460			1.5	203
XXXVII $-OCH_2CH(OH)CH_2SO_3^-Na^+$	Acrylic resin	N,N'-Dimethylaniline		21,886			0.93	203
XXXVIII $-OCH_2CH_2CH_2SO_3^-Na^+$	Acrylic resin	N,N'-Dimethylaniline		20,372			0.73	203

[a] Φ_m are expressed as the number of grams of monomer polymerized by mole of absorbed photons.

[b]   Transient absorption maxima (520—550 nm).

[c]   In the presence of 2-N,N-diethanolamine, in water.

[d]   Monomer, hydroxyethylmethacrylate (HEMA), monomer/water 1:1.

[e]   Monomer/water (1:4) (v/w); 1% of amine.

TABLE 22

Quenching of Excited States and Photoreduction of Fluorenone by Amines

Amine	Solvent	Φ_{red}[a]	$k_{37} \times 10^{-8}$[b] (1/M s)	$k_{31} \times 10^{-9}$[c] (1/M s)	Ref.
Triethylamine	Bz	0.8 (10.01—0.1 M)	0.23		210
	Bz	0.68 (0.1 M)			212
	Bz/MMA (1:1)	0.18 (0.1 M)			212
	Bz			10	208
Dimethylethanolamine	Bz	1.22 (0.33 M)			212
	Bz/MMA (1:1)	0.37 (0.33 M)			212
	Cyclohexane		0.76		210
Diethylethanolamine	Bz	1.9 (0.01 M)	0.54		210
Triethanolamine	Bz	2.0 (0.01 M)	2.2		210
Diethylamine	Bz			6.5	208
Diisopropylamine	Bz	0.01 (0.012 M)			210
	Bz			3.8	208
sec-Butylamine	Bz	0.04 (0.1 M)	3.9		210
Aniline	Cyclohexane	0.01 (0.02 M)			210
N-Methylaniline	Cyclohexane		2.2		211

[a]   Fluorenone concentration ≤ 0.01 M, in parenthesis is given the amine concentration.
[b]   Determined from the retardation of the photoreduction by stilbene.
[c]   Determined from Davis et al.,[208] $\tau = 2.8$ ns.

Fluorenone photolysis in the presence of tertiary amines has been extensively employed in polymerization studies.[2,45,85,212,214] MMA photopolymerization initiated by fluorenone photolysis in the presence of triethylamine and dimethylethanolamine presents a maximum rate at a given amine concentration, with smaller rates both at higher and lower amine concentrations.[2,212] Similar results were obtained in acrylonitrile polymerization initiated by fluorenone in the presence of aliphatic amines.[85] These results can be interpreted in terms of the occurrence of Processes 31 and 37, with $\alpha_{37} \gg \alpha_{31}$.

MMA has been reported to quench both the excited singlet and triplet states of fluorenone in acetonitrile, with rate constants of 4.3×10^6 and 1×10^3 M^{-1} s^{-1}, respectively.[215] From these rate constants and the singlet and triplet lifetimes, it is concluded that the ratio between the triplet and singlet processes is 0.8 when the MMA concentration is 3.8 M. These reactions were invoked to explain the small ($\Phi = 2.2 \times 10^{-4}$) polymerization observed in the absence of additives. Competition of Processes 31 and 37 with Processes 30 and 36 leads to a reduction of fluorenone photoreduction yields by amines in the presence of MMA.

Φ values for the polymerization of MMA in the presence of amines reach up to *circa* 0.3. However, the $k_p/k_t^{1/2}$ calculated value is significantly smaller than those found in other systems,[212] suggesting significant termination by primary radicals. In this regard, the fluorenone-derived free radical XXXIX behaves similarly to other radicals derived from hydrogen transfer to aromatic carbonyl compounds. Also, the high values for the order in monomer (1.9 to 2.2) found in the acrylonitrile polymerization sensitized by fluorenone and aliphatic alkyl amines could be interpreted in terms of competition between the addition of radical XXXIX to the monomer and termination processes involving the primary radical.[85]

Copolymers of MMA and 2-vinylfluorenone have been employed to photoinitiate MMA polymerization in the presence of tertiary amines.[212] The macrosensitizer behaves as the small model compound, being the polymerization rates, when copolymers with fluorenone content between 1 and 68% are employed, slightly higher than those obtained for the

TABLE 23
Initiation Efficiencies in the Vinyl Polymerization Sensitized by the Benzil-Coinitiator System[a]

Monomer	Coinitiator	Benzil concentration (mM)	Φ
MMA	TEA 0.014 M	1.25	0.45
	TEA 0.0037 M	1.55	0.2
	TEA 0.014 M	1.55	0.49
	TEA 0.025 M	1.55	0.49
	TEA 0.014 M	3.15	0.23
	TEA 0.014 M	6.6	0.12
Vinyl acetate	TEA 0.014 M	1.55	0.53
	TEA 0.014 M	3.15	0.20
	TEA 0.014 M	6.6	0.13
MMA	THF 1.84 M	1.55	0.16
	THF 1.84 M	6.6	0.14
	N,N-Dimethylaniline 0.01 M	1.55	0.41
	N,N-Dimethylaniline 0.01 M	3.15	0.40
	N,N-Dimethylaniline 0.01 M	5.25	0.39
	N,N-Dimethylaniline 0.01 M	6.6	0.38

[a]   In monomer/benzene 1:1 (v/v), $\gamma_{\text{irradiation}}$ = 366 nm.

From Encinas et al.[187]

monomer chromophore. The polymeric photoinitiation leads to a mixture of both homo and graft polymer, this last polymer arises from chain initiation and/or termination processes involving the fluorenyl macroradical.

7. α-Dicarbonyl Compounds and Amines

Benzil behaves as a mild photoinitiator due to its capacity to produce active free radicals through the formation of a charge transfer intermediate.[152] Interaction of the excited triplet with aliphatic amines further increases the photoinitiation efficiency.[187,216] The quenching process involved takes place by a charge transfer mechanism with overall rate constants in the $10^9\,M^{-1}\,\text{s}^{-1}$ range for tertiary amines, both in nonpolar (heptane) and polar (acetonitrile) solvents.[217]

Encinas et al.[187] have carried out an extensive study of the factors determining the initiation efficiencies in the benzil/triethylamine and benzil/N,N-dimethylaniline systems, where quenching by the amine competes with quenching by the monomer. If only triplet processes are considered, the value of Φ is related to the amine concentration by

$$\Phi \approx \frac{k_{37}[\text{Amine}]}{k_{37}[\text{Amine}] + k_{36}[\text{M}] + \Sigma k} \tag{94}$$

As expected from Equation 94, at a fixed benzil concentration the value of Φ increases with the amine concentration until a plateau is reached. Typical photoinitiation efficiency values are given in Table 23.

A noticeable difference was observed when triethylamine or N,N-dimethyl aniline was employed as coinitiator. With the former compound, the initiation efficiency decreases when the benzil concentration increases, while with N,N-dimethylaniline, the value of Φ is independent on the benzil concentration.

Initiation in the benzil/triethylamine system is due to the occurrence of Processes 95 to 97

$$[\text{Benzil}]^{\delta-}\; {}^{\delta+}\text{TEA}] \rightarrow \underset{\text{XL}}{\text{Ph--CO--}\overset{\overset{\displaystyle\text{OH}}{|}}{\underset{\text{XLI}}{\text{C}}}\text{--Ph}} \;+\; CH_3\text{--}\dot{C}H\text{--}N(C_2H_5)_2 \tag{95}$$

$$CH_3\text{--}\dot{C}H\text{--}N(C_2H_5)_2 \;+\; M \rightarrow \text{addition} \tag{96}$$

with the ketyl radical XLI being unable to initiate the polymerization. The decrease in Φ produced by the increase in benzil concentration is due to competition of Reaction 96 with ground-state benzil reduction by the α-aminoalkyl radical

$$CH_3\text{--}\dot{C}H\text{--}N(C_2H_5)_2 \;+\; Ph\text{--CO--CO--Ph} \rightarrow Ph\text{--CO--}\overset{\overset{\textstyle\;}{|}}{\underset{\underset{\textstyle\text{OH}}{|}}{\dot{C}}}\text{--Ph} \;+\; CH_2\text{=}CH\text{--}N(C_2H_5)_2 \tag{97}$$

Reaction 97 is a very fast process for which has been reported a rate constant of $1.2 \times 10^9\ M^{-1}\ s^{-1}$.[186]

Extrapolation of Φ to high amine and low benzil concentrations renders a value of nearly one. It it is considered that the ketyl radicals do not initiate the polymerization, this result implies that the charge transfer complex produced in the excited-benzil/triethylamine (XL) interaction generates active free radicals almost quantitatively. In fact, under conditions of almost total quenching by the amine, the value of Φ will be given by

$$\Phi \;=\; \alpha_{95}\,\frac{k_{96}[M]}{k_{96}[M] \;+\; k_{97}[\text{Benzil}]} \tag{98}$$

Plots of $1/\Phi$ against the benzil concentration allows the evaluation of α_{95} and k_{97}/k_{96}. Values of α_{95} of one were obtained both for MMA and vinyl acetate in benzene. A considerably smaller value (0.2) was obtained for MMA in acetonitrile. Taking $k_{97} = 1.8 \times 10^9$,[186] the values of k_{96} obtained for this procedure were 3.1×10^5 for MMA and vinyl acetate in benzene, and $9.6 \times 10^5\ M^{-1}\ s^{-1}$ for MMA in acetonitrile. These values are of similar order of magnitude to those obtained for the addition to monomers of the radicals generated in the decomposition of benzoin derivatives.[95]

When N,N-dimethylaniline is employed as coinitiator, no dependence of Φ with benzil concentration is observed due to the lack of α-hydrogens in the amine-derived free radical. A kinetic analysis of MMA polymerization photoinitiated by the benzil/N,N-dimethylaniline has been carried out by Sengupta and Modak,[216] who found low values for the initiator exponent and $k_p/k_t^{1/2}$ ratios. These results were interpreted in terms of the termination by the primary ketyl radicals XLI.

In most systems, and due to values of α_{37} larger than α_{36}, the value of Φ increases in the presence of amines. Aliphatic α-dicarbonyl compounds, at least when MMA is employed as monomer, constitute an exception. In these systems, the initiation efficiency markedly decreases in the presence of either diethylamine or triethylamine. This result is the consequence of a rather high value of α_{36} and low values of both α_{31} and/or α_{37}. In particular, at high amine concentrations where a large fraction of the singlets are quenched by the amines, very low Φ values are obtained, pointing to very low α_{31} values in these systems. CIDEP studies of the reaction between excited 2,3-butanedione and triethylamine show that free radicals are produced only when the triplet state of the carbonyl compound is quenched.[118]

TABLE 24
Rate Constants for the Quenching of Triplet Thioxanthone Derivatives by Amines[a]

Amine	$k_{37} \times 10^{-9}$ (1/M s)				
	TX	**2-Cl-TX**	**2-Methyl-TX**	**2-Isopropyl-TX**	**Ref.**
Ethyl 4-(dimethylamino)benzoate	6	5	4	4	225
Bis(2-hydroxyethyl)methylamine	8	6	6	6	225
Triethylamine		2.5			225
Dimethylaniline		8			225
	9.1				184
2-(Dimethyl)ethyl benzoate		0.85			225
Dibutylamine	2.7				184
tert-Butylamine	0.0046				184

[a]   In benzene as solvent.

Values of α_{37} considerably larger than α_{31} have also been reported for the interaction of excited acetone with both diethylamine and triethylamine.[12]

8. Thioxanthone Derivatives

Thioxanthones (TXs) constitute a particular class of photoinitiators with high photoinitiation efficiencies when irradiated in the near-UV with wavelengths up to 400 nm, in conjunction with activators. This feature makes them particularly useful for the UV curing of coatings, when they are employed in conjunction with activators.[218-220]

TXs present, as fluorenone, a photobehavior that is strongly solvent dependent. The position of the fluorescence band is considerably red shifted when the solvent polarity increases, and the fluorescence quantum yield is considerably higher in polar solvents, i.e., it is nearly zero in n-hexane, 5×10^{-3} in acetonitrile, and 0.45 in trifluoroethanol.[221,222]This effect is due to the proximity of the nπ* and $\pi\pi$* states that modify, through vibronic coupling, the internal conversion rates.[222] Similar proximity effects have also been reported for the triplet state.[223,224] The transient UV absorption spectra of thioxanthone triplet are well characterized, ranging from 600 to 700 nm, the absorption maxima being very dependent on the solvent polarity.[223]

Thioxanthone and its derivatives are efficient photoinitiators only in the presence of electron and/or hydrogen donors.[225,226] Most studies have been carried out employing tertiary amines as coinitiators. Interaction between the excited triplet of TXs and the amines proceeds through a charge transfer intermediate:[184]

$$\text{(reaction scheme)} \tag{99}$$

XLII

XLIII

TABLE 25

Rate Constants for the Reaction of Thioxanthone Triplets with Monomers in Benzene[a]

	$k_{36} \times 10^{-6}$ (1/M s)			
Monomer/initiator	TX	2-Cl-TX	2-Methyl-TX	2-Isopropyl-TX
Styrene	3000	6000	6000	6000
MMA	15	2	3	3
Acrylonitrile	4	0.4	1	1
Vinyl acetate	0.2	0.02	0.03	0.04
N-Vinyl-2-pyrrolidone	40	30	5	6
Butyl vinyl ether	1	1	0.3	0.3

[a] In benzene as solvent.

From Amirzadeh and Schnabel.[225]

generating the ketone radical anion XLII and/or the amino cation radical and the ketyl radical XLIII by proton transfer. The overall rate of the quenching process is dependent on the amine ionization potential (Table 24), reaching the diffusion-controlled limit for tertiary amines.[184,225]

Both the ketyl radical XLIII and the radical anion XLII have been characterized by transient spectroscopy,[226,227] although the optical density of the transient associate to the anion is rather weak. Flash photolysis of several substituted TXs in 2-propanol leads to the formation, through hydrogen abstraction from the solvent, of the corresponding ketyl radicals.[226-228] However, in 2-propanol as solvent the photoinitiation efficiency Φ of TX is extremely low. In the presence of tertiary amines, the TX-derived radical anion is also detected.

Both photoreduction by the solvent and interaction with the amine compete with quenching by the monomer (Table 25). Comparison of the data given in Tables 24 and 25 indicates that, in benzene as solvent, quenching by the monomer will prevail when styrene is employed as solvent, while for the other monomers substantial quenching by the amines could remain even in bulk monomer. The efficiency of the initiation will be determined by the fraction of triplets quenched by the amine and by the fate of the produced radicals. In this regard, it is important to note that the ketyl radicals are fairly unreactive toward the monomers. Values of 4×10^2 and 40 M^{-1} s^{-1} have been reported for the rate constant of the ketyl radical addition to N-vinyl-2-pirrolidone and MMA, respectively.[225]

The spectroscopic properties of different substituted TXs have been related to the values of Φ obtained in the polymerization of n-butyl methacrylate in 2-butanone as solvent when N-diethylmethylamine is employed as cocatalyst.[226,227,229,230] No correlation was observed between luminescence quantum yields and photopolymerization activities. On the other hand, photoinitiation activities correlate with the transient absorption intensity; this result has been interpreted in terms of the participation of the triplet exciplex in the initiation mechanism.[226,227,230] The influence of several amines on the photoinitiation activity of 4-n-propoxy thioxanthone,[228] and the effect of different substituents on the activity of 2-acetoxy-thioxanthone[229] have been reported for the polymerization of n-butyl methacrylate (Table 26). For a given thioxanthone and different amines, there are good correlationships between polymerization rates, amine ionization potentials, and photoreduction quantum yields.[228,229] For a given amine, photoinitiation quantum yields correlate with photoreduction yields, and the introduction of donating groups in the ketone increases the rate of the processes.[229] Substitution at 1-position, to the carbonyl group, enhances the radiationless conversion rate through intramolecular hydrogen bonding in the excited state, reducing the photoinitiation capacity of the ketone. For the acetoxy derivatives, initiation efficiencies of *circa* 0.11 were calculated from polymerization rates, light intensity, and $k_p/k_t^{1/2}$ values.[229]

TABLE 26

Photopolymerization of n-Butyl Methacrylate Initiated by Thioxanthone Derivatives in the Presence of Amines

Ketone	Amine	[Amine] (M)	IP (eV)	%C[a]	Φ_m^b	Φ_{red}^c	Ref.
4-n-Propoxy-TX	Dicyclohexylamine	0.006	9.2	0.004		0.05[d]	228
	Diethylamine	0.015	8.4	0.175			228
	N-Diethylmethylamine	0.013	8.1	0.304			228
	Triethylamine	0.011	7.85	0.583		0.29[d]	228
	Tri-n-butylamine	0.006	7.4	0.658		0.29[d]	228
	N-Diethylmethylamine	0.013		0.09[e]			228
	N-Diethylmethylamine	0.026		1.75[e]			228
	N-Diethylmethylamine	0.039		2.12[e]			228
	N-Diethylmethylamine	0.052		2.40[e]			228
2-Acetoxy-TX	Tri-n-butylamine	0.006			19	0.6	229
1-Methyl-acetoxy-TX	Tri-n-butylamine	0.006			2.6	0.001	229
3-Methyl-2-acetoxy-TX	Tri-n-butylamine	0.006			23	0.54	229
4-Methyl-2-acetoxy-TX	Tri-n-Butylamine	0.006			13	0.73	229
1,3-Dimethyl-2-acetoxy-TX	Tri-n-butylamine	0.006			3.3	0.005	229
1,4-Dimethyl-2-acetoxy-TX	Tri-n-butylamine	0.006			2.6	0.003	229
3,4-Dimethyl-2-acetoxy-TX	Tri-n-butylamine	0.006			12	0.56	229
1,3,4-Trimethyl-2-acetoxy-TX	Tri-n-butylamine	0.006			2.6	0.001	229

[a] Conversion of n-butyl methacrylate after 60 min irradiation.
[b] Butyl acrylate in bulk, $\lambda_{irradiation} = 366$ nm.
[c] In benzene solution, using diethanolamine 0.001 M.
[d] In 2-propanol, amine 5×10^{-5} M.
[e] After 30 min irradiation.

In order to overcome problems of poor solubility, polymeric photoinitiators such as XLIV have been developed:[231,232]

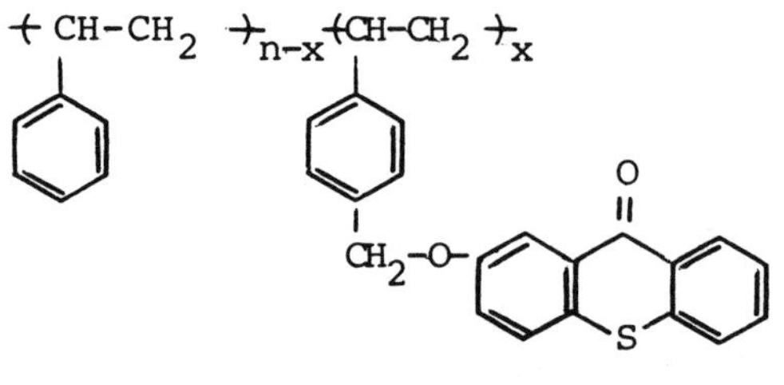

XLIV

Spectroscopic properties and the initiation efficiency of photoinduced MMA polymerization were similar for 2-benzoyloxythioxanthone XLV

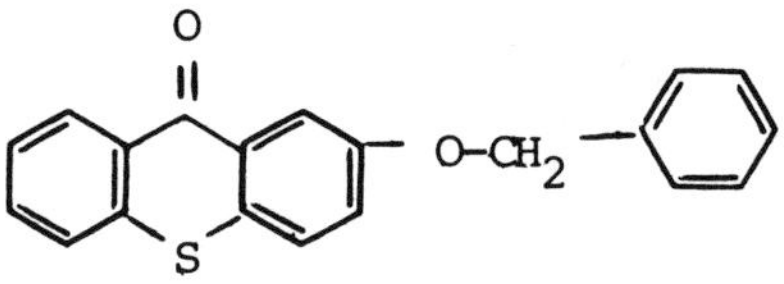

XLV

and the macroinitiator XLIV with 15% of thioxanthone groups. A value of $\Phi = 0.07$ has been reported when MMA is employed as monomer.[232]

Water-soluble thioxanthone derivative photoinitiators have received increasing interest due to their applications in the grafting of monomers onto natural polymers and the UV curing of water-based ink systems.[233-235] Water solubility is achieved by introducing ionic substituents in one of the benzene rings. A series of compounds of general formula

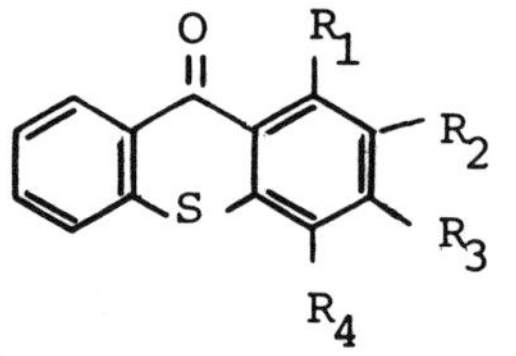

have been prepared where in one of the R_2, R_3, or R_4 is one of the following substituents:

$$A = -O\,CH_2\,CH_2\,CH_2N^+\,(CH_3)_3\,SO_3^-\,CH_3$$

$$B = -O\,CH_2\,CH\,(OH)\,CH_2N^+\,(CH_3)_3\,Cl^-$$

$$C = -O\,CH_2\,CH_2\,CH_2\,SO_3^-\,Na^+$$

$$D = -O\,CH_2\,COOH$$

and the other R substituents are either a hydrogen atom or a methyl group:

XLVI $R_2 = B$; $R_1 = R_3 = R_4 = H$

XLVII $R_2 = B$; $R_1 = R_4 = H$; $R_3 = $ methyl

XLVIII R$_2$ = B; R$_4$ = H; R$_1$ = R$_3$ = methyl

XLIX   R$_2$ = B; R$_1$ = R$_3$ = H; R$_4$ = methyl

L   R$_2$ = B; R$_1$ = R$_3$ = R$_4$ = methyl

LI   R$_2$ = B; R$_1$ = H: R$_3$ = R$_4$ = methyl

LII   R$_2$ = A; R$_1$ = R$_3$ = R$_4$ = H

LIII   R$_3$ = A; R$_1$ = R$_2$ = R$_4$ = H

LIV   R$_4$ = A; R$_1$ = R$_2$ = R$_3$ = H

LV   R$_3$ = B; R$_1$ = R$_2$ = R$_4$ = H

LVI   R$_4$ = B; R$_1$ = R$_2$ = R$_3$ = H

LVII   R$_4$ = B; R$_2$ = R$_3$ = H; R$_1$ = methyl

LVIII   R$_4$ = B; R$_2$ = H; R$_1$ = R$_3$ = methyl

LIX   R$_4$ = B; R$_3$ = H; R$_1$ = R$_2$ = methyl

LX   R$_2$ = C; R$_1$ = R$_3$ = R$_4$ = H

LXI   R$_2$ = D; R$_1$ = R$_3$ = R$_4$ = H

Fluorescence emission of these compounds is highly shifted to the red and the fluorescence quantum yields in water solutions are considerably higher than those measured in 2-propanol. The presence of a methyl substituent in the *ortho* position with respect to the carbonyl group (compounds LVII, XLVIII, and LIX) notably reduces the emission yields due, as in the oil-soluble thioxanthone derivatives, to intramolecular hydrogen bonding.[234,235]

Due to their rather large singlet lifetimes, thioxanthone derivatives in their first excited singlet states are quenched by electron donors.[173] Thioxanthone derivatives bearing both cationic and anionic substituents are quenched by methyldiethanolamine with rate constants in the 2 to 4 × 10^9 M^{-1} s^{-1} range. Quenching rate constants correlate with the amine ionization potential, although hydroxy amines slightly depart from the correlationship.[236] Quenching of the triplet state shows a similar correlation with the amine ionization potential but the rate of the process is nearly two orders of magnitude slower[236] (Table 27). Transient absorption spectroscopy reveals the presence of both the ketyl and anion radicals.[233,235]

Photoinitiation efficiencies measured in the polymerization of 2-hydroxyethyl methacrylate and acrylamide in aqueous solutions are given in Table 28. Substitution increases the photopolymerization rates, being the higher values obtained for the compounds substituted at the 4-position. Even, high efficiencies are obtained for 1-methyl-substituted compounds in spite of significant intramolecular hydrogen abstraction, evidenced by the lower emission quantum yields. For a given thioxanthone derivative, there is a good correlation between Φ values and the amine ionization potential.[234-236]

Early reports ascribed the polymerization activity to the excited singlet/amine interaction.[233-235] However, the effect of the amine concentration upon the polymerization rate indicates that the active free radicals are produced predominantly from the triplet/amine interaction. In fact, polymerization rates increase with the amine concentration until a maximum value is reached, with a subsequent decrease at higher amine concentration,[236] a behavior that indicates that $\alpha_{31} \ll \alpha_{37}$.

Water-soluble thioxanthone derivatives are quenched by acrylamide. Triplet quenching rates are in the 5 to 12 × 10^4 M^{-1} s^{-1} range.[236] This process can become significant only at low amine concentrations and/or for those amines of low reactivity.

TABLE 27
Quenching Rate Constants of the Excited Singlet and Triplet
States of Water-Soluble Thioxanthones by Amines[a]

Ketone structure	Amine	IP (eV)	$k_{31} \times 10^{-9}$ (1/M s)	$k_{37} \times 10^{-7}$ (1/M s)
LII	Methyldiethanolamine (MDEA)		2.4	3.2
LIII	MDEA		4.6	
XLVI	MDEA		2.3	
XLVII	MDEA		2.9	5.0
L	MDEA		2.7	
LI	MDEA		3.1	3.0
LV	MDEA		4.2	3.7
LVI	MDEA		2.8	
LVII	MDEA		1.6	
LIX	MDEA			7.8
LXI	MDEA		1.8	6.5
LX	MDEA		2.4	4.4
	Diazabicyclooctane	7.52	3.9	
	Triethylamine	8.08	2.9	
	Diethylamine	8.67	0.6	
	Triethanolamine	8.7	3.0	
	Dimethylethanolamine	8.85	1.3	
	n-Butylamine	9.29	0.05	
	Ethanolamine	9.85	0.02	

[a]   In water.

From Lougnot et al.[236]

TABLE 28
Polymerization of Vinyl Monomers Photoinitiated by Water-Soluble Thioxanthones

TX structure	Amine	IP (eV)	%C[a]	Φ_m[b]	Ref.
LII	N-Diethylmethylamine (DEMA)		1.13	675	235,236
LIII	DEMA		0.7	1502	235, 236
LIV	DEMA		4.23	1140	235, 236
XLVI	DEMA		1.57	712	235, 236
LV	DEMA		0.71	1640	235, 236
LVI	DEMA		Gelation	1200	235, 236
LVIII	DEMA		1.66	1423	235, 236
XLVII	DEMA		1.72	790	234, 236
XLIX	DEMA		2.99	926	234, 236
L	DEMA		2.93	1800	234, 236
LVI	Tri-n-butylamine	7.6	Gelation		235
	Triethylamine	8.08	5.5		235
	Diethylmethylamine	8.30	4.23		235
	Diethylamine	8.67	1.13		235
	Dicyclohexylamine	9.20	0.8		235

[a]   2-Hydroxyethyl methacrylate in water (2:1), DEMA = 0.17 M, Allen et al.[234,235]
[b]   Acrylamide in aqueous solutions, methyldiethanolamine 0.05 M, Lougnot et al.[236]

V. SOLVENT EFFECTS ON PHOTOINITIATION EFFICIENCY

The solvent can modify the photoinitiation rate by direct interaction with the excited states (Processes 29 and 35) and/or by modification of the yields and α values of the other photochemical and photophysical processes.

Participation of the solvent as coreactive can be considered as a particular example of the effect of additives, and some particular examples have already been mentioned in previous sections. Extensive studies have been carried out employing anthraquinone (AQ) as sensitizer.

Solvents bearing labile hydrogen atoms can photoreduce AQ from its lowest, $n\pi^*$, triplet.[237-243] The produced anthrasemiquinone radicals (AQH$-$) yield anthrahydroquinone (AQH$_2$) and AQ,[238] according to

$$^3AQ^* \rightarrow AQ \tag{100}$$

$$^3AQ^* + RH \rightarrow AQH\cdot + R\cdot \tag{101}$$

$$R\cdot + AQ \rightarrow AQH\cdot + \dot{R}(-H) \tag{102}$$

$$2AQH\cdot \rightarrow AQH_2 + AQ \tag{103}$$

The AQH$\cdot$ radical has been well characterized by its absorption spectrum at long wavelengths (580 to 700 nm).[183,240,244] The AQ photoreduction quantum yield is solvent dependent, and values of 0.98 in ethanol,[238] 0.11 in hexane,[238] and 0.3 in toluene[183,244] have been reported. The solvent effect is mainly determined by the rate of Reaction 102. The relatively slow rate of this process, when hexane is employed as solvent, would favor the occurrence of

$$AQH\cdot + R\cdot \rightarrow AQ + RH \tag{104}$$

that decreases the photoreduction yield.

Free radical vinyl polymerization can be initiated by AQ irradiation in solvents bearing labile hydrogen atoms. A detailed study using THF as solvent shows that the initiating species is the THF-derived radical, and that the AQH$\cdot$ radical is an effective terminating species.[242]

Ledwith et al.[242] have reported MMA polymerization rates in several solvents and the results indicate that the process is determined by the rate of hydrogen abstraction by the triplet AQ. Similar results have also been reported by Li et al.,[245] who found that alcohols are better solvents for the polymerization than hydrocarbons, a result related to the relative readiness to release a hydrogen atom from the solvent.

Peculiar results are obtained when photopolymerizations are carried out in alcohol-hydrocarbon mixtures. For example, the polymerization in mixtures of cyclohexane and alcohols shows a maximum rate at a certain concentration of alcohol[245-248] (Table 29). The presence of the maximum depends on the hydrocarbon employed and, for a given alcohol/hydrocarbon pair, on the monomer being polymerized. For example, the maximum is present when MMA is employed, but it does not appear in acrylates or other methacrylates such as ethyl acrylate, ethyl methacrylate, or *n*-butylmethacrylate.[247] When isopropanol is employed as alcohol and MMA as monomer, maximum rates at a given solvent composition are observed when the mixtures comprise saturated hydrocarbons, but not when aromatic or unsaturated hydrocarbons are considered[248] (Table 29). The presence of the maximum is also observed when other quinones such as 2-*tert*-butylanthraquinone or α-naphthoquinone are used, but is absent when initiators such as benzophenone[246] or benzoyl peroxide[245] are

TABLE 29

**Anthraquinone-Sensitized Photopolymerization of Methyl
Methacrylate in Hydrocarbon-Alcohol Solvent Mixtures**

Solvent mixtures	Alcohol content[a]	Maximum conversion	Ref.
Cyclohexane-alcohols[b]			
Methanol	4.1	34	246
Ethanol	7.4	40.1	246
n-Propanol	10.5	36.5	246
2-Propanol	13.5	35.8	246
n-Butanol	17.5	38	246
sec-Butanol	17.5	30.2	246
tert-Butanol	33.5	27	246
2-Propanol-hydrocarbons[c]			
n-Hexane	25		248
n-Octane	50		248
n-Decane	70		248
n-Dodecane	80		248
Cyclohexane	10		248
Decalin	20		248
2,4-Dimethylpentane	40		248
1-Octene	Not found		248
Toluene	Not found		248

[a]   At the maximum conversion (% mol).
[b] $[MMA] = 0.94\ M$.
[c]   Initiator β-methylanthraquinone, $[MMA] = 1.57\ M$.

employed. Saturated hydrocarbons are not able to photoreduce AQ to AQH_2, but this compound is formed in alcohols and mixed solvents comprising alcohols and saturated hydrocarbons.[245,248] AQ is consumed when photolyzed both in hydrocarbons or alcohols, the consumption quantum yields being highest in alcohols bearing labile hydrogen atoms such as isopropanol. However, the photoreduction quantum yields do not correlate with photoinitiation efficiencies,[245,248] suggesting that in alcohols larger amounts of AQH_2 are produced through the occurrence of reactions

$$R_2CHOH + AQH\cdot \rightarrow R_2\dot{C}OH + AQH_2 \tag{105}$$

and

$$R_2\dot{C}OH + AQH\cdot \rightarrow RCOR + AQH_2 \tag{106}$$

The effect of the solvent can be particularly important in reactions where the photoinitiator system produces the active radicals through a charge transfer mechanism, since both the rate of the processes and the free radical yields can be modified by the solvent.

Triethylamine (TEA) photoreduces AQ via formation of an exciplex (LXII) between a ground-state TEA molecule and the AQ triplet.[183,244,249] The exciplex changes to a contact pair (LXIII), from which can take place proton transfer to produce the AQH· and amine-derived (TEA·) radicals:

TABLE 30
Polymerization Data of Methyl
Methacrylate Photoinitiated by
Anthraquinone-Triethylamine

Solvent[a]	[TEA] (*M*)	Φ^b
MMA[c]	0.086	0.144
Benzene	—	0.016
	0.004	0.14
	0.012	0.18
	0.043	0.44
	0.17	0.44
Ethanol	—	0.16
	0.086	0.06
Isopropanol	—	0.2
	0.086	0.10
Acetonitrile	0.086	0.01
Acetoni-trile:benzene (1:2)	0.086	0.06

[a]   Monomer/solvent (1:1) (v/v).
[b]   DTBP as reference photoinitiator.
[c]   Bulk monomer.

From Encinas et al.[250]

$$AQ + TEA \tag{107}$$

$$(AQ^- \; TEA^+) \rightarrow (AQ^-...TEA^+) \rightarrow AQH\cdot + TEA\cdot \tag{108}$$

$$LXII \qquad\qquad LXIII$$

$$+ TEA$$
$$Triplex \tag{109}$$

The exciplex (LXII) and the contact ion pair (LXIII) have been characterized by time-resolved spectroscopy.[183,244] In toluene as solvent, the decay constant of the exciplex and the photoreduction yield, under total quenching conditions, are independent on the amine concentration. On the other hand, in ethanol as solvent, the decay constant of the exciplex increases and the photoreduction yield decreases when the amine concentration increases, suggesting a triplex formation as a competitive deactivation pathway of the exciplex (Reaction 109). The contact ion pair (LXIII) is relatively stable in polar solvents such as ethanol, with a decay constant of 12 s^{-1}. In nonpolar solvents such as toluene, the decay is considerably faster, with a decay rate constant of 2×10^5 s^{-1}.[183]

The polymerization of MMA photoinitiated by the AQ/TEA system has been investigated in solvents of different polarity,[250] and the derived Φ values are given in Table 30. In absence of TEA, even in solvents of high hydrogen-donating capacity, AQ is a rather poor photoinitiator. However, high efficiencies are obtained in the presence of amines in nonpolar solvents. In polar solvents, such as acetonitrile and ethanol, initiation efficiencies remain low, even in the presence of amines, and decrease when the monomer concentration increases.

The dependence of AQ photoconsumption and AQH$_2$ formation yields with MMA and amine concentrations is different when benzene or ethanol is employed as solvent. Under conditions of almost total triplet quenching by the amine, an increase in amine concentration reduces both yields when the solvent is ethanol, but does not modify them when the photolysis is carried out in benzene. The different behavior in both solvents is observed in the presence and in the absence of MMA. In both solvents, an increase in MMA concentration reduces the AQ consumption and the fraction of it that appears as AQH_2. Given the large differences between the quenching rate constants by the amine and MMA, these results cannot be due to triplet AQ quenching by the monomer and have been interpreted in terms of the occurrence of a process such as

$$(AQ^- \ TEA^+) + MMA \rightarrow \text{Deactivation} \qquad (110)$$

taking place with partial consumption of AQ without giving AQH_2 or active free radicals. The low polymerization rates obtained in ethanol as solvent are a consequence of the high efficiency of Processes 109 and 110 in this solvent. Higher polymerization rates are obtained in benzene due to the smaller rate of Process 109 in this solvent. However, the occurrence of Process 110 competes with the free radical formation from the exciplex, and leads to a decrease in the photoinitiation efficiency when the monomer concentration increases.[250]

In the photopolymerization of MMA sensitized by the benzil/TEA system, higher efficiencies are also obtained in nonpolar solvents.[187] In a solution 1.55 mM of benzil and with amine concentrations large enough to quench almost totally the excited triplets, the Φ values are 0.49 and 0.16 in benzene and acetonitrile, respectively. The lower values of Φ obtained in acetonitrile can be explained in terms of the behavior expected for the charge transfer complex. This complex can decay by back electron transfer or by producing a pair of free radicals or radical ions. It has been well established from nanosecond studies in the benzophenone-amine systems that solvents of high dielectric constants favor the formation of radical ions over ketyl radicals.[179] The photopolymerization results, extrapolated to low benzil concentrations, show that ketyl radical formation is almost quantitative in benzene, while in acetonitrile it drops to 0.2 due to the predominance of the ion pair formation.

Ghosh et al.[196-198] have analyzed the role of the solvent in polymerizations initiated by irradiation of benzophenone in the presence of several amines. In MMA photopolymerization, the order in monomer depends on the type of solvent, being 1.0 in inert solvents such as benzene, toluene, acetone, and methyl ethyl ketone, <1 in halomethanes (chloroform and carbon tetrachloride), and >1 in alkanols. Initiation rates are independent on the solvent concentration for the inert solvents, increase with the solvent concentration for the halomethanes, and decrease with the alkanol concentration. These results can be interpreted in terms of the influence of the solvent on the quantum yield of free radical formation from the charge transfer complex. Also, when halomethanes are employed, direct quenching of the ketone triplet by the solvent leading to an efficient production of free radicals[169] cannot *a priori* be disregarded.

p-Alkylamino benzophenones are photoinitiators whose efficiencies are strongly affected by the solvent as a consequence of changes in their photochemical and photophysical behavior.[251-254] These compounds present an intense charge-transfer band in the 300 to 400 nm region. The relative position of the charge transfer and the $n\pi^*$ states change with the solvent, rendering the excited-state behavior extremely solvent dependent. The intersystem crossing and photoreduction quantum yields notably decrease when the solvent polarity increases (Table 31), due to the more favorable energy gap between the charge transfer and $n\pi^*$ triplet states in nonpolar solvents.[251] Furthermore, efficient self-quenching reduces the photoreduction yields when the ketone concentration increases. The deactivation kinetic scheme is given by Equations 111 to 118:

$$A + h\nu \rightarrow {}^1A^* \qquad (111)$$

$$^1A^* \rightarrow A \qquad (112)$$

TABLE 31
Photoreduction, Intersystem Crossing, and Polymerization Efficiency of *p*-Aminobenzophenones

Ketone	Ketone × 10³ (*M*)	Solvent	Φ_{isc}	Φ_{red}	Rp × 10⁵ᵃ (*M/s*)	Ref.
LXIV	0.012	Cyclohexane	0.91			251
	0.012	Cyclohexane		0.40		252
	0.05	Benzene	1			251
	0.05	2-Propanol	0.24	$<10^{-3}$		251
	0.05	Ethanol	0.08	$<10^{-4}$		251
LXV	0.04	Cyclohexane		0.58		254
	1	Cyclohexane		0.17		254
	0.011	Benzene	1	0.055		254
	1	Benzene		0.025		254
	0.055	Dimethylformamide	0.11	$<10^{-3}$		254
	1.4	Benzene			2.25	257
	0.1	Benzene			4.45	257
	0.061	Benzene			5.48	257
	0.0084	Benzene			3.51	257
	0.0044	Benzene			2.45	257

ᵃ   Methyl acrylate 3.28 *M*.

$$^1A^* \rightarrow {}^3A^* \tag{113}$$

$$^3A^* \rightarrow A \tag{114}$$

$$^3A^* + RH \rightarrow \ \underset{/}{\overset{\backslash}{C}}-OH + R\cdot \tag{115}$$

$$^3A^* + A \rightarrow {}^3[A_2]^* \ (\text{triplet excimer}) \tag{116}$$

$$^3[A_2]^* \rightarrow 2A \tag{117}$$

$$^3[A_2]^* \rightarrow \ \dot{C}-OH + \dot{C}H_2-N\diagup \tag{118}$$

Amino benzophenones LXIV to LXVI

$$R_1 - C_6H_4 - \overset{\overset{O}{\|}}{C} - C_6H_4 - R_2$$

LXIV $R_1 = R_2 = -N\diagdown{}_{CH_3}^{CH_3}$

LXV $R_1 = -N\diagdown{}_{CH_3}^{CH_3}$ $R_2 = -CH\diagdown{}_{CH_3}^{CH_3}$

LXVI $R_1 = R_2 = -N\diagdown{}_{CH_2CH_3}^{CH_2CH_3}$

have been used as photoinitiators of MMA and methyl acrylate free radical polymerization. The solvent dependence of their photochemical behavior is reflected in their efficiency as photoinitiators.[255-258] When LXVI is employed as a sensitizer, polymerization rates are large in benzene, intermediate in cyclohexane, and null in ethanol.[255] The lower polymerization in cyclohexane is due to the contribution of Process 115, and the lack of polymerization in ethanol is a consequence of the higher rate of Process 112 that reduces the triplet quantum yield. Polymerization rates obtained employing compound LXV as photosensitizer are included in Table 31. Polymerization rates vary linearly with monomer concentration and the square root of light intensity. For small absorbances, the initial polymerization rate varies linearly with the square root of the photoinitiator concentration, while for high absorbances is independent on the photoinitiator concentration in a well-stirred system and decreases as the photoinitiator concentration increases in an unstirred system.[257] These results are then fully compatible with those predicted by Equations 22 and 23. Recently Mateo et al.[258] have reported Φ values of 0.0015 and 0.013 for MMA polymerization by LXV in cyclohexane and benzene, respectively. This difference has been explained in terms of the nature of the polymerization-initiating radicals. In cyclohexane solution and at low LXV ($<5 \ 10^{-5} \ M$), both radicals, the cyclohexyl and the aminoalkyl derived from ketone, participate in the initiation step. When benzene is used as solvent both phenyl and amino alkyl radicals participate in the initiation step at any ketone concentration employed.

Photoinitiation has been also observed when aromatic hydrocarbons are irradiated in the presence of amines.[259,260] Singlet excited states of aromatic hydrocarbons are efficiently quenched by amines through a mechanism that involves as a first step the formation of an exciplex with significant charge transfer contribution,[261-266] whose behavior is strongly dependent upon the solvent polarity. The solvent plays an important role on the dynamic of the radical ion pairs,[265-267] and, in polar solvents, the formation of the free, solvated ions becomes an important nonradiative deactivation pathway.[267-269]

The interaction of singlet excited pyrene with TEA has been studied using nano- and picosecond absorption spectroscopy and transient photoconductivity.[266-269] The mechanism of the process is represented by

$$\text{Py*} + \text{TEA}$$
$$\downarrow$$
$$(\text{Py}^- \ \text{TEA}^+) \rightarrow \text{decay}$$
$$\text{LXVII}$$
$$\swarrow$$
$$(\text{Py}^-...\text{TEA}^+) \leftrightarrows (\text{Py}^- + \text{TEA}^+) \rightarrow \text{PyH}\cdot + \text{TEA}\cdot$$
$$\text{LXVIII} \qquad\qquad \text{LXIX} \qquad\qquad \text{LXX}$$
$$\downarrow \qquad\qquad\qquad \downarrow$$
$$\text{Py} + \text{TEA} \qquad \text{Products}$$

and comprises the exciplex (LXVII), the solvated ion pair (LXVIII), the separated ion pair (LXIX), and the free pyrenyl (LXX) and amino-derived free radicals. The rate constant of the complex dissociation into the separated radical ion pair is notably dependent on the dielectric constant of the solvent ($4.9 \times 10^8 \ s^{-1}$ in acetonitrile and $\ll 10^6 \ s^{-1}$ in diethylether).[266]

Photobleaching of pyrene by TEA is extremely solvent dependent (0.22 in acetonitrile and <0.001 in benzene) and decreases in the presence of monomers,[260] the protection being higher for MMA, intermediate for styrene, and negligible for vinyl acetate.[270] Time-resolved spectroscopic measurements show that MMA and styrene addition quench the pyrene-free radical anion. The bimolecular rate constants are 5.3 and $1 \times 10^5 \ 1/M$ s for MMA and

TABLE 32
Polymerization Data of Methyl
Methacrylate Photoinitiated by
Pyrene-Triethylamine

Solvent[a]	[TEA] (*M*)	Φ[b]
Acetonitrile	—	0.01
	0.00015	0.042
	0.00045	0.22
	0.0015	0.4
	0.005	0.48
	0.01	0.48
	0.05	0.52
	0.1	0.58
Benzene	—	0.01
	0.003	0.036
	0.01	0.052
	0.1	0.074

[a] MMA/solvent (1:1) v/v.
[b] Taken DTBP as reference photoinitiator.

From Encinas et al.[260]

styrene, respectively. On the other hand, monomer addition decreases the initial absorption due to the pyrenyl radical (LXX) without changing its decay rate.[270]

When MMA polymerization is initiated by irradiation of pyrene in the presence of TEA, polymerization rates are considerably higher in polar solvents (Table 32), a result opposite to that observed with carbonyl compounds.[187] The observed differences in polymerization rates are due to differences in Φ since similar rates in both solvents are obtained with di-*tert*-butylperoxide, whose photocleavage yield is solvent independent.[260] Initiation efficiencies as a function of the amine concentration are included in Table 32. The values of Φ show a good correlation with the fraction of excited pyrene singlets quenched by the amine, as determined by the decrease in the hydrocarbon fluorescence. Active free radicals must then be produced along the singlet quenching by the amine.

The trend of photoinitiation efficiencies is similar to that found for the interaction of Py^- with the monomers and the pyrene photobleaching protection. All the results are compatible with a mechanism involving the interaction of a pyrene anion, either free or the ion pair, with the monomer:

$$Py^- + MMA \rightarrow Py + MMA^- \tag{119}$$

$$MMA^- + TEA^+ \rightarrow MMA(H)\cdot + TEA\cdot \tag{120}$$

or

$$(Py^-\ TEA^+) + MMA \rightarrow Py + TEA\cdot + MMA(-H)\cdot \tag{121}$$

followed by initiation by the amino-derived radicals. This mechanism explains the polymerization data as well as the effect of the monomer on the pyrene anion transient, the decrease in PyH· yield, the dependence of the pyrene bleaching rate on the monomer concentration, and the invariance of the PyH· decay with the olefin concentration.[270] The value of Φ for a given monomer reflects its capacity to interact with the pyrene radical anion. The rate constant of this process will be determined either by the monomer electron affinity (Process 119) or the stability of the monomer-derived radical (Process 121). The high electron affinity of MMA leads to high efficiencies when this monomer is employed. On the other hand, neither Reaction 119 nor 121 will be favored when vinyl acetate is employed as monomer.

ACKNOWLEDGMENT

The authors acknowledge FONDECYT, Grant #912/90 for supporting this work.

REFERENCES

1. **Oster, G. and Yang, N. L.**, Photopolymerization of vinyl monomers, *Chem. Rev.*, 68, 125, 1968.
2. **Ledwith, A.**, Photoinitiation of polymerization, *Pure Appl. Chem.*, 4, 431, 1977.
3. **Hageman, H. J.**, Photoinitiators for free radical polymerization, *Prog. Org. Coat.*, 13, 123, 1985.
4. **Pappas, S. P.**, Photoinitiated radical polymerization, *J. Radiat. Curing*, 14, 6, 1987.
5. **Rabek, J. F.**, *Mechanism of Photophysical Processes and Photochemical Reactions in Polymers*, John Wiley & Sons, New York, 1988.
6. **Allen, N. S.**, *Photopolymerization and Photoimaging Science and Technology*, Elsevier, Amsterdam, 1989.
7. **Shultz, A. R. and Joshi, M. G.**, Kinetics of photoinitiated free-radical polymerization, *J. Polym. Sci. Polym. Chem. Ed.*, 22, 1753, 1984.
8. **Lissi, E. A. and Zanocco, A.**, Photoinitiated polymerization: effect of the initiator absorbance, *J. Polym. Sci. Polym. Chem. Ed.*, 21, 2197, 1983.
9. **Lissi, E. A., Garrido, J., and Zanocco, A.**, Photopolymerization of methyl methacrylate by benzoin methyl ether irradiated at 366 nm: effect of the initiator absorbance upon the polymerization rate, *J. Polym. Sci. Polym. Chem. Ed.*, 22, 391, 1984.
10. **Zanocco, A., Garrido, J., and Lissi, E. A.**, Effect of stirring on photoinitiated polymerization, *J. Polym. Sci. Polym. Chem. Ed.*, 26, 2827, 1988.
11. **Lissi, E.A., Encinas, M. V., and Abarca, M. T.**, Photopolymerization of methyl methacrylate sensitized by aliphatic ketones, *J. Polym. Sci. Polym. Chem. Ed.*, 17, 19, 1979.
12. **Lissi, E. A. and Encinas, M. V.**, Polymerization photosensitized by carbonyl compounds, *J. Polym. Sci. Polym. Chem. Ed.*, 17, 2791, 1979.
13. **Merlin, A. and Fouassier, J. P.**, Benzoin derivatives as photo-initiators for vinyl polymerization in micellar systems, *Polymer*, 21, 1363, 1980.
14. **Ghosh, P., Mitra, P. S., and Banerjee, A. N.**, Photopolymerization of methyl methacrylate with the use of bromide as photoinitiator, *J. Polym. Sci. Polym. Chem. Ed.*, 11, 2021, 1973.
15. **Ghosh, P. and Banerjee, A. N.**, Photopolymerization of methyl methacrylate with the use of iodine as the photoinitiator, *J. Polym. Sci. Polym. Chem. Ed.*, 12, 375, 1974.
16. **Lissi, E. A. and Aljaro, J.**, Methyl methacrylate polymerization in the presence of iodine, *J. Polym. Sci. Polym. Lett. Ed.*, 14, 499, 1976.
17. **McCloskey, C. M. and Bond, J.**, Photosensitizers for polyester-vinyl polymerization, *Ind. Eng. Chem.*, 47, 2125, 1955.
18. **Otsu, T., Tanaka, H., and Wasaki, H.**, Photodegradation of chloromethyl vinyl ketone polymer and copolymers with styrene and α-methyl styrene, *Polymer*, 20, 55, 1979.
19. **Ozawa, T., Sukegawa, S., and Masaki, K.**, Polymerization of vinyl acetate by ultraviolet light with bromoacetophenone. Ultraviolet absorption and decomposition of bromoacetophenone and bulk polymerization of vinyl acetate of ultraviolet light, *Chem. High Polym.*, 17, 367, 1960.
20. **Barson, C. A., Henbest, R. G. C., and Robb, J. C.**, Styrene dibromide as a photoinitiator in the polymerization of styrene, *Trans. Faraday Soc.*, 66, 1688, 1970.
21. **Bunce, N. J., Ingold, K. U., Landers, J. P., Lusztyk, J., and Scaiano, J. C.**, Kinetics study of the photochlorination of 2,3-dimethylbutane and other alkanes in solution in the presence of benzene. First measurements of the absolute rate constants for hydrogen abstraction by the free chloride atom and the chloride atom-benzene π-complex. Identification of these species as the only hydrogen abstractors in these systems, *J. Am. Chem. Soc.*, 107, 5464, 1985.
22. **Encinas, M. V., Lissi, E.A., Koch, V., and Elorza, E.**, Polymerization photoinitiated by carbonyl compounds. IV. α-Diketones and α-substituted alkanones, *J. Polym. Sci. Polym. Chem. Ed.*, 20, 73, 1982.

23. **Morrison, H. and Cardenas, L.,** Stereoelectronic control in the photolytic cleavage of α-chloro ketones, *J. Org. Chem.,* 52, 2590, 1987.
24. **McGimpsey, W. G. and Scaiano, J. C.,** Photochemistry of α-chloro- and α-bromoacetophenone. Determination of extinction coefficients for halogen-benzene complexes, *Can. J. Chem.,* 66, 1474, 1988.
25. **Evleth, E. M.,** Ground and excited state correlation in the rupture of oxygen-oxygen bonds in peroxides, *J. Am. Chem. Soc.,* 98, 1637, 1976.
26. **Lissi, E. A.,** Photolysis of di-tert-butyl peroxide, *Can. J. Chem.,* 52, 2491, 1974.
27. **Scaiano, J. C. and Wubbels, G. G.,** Photosensitized dissociation of di-tert-butyl peroxide. Energy transfer to a repulsive state, *J. Am. Chem. Soc.,* 103, 640, 1981.
28. **Mendenhall, G. D., Stewart, L. C., and Scaiano, J. C.,** Laser photolysis study of the reactions of alkoxy radicals generated in the photosensitized decomposition of organic hyponitrites, *J. Am. Chem. Soc.,* 104, 5109, 1982.
29. **Engel, P. S., Woods, T. L., and Page, M. A.,** Quenching of excited triplet sensitizers by organic peroxides, *J. Phys. Chem.,* 87, 10, 1983.
30. **Thijs, L., Gupta, S. N., and Neckers, D. C.,** Photochemistry of perester initiators, *J. Org. Chem.,* 44, 4123, 1979.
31. **Ghosh, P., Mukhopadhyay, G., and Ghosh, R.,** Effect of benzophenone on the H_2O_2-induced photopolymerization of methyl methacrylate, *Eur. Polym. J.,* 16, 457, 1980.
32. **Abuin, E., Mujica, C., and Lissi, E. A.,** Reactivity of methoxy radicals: hydrogen abstraction and double bond addition, *Rev. Latinoam. Quim.,* 11, 78, 1980.
33. **Encinas, M. V., Rivera, M., and Lissi, E. A.,** Reactions of tert-butoxy radicals with vinyl monomers, *J. Polym. Sci. Polym. Chem. Ed.,* 16, 1709, 1978.
34. **Moad, G., Rizzardo, E., and Solomon, D. H.,** Selectivity of the reaction of free radicals with styrene, *Macromolecules,* 15, 909, 1982.
35. **Rizzardo, E. and Solomon, D. H.,** Determination of tert-butoxy endgroups in polymers: the mode of reaction of tert-butoxy radicals with methyl methacrylate and styrene, *J. Macromol. Sci. Chem.,* A13, 1005, 1979.
36. **Encinas, M. V. and Lissi, E. A.,** Quenching of excited singlets by peroxides and hydroperoxides, *J. Photochem.,* 20, 153, 1982.
37. **Abuin, E. B., Lissi, E. A., and Avaria, L.,** Effect of pH on the behavior of hydrogen peroxide as an inhibitor or chain transfer agent, *J. Polym. Sci. Polym. Lett.,* 26, 501, 1988.
38. **Ghosh, P., Sengupta, P. K., and Mukherjee, N.,** Photopolymerization of methyl methacrylate using hydrogen peroxide as the photoinitiator, *J. Polym. Sci. Polym. Chem. Ed.,* 17, 2119, 1979.
39. **Tang, D. K. and Ho, S. Y.,** The photopolymerization of isoprene with the use of hydrogen peroxide as photoinitiator, *J. Polym. Sci. Polym. Chem. Ed.,* 22, 1357, 1984.
40. **Chang, J. F., Chang, T. C., and Ho, S. Y.,** Rate constants in the photopolymerization of H_2O_2-isoprene systems, *J. Polym. Sci. Polym. Chem. Ed.,* 23, 3045, 1985.
41. **Montero, C., Abuin, E., and Lissi, E. A.,** Thermal and photochemical decomposition of a styrene-oxygen copolymer, *Contrib. Cient. Technol.,* 51, 25, 1981.
42. **Abu-Abdoun, I. I., Thijs, L., and Neckers, D. C.,** Nonketonic perester photoinitiators, *Macromolecules,* 17, 282, 1984.
43. **Abu-Abdoun, I. I., Thijs, L., and Neckers, D. C.,** Photoinitiated polymerization of vinyl monomers by benzophenone t-butyl peresters, *J. Polym. Sci. Polym. Chem. Ed.,* 21, 3129, 1983.
44. **Gupta, S. N., Gupta, I., and Neckers, D. C.,** Studies on the vinyl polymerization photoinitiated by p-benzoylperoxybenzoic acid tert-butyl-ester, *J. Polym. Sci. Polym. Chem. Ed.,* 19, 103, 1981.
45. **Allen, N. S., Hardy, S. J., Jacobine, A., Glaser, D. M., and Catalina, F.,** Photochemistry and photopolymerization activity of novel perester derivatives of fluorenone, *Eur. Polym. J.,* 25, 1219, 1989.
46. **Kuhlman, R. and Schnabel, W.,** Laser flash photolysis investigation on primary processes of the sensitized polymerization of vinyl monomers: experiments with benzophenone, *Polymer,* 17, 419, 1976.
47. **Gupta, S. N., Thijs, L., and Neckers, D. C.,** Synthesis and polymerization of p-(p′-vinylbenzoyl)-peroxybenzoic acid tert-butylester a monomer containing a photodissociable radical source, *J. Polym. Sci. Polym. Chem. Ed.,* 19, 855, 1981.
48. **Gupta, I., Gupta, S. N., and Neckers, D. C.,** Photocrosslinking and photografting of vinyl polymers using poly(styrene-co-p-vinylbenzophenone-p′-tert-butyl perzoate) as a comonomer, *J. Polym. Sci. Polym. Chem. Ed.,* 20, 147, 1982.
49. **Andrews, L. J., Deroulede, A., and Linschitz, H.,** Photophysical processes in fluorenone, *J. Phys. Chem.,* 82, 2304, 1978.
50. **Naito, I., Ueki, T., Tabara, S., Tomiki, M., and Kinoshita, A.,** Photosensitized polymerizations of some vinyl monomers using aliphatic carbonyl polymers as initiators, *J. Polym. Sci. Polym. Chem. Ed.,* 24, 875, 1986.
51. **Ghosh, P. and Mukherji, N.,** Photopolymerization of methyl methacrylate using triethylamine-benzoyl-peroxide redox initiator system, *Eur. Polym. J.,* 15, 797, 1979.

52. **Ghosh, P. and Maity, S. N.,** Polymerization of methylmethacrylate with cetyl pyridinium bromide (CPB)-benzoyl peroxide (Bz_2O_2) combination as a redox initiator and with CPB as a lone photoinitiator, *Eur. Polym. J.,* 15, 787, 1979.

53. **Ghosh, P. and Maity, S. N.,** Polymerization of methyl methacrylate with the use of cetyl trimethyl ammonium bromide-benzoyl peroxide combination as the initiating system, *Eur. Polym. J.,* 14, 855, 1978.

54. **Ghosh, P., Biswas, S., and Niyogi, U.,** Photopolymerization of methyl methacrylate using a combination of Michler's ketone and benzoyl peroxide as photoinitiator, *Eur. Polym. J.,* 25, 1285, 1989.

55. **Leplyanin, G. V., Rafikov, S. R., Barisova, E. G., Korchev, O. I., and Galin, F. Z.,** On the kinetics of polymerization photoinitiated by disulphides participating in chain termination, *Vysokomol. Soedin. Ser. A,* 18, 597, 1976.

56. **Nambu, Y., Acar, M. H., Suzuki, T., and Endo, T.,** Thermal and photoinitiated copolymerization of the cyclic disulfide lipoamide with styrene, *Makromol. Chem.,* 189, 495, 1988.

57. **Niwa, M., Matsumoto, T., and Izumi, H.,** Kinetics of the photopolymerization of vinyl monomers by bis(isopropylxanthogen) disulfide. Design of block copolymers, *J. Macromol. Sci. Chem.,* A24, 567, 1987.

58. **Engel, P. S.,** Mechanism of the thermal and photochemical decomposition of azoalkanes, *Chem. Rev.,* 80, 99, 1980.

59. **Engel, P. S. and Bartlett, P. D.,** The sensitizer photolysis of acyclic azo compounds. Singlet energy transfer, *J. Am. Chem. Soc.,* 92, 5883, 1970.

60. **Burkhart, R. D. and Merrill, J. C.,** Photo and thermal initiator efficiency of 2,2'-azobisisobutyronitrile at 25°C, *J. Phys. Chem.,* 73, 2699, 1969.

61. **Nuyken, O., Dyckerhoff, L., Schuster, H., and Kerber, R.,** Radical efficiency and ionic reactions of same unsymmetrical azo compounds, *Makromol. Chem.,* 184, 2251, 1983.

62. **Nuyken, O., Schuster, H., and Kerber, R.,** Novel monomeric and polymeric azo initiators, *Makromol. Chem.,* 184, 2285, 1983.

63. **Encinas, M. V. and Lissi, E. A.,** Type I-type II competition in the photolysis of aliphatic ketones, *J. Photochem.,* 8, 131, 1978.

64. **Encinas, M. V., Lissi, E. A., Lemp, E., Zanocco, A., and Scaiano, J. C.,** Temperature dependence of the photochemistry of aryl alkyl ketones, *J. Am. Chem. Soc.,* 105, 1856, 1983.

65. **Encinas, M. V., Lissi, E. A., and Olea, F. A.,** Photochemistry of aliphatic aldehydes, *J. Photochem.,* 14, 233, 1980.

66. **Encinas, M. V. and Lissi, E. A.,** 2,4-Dimethyl-3-pentanone-sensitized photopolymerization of vinyl monomers, *J. Polym. Sci. Polym. Chem. Ed.,* 17, 1645, 1979.

67. **Encinas, M. V., Lissi, E. A., and Scaiano, J. C.,** Photochemistry of aliphatic ketones in polar solvents, *J. Phys. Chem.,* 84, 948, 1980.

68. **Forgeteg, S. and Bérces, T.,** Structural and solvent effects in the photochemistry of aliphatic ketones, *J. Photochem.,* 36, 49, 1987.

69. **Yang, N. C., Feit, E. D., Hui, M. H., Turro, N. J., and Dalton, J. C.,** Photochemistry of di-tert-butyl ketone and structural effects on the rate and efficiency of intersystem crossing of aliphatic ketones, *J. Am. Chem. Soc.,* 92, 6974, 1970.

70. **Encinas, M. V., Rufs, A. M., and Lissi, E. A.,** Photochemistry of hydroxyalkanones in solution, *J. Chem. Soc. Perkin Trans. 2* p. 457, 1985.

71. **Gould, I. R., Baretz, B. H., and Turro, N. J.,** Primary processes in the type I photocleavage of dibenzyl ketones. A pulsed laser and photochemically induced dynamic nuclear polarization study, *J. Phys. Chem.,* 91, 925, 1987.

72. **Kochevar, I. E. and Wagner, P. J.,** Quenching of triplet phenyl ketones by olefins, *J. Am. Chem. Soc.,* 94, 3859, 1972.

73. **Yang, N. C., Hui, M H., Shold, D. M., Turro, N. J., Hautala, R. R., Dawes, K., and Dalton, J. C.,** Quenching of the $^1n\pi^*$ of alkanones by unsaturated compounds, *J. Am. Chem. Soc.,* 99, 3023, 1977.

74. **Loutfy, R. O., Dogra, S. K., and Yip, R. W.,** The interaction between the excited triplet state of ketones and olefins: the role of triplet exciplexes, *Can. J. Chem.,* 57, 342, 1979.

75. **Maharaj, V. and Winnik, M. A.,** Electron transfer contribution to the quenching of aromatic ketone phosphorescence by electron-poor olefins, *Tetrahedron Lett.,* 23, 3035, 1982.

76. **Scaiano, J. C., Lissi, E. A., and Stewart, L. C.,** Quenching of triplet macromolecules by small molecules. The role of energy migration, *J. Am. Chem. Soc.,* 106, 1539, 1984.

77. **Takemura, F., Iwai, K., Ikesu, M., Itoh, K., and Matoike, A.,** Effects of inorganic anions on the photopolymerization of acrylamide sensitized by 2,3-butanedione in aqueous solutions, *Polym. J.,* 20, 565, 1988.

78. **Bottom, R. A., Guthrie, J. T., and Green, P. N.,** The influence of H-donors on the photodecomposition of selected water-soluble photoinitiators, *Polym. Photochem.,* 6, 59, 1985.

79. **Allen, N. S., Catalina, F., Green, P. N., and Green, W. A.,** Photochemistry of carbonyl photoinitiators. Photopolymerization, flash photolysis and spectroscopic study, *Eur. Polym. J.,* 22, 49, 1986.

80. **Cáceres, T., Encinas, M. V., and Lissi, E. A.,** Photocleavage of benzil, *J. Photochem.*, 27, 109, 1984.
81. **McGimpsey, W. G. and Scaiano, J. C.,** A two-photon study of the reluctant Norrish type I reaction of benzil, *J. Am. Chem. Soc.*, 109, 2179, 1987.
82. **Hong, S. I., Kurosaki, T., and Okawara, M.,** Photopolymerizations initiated by oxime derivatives, *J. Polym. Sci. Polym. Chem. Ed.*, 12, 2553, 1974.
83. **Encinas, M. V., Garrido, J., and Lissi, E. A.,** Photopolymerization initiated by carbonyl compounds. VII. Solvent effects on biacetyl photoinitiation, *J. Polym. Sci. Polym. Chem. Ed.*, 23, 2481, 1985.
84. **Miyata, K., Nakashima, K., and Koyanagi, M.,** Spectroscopic evidence of α,α-dihydroxy ketone in biacetyl aqueous solutions, *Bull. Chem. Soc. Jpn.*, 62, 367, 1989.
85. **Kubota, H. and Ogiwara, Y.,** Photopolymerization of acrylonitrile sensitized with a combination of aromatic ketone and amine, *J. Appl. Polym. Sci.*, 27, 2683, 1982.
86. **Lewis, F. D. and Magyar, J. G.,** Photoreduction and α cleavage of aryl alkyl ketones, *J. Org. Chem.*, 37, 2102, 1972.
87. **Sandner, M. R. and Osborn, C. L.,** Photochemistry of 2,2-dimethoxy-2-phenylacetophenone triplet detection via spin memory, *Tetrahedron Lett.*, p. 415, 1974.
88. **Timpe, H. J., Wagner, R., and Paleta, O.,** Light-induced polymer and polymerization reactions. Benzoin derivative/halogen compound systems as photoinitiators for polymerization reactions, *Acta Polym.*, 38, 641, 1987.
89. **Pappas, S. P.,** Photochemical aspects of UV curing, *Prog. Org. Coat.*, 2, 333, 1973.
90. **Oster, G., Oster, G. K., and Moroson, H.,** Ultraviolet induced crosslinking and grafting of solid high polymers, *J. Polym. Sci.*, 34, 671, 1959.
91. **Delzenne, G. A.,** Photopolymerization systems. An integral approach, *J. Radiat. Curing*, 2, 1979.
92. **Heine, H. G., Hartmann, W., Kory, D. C., Magyar, J. G., Hoyle, C. E., McVey, J. K., and Lewis, F. D.,** Photochemical α cleavage and free-radical reactions of some deoxybenzoins, *J. Org. Chem.*, 39, 691, 1974.
93. **Amirzaden, G., Kuhlmann, R., and Schnabel, W.,** Photolysis of deoxybenzoin, *J. Photochem.*, 10, 133, 1979.
94. **Lewis, F. D., Lauterbach, R. T., Heine, H. G., Hartmann, W., and Rudolph, H.,** Photochemical α cleavage of benzoin derivatives. Polar transition states for free radical formation, *J. Am. Chem. Soc.*, 97, 1519, 1975.
95. **Schnabel, W.,** Flash photolysis of benzoin and benzophenone containing systems, *Photogr. Sci. Eng.*, 23, 154, 1979.
96. **Kuhlmann, R. and Schnabel, W.,** Flash photolysis investigation on primary processes of the sensitized polymerization of vinyl monomers, *Angew. Makromol. Chem.*, 70, 145, 1978.
97. **Carlblom, L. H. and Pappas, S. P.,** Photoinitiated polymerization of methyl methacrylate and methyl acrylate with ^{14}C-labeled benzoin methyl ethers, *J. Polym. Sci. Polym. Chem. Ed.*, 15, 1381, 1977.
98. **Lougnot, D. J. and Fouassier, J. P.,** Comparative study of the efficiency of photoinitiators under light excitation by conventional sources and laser beams, *Makromol. Chem. Rapid Commun.*, 4, 11, 1983.
99. **Fouassier, J. P. and Riviere, D.,** A comparative investigation of the photoinitiation process in bulk and micelle polymerization. I. Steady-state experiments, *Polym. Photochem.*, 3, 29, 1983.
100. **Salmassi, A., Eichler, J., Herz, C. P., and Schnabel, W.,** On the photolysis of 1-phenyl-2-hydroxy-2-methyl-propanone-1 in the presence of methylmethacrylate, styrene, and acrylonitrile: laser flash photolysis studies, *Polym. Photochem.*, 2, 209, 1982.
101. **Fouassier, J. P. and Lougnot, D. J.,** Excited-state reactivity in a series of polymerization photoinitiators based on the acetophenone nucleus, *J. Chem. Soc. Faraday Trans. 1*, 83, 2935, 1987.
102. **Gaur, H. A., Groenenboom, C. J., Hageman, H. J., Hakvoort, G. T. M., Oosterhoff, P., Overeem, T., Polman, R. J., and van der Werf, S.,** Photoinitiators and photoinitiation. V. Photodecomposition of some α-hydroxymethylbenzoin derivatives: sulfonic esters, *Makromol. Chem.*, 185, 1795, 1984.
103. **Heine, H. G.,** Photochemische α-spaltung von ketonen in losung. IV. Photolyse der benzoin ether, *Tetrahedron Lett.*, p. 4755, 1972.
104. **Bradshaw, J. S., Knudsen, R. D., and Parrish, W. R.,** Photolysis of benzaldehyde in solution: the products, *J. Chem. Soc. Chem. Commun.*, p. 1321, 1972.
105. **Kornis, G. and De Mayo, P.,** Photochemical synthesis. The conversion of dibenzoyl methane to tribenzoylethane, *Can. J. Chem.*, 42, 2822, 1964.
106. **Pappas, S. P. and Chattopadhyay, A.,** Photochemistry of benzoin ethers. Type I cleavage by low energy sensitization, *J. Am. Chem. Soc.*, 95, 6484, 1973.
107. **Sheeham, J. C., Wilson, R. M., and Oxford, W. A.,** The photolysis of methoxy-substituted benzoin esters. A photosensitive protecting group for carboxylic acids, *J. Am. Chem. Soc.*, 93, 7222, 1971.
108. **Osborn, C. L. and Sandner, M. R.,** Photocurable systems effect of various photoinitiators on the polymerization rate of acrylate esters, *Org. Coat. Plast. Preprints*, 34, 660, 1974.
109. **Cocivera, M. and Trozollo, A. M.,** Photolysis of benzaldehyde in solution studied by nuclear magnetic resonance spectroscopy, *J. Am. Chem. Soc.*, 92, 1772, 1970.

110. **Closs, G. L. and Paulsen, D. R.,** Application of the radical-pair theory of chemically induced dynamic nuclear spin polarization (CIDNP) to photochemical reactions of aromatic aldehydes and ketones, *J. Am. Chem. Soc.,* 92, 7229, 1970.

111. **Maruyama, K., Furuta, H., and Otsuki, T.,** Photo-CIDNP observed in O-methylbenzoin carbon tetrachloride system, *Bull. Chem. Soc. Jpn.,* 53, 2421, 1980.

112. **Baumann, H., Mueller, U., Pfeifer, D., and Timpe, H. J.,** Light-initiated polymer and polymerization reactions. Photoinduced decomposition of arenediazonium salts by benzoin derivatives, *J. Prakt. Chem.,* 324, 217, 1982.

113. **Paul, H. and Fisher, H.,** Electron spin resonance of free radicals in photochemical reactions of ketones in solutions, *Helv. Chem. Acta,* 56, 1575, 1973.

114. **Ledwith, A., Rusell, P. J., and Sutcliffe, L. H.,** Radical intermediates in chemical decomposition of benzoin and related compounds, *J. Chem. Soc. Perkin Trans. 2,* p. 1925, 1972.

115. **Hageman, H. J. and Overeem, T.,** Photoinitiators and photoinitiation. IV. Trapping of primary radicals from photoinitiators by 2,2,6,6-tetramethylpiperidinoxyl, *Makromol. Chem. Rapid Commun.,* 2, 719, 1981.

116. **Fouassier, J. P. and Merlin A.,** Laser investigation of Norrish type II photoscission in the photoinitiator Irgacure (2,2-dimethoxy 2-phenylacetophenone), *J. Photochem.,* 12, 17, 1980.

117. **Lewis, F. D. and Magyar, J. G.,** Cage effects in the photochemistry of (S)-(+)-2-phenylpropiophenone, *J. Am. Chem. Soc.,* 95, 5973, 1973.

118. **Baumann, H., Timpe, H. J., Zubarev, V. E., Fok, N. V., and Mel'nikov, M. Y.,** Light-initiated polymer and polymerization reaction. ESR spin trapping in bivalent photoinitiator systems consisting of α-phenylbenzoin and arylonium compounds, *Z. Chem.,* 25, 181, 1985.

119. **Hutchison, J. and Ledwith, A.,** Mechanism and relative efficiencies in radical polymerization photoinitiated by benzoin, benzoin methyl ether and benzil, *Polymer,* 14, 405, 1973.

120. **Pappas, S. P. and Chattopadhyay, A. K.,** Benzoin ether photoinitiated polymerization, *J. Polym. Sci. Polym. Lett. Ed.,* 13, 483, 1975.

121. **Hageman, H. J., van der Maeden, F. P., and Janssen, P. C. G.,** Photoinitiators and photoinitiation. I. Vinyl polymerization photoinitiated by benzoin methyl ether, *Makromol. Chem.,* 180, 2531, 1979.

122. **Groenenboom, C. J., Hageman, H. J., Overeem, T., and Weber, A. J. M.,** Photoinitiators and photoinitiation. III. Comparison of the photodecomposition of α-methoxy- and α,α-dimethoxybenzoin in 1,1-diphenylethylene as model substrate, *Makromol. Chem.,* 183, 281, 1982.

123. **Kuhlmann, R. and Schnabel, W.,** Flash photolysis investigation on primary processes of the sensitized polymerization of vinyl monomers. Experiments with benzoin and benzoin derivatives, *Polymer,* 18, 1163, 1977.

124. **Pappas, S. P. and Asmus, R. A.,** Photoinitiated polymerization of methyl methacrylate with benzoin methyl ether. III. Independent photogeneration of the ether radical, *J. Polym. Sci. Polym. Chem. Ed.,* 20, 2643, 1982.

125. **Phan, X. T.,** Effect of molecular oxygen on the UV-polymerization of methyl methacrylate initiated by 2,2-dimethoxy-2-phenylacetophenone and 1-hydroxycyclohexylphenyl ketone in solution, *J. Radiat. Curing,* 13, 18, 1986.

126. **Lipscomb, N. T. and Tarshiani, Y.,** Kinetics and mechanism of the benzoin isobutyl ether photoinitiated polymerization of styrene, *J. Polym. Sci. Polym. Chem. Ed.,* 26, 529, 1988.

127. **Phan, X. T.,** Photochemical α-cleavage of 2,2-dimethoxy-2-phenylacetophenone and 1-hydroxycyclohexyl phenyl ketone photoinitiators. Effect of molecular oxygen on free reaction in solution, *J. Radiat. Curing,* 13, 11, 1986.

128. **Ahn, K. D., Ihn, K. J., and Kwon, I. C.,** A photosensitive polymer benzoin ether side chains: poly(α-methylolbenzoin methyl ether acrylate), *J. Macromol. Sci. Chem.,* A23, 355, 1986.

129. **Hageman, H. J. and Jansen, L. G. J.,** Photoinitiators and photoinitiation. IX. Photoinitiators for radical polymerization which counter oxygen-inhibition, *Makromol. Chem.,* 189, 2781, 1988.

130. **Nigan, S., Asmus, K. D., and Willson, R. L.,** Electron transfer and addition reactions of the free nitroxyl radicals with radiation-induced radicals, *J. Chem. Soc. Faraday Trans. 1,* p. 2324, 1976.

131. **Eichler, J., Herz, C. P., Naito, I., and Schnabel, W.,** Laser flash photolysis investigation of primary processes in the sensitized polymerization of vinyl monomers. IV. Experiments with hydroxy alkylphenones, *J. Photochem.,* 12, 225, 1980.

132. **Osborn, C. L.,** Photoinitiation systems and their role in UV-curable coatings and inks, *J. Radiat. Curing,* 3, 5, 1976.

133. **Borer, A., Kirchmayer, R., and Rist, G. H.,** CIDNP Investigation of photoinduced polymerization, *Helv. Chim. Acta,* 61, 305, 1978.

134. **Majima, T. and Schnabel, W.,** On the reactivity of phosphinoyl and thiophosphinoyl radicals: flash photolysis studies, *J. Photochem. Photobiol. A,* 50, 31, 1989.

135. **Baxter, J. E., Davidson, R. S., and Hageman, H. J.,** Acylphosphine oxides as photoinitiators for acrylate and unsaturated polyester resins, *Eur. Polym. J.,* 24, 419, 1988.

136. **Sumiyoshi, T., Henne, A., Lecktken, P., and Schnabel, W.,** Optical absorption spectra of phosphonyl radicals, *Z. Naturforsch. Teil A,* 39, 434, 1984.

137. **Sumiyoshi, T. and Schnabel, W.,** On the reactivity of phosphonyl radicals towards olefinic compounds, *Makromol. Chem.,* 186, 1811, 1985.

138. **Sumiyoshi, T., Schnabel, W., Henne, A., and Lechtken, P.,** On the photolysis of acylphosphine oxides. I. Laser flash photolysis studies with 2,4,6-trimethylbenzoyldiphenylphosphine oxide, *Polymer,* 26, 141, 1985.

139. **Schnabel, W.,** Mechanistics and kinetics aspects concerning the initiation of free radical polymerization by thioxanthones and acylphosphine oxides, *J. Radiat. Curing,* 13, 26, 1986.

140. **Baxter, J. E., Davidson, R. S., Hageman, H. J., and Overeem, T.,** Photoinitiators and photoinitiation. VII. The photo-induced α-cleavage of acylphosphine oxides. Trapping of primary radicals by a stable nitroxyl, *Makromol. Chem. Rapid Commun.,* 8, 311, 1987.

141. **Baxter, J. E., Davidson, R. S., and Hageman, H. J.,** Use of acylphosphine oxides and acylphosphonates as photoinitiators, *Polymer,* 29, 1569, 1988.

142. **Baxter, J. E., Davidson, R. S., Hageman, H. J., McLauchlan, K. A., and Stevens, D. G.,** The photo-induced cleavage of acylphosphine oxides, *J. Chem. Soc. Chem. Commun.,* p. 73, 1987.

143. **Baxter, J. E., Davidson, R. S., Hageman, H. J., and Overeem, T.,** Photoinitiators and photoinitiation. VIII. The photoinduced α-cleavage of acylphosphine oxides. Identification of the initiating radicals using a model substrate, *Makromol. Chem.,* 189, 2769, 1988.

144. **Baysal, B.,** Initiation of polymerization by cyclic peroxides, *J. Polym. Sci.,* 33, 381, 1958.

145. **Hamity, M. and Scaiano, J. C.,** Trapping by methyl methacrylate of the biradicals derived from the photolysis of some alkyl aryl ketones: a novel initiation of polymerization, *J. Photochem.,* 4, 229, 1975.

146. **Naito, I.,** The laser-flash photolysis of 2-hexanone and 5-methyl-2-hexanone for the direct detection of the triplet 1,4-biradical, *Bull. Chem. Soc. Jpn.,* 56, 2851, 1983.

147. **Scaiano, J. C.,** Laser flash photolysis studies of the reactions of some 1,4-biradicals, *Acc. Chem. Res.,* 15, 252, 1982.

148. **Encinas, M. V. and Lissi, E. A.,** Photochemical processes involving the $^{1}n\pi^{*}$ excited state of aliphatic ketones bearing γ-hydrogens, *J. Photochem.,* 4, 321, 1975.

149. **Encinas, M. V. and Lissi, E. A.,** Photochemistry of alkyl ketones bearing γ-hydrogens, *J. Photochem.,* 6, 173, 1976/1977.

150. **Scaiano, J. C., Lissi, E. A., and Encinas, M. V.,** Chemistry of the biradicals produced in the Norrish type II reaction, *Rev. Chem. Intermed.,* 2, 139, 1978.

151. **Johnston, L. J. and Scaiano, J. C.,** Time-resolved studies of biradical reactions in solution, *Chem. Rev.,* 89, 521, 1989.

152. **Encinas, M. V., Lissi, E. A., Gargallo, L., Radic, D., and Sigdman, R.,** Polymerization photoinitiated by carbonyl compounds. VI. Mechanism of benzil photoinitiation, *J. Polym. Sci. Polym. Chem. Ed.,* 22, 2469, 1984.

153. **Greenberg, S. A. and Forster, L. S.,** The photolysis of biacetyl solutions, *J. Am. Chem. Soc.,* 83, 4339, 1961.

154. **Turro, N. J. and Engel, R.,** Quenching of biacetyl fluorescence and phosphorescence, *J. Am. Chem. Soc.,* 91, 7113, 1969.

155. **Jones, G., Santhanam, M., and Chiang, S. H.,** Photoaddition of biacetyl and alkenes. Reaction stereochemistry, multiplicity and photokinetics, *J. Am. Chem. Soc.,* 102, 6088, 1980.

156. **Turro, N. J. and Lee, T. J.,** Intramolecular photoreduction of alkyl α-diketones, *J. Am. Chem. Soc.,* 91, 5651, 1969.

157. **Gersdorf, J., Mattay, J., and Gorner, H.,** Photoreactions of biacetyl, benzophenone, and benzil with electron-rich alkenes, *J. Am. Chem. Soc.,* 109, 1203, 1987.

158. **Carless, H. A. J. and Fekarurhobo, G. K.,** Selective photocycloadditions of biacetyl to allyl vinyl ethers: the synthesis of 2-alkoxyoxetanes as thromboxane A_2 analogs, *Tetrahedron Lett.,* 26, 4407, 1985.

159. **Wagner, P. J.,** Type II photoelimination and photocyclization of ketones, *Acc. Chem. Res.,* 4, 168, 1971.

160. **Scaiano, J. C.,** Intermolecular photoreduction of ketones, *J. Photochem.,* 2, 81, 1973/1974.

161. **Suppan, P.,** The mechanism of the primary process in the photoreduction of aromatic carbonyl compounds, *J. Chem. Soc. Faraday Trans. 2,* 82, 2167, 1986.

162. **Fouassier, J. P. and Lougnot, D. J.,** A comparative investigation of the photoinitiation processes in bulk and micelle polymerization. II. Excited states dynamics, *Polym. Photochem.,* 3, 79, 1983.

163. **Schuster, D. J. and Karp, P. B.,** Photochemistry of ketones in solution. LVIII. Mechanism of photoreduction of benzophenone by benzhydrol, *J. Photochem.,* 12, 333, 1980.

164. **Sandner, M. R., Osborn, C. L., and Trecker, D. J.,** Benzophenone/triethylamine-photoinitiated polymerization of methyl acrylate, *J. Polym. Sci. Polym. Chem. Ed.,* 10, 3173, 1972.

165. **Block, H., Ledwith, A., and Taylor, A. R.,** Polymerization of methyl methacrylate photosensitized by benzophenone, *Polymer,* 12, 271, 1971.

166. **Inbar, S., Linschitz, H., and Cohen, S. G.,** Nanosecond flash studies of reduction of benzophenone by aliphatic amines. Quantum yields and kinetic isotope effects, *J. Am. Chem. Soc.,* 103, 1048, 1981.

167. **Hutchison, J., Lambert, M. C., and Ledwith, A.,** Role of semi-pinacol radicals in the benzophenone-photoinitiated polymerization of methyl methacrylate, *Polymer,* 14, 250, 1973.

168. **Ghosh, P. and Mukherjee, G. S.,** Comparative kinetics of photopolymerization of methyl methacrylate using chloro-derivatives of acetic acid in combination with dimethylaniline as photoinitiators, *Eur. Polym. J.,* 10, 1049, 1989.

169. **Lissi, E. A. and Encinas, M. V.,** The reaction of 2-pentanone triplets with carbon tetrachloride, *J. Photochem.,* 3, 237, 1974.

170. **Loutfy, R. O. and Somersall, A. C.,** An exciplex mechanism for the quenching of singlet excited states of aliphatic ketones by carbon tetrachloride, *Can. J. Chem.,* 54, 760, 1976.

171. **Shimada, S., Nakagawa, K., and Tabuchi, K.,** Photopolymerization of methyl methacrylate with 1-benzyl-1,4-dihydronicotinamide in the presence of carbon tetrachloride, *Polym. J.,* 21, 275, 1989.

172. **Li, T., Cao, W. X., and Feng, X.,** Photoinduced charge-transfer polymerization of vinyl monomers, *Rev. Macromol. Chem. Phys. C,* 29, 153, 1989.

173. **Cohen, S. G., Parola, A., and Parsons, G. H.,** Photoreduction by amines, *Chem. Rev.,* 73, 141, 1973.

174. **Parola, A. H., Rose, A. W., and Cohen, S. G.,** Effects of concentration of amine and of medium in photoreduction of ketones by amines, *J. Am. Chem. Soc.,* 97, 6202, 1975.

175. **Parola, A. H. and Cohen, S. G.,** Effect of solvent in the photoreduction and quenching of benzophenone by triethylamine, 1-azabicyclo[2.2.2]octane and 1,4-diazabicyclo[2.2.2]octane, *J. Photochem.,* 12, 41, 1980.

176. **Shaefer, C. G. and Peter, K. S.,** Picosecond dynamics of the photoreduction of benzophenone by triethylamine, *J. Am. Chem. Soc.,* 102, 7566, 1980.

177. **Bhattachryya, K. and Das, P. K.,** Nanosecond transient processes in the triethylamine quenching of benzophenone triplets in aqueous alkaline media. Substituent effect, ketyl radical deprotonation, and secondary photoreduction kinetics, *J. Phys. Chem.,* 90, 3987, 1986.

178. **Simon, J. D. and Peter, K. S.,** Solvent effects on the picosecond dynamics of the photoreduction of benzophenone by aromatic amines, *J. Am. Chem. Soc.,* 103, 6403, 1981.

179. **Simon, J. D. and Peter, K. S.,** Picosecond dynamics of ion pairs: the effect of hydrogen bonding on ion-pair intermediates, *J. Am. Chem. Soc.,* 104, 6542, 1982.

180. **Merlin, A., Lougnot, D. J., and Fouassier, J. P.,** The benzophenone-amine photoinitiator in vinyl polymerization, *Polym. Bull.,* 3, 1, 1980.

181. **Guttenplan, J. B. and Cohen, S. G.,** Triplet energies, reduction potentials and ionization potentials in carbonyl-donor partial charge-transfer interaction, *Tetrahedron Lett.,* p. 2163, 1972.

182. **Gorman, A. A., Parekh, C. T., Rodgers, M. A. J., and Smith, P. G.,** The quenching of the triplet states of aromatic carbonyl compounds by triethylamine: solvent effects on correlations with the function $(^3\Delta E_{oo} + E(A - /A))$, *J. Photochem.,* 9, 11, 1978.

183. **Hamanoue, K., Nakayama, T., Yamamoto, Y., Sawada, K., Yuhara, Y., and Teranishi, H.,** Photoreduction of anthraquinone by triethylamine in ethanol and toluene studied by steady-state photolysis and laser photolysis, *Bull. Chem. Soc. Jpn.,* 61, 1121, 1988.

184. **Yates, S. F. and Schuster, G. B.,** Photoreduction and triplet thioxanthone by amines: charge transfer generated radicals that initiate polymerization of olefins, *J. Org. Chem.,* 49, 3350, 1984.

185. **Cohen, S. G., Stein, N. J., and Chao, H. M.,** Effect of water on photoreduction of aromatic ketones by tertiary amines, *J. Am. Chem. Soc.,* 90, 521, 1968.

186. **Scaiano, J. C.,** Photochemical and free-radical processes in benzil-amine systems. Electron-donor properties of α-aminoalkyl radicals, *J. Phys. Chem.,* 85, 2851, 1981.

187. **Encinas, M. V., Garrido, J., and Lissi, E. A.,** Polymerization photoinitiated by carbonyl compounds. VIII. Solvent and photoinitiator concentration effects, *J. Polym. Sci. Polym. Chem. Ed.,* 27, 139, 1989.

188. **McLauchlan, K. A., Sealy, R. C., and Wittmann, J. M.,** Electron spin polarization (CIDEP) in photosensitized reactions, *J. Chem. Soc. Faraday Trans. 2,* 73, 926, 1977.

189. **Carlini, C., Toniolo, L., Rolla, P. A., Barigelletti, F., Bortolus, P., and Flamigni, L.,** Polymeric photoinitiators containing side chain benzophenone chromophores: relationships between structure and activity, *New Polym. Mater.,* 1, 63, 1987.

190. **Pappas, S. P.,** UV curing by radical cationic and concurrent radical-cationic polymerization, *Radiat. Phys. Chem.,* 25, 633, 1985.

191. **Clarke, S. R. and Shanks, R. A.,** A study of sensitizer systems for photoinitiated polymerization, *J. Macromol. Sci. Chem.,* A14, 69, 1980.

192. **Clarke, S. R. and Shanks, R. A.,** Factors affecting the ultraviolet-initiated polymerization of vinyl monomers, *J. Macromol. Sci. Chem.,* A17, 77, 1982.

193. **Davidson, R. S. and Goodin, J. W.,** The polymerization of acrylates using a combination of a carbonyl compound and an amine as a photo-initiator system, *Eur. Polym. J.,* 18, 597, 1982.

194. **Ghosh, P. and Bandyopadhyay, A. R.,** Photopolymerization of methyl methacrylate using triethylene tetramine-benzophenone combination as the photoinitiator, *Eur. Polym. J.*, 20, 1117, 1984.
195. **Ghosh, P., Chowdhury, D. K., and Bandyopadhyay, A. R.,** Use of some organic acids and amines as photoinitiators of vinyl polymerization in the absence and in the presence of benzophenone photosensitizer, *J. Macromol. Sci. Chem.*, A20, 549, 1983.
196. **Ghosh, P. and Bandyopadhyay, A. R.,** Photopolymerization of methyl methacrylate using diethylene triamine-benzophenone combination as the photoinitiator, *J. Polym. Mater.*, 2, 1, 1985.
197. **Ghosh, P. and Ghosh, R.,** Photopolymerization of methyl methacrylate using benzophenone-dimethylaniline combination as photoinitiator, *Eur. Polym. J.*, 17, 545, 1981.
198. **Ghosh, P. and Ghosh, R.,** Comparative studies of the effects of dimethylaniline, methylaniline and aniline in the benzophenone sensitized photopolymerization of methyl methacrylate in bulk and in solution, *Eur. Polym. J.*, 17, 817, 1981.
199. **Lougnot, D. J., Jacques, P., Fouassier, J. P., Casal, H. L., Kim.-Thuan, N., and Scaiano, J. C.,** New functionalized water-soluble benzophenone: a laser flash photolysis study, *Can. J. Chem.*, 63, 3001, 1985.
200. **Fouassier, J. P., Lougnot, D. J., Zuchowicz, I., Green, P. N., Timpe, H. J., Kronfeld, K. P., and Muller, U.,** Photoinitiation mechanism of acrylamide polymerization in the presence of water-soluble benzophenones, *J. Photochem.*, 36, 347, 1987.
201. **Lougnot, D. J. and Fouassier, J. P.,** Comparative reactivity of water soluble photoinitiators as viewed in terms of excited state processes, *J. Polym. Sci. Polym. Chem. Ed.*, 26, 1021, 1988.
202. **Allen, N. S., Howells, E. M., Lam., E., Catalina, F., Green, P. N., Green, W. A., and Chen, W.,** Photopolymerization and flash photolysis of a water soluble benzophenone photoinitiator: influence of tertiary amine, *Eur. Polym. J.*, 24, 591, 1988.
203. **Allen, N. S., Catalina, F., Mateo, J. L., Sastre, R., Green, P. N., and Green, W. A.,** Photochemistry of novel water-soluble parasubstituted benzophenone photoinitiators: a photocalorimetric and photoreduction study, *J. Photochem. Photobiol. A*, 44, 171, 1988.
204. **Allen, N. S., Chen, W., Catalina, F., Green, P. N., and Green, A.,** Photochemistry of novel water-soluble para-substituted benzophenone photoinitiators: a polymerization, spectroscopic and flash photolysis study, *J. Photochem. Photobiol. A*, 44, 349, 1988.
205. **Allen, N. S., Catalina, F., Mateo, J. L., Sastre, R., Chen, W., Green, P. N., and Green, W. A.,** Photochemistry and photopolymerization activity of water soluble benzophenone initiators, *Polym. Mater. Sci. Eng.*, 60, 10, 1989.
206. **Yamaguchi, H., Ninomiya, K., and Ogata, M.,** A study of the electronic spectra of 9-fluorenone, *Chem. Phys. Lett.*, 75, 593, 1986.
207. **Biczók, L. and Berces, T.,** Temperature dependence of the rates of photophysical processes of fluorenone, *J. Phys. Chem.*, 92, 3842, 1988.
208. **Davis, G. A., Carapellucci, P. A., Szoc, K., and Gresser, J. D.,** The photoreduction of fluorenone, *J. Am. Chem. Soc.*, 91, 2264, 1969.
209. **Cohen, S. G. and Parson, G.,** Effects of polar substituents on photoreduction and quenching of fluorenone by dimethylanilines, *J. Am. Chem. Soc.*, 92, 7603, 1970.
210. **Stone, P. G. and Cohen, S. G.,** Effects of structure on rates and quantum yields in photoreduction of fluorenone by amines. Catalysis and inhibition by thiols., *J. Am. Chem. Soc.*, 104, 3435, 1982.
211. **Parsons, G. H., Mendelson, L. T., and Cohen, S. G.,** Quenching of photoexcited fluorenone by m- and p-substituted anilines and mono-N-substitutes anilines. Linear free-energy relations, *J. Am. Chem. Soc.*, 96, 6643, 1974.
212. **Ledwith, A., Bosley, A., and Purbrick, M. D.,** Exciplex interactions in photoinitiation of polymerization by fluorenone-amine systems, *J. Radiat. Curing*, 2, 1979.
213. **Davidson, R. S. and Santhanam, M.,** The photoreactions of aromatic carbonyl compounds with amines. III. The photoreactions of fluorenone with tertiary amines, *J. Chem. Soc. Perkin Trans. 2*, p. 2355, 1972.
214. **Ledwith, A. and Purbrick, M. D.,** Initiation of free radical polymerization by photoinduced electron transfer processes, *Polymer*, 14, 521, 1973.
215. **Timpe, H. J. and Kronfeld, K. P.,** Light-induced polymer and polymerization reactions: direct photoinitiation of methyl methacrylate polymerization by excited states of ketones, *J. Photochem. Photobiol A*, 46, 253, 1989.
216. **Sengupta, P. K. and Modak, S. K.,** Photopolymerization of methyl methacrylate using the benzil-dimethylaniline combination as the photoinitiator, *J. Macromol. Sci. Chem.*, A20, 789, 1983.
217. **Encinas, M. V. and Scaiano, J. C.,** Laser photolysis study of the exciplex between triplet benzil and triethylamine, *J. Am. Chem. Soc.*, 101, 7740, 1979.
218. **Allen, N. S., Catalina, F., Green, P. N., and Green, W. A.,** Photochemical study of carbonyl initiators: aromatic ketones and thioxanthones, *Rev. Plast. Mod.*, 54, 370, 1987.
219. **Meier, K. and Zweifel, H.,** Thioxanthone ester derivatives: efficient triplet sensitizers for photopolymer applications, *J. Photochem.*, 35, 353, 1986.

220. **Green, P. N.**, Photoinitiators, types and properties, *Polym. Paint Colour J.*, 175, 246, 1985.

221. **Dalton, J. C. and Montgomery, F. C.**, Solvent effects on thioxanthone fluorescence, *J. Am. Chem. Soc.*, 96, 6230, 1974.

222. **Lai, T. and Lim, E. C.**, Photophysical behavior of aromatic carbonyl compounds related to proximity effect: thioxanthone, *Chem. Phys. Lett.*, 73, 244, 1980.

223. **Fouassier, J. P., Jacques, P., and Encinas, M. V.**, Solvent effects on the thioxanthone triplet quenching by vinyl monomers, *Chem. Phys. Lett.*, 148, 309, 1988.

224. **Fouassier, J. P., Lougnot, D. J., Paverne, A., and Wieder, F.**, Time-resolved laser-pumped dye laser spectroscopy: energy and electron transfer in ketones, *Chem. Phys. Lett.*, 135, 30, 1987.

225. **Amirzadeh, G. and Schnabel, W.**, On the photoinitiation of free radical polymerization-laser flash photolysis investigations on thioxanthone derivatives, *Makromol. Chem.*, 182, 2821, 1981.

226. **Allen, N. S., Catalina, F., Moghaddam, B., Green, P. N., and Green, W. A.**, Photochemistry of thioxanthones. III. Spectroscopic and flash photolysis study on hydroxy and methoxy derivatives, *Eur. Polym. J.*, 22, 691, 1986.

227. **Allen, N. S., Catalina, F., Green, P. N., and Green, W. A.**, Photochemistry of thioxanthones. I. Spectroscopic and flash photolysis study on oil soluble structures, *Eur. Polym. J.*, 21, 841, 1985.

228. **Allen, N. S., Catalina, F., Luc-Gardette, J., Green, W. A., Green, P. N., Chen, W., and Fatinikun, K. O.**, Spectroscopic properties and photopolymerization activity of 4-n-propoxythioxanthone, *Eur. Polym. J.*, 24, 435, 1988.

229. **Catalina, F., Tercero, J. M., Peinado, C., Sastre, R., Mateo, J. L., and Allen, N. S.**, Photochemistry and photopolymerization study on 2-acetoxy and methyl-2-acetoxy derivatives of thioxanthone as photoinitiators, *J. Photochem. Photobiol. A*, 50, 249, 1989.

230. **Allen, N. S., Catalina, F., Green, P. N., and Green, W. A.**, Photochemistry of thioxanthones. IV. Spectroscopic and flash photolysis study on novel n-propoxy and methyl, n-propoxy derivatives, *Eur. Polym. J.*, 22, 793, 1986.

231. **Allen, N. S., Catalina, F., Peinado, C., Sastre, R., Mateo, J. L., and Green, P. N.**, Synthesis, characterization and photopolymerization activity of a novel thioxanthone monomer and photopolymers, *Eur. Polym. J.*, 23, 985, 1987.

232. **Catalina, F., Peinado, C., Sastre, R., and Mateo, J. L.**, Thioxanthone photopolymers. I. A kinetics study of methyl methacrylate polymerization using 2-benzyloxy thioxanthone, free and polymer bound, as photoinitiator, *J. Photochem. Photobiol. A*, 47, 365, 1989.

233. **Allen, N. S., Catalina, F., Green, P. N., and Green, W. A.**, Photochemistry of thioxanthones. II. A spectroscopic and flash photolysis study on water soluble structures, *Eur. Polym. J.*, 22, 347, 1986.

234. **Allen, N. S., Catalina, F., Green, P. N., and Green, W. A.**, Photochemistry of thioxanthones. V. A polymerization, spectroscopic and flash photolysis study on novel water soluble methyl substituted 3-(9-oxo-9H-thioxanthene-2-yloxyl)N,N,N-trimethyl-1-propanaminium salts, *Eur. Polym. J.*, 22, 871, 1986.

235. **Allen, N. S., Catalina, F., Green, P. N., and Green, W. A.**, Photochemistry of thioxanthones. VI. A polymerization, spectroscopic and flash photolysis study on novel water-soluble methyl substituted 3-(9-oxo-9H-thioxanthene-2,3-γ-4-yloxy)N,N,N-trimethyl-1-propanaminium salts, *J. Photochem.*, 36, 99, 1987.

236. **Lougnot, D. J., Turck, C., and Fouassier, J. P.**, Water-soluble polymerization initiators based on the thioxanthone structure. A spectroscopic and laser photolysis study, *Macromolecules*, 22, 108, 1989.

237. **Tickle, K. and Wilkinson, F.**, Photoreduction of anthraquinone in isopropanol, *Trans. Faraday Soc.*, 61, 1981, 1965.

238. **Carlson, S. A. and Hercules, D. M.**, Studies on some intermediates and products of the photoreduction of 9,10-anthraquinone, *Photochem. Photobiol.*, 17, 123, 1973.

239. **Moger, G., Simon, P., and Rockenbauer, A.**, Rate constants for photochemically induced hydrogen abstraction calculated from spin-trapping rates, *React. Kinet. Catal. Lett.*, 14, 301, 1980.

240. **Hamanoue, K., Sawada, K., Yokoyama, K., Nakayama, T., Hirase, S., and Teranishi, H.**, Photoreduction of α-chloroanthraquinones followed by photochemical dehydrochlorination in ethanol at room temperature, *J. Photochem.*, 33, 99, 1986.

241. **Hamanoue, K., Nakayama, T., Kajiwara, Y., Yamaguchi, T., and Teranishi, H.**, The lowest triplet states of anthraquinone and chloroanthraquinones. The 1-chloro, 2-chloro, 1,5-dichloro, 1,8-dichloro compounds, *J. Chem. Phys.*, 86, 6654, 1987.

242. **Ledwith, A., Ndaalio, G., and Taylor, A. R.**, Polymerization of methyl methacrylate photoinitiated by anthraquinone and 2-tert-butylanthraquinone, *Macromolecules*, 8, 1, 1975.

243. **Palit, D. K., Pal, H., Mukherjee, T., and Mittal, J. P.**, Triplet excited states and semiquinone radicals of 1,4-disubstituted anthraquinones, *J. Photochem. Photobiol. A*, 52, 375, 1990.

244. **Hamanoue, K., Yokoyama, K., Kajiwara, Y., Kimoto, M., Nakayama, T., and Teranishi, H.**, The absorption kinetics of photoreductions of anthraquinone and 1-chloroanthraquinone by triethylamine in toluene and ethanol, *Chem. Phys. Lett.*, 113, 207, 1985.

245. **Li, Z. T., Kubota, H., and Ogiwara, Y.**, Effect of solvent on anthraquinone-sensitized photopolymerization of methyl methacrylate, *J. Appl. Polym. Sci.*, 27, 1465, 1982.

246. **Li, Z. T., Kubota, H., and Ogiwara, Y.,** Some factors affecting the photosensitized polymerization of methyl methacrylate, *J. Macromol. Sci. Chem.,* A18, 493, 1982.
247. **Li, Z. T., Kubota, H., and Ogiwara, Y.,** Anthraquinone sensitized photopolymerization: effect of monomer in isopropanol-cyclohexane solvent, *Polym. Photochem.,* 3, 435, 1983.
248. **Kubota, H. and Ogiwara, Y.,** β-Methylanthraquinone-sensitized photopolymerization of methyl methacrylate in iso-propanol/hydrocarbon solvents, *Eur. Polym. J.,* 25, 301, 1989.
249. **Hamanoue, K., Yamamoto, Y., Nakayama, T., and Teranishi, H.,** The excited state dynamics of triplet 1,8-dichloroanthraquinone and its complex formation with triethylamine and 2,5-dimethyl-2,4-hexadiene, in *Physical Organic Chemistry,* Kobayashi, M., Ed., 1986, 235.
250. **Encinas, M. V., Majmud, C., and Lissi, E. A.,** Photopolymerization photoinitiated by carbonyl compounds. IX. Methyl methacrylate polymerization photoinitiated by anthraquinone in presence of triethylamine, *J. Polym. Sci. Polym. Chem. Ed.,* 28, 2465, 1990.
251. **Schuster, D. I., Goldstein, M. D., and Bane, P.,** Photochemistry of unsaturated ketones in solution. Photophysical and photochemical studies of Michler's ketone, *J. Am. Chem. Soc.,* 99, 187, 1977.
252. **Brown, R. G. and Porter, G.,** Photochemistry of Michler's ketone in cyclohexane and alcohol solvents, *J. Chem. Soc. Faraday Trans. 1,* 73, 1569, 1977.
253. **Suppan, P.,** Photochemistry of Michler's ketone in solution, *J. Chem. Soc. Faraday Trans. 1,* 71, 539, 1975.
254. **Mateo, J. L., Manzarbeitia, J. A., Sastre, R., and Martinez-Utrilla, R.,** Photoreactivity of linear p-dimethylaminobenzoylated polystyrene and its model compounds, 4-dimethyl-4′-isopropylbenzophenone, *J. Photochem. Photobiol. A,* 40, 169, 1987.
255. **McGinniss, V. D., Provder, T., Kuo, C., and Gallopo, A.,** Polymerization of methyl methacrylate photoinitiated by 4,4′-bis(N,N-diethylamino)benzophenone, *Macromolecules,* 11, 393, 1978.
256. **McGinniss, V. D., Provder, T., Kuo, C., and Gallopo, A.,** Polymerization of methyl methacrylate photoinitiated by 4,4′-bis(N,N-diethylamino)benzophenone. Oxygen effects, *Macromolecules,* 11, 405, 1978.
257. **Mateo, J. L., Bosh, P., Vazquez, E., and Sastre, R.,** Polymerization of methyl acrylate photoinitiated by 4-dimethylamino-4′-isopropylbenzophenone, *Makromol. Chem.,* 189, 1219, 1988.
258. **Mateo, J. L., Bosch, P., Catalina, F., and Sastre, R.,** 4-N,N-Dimethylamino-4′-isopropylbenzophenone as polymerization photoinitiator. Effect of solvent and photoinitiator concentration on its photoreactivity and on the polymerization process, *J. Polym. Sci. Polym. Chem. Ed.,* 28, 1445, 1990.
259. **Kubota, H. and Ogiwara, Y.,** Photopolymerization using sensitizers of binary system of aromatic hydrocarbon and amines, *J. Appl. Polym. Sci.,* 28, 2425, 1983.
260. **Encinas, M. V., Majmud, C., and Lissi, E. A.,** Methyl methacrylate polymerization photoinitiated by pyrene in the presence of triethylamine, *Macromolecules,* 22, 563, 1989.
261. **Yang, N. C. and Libman, J.,** Chemistry of exciplexes, photochemical addition of secondary amines to anthracene, *J. Am. Chem. Soc.,* 95, 5783, 1973.
262. **Van, S. P. and Hammond, G. S.,** Amine quenching of aromatic fluorescence and fluorescent exciplexes, *J. Am. Chem. Soc.,* 100, 3895, 1978.
263. **Meeus, F., Van der Auweraer, M., and De Schryver, F. C.,** Thermodynamic and kinetic aspects of the intermolecular exciplex formation between 2-methylnaphthalene and aliphatic amines, *J. Am. Chem. Soc.,* 102, 4017, 1980.
264. **Lewis, F. D., Zebrowski, B. E., and Correa, P. E.,** Photochemical reactions of arenecarbonitriles with aliphatic amines. Effect of arene structure on aminyl vs. α-aminoalkyl radical formation, *J. Am. Chem. Soc.,* 106, 187, 1984.
265. **Mataga, N.,** Photochemical charge transfer phenomena-picosecond laser photolysis study, *Pure Appl. Chem.,* 56, 1255, 1984.
266. **Weller, A.,** Mechanism and spindynamics of photoinduced electron transfer reactions, *Z. Phys. Chem. N. F.,* 130, 129, 1982.
267. **Hirata, Y., Saito, T., and Mataga, N.,** Photochemistry of pyrene-trialkylamine systems in polar solvents: hydropyrenyl formation from a new type of long-lived ion pair formed by recombination of free ions, *J. Phys. Chem.,* 91, 3119, 1987.
268. **Mataga, N., Okada, T., Kanda, Y., and Shioyama, H.,** Behavior of radical ion pairs produced by photoinduced electron transfer in polar solutions, *Tetrahedron Lett.,* 42, 6143, 1986.
269. **Hirata, Y., Kanda, Y., and Mataga, N.,** Picosecond laser photolysis and transient photocurrent studies of the ionic dissociation mechanism of heteroexcimers: pyrene-N,N-dimethylaniline and pyrene-p-dicyanobenzene systems in polar solvents, *J. Phys. Chem.,* 87, 1659, 1983.
270. **Encinas, M. V., Majmud, C., Lissi, E. A., and Scaiano, J. C.,** Polymerization photoinitiated by pyrene in the presence of triethylamine: interaction between monomers and pyrene-derived reaction intermediates, *Macromolecules,* in press.

INDEX

A

M

Q

R